T0179798

# Applied Bayesian Modelling

# Applied Bayesian Modelling

## Second Edition

**Peter Congdon**

*Centre for Statistics and Department of Geography,*
*Queen Mary, University of London, UK*

This edition first published 2014
© 2014 John Wiley & Sons, Ltd

*Registered office*
John Wiley & Sons Ltd, The Atrium, Southern Gate, Chichester, West Sussex, PO19 8SQ, United Kingdom

For details of our global editorial offices, for customer services and for information about how to apply for permission to reuse the copyright material in this book please see our website at www.wiley.com.

*Library of Congress Cataloging-in-Publication Data*

Congdon, P.
  Applied Bayesian modelling / Peter Congdon. – Second edition.
     pages cm
  Includes bibliographical references and index.
  ISBN 978-1-119-95151-3 (cloth)
 1. Bayesian statistical decision theory. 2. Mathematical statistics. I. Title.
  QA279.5.C649 2014
  519.5′42–dc23

                                        2014004862


A catalogue record for this book is available from the British Library.

ISBN: 978-1-119-95151-3

Set in 10/12pt TimesLTStd by Laserwords Private Limited, Chennai, India

Printed and bound in Singapore by Markono Print Media Pte Ltd

1   2014

# Contents

Contents

# Preface

My gratitude is due to Wiley for proposing a revised edition of *Applied Bayesian Modelling*, first published in 2003. Much has changed since then for those seeking to apply Bayesian principles or to exploit the growing advantages of Bayesian estimation.

The central program used throughout the text in worked examples is BUGS, though R packages such as R-INLA, R2BayesX and MCMCpack are also demonstrated. Reference throughout the text to BUGS can be taken to refer both to WinBUGS and the ongoing Open-BUGS program, on which future development will concentrate (see http://www.openbugs. info/w/). There is a good deal of continuity between the final WinBUGS14 version and Open-BUGS (for details of differences see http://www.openbugs.info/w.cgi/OpenVsWin), though OpenBUGS has a wider range of sampling choices, distributions and functions. BUGS code can also be simply adapted to JAGS applications and the JAGS interfaces with R such as rjags.

Although R interfaces to BUGS or encapsulating the program are now widely used, the BUGS programming language itself remains a central aspect. Direct experience in WinBUGS or OpenBUGS programming is important as a preliminary to using R Interfaces such as BRUGS and rjags.

For learning Bayesian methods, especially if the main goal is data analysis *per se*, BUGS has advantages both practical and pedagogical. It can be seen as a half-way house between menu driven Bayesian computing (still not really established in any major computing package, though SAS has growing Bayesian capabilities) on the one hand, and full development of independent code, including sampling algorithms, on the other.

Many thanks are due to the following for comments on chapters or programming advice: Sid Chib, Cathy Chen, Brajendra Sutradhar and Thomas Kneib.

Please send comments or questions to me at p.congdon@qmul.ac.uk.

Peter Congdon, London

# 1

# Bayesian methods and Bayesian estimation

## 1.1  Introduction

Bayesian analysis of data in the health, social and physical sciences has been greatly facilitated in the last two decades by improved scope for estimation via iterative sampling methods. Recent overviews are provided by Brooks *et al.* (2011), Hamelryck *et al.* (2012), and Damien *et al.* (2013). Since the first edition of this book in 2003, the major changes in Bayesian technology relevant to practical data analysis have arguably been in distinct new approaches to estimation, such as the INLA method, and in a much extended range of computer packages, especially in R, for applying Bayesian techniques (e.g. Martin and Quinn, 2006; Albert, 2007; Statisticat LLC, 2013).

Among the benefits of the Bayesian approach and of sampling methods of Bayesian estimation (Gelfand and Smith, 1990; Geyer, 2011) are a more natural interpretation of parameter uncertainty (e.g. through credible intervals) (Lu *et al.*, 2012), and the ease with which the full parameter density (possibly skew or multi-modal) may be estimated. By contrast, frequentist estimates may rely on normality approximations based on large sample asymptotics (Bayarri and Berger, 2004). Unlike classical techniques, the Bayesian method allows model comparison across non-nested alternatives, and recent sampling estimation developments have facilitated new methods of model choice (e.g. Barbieri and Berger, 2004; Chib and Jeliazkov, 2005). The flexibility of Bayesian sampling estimation extends to derived 'structural' parameters combining model parameters and possibly data, and with substantive meaning in application areas, which under classical methods might require the delta technique. For example, Parent and Rivot (2012) refer to 'management parameters' derived from hierarchical ecological models.

New estimation methods also assist in the application of hierarchical models to represent latent process variables, which act to borrow strength in estimation across related units and outcomes (Wikle, 2003; Clark and Gelfand, 2006). Letting $[A, B]$ and $[A|B]$ denote joint and

*Applied Bayesian Modelling*, Second Edition. Peter Congdon.
© 2014 John Wiley & Sons, Ltd. Published 2014 by John Wiley & Sons, Ltd.

conditional densities respectively, the paradigm for a hierarchical model specifies

$$[\text{Process, Parameters|Observations}] \propto [\text{Observations|Process, Parameters}]$$

$$[\text{Process|Parameters}]\,[\text{Parameters}] \qquad (1.1)$$

based on an assumption that observations are imperfect realisations of an underlying process and that units are exchangeable. Usually the observations are considered conditionally independent given the process and parameters.

Such techniques play a major role in applications such as spatial disease patterns, small domain estimation for survey outcomes (Ghosh and Rao, 1994), meta-analysis across several studies (Sutton and Abrams, 2001), educational and psychological testing (Sahu, 2002; Shiffrin et al., 2008) and performance comparisons (e.g. Racz and Sedransk, 2010; Ding et al., 2013).

The Markov chain Monte Carlo (MCMC) methodology may also be used to augment the data, providing an analogue to the classical EM method. Examples of such data augmentation (with a missing data interpretation) are latent continuous data underlying binary outcomes (Albert and Chib, 1993; Rouder and Lu, 2005) and latent multinomial group membership indicators that underlie parametric mixtures. MCMC mixing may also be improved by introducing auxiliary variables (Gilks and Roberts, 1996).

### 1.1.1   Summarising existing knowledge: Prior densities for parameters

In classical inference the sample data $y$ are taken as random while population parameters $\theta$, of dimension $p$, are taken as fixed. In Bayesian analysis, parameters themselves follow a probability distribution, knowledge about which (before considering the data at hand) is summarised in a prior distribution $\pi(\theta)$. In many situations it might be beneficial to include in this prior density cumulative evidence about a parameter from previous scientific studies. This might be obtained by a formal or informal meta-analysis of existing studies. A range of other methods exist to determine or elicit subjective priors (Garthwaite et al., 2005; Gill and Walker, 2005). For example, the histogram method divides the range of $\theta$ into a set of intervals (or 'bins') and uses the subjective probability of $\theta$ lying in each interval; from this set of probabilities, $\pi(\theta)$ may be represented as a discrete prior or converted to a smooth density. Another technique uses prior estimates of moments, for instance in a normal $N(m, V)$ density with prior estimates $m$ and $V$ of the mean and variance, or prior estimates of summary statistics (median, range) which can be converted to estimates of $m$ and $V$ (Hozo et al., 2005).

Often, a prior amounts to a form of modelling assumption or hypothesis about the nature of parameters, for example, in random effects models. Thus small area death rate models may include spatially correlated random effects, exchangeable random effects with no spatial pattern, or both. A prior specifying the errors as spatially correlated is likely to be a working model assumption rather than a true cumulation of knowledge.

In many situations, existing knowledge may be difficult to summarise or elicit in the form of an informative prior, and to reflect such essentially prior ignorance, resort is made to non-informative priors. Examples are flat priors (e.g. that a parameter is uniformly distributed between $-\infty$ and $+\infty$) and Jeffreys prior

$$\pi(\theta) \propto \det\{I(\theta)\}^{0.5},$$

where $I(\theta)$ is the expected information[1] matrix. It is possible that a prior is improper (does not integrate to 1 over its range). Such priors may add to identifiability problems (Gelfand and Sahu, 1999), especially in hierarchical models with random effects intermediate between hyperparameters and data. An alternative strategy is to adopt vague (minimally informative) priors which are 'just proper'. This strategy is considered below in terms of possible prior densities to adopt for the variance or its inverse. An example for a parameter distributed over all real values might be a normal with mean zero and large variance. To adequately reflect prior ignorance while avoiding impropriety, Spiegelhalter *et al.* (1996) suggest a prior standard deviation at least an order of magnitude greater than the posterior standard deviation.

## 1.1.2   Updating information: Prior, likelihood and posterior densities

In classical approaches such as maximum likelihood, inference is based on the likelihood of the data alone. In Bayesian models, the likelihood of the observed data $y$, given a set of parameters $\theta = (\theta_1, \ldots, \theta_d)$, denoted $p(y|\theta)$ or equivalently $L(\theta|y)$, is used to modify the prior beliefs $\pi(\theta)$. Updated knowledge based on the observed data and the information contained in the prior densities is summarised in a posterior density, $\pi(\theta|y)$. The relationship between these densities follows from standard probability relations. Thus

$$p(y, \theta) = p(y|\theta)\pi(\theta) = \pi(\theta|y)p(y)$$

and therefore the posterior density can be written

$$\pi(\theta|y) = p(y|\theta)\pi(\theta)/p(y).$$

The denominator $p(y)$ is a known as the marginal likelihood of the data, and found by integrating the likelihood over the joint prior density

$$p(y) = \int p(y|\theta)\pi(\theta)\mathrm{d}\theta.$$

This quantity plays a central role in formal approaches to Bayesian model choice, but for the present purpose can be seen as an unknown proportionality factor, so that

$$\pi(\theta|y) \propto p(y|\theta)\,\pi(\theta),$$

or equivalently

$$\pi(\theta|y) = k\,p(y|\theta)\,\pi(\theta). \tag{1.2}$$

The product $\pi_u(\theta|y) = p(y|\theta)\pi(\theta)$ is sometimes called the un-normalised posterior density. From the Bayesian perspective, the likelihood is viewed as a function of $\theta$ given fixed data $y$ and so elements in the likelihood that are not functions of $\theta$ become part of the proportionality constant in (1.2). Similarly, for a hierarchical model as in (1.1), let $Z$ denote latent variables depending on hyperparameters $\theta$. Then one has

$$\pi(Z, \theta|y) = p(y|Z, \theta)p(Z|\theta)\,\pi(\theta)/p(y),$$

---

[1] If $\ell(\theta) = \log(L(\theta))$ then $I(\theta) = -\mathrm{E}\left\{\dfrac{\delta^2 \ell(\theta)}{\delta\ell(\theta_i)\delta\ell(\theta_j)}\right\}$.

or equivalently

$$\pi(Z, \theta|y) = kp(y|Z, \theta)p(Z|\theta)\pi(\theta). \tag{1.3}$$

Equations (1.2) and (1.3) express mathematically the process whereby updated beliefs are a function of prior knowledge and the sample data evidence.

It is worth introducing at this point the notion of the full conditional density for individual parameters (or parameter blocks) $\theta_j$, namely

$$p(\theta_j|\theta_{[j]}, y) = p(\theta|y)/p(\theta_{[j]}/y),$$

where $\theta_{[j]} = (\theta_1, \ldots, \theta_{j-1}, \theta_{j+1}, \ldots, \theta_p)$ denotes the parameter set excluding $\theta_j$. These densities are important in MCMC sampling, as discussed below. The full conditional density can be abstracted from the un-normalised posterior density $\pi_u(\theta|y)$ by regarding all terms except those involving $\theta_j$ as constants.

For example, consider a normal density $N(\mu, 1/\tau)$ for observations $(y_1, \ldots, y_n)$ with likelihood

$$(2\pi)^{-n/2}\tau^{n/2} \exp\left[-0.5\tau \sum_{i=1}^{n} (y_i - \mu)^2\right].$$

Assume a gamma $Ga(\alpha, \beta)$ prior on $\tau$, and a $N(m, V)$ prior on $\mu$. Then the joint posterior density, concatenating constant terms (including the inverse of the marginal likelihood) into the constant $k$, is

$$p(\mu, \tau|y) = k\tau^{n/2}\exp^{-0.5\tau \sum_{i=1}^{n} (y_i - \mu)^2} V^{-0.5}e^{-0.5(\mu-m)^2/V} \frac{\beta^\alpha}{\Gamma(\alpha)}\tau^{\alpha-1}e^{-\beta\tau}. \tag{1.4}$$

The full conditional density for $\mu$ is expressed analytically as

$$p(\mu|\tau, y) = p(\mu, \tau|y)/p(\tau|y),$$

and can be obtained from (1.4) by focusing only on terms that are functions of $\mu$. Thus

$$p(\mu|\tau, y) \propto e^{-0.5\tau \sum_{i=1}^{n} (y_i-\mu)^2} e^{-0.5(\mu-m)^2/V}.$$

By algebraic re-expression, and with $h = 1/V$, one may show

$$p(\mu|\tau, y) = N\left(\frac{mh + \tau n\bar{y}}{h + n\tau}, \frac{1}{h + n\tau}\right).$$

Similarly

$$p(\tau|\mu, y) \propto \tau^{n/2}e^{-0.5\tau \sum_{i=1}^{n} (y_i-\mu)^2} \tau^{\alpha-1}e^{-\beta\tau},$$

which can be re-expressed as

$$p(\tau|\mu, y) = Ga(\alpha + 0.5n, \beta + 0.5 \sum_{i=1}^{n} (y_i - \mu)^2),$$

where $Ga(a, b)$ denotes a gamma density with mean $a/b$ and variance $a/b^2$.

### 1.1.3 Predictions and assessment

The principle of updating extends to replicate values or predictions. Before the study a prediction would be based on random draws from the prior density of parameters and is likely to have little precision. Part of the goal of a new study is to use the data as a basis for making improved predictions or evaluation of future options. Thus in a meta-analysis of mortality odds ratios (e.g. for a new as against conventional therapy), it may be useful to assess the likely odds ratio $y_{\text{rep}}$ in a hypothetical future study on the basis of findings from existing studies. Such a prediction is based on the likelihood of $y_{\text{rep}}$ averaged over the posterior density based on $y$:

$$p(y_{\text{rep}}|y) = \int p(y_{\text{rep}}|\theta)\,\pi(\theta|y)\,d\theta,$$

where the likelihood of $y_{\text{rep}}$, $p(y_{\text{rep}}|\theta)$, usually takes the same form as adopted for the observations themselves. In a hierarchical model, one has

$$p(y_{\text{rep}}|y) = \int_\theta \int_Z p(y_{\text{rep}}|\theta,Z)p(\theta,Z|y)d\theta dZ.$$

One may also take predictive samples order to assess the model performance, namely in model criticism (Vehtari and Ojanen, 2012). A particular instance of this (see Chapters 2 and 3) is in cross-validation based on omitting a single case. Data for case $i$ is observed, but a prediction of $y_i$ is nevertheless made on the basis of the remaining data $y_{[i]} = \{y_1, y_2, \ldots, y_{i-1}, y_{i+1}, \ldots, y_n\}$. Thus in a regression example, the prediction $y_{\text{rep},i}$ would use observed covariates $X_i$ for case $i$, but the regression coefficients would be from a model fitted to $y_{[i]}$.

One may also derive

$$p(y_i|y_{[i]}) = \int p(y_i|\theta)\,\pi(\theta|y_{[i]})\,\pi(\theta)\,d\theta,$$

namely the probability of $y_i$ given the rest of the data (Gelfand *et al.*, 1992). This is known as the conditional predictive ordinate (CPO) and is equivalent to the leave-one-out posterior predictive distribution

$$p(y_{\text{rep},i}|y_{[i]}) = \int p(y_{\text{rep},i}|\theta)\,\pi(\theta|y_{[i]})\,\pi(\theta)\,d\theta$$

evaluated at the observed value $y_i$. Observations with low CPO values are not well fitted by the model. Predictive checks may be made comparing $y_i$ and $y_{\text{rep},i}$, providing cross-validatory posterior $p$-values (Marshall and Spiegelhalter, 2007)

$$\Pr(y_i > y_{\text{rep},i}|y_{[i]}),$$

to assess whether predictions tend to be larger or smaller than the observed values.

However, full $n$-fold cross-validation may be computationally expensive except in small samples. Another option is for a large dataset to be randomly divided into a small number $k$ of groups; then cross-validation may be applied to each partition of the data, with $k-1$ groups as the training sample and the remaining group as the validation sample (Alqalaff and Gustafson, 2001; Vehtari and Ojanen, 2012). For large datasets one might take 50% of the data as the training sample and the remainder as the validation sample (i.e. $k = 2$).

One may also sample replicate data based on a model fitted to all observed cases to carry out posterior predictive checks. For instance, in a normal linear regression application, a prediction $y_{\text{rep},i} \sim N(\mu_i, \sigma^2)$ would make use of regression means $\mu_i = X_i \beta$ based on the complete data. These predictions may be used in predictive loss model selection (e.g. Laud and Ibrahim, 1995; Gelfand and Ghosh, 1998; Daniels *et al.*, 2012), or in predictive checks and significance tests. For example, the above comparison becomes

$$\Pr(y_i > y_{\text{rep},i}|y).$$

However, such tests may lead to conservative inferences as the data are used twice (Bayarri and Castellanos, 2007), once for model fitting and again for model assessment. This conservatism may be reduced by calibration (Hjort *et al.*, 2006) or by a mixed predictive approach (Marshall and Spiegelhalter, 2007).

### 1.1.4   Sampling parameters

To update knowledge about the parameters requires that one can sample from the posterior density. From the viewpoint of sampling from the density of a particular parameter $\theta_j$, it follows from (1.2) and (1.3) that aspects of the likelihood which are not functions of $\theta_j$ may be omitted. Thus consider a binomial outcome $y$ with $r$ successes from $n$ trials, and with unknown parameter $\rho$ representing the binomial probability, with a beta prior $\text{Be}(a, b)$, where the beta density is

$$\frac{\Gamma(a+b)}{\Gamma(a)\Gamma(b)} \rho^{a-1} (1-\rho)^{b-1}.$$

The likelihood is then, viewed as a function of $\rho$, proportional to a beta density, namely

$$L(\rho|y) \propto \rho^r (1-\rho)^{n-r},$$

and the posterior density for $\rho$ is obtained as a beta density with parameters $r + a$ and $n + b - r$

$$p(\rho|y) = \text{Be}(r + a, n + b - r). \tag{1.5}$$

Therefore the parameter's posterior density may be obtained by sampling from the relevant beta density. Incidentally, this example shows how the prior may in effect be seen to provide a prior sample, here of size $a + b - 2$, the size of which increases with the confidence attached to the prior belief. For instance if $a = b = 2$ then the prior is equivalent to a prior sample of 1 success and 1 failure.

In (1.5), a simple analytic result provides a method for sampling of the unknown parameter. This is an example where the prior and the likelihood are conjugate since both the prior and posterior density are of the same type. In more general situations, with many parameters in $\theta$ and with possibly non-conjugate priors, analytic forms of the posterior density are typically unavailable, but the goal is still to estimate the marginal posterior of a particular parameter $\theta_j$ given the data. This involves integrating out all the parameters but this one

$$p(\theta_j|y) = \int p(\theta_j|\theta_{[j]})p(\theta_{[j]}|y)d\theta_{[j]}.$$

Such integrations in the past involved demanding methods such as numerical quadrature.

MCMC methods (Section 1.2) use various techniques to sample repeatedly from the joint posterior of all the parameters

$$p(\theta_1, \theta_2, \dots, \theta_d|y),$$

without undertaking such integrations. Note, however, that unlike simple Monte Carlo sampling, estimation is complicated by features such as correlation between samples.

Suppose an initial burn-in of $B$ iterations is taken to ensure sampling is concentrated in the area of highest posterior density, and $T$ samples are taken thereafter from the joint posterior via MCMC sampling, possibly by combining samples from multiple chains (see Section 1.4). Then marginal posteriors for particular parameters $\theta_j$ may be estimated by summarising the information contained in the $T$ samples $\theta_j^{(1)}, \theta_j^{(2)}, \ldots, \theta_j^{(T)}$. For example, the mean and variance of the posterior density may be estimated from the average and variance of the sampled values, and the quantiles of the posterior density may be estimated by the relevant points from the ranked sample values. Thus

$$\hat{E}(\theta_j|y) = \sum_{t=1}^{T} \theta_j^{(t)}/T.$$

is an estimator of the integral

$$E(\theta_j|y) = \int \theta_j p(\theta|y) d\theta.$$

The overall posterior density may be estimated by kernel density methods using the samples $\theta_j^{(1)}, \theta_j^{(2)}, \ldots, \theta_j^{(T)}$.

## 1.2 MCMC techniques: The Metropolis–Hastings algorithm

Assume a preset initial parameter value $\theta^{(0)}$. Then MCMC methods involve generating a correlated sequence of sampled values $\theta^{(t)}(t = 1, \ldots, T)$, with updated values $\theta^{(t)}$ drawn from a transition sequence

$$K(\theta^{(t)}|\theta^{(0)}, \ldots, \theta^{(t-1)}) = K(\theta^{(t)}|\theta^{(t-1)})$$

that is Markovian in the sense of depending only on $\theta^{(t-1)}$. The transition kernel $K(\theta^{(t)}|\theta^{(t-1)})$ is required to satisfy certain conditions (irreducibility, aperiodicity, positive recurrence) to ensure that the sequence of sampled parameters has the joint posterior density $p(\theta|y)$ as its stationary distribution (Brooks, 1998; Andrieu and Moulines, 2006). These conditions amount to requirements on the proposal distribution and acceptance rule used to generate new parameters.

The Metropolis–Hastings (M-H) algorithm is the baseline for MCMC sampling schemes (Griffin and Stephens, 2013). Let $\theta'$ be a candidate parameter value generated by a proposal density $q(\theta'|\theta^{(t)})$. The candidate value has probability of acceptance

$$\alpha(\theta', \theta^{(t)}|y) = \min\left(1, \frac{p(\theta'|y)\, q(\theta^{(t)}|\theta')}{p(\theta^{(t)}|y)q(\theta'|\theta^{(t)})}\right),$$

with transition kernel $K(\theta^{(t)}|\theta^{(t-1)}) = \alpha(\theta', \theta^{(t)}|y)q(\theta'|\theta^{(t)})$ (Chib, 2013). If the chosen proposal density is symmetric, so that $q(\theta'|\theta^{(t)}) = q(\theta^{(t)}|\theta')$, then the M-H algorithm reduces to the Metropolis algorithm whereby

$$\alpha(\theta', \theta^{(t)}|y) = \min\left[1, \frac{p(\theta'|y)}{p(\theta^{(t)}|y)}\right].$$

A particular symmetric density in which $q(\theta'|\theta^{(t)}) = q(|\theta^{(t)} - \theta'|)$ leads to random walk Metropolis updating (Chib and Greenberg, 1995; Sherlock et al., 2010). Typical Metropolis updating schemes use uniform and normal densities. A normal proposal with variance $\sigma_q^2$ involves standard normal samples

$$h_t \sim N(0, 1),$$

with candidate values

$$\theta' = \theta^{(t)} + \sigma_q h_t,$$

where $\sigma_q$ determines the size of the potential shift from current to future value. A uniform random walk samples uniform variates $h_t \sim U(-1, 1)$ and scales these to form a proposal

$$\theta' = \theta^{(t)} + \kappa h_t,$$

with the value of $\kappa$ determining the potential shift. Parameters $\sigma_q$ and $\kappa$ may be varied to achieve a target acceptance rate for proposals. A useful modification of random walk Metropolis for constrained (e.g. positive or probability) parameters involves reflexive random walks. For example, suppose $\theta$ is a probability and a value $\theta' = \theta^{(t)} + u^{(t)}$, where $u^{(t)} \sim U(-\kappa, \kappa)$, is sampled. Then if $-1 < \theta' < 0$, one sets $\theta' = |\theta'|$, and if $|\theta'| > 1$, one sets $\theta' = 1$. Truncated normal sampling can also be used, as in

$$\theta' \sim N(\theta^{(t)}, \sigma_q^2)I(0, 1).$$

Evaluating the ratio of $p(\theta'|y)/p(\theta^{(t)}|y)$ in practice involves a comparison of the unstandardised posterior density, namely the product of likelihood and prior ordinates, as the normalising constant in $p(\theta|y) = kL(\theta|y)\pi(\theta)$ cancels out. In practice also the parameters are updated individually or in sub-blocks of the overall parameter set. In fact, for updating a particular parameter, with proposed value $\theta'_j$ from a proposal density specific to $\theta_j$, all other parameters than the jth can be regarded as fixed. So all terms in the ratio $p(\theta'|y)/p(\theta^{(t)}|y)$ cancel out, apart from those in the full conditional densities $p(\theta_j|y, \theta_{[j]})$. So for updating parameter j, one may consider the ratio of full conditional densities evaluated at the candidate and current values respectively

$$p(\theta'_j|y, \theta_{[j]})q(\theta^{(t)}|\theta^*)/[p(\theta_j^{(t)}|y, \theta_{[j]})q(\theta^*|\theta^{(t)})].$$

where $\theta^* = (\theta_1, \ldots, \theta_{j-1}, \theta'_j, \theta_{j+1}, \ldots, \theta_p)$. It may in practice be easier (if not strictly necessary) to program using the ratio

$$L(\theta^*|y)\pi(\theta^*)q(\theta|\theta^*)/[L(\theta|y)\pi(\theta)q(\theta^*|\theta)].$$

If the proposal $\theta'$ is rejected, the parameter value at iteration $t + 1$ is the same as at iteration $t$. The acceptance rate for proposed parameters depends on how close $\theta'$ is to $\theta^{(t)}$, which is influenced by the variance of the proposal density. For a normal proposal density, $q(\theta'|\theta^{(t)}) = N(\theta^{(t)}, \sigma_q^{(2)})$, a higher acceptance rate follows from reducing $\sigma_q$, but this implies slower exploration of the posterior density.

## 1.2.1   Gibbs sampling

The Gibbs sampler (Gelfand and Smith, 1990) is a special componentwise M-H algorithm whereby the proposal density for updating $\theta_j$ is the full conditional $p(\theta_j|y, \theta_{[j]})$, so that proposals are accepted with probability 1. Successive samples from full conditional densities may involve sampling from analytically defined densities (gamma, normal, Student $t$, etc.), as in the normal likelihood example above, or by sampling from non-standard densities. If the full conditionals are non-standard, but maybe of a certain mathematical form (e.g. log-concave), then other forms of sampling, such as slice sampling (Neal, 2003), adaptive rejection sampling (Gilks and Wild, 1992) or griddy Gibbs sampling (Ritter and Tanner, 1992) may be used. The BUGS program (Section 1.3) may be applied with some or all parameters sampled from formally coded conditional densities; however, provided with prior and likelihood, BUGS will infer the correct conditional densities using directed acyclic graphs.

In some instances the full conditionals may be converted to simpler forms by introducing latent data $w_i$, either continuous or discrete, a device known as data augmentation. An example is the probit regression model for binary responses (Albert and Chib, 1993), where continuous latent variables $w_i$ underlie the observed binary outcome $y_i$. Thus the formulation

$$w_i = X_i\beta + u_i$$

$$u_i \sim N(0, 1)$$

$$y_i = I(w_i > 0)$$

is equivalent to the probit model.[2] The parameter $\beta$ may be estimated in the same way (i.e. using the same form of MCMC updating) as for normal linear regression applied to metric data.

## 1.2.2    Other MCMC algorithms

The reversible jump MCMC (RJMCMC) algorithm (Green, 1995; Sisson, 2005; Fan and Sisson, 2011; Griffin and Stephens, 2013) generalises the M-H algorithm to include a model indicator. Let $m$ and $m'$ be models with parameters $\theta$ and $\theta'$, of dimension $d$ and $d'$ respectively. Moves from model $m$ to model $m'$ are proposed with probability $J(m'|m)$, and candidate values $u$ (for parameters present in model $m'$ but not in model $m$) are proposed according to a density $q(u|\theta, m)$. The reverse move involves generating $u'$ from a density $q(u'|\theta', m')$. The dimension of $u$ and $u'$ are such that $d + \dim(u) = d' + \dim(u')$. For a move from $m$ to $m'$, one sets $(\theta') = g_{m,m'}(\theta, u)$ where $g_{m,m'}$ is a bijective function (one to one and onto), such that $g_{m,m'} = 1/g_{m',m}$.

The acceptance probability is the minimum of 1 and

$$\alpha = \left[ p\left(\theta', m'|y\right) J(m|m') q(u'|\theta', m') \left| \frac{\partial(\theta')}{\partial(\theta, u)} \right| \right] \bigg/ \left[ p\left(\theta, m|y\right) J(m'|m) q(u|\theta, m) \right],$$

where the Jacobean $\partial(\theta')/\partial(\theta, u)$ accounts for the change in parameters between models. A simplified version of the acceptance probability applies when moves are between nested models (Fan and Sisson, 2011), when $\alpha$ reduces to

$$\alpha = \left[ p\left(\theta', m'|y\right) J(m|m') \left| \frac{\partial(\theta')}{\partial(\theta, u)} \right| \right] \bigg/ \left[ p\left(\theta, m|y\right) J(m'|m) q(u|\theta, m) \right].$$

In practice the posterior densities $p(\theta', m'|y)$ and $p(\theta, m|y)$ are expressed as product of likelihood, prior and model probability, as in $p(\theta, m|y) = k L(\theta|y) p(\theta|m) p(m)$, since the normalising constant cancels out (Green, 2003).

---

[2] $I(u)$ is 1 if $u$ holds and zero otherwise.

An example would be when $d = 1$ with parameter $\theta_1$, while $d' = 2$ with parameters $(\theta_1', \theta_2')$. If the current state is $(\theta_1, m)$ the candidate parameters could be generated as

$$\theta_1' = \theta_1,$$

$$\theta_2' = u,$$

where $u \sim q(u|\theta_1, m)$ has the same support as $\theta_2'$. For this example,

$$\left| \frac{\partial(\theta')}{\partial(\theta, u)} \right| = \begin{vmatrix} \frac{\partial \theta_1'}{\partial \theta_1} & \frac{\partial \theta_2'}{\partial \theta_1} \\ \frac{\partial \theta_1'}{\partial u} & \frac{\partial \theta_2'}{\partial u} \end{vmatrix} = \begin{vmatrix} 1 & 0 \\ 0 & 1 \end{vmatrix} = 1.$$

For the reverse move (from a larger to a smaller model), one could set

$$u = \theta_2'$$

$$\theta_1 = \theta_1'$$

with the inverse of the function governing the move from $m$ to $m'$. This proposal is accepted with probability $\min(1, 1/\alpha)$.

If the function linking current and proposed parameters is the identity function (as in the above example, and in the regression predictor selection in Example 1.2), the Jacobean equals unity (Griffin and Stephens, 2013; King *et al.*, 2010). Choice of proposal densities for the between model step (as distinct from within model updating by the usual M-H methods) involves distinct issues, as discussed by Brooks *et al.* (2003) and Al-Awadhi *et al.* (2004).

## 1.2.3   INLA approximations

Integrated nested Laplace approximations (INLA) are an alternative to estimation and inference via MCMC in latent Gaussian models, which are a particular form of hierarchical model (Rue *et al.*, 2009). This includes a wide class of models, including generalised linear models, spatial data applications, survival data, and time series.

Denote the observations as $y$, and Gaussian distributed random effects (or latent Gaussian field) as $x$. Then with $\theta$ denoting hyperparameters, the assumed hierarchical model is

$$y_i|x_i \sim p(y_i|x_i, \theta_1),$$

$$x_i|\theta_2 \sim \pi(x|\theta_2) = N(., Q^{-1}(\theta_2)),$$

$$\theta \sim \pi(\theta),$$

with posterior density

$$\pi(x, \theta|y) \propto \pi(\theta)\pi(x|\theta_2) \prod_i p(y_i|x_i, \theta_1).$$

For example, consider a binary time sequence,

$$y_t \sim \text{Bern}(p_t),$$

$$\text{logit}(p_t) = \eta_t,$$

with linear predictor

$$\eta_t = G_t\beta + u_t + v_t,$$

where the regression coefficient vector $\beta$ is assigned a normal prior, $v_t$ is an unstructured normal random effect with variance $\sigma_v^2$, and

$$u_t = \phi u_{t-1} + e_t,$$

with $e_t$ normal with variance $\sigma_e^2$. Then $x = (\eta, u, v, \beta)$ is jointly Gaussian with hyperparameters

$$\theta = (\sigma_v^2, \sigma_e^2, \phi).$$

Integrated nested Laplace approximation (or INLA) is a deterministic algorithm, as opposed to a stochastic algorithm such as MCMC, specifically designed for latent Gaussian models. The focus in the algorithm is on posterior density of the hyperparameters,

$$\pi(\theta|y),$$

and on the conditional posterior of the latent field

$$\pi(x_i|\theta, y).$$

The algorithm uses a Laplace approximation for the posterior density of the hyperparameters, denoted

$$\tilde{\pi}(\theta|y),$$

and a Taylor approximation for the conditional posterior of the latent field, denoted

$$\tilde{\pi}(x_i|\theta, y).$$

From these approximations, marginal posteriors are obtained as

$$\tilde{\pi}(x_i|y) = \int \tilde{\pi}(\theta|y)\tilde{\pi}(x_i|\theta, y)d\theta,$$

$$\tilde{\pi}(\theta_j|y) = \int \tilde{\pi}(\theta|y)d\theta_{[j]},$$

where $\theta_{[j]}$ denotes $\theta$ excluding $\theta_j$, and integrations are carried out numerically. An estimate for the marginal likelihood is provided by the normalising constant for

$$\tilde{\pi}(\theta|y).$$

## 1.3   Software for MCMC: BUGS, JAGS and R-INLA

BUGS (encompassing WinBUGS and OpenBUGS) is a general purpose Bayesian estimation package using MCMC, and despite the acronym employs a range of M-H parameter sampling options beyond Gibbs sampling. RJMCMC may currently be implemented for certain types of model (e.g. normal linear regression), using the JUMP interface in WinBUGS14 (Lunn *et al.*, 2008).

JAGS is also a general purpose estimation package with a very similar coding scheme to BUGS, but may have advantages in interfacing with R, and for this purpose can be implemented using the library rjags. Among differences in coding between BUGS and JAGS are more economical likelihood statements: so a linear regression with a single predictor, and parameters beta0 and beta1, can be expressed

```
y[i] ~ dnorm(beta0 + beta1 * x[i], tau)
```

rather than

```
y[i] ~ dnorm(mu[i], tau)
mu[i] <- beta0 + beta1 * x[i].
```

Also distinct is an option for sorting $K$ sampled parameters, rather than constrained sampling on each parameter, using the command sort(theta[1:K]). This may assist in sampling ordered parameters, as illustrated in the ordinal regression discussion in Chapter 3.

Estimation via BUGS in the standalone form (not involving R) involves checking the syntax of the program code (enclosed in a model file), reading in data, and then compiling. Each statement in the program code involves either a relation ~ (distributed as) corresponding to solid arrows in a directed acyclic graph, or a deterministic relation < -, corresponding to a hollow arrow in a DAG. Model checking, data input and compilation involve the model menu in BUGS – though models may also be constructed directly by graphical means. Syntax checking involves highlighting the entire model code, or just the first few letters of the word model, and choosing the sequence model/specification/check model. The number of chains (if in excess of one) needs to be specified before compilation.

Data also need to be loaded before compilation. To load a list data file either the whole file is highlighted or just the first few letters of the word 'list'. For ascii data files, the first few letters of the first vector name need to be highlighted. Several separate data files may be input if needed.

If the compilation is successful the initial parameter value file or files ('inits files') are read in. These need not necessarily contain initial values for all the parameters, and some may be randomly generated from the priors (as specified in the model code) using 'gen inits'. Sometimes doing this may produce aberrant values which lead to numerical overflow, and generating inits is generally excluded for precision parameters.

An expert system chooses the sampling method, opting for standard Gibbs sampling if conjugacy is identified. For non-conjugate problems without log-concavity, M-H updating options are invoked, for example slice sampling (Neal, 2003) or adaptive sampling (Gilks *et al.*, 1998).

To monitor parameters (with the goal of obtaining estimated parameter summaries from averaging over sampled values) one selects inference/samples, and enters the relevant parameter name. For parameters which would require extensive storage to be monitored fully, an abbreviated summary (for, say, the model means of all observations in large samples, as required for subsequent calculation of model fit formulas) is obtained by inference/summary, and then entering the relevant parameter name.

There are a number of R interfaces with BUGS, such as BRugs, an R interface to Open-BUGS. This module allows OpenBUGS analyses to be completely performed within R, without needing to launch the standalone OpenBUGS program. Each OpenBUGS command (e.g. for compiling or loading data) has its own R function. By contrast, the R2OpenBUGS and R2WinBUGS modules in R construct a BUGS script to perform the analysis, but necessitate launching WinBUGS or OpenBUGS in the background to run the script. All interface options involve a program code (in BUGS/JAGS format) which can be saved to the R working directory, or incorporated in the code by using the R functions cat() or modelString.

To use INLA in R requires first installing the packages named pixmap and sp, then INLA may be downloaded via the command source ("http://www.math.ntnu.no/inla/givemeINLA.R"), rather than the install.packages command. Updates may be obtained using the command

$$inla.upgrade(testing = TRUE).$$

**Example 1.1    MCMC sampling for a normal likelihood**

Consider $n = 100$ observations $y_i$ generated in R from a normal density with mean 10 and variance 9. The observed mean and variance is likely to differ from the simulation values. Consider estimation of the normal parameters $\mu$ and $\sigma^2$ for the likelihood $y_i \sim N(\mu, \sigma^2)$. Firstly assuming an independent normal-gamma prior for $\mu$ and $\tau = 1/\sigma^2$, one may apply Gibbs sampling, namely repetitive sampling from the full conditionals for $\mu$ and $\tau = 1/\sigma^2$, set out above. In particular, assume a $Ga(1, 1)$ prior and a $N(0, 100)$ prior for $\mu$.

Then a code in R for simulating the data, and subsequent Gibbs sampling to estimate $\mu$ and $\sigma^2$, with $T = 10000$ iterations and $B = 1000$ for burn-in, is

```
# Observations: sample of n=100 from N(10,9)
x <- rnorm(100,10,3); n <- 100; mn.x <- mean(x); var.x <- var(x)
cat("Observed mean ",mn.x, "\n ")
cat("Observed variance ",var.x, "\n ")
# Gibbs Sampling in conjunction with Normal/Gamma Priors
# Parameters for Normal priors on mu, and gamma prior on tau
  m <- 0; V  <- 100; h <- 1/V; alph  <- 1; beta  <- 1
# MCMC sample settings, arrays for holding samples, initial
# values (0 for mu, 1 for tau)
  T <- 10000; B <- 1000; TB <- T-B
  mu <-   tau <- sigma2  <- numeric(T);
  mu[1]  <- 0; tau[1] <- 1
# start Gibbs sampler loop
for(t in 2:T){  # full conditional for mu
  m1 <- (h*m + n*tau[t-1]*mean(x))/(h+n*tau[t-1])
  V1 <- 1/(h+n*tau[t-1])
  mu[t] <- rnorm(1,m1,sqrt(V1))
  # full conditional for tau
  alph1 <- alph + (n/2);      beta1 <- (sum((x-mu[t])^2)/2) + beta
  tau[t]  <- rgamma(1,alph1,beta1)
  sigma2[t] <- 1/tau[t]}
# end loop
```

Note that for the particular sample of $y$-values considered here, we have $\bar{y} = 9.65$ and $\text{var}(y) = 7.67$. Numerical and graphical summaries may be obtained (for iterations $t > B$) using the commands

```
# Retain samples after Burn-in
mu <- mu[B+1:T]; sigma2 <- sigma2[B+1:T]
# parameter summaries
summary(mu);       quantile(mu[1:TB], c(0.025, 0.05, 0.90, 0.975))
summary(sigma2); quantile(sigma2[1:TB], c(0.025, 0.05, 0.90, 0.975))
# Set up subplots
par(mfrow=c(2,2))
# Trace Plots
plot(mu,type="l");   plot(sigma2,type="l")
# Marginal Posterior Densities
plot(density(mu),col="red",main=expression(paste("Posterior
                        Density of ",mu)))
plot(density(sigma2),col="blue",main=expression(paste("Posterior
                        Density of ",sigma^2)))
```

We find posterior means and 95% credible intervals for $\mu$ and $\sigma^2$ are 9.63 (9.08, 10.18) and 7.69 (5.83, 10.19) respectively. Figure 1.1 contains kernel plots of the marginal posterior densities based on the last 9000 iterations.

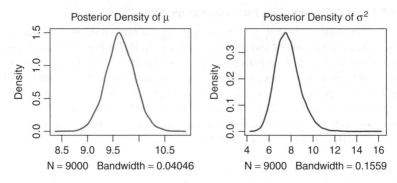

**Figure 1.1**   Trace plots and kernel densities, Example 1.1

One may also apply more general M-H sampling to these data if different prior assumptions are adopted. Suppose diffuse priors are adopted, specifically a flat prior on $\mu$, and a Jeffrey's prior on the residual variance, so that

$$\pi(\mu, \sigma) \propto 1/\sigma.$$

The posterior density in this case is

$$p(\mu, \sigma^2 | y) = k\sigma^{(-n-1)} \exp\left[-0.5 \sum_{i=1}^{n} (y_i - \mu)^2 / \sigma^2\right].$$

Uniform random walk Metropolis sampling may then be applied to update $\mu$ and $\sigma$. Generically $\theta' = \theta^{(t)} + \kappa h_t$, where $h_t \sim U(-1, 1)$. For the proposal for $\sigma$, the absolute value of $\sigma^{(t)} + \kappa h_t$ is retained, illustrating a reflecting random walk to avoid negative values for $\sigma$. The setting $\kappa = 0.5$ is adopted for proposals of both parameters, again with $T = 10\,000$ iterations and $B = 1000$ for burn-in.

The code, including calculation of proposal rejection rates, is

```
# MCMC sample settings, arrays for holding samples, initial values
T <- 10000; B <- 1000;
mu <- sigma2 <- sig <- numeric(T) ; u.mu <-  u.sig <- runif(T)
sig[1] <- 1; mu[1] <- 0
# totals for rejections of proposals
REJmu <- 0; REJsig <- 0
# log posterior density (up to a constant)
logpost=function(x,mu,sig){logpost= sum(dnorm(x,mu,sig,log=T)) - log(sig)}
# MCMC sampling loop
for (t in 2:T) {mucand <- mu[t-1] + runif(1,-0.5,0.5)
                sigcand <- abs(sig[t-1] + runif(1,-0.5,0.5))
# accept (or not) proposal for mu
log.alph.mu = logpost(x,mucand,sig[t-1])-logpost(x,mu[t-1],sig[t-1])
if (log(u.mu[t]) < log.alph.mu) mu[t] <- mucand
else { mu[t] <- mu[t-1]; if(t > B) REJmu <- REJmu+1 }
# accept (or not) proposal for sigma
log.alph.sig = logpost(x,mu[t],sigcand)-logpost(x,mu[t],sig[t-1])
if (log(u.sig[t]) < log.alph.sig) sig[t] <- sigcand
else { sig[t] <- sig[t-1]; if (t>B) REJsig <- REJsig+1 }
sigma2[t] <- sig[t]*sig[t]}
# Rejection Rates
cat("Rejection Rate mu = ",REJmu/(T-B), "\n ")
cat("Rejection Rate sigma = ",REJsig/(T-B), "\n ")
```

The rejection rates for proposed updates for $\mu$ and $\sigma$ are 33% and 45% respectively. Posterior means and 95% credible intervals for $\mu$ and $\sigma^2$ are 9.65 (9.09, 10.18) and 7.83 (5.86, 10.42).

Finally, a M-H update is considered, involving a gamma proposal $q(\tau'|\tau)$ for updating values of $\tau = 1/\sigma^2$, and so a Hastings correction $q(\theta^{(t)}|\theta')/q(\theta'|\theta^{(t)})$ is involved in the calculations for the acceptance/rejection step. The update on $\mu$ involves a random walk Metropolis. Priors are as in the preceding analysis. The proposal density for $\tau$ has the form

$$\tau' \sim Ga(\kappa, \kappa/\tau^{(t)}),$$

with $\kappa$ as a tuning parameter, and with higher values for $\kappa$ resulting in proposals closer to the current value $\tau^{(t)}$. We take $\kappa = 5$, with the R code being

```
T <- 10000; B <- 1000;
mu <- sigma2   <- tau <- numeric(T) ; u.mu <-   u.tau <- runif(T)
tau[1] <- 1; mu[1] <- 0; kap <- 5
# totals for rejections of proposals
REJmu <- 0; REJtau <- 0
# log posterior density (up to a constant)
logpost = function(x,mu,tau){sig <- 1/sqrt(tau)
logpost = sum(dnorm(x,mu,sig,log=T))-log(sig)}
# MCMC sampling loop
for (t in 2:T) {mucand <- mu[t-1] + runif(1,-0.5,0.5)
                        taucand <- rgamma(1,kap,kap/tau[t-1])
# accept (or not) proposal for mu
log.alph.mu = logpost(x,mucand,tau[t-1])-logpost(x,mu[t-1],tau[t-1])
if (log(u.mu[t]) < log.alph.mu) mu[t] <- mucand
else { mu[t] <- mu[t-1]; if(t > B) REJmu <- REJmu+1 }
# accept (or not) proposal for tau
Hastcorr <- log(dgamma(tau[t-1],kap,kap/taucand)/dgamma(taucand,
                                kap,kap/tau[t-1]))
log.alph.tau = logpost(x,mu[t],taucand)-logpost(x,mu[t],tau[t-1])
                                +Hastcorr
if (log(u.tau[t]) < log.alph.tau) tau[t] <- taucand
else { tau[t] <- tau[t-1]; if (t>B) REJtau <- REJtau+1 }
sigma2[t] <- 1/tau[t]}
# Rejection Rates
cat("Rejection Rate mu = ",REJmu/(T-B), "\n ")
cat("Rejection Rate tau = ",REJtau/(T-B), "\n ")
# Retain samples after Burn-in
mu <- mu[B+1:T]; sigma2 <- sigma2[B+1:T]
# parameter summaries
summary(mu); quantile(mu[1:TB], c(0.025, 0.05, 0.90, 0.975))
summary(sigma2); quantile(sigma2[1:TB], c(0.025, 0.05, 0.90, 0.975))
```

Rejection rates for proposals of $\mu$ and $\tau$ are 34% and 65% respectively. Posterior means and 95% credible intervals for $\mu$ and $\sigma^2$ are 9.65 (9.12, 10.20) and 7.62 (5.80, 10.11) respectively.

## Example 1.2   Logistic regression, M-H estimation with and without predictor selection

This example considers logistic regression applied to the prostatic nodal involvement data considered in Collett (1991) (see Chapter 3). There are $n = 53$ subjects and response $y = 1$ if cancer had spread to the surrounding lymph nodes and value zero otherwise. Predictors are level of serum acid phosphate ($x_1$, log transformed); X-ray exam result, coded 0 if negative and

1 if positive ($x_2$); tumour size, coded 0 if small and 1 if large ($x_3$); and pathological grade of the tumour, coded 0 if less serious and 1 if more serious ($x_4$). Priors on regression coefficients are as in Chib (1995).

First consider M-H estimation without predictor selection. Metropolis random walk updates are based on a uniform proposal density for regression parameters $\theta = (\theta_1, \ldots, \theta_5)$, where $\theta_1$ is the intercept, namely

$$\theta' = \theta^{(t)} + \kappa h_t,$$

with $\kappa = 1$. The chain is run for $T = 10000$ iterations with $B = 1000$ burn-in. The code, with predictor matrix $X$ including an intercept, is as follows

```
n =53;    npar = 5;    T =10000; B=1000; B1 <- B+1
# Input data
y <- c(0,0,0,0,0,0,0,0,1,0,0,0,0,1,0,0,0,0,0,0,0,0,1,0,1,1,0,0,0,
0,0,0,1,1,1,0,0,0,0,0,0,1,1,1,1,1,1,1,1,1,1,1,1)
 X <- matrix(c(1,-0.73,0,0,0,
1,-0.58,0,0,0,
. . . .
1,-0.12,1,1,0,
1,0.23,1,1,1),nrow=53,byrow=T)
# Set up the matrices to calculate posterior mean and sd
    mn <- var <- std <- accrate <- array(0,npar)
 # Normal priors on regression coefficients
    mu <- c(0,0.75,0.75,0.75,0.75); sig2 <- c(100,25,25,25,25);
    sig <- sqrt(sig2)
# Parameters for MH updates (Uniform random walk)
    kappa <- c(1,1,1,1,1)
# Initial parameter values
    theta <- c(0,0,0,0,0)
# sampled parameters
    sample <- samp2 <- acc <- array(0, dim=c(T, npar))
# calculate log likelihood
calcLK <- function(n, y, theta, npar, X){ p <- array(0,n); likhood <- 0
    for (i in 1:n) { eta <- sum(theta[1:npar]*X[i,1:npar])
                     p[i] <- 1/(1+exp(-eta))
                     likhood <- likhood + y[i]*log(p[i])+(1-y[i])*log(1-p[i])}
                     likhood}
# log-likelihood for initial parameters
    LK <- calcLK(n, y, theta, npar, X)
# MH updates Start Loop
    for (t in 1:T) { for (i in 1:npar) {    oldtheta <- theta[i]
# Candidate using Uniform proposal density
            theta[i] <- runif(1, theta[i]-kappa[i], theta[i]+kappa[i])
# Likelihood of proposed value
            newLK <- calcLK(n, y, theta, npar, X)
            num <- newLK + log(dnorm(theta[i],mu[i],sig[i]))
            den <- LK + log(dnorm(oldtheta,mu[i],sig[i]))
            A <- min(1,exp(num-den))
            u <- runif(1)
# Accept move with probability A, or stay at existing value
            if (u <= A) {   LK <- newLK; if(t>B) {acc[t,i] <- 1 }}
                else       { theta[i] <- oldtheta ;    if(t>B)
                                        {acc[t,i] <- 0  }}}
# Record parameter values:
        for (i in 1:npar) {sample[t,i] <- theta[i]; samp2[t,i] <- theta[i]^2}}
# End Loop
```

```
# posterior means and sd, acceptance rates
for (i in 1:npar) {mn[i] <- sum(sample[B1:T,i])/(T-B)
                    accrate[i] <- sum(acc[B1:T,i])/(T-B)
        std[i] <- sqrt((sum(samp2[B1:T,i])-(T-B)*mn[i]^2)/(T-B-1))}
# examples of numerical/graphical summaries
summary(sample[B1:T,1]);
plot(density(sample[B1:T,1],bw=0.15))
library(LaplacesDemon)
# press enter after message regarding page change
p.interval(sample[B1:T,1], HPD=TRUE, MM=FALSE, plot=TRUE)
```

Proposal acceptance rates for the five parameters are between 54% and 74%, with posterior means (sd) of $-1.44$ (0.79), 2.76 (1.23), 2.26 (0.84), 1.74 (0.80), and 0.91 (0.78). The high standard deviations of the regression parameters indicate possible predictor redundancy, and a RJMCMC selection is applied.

In the extra model selection step then involved, coefficients are randomly selected from $\theta_2$ to $\theta_5$ for possible inclusion or exclusion (the intercept is retained by default). So let $c \in$ $(2, \ldots, 5)$ be randomly selected. Then consider the case where $\theta_c$ is not currently in the regression model $m$ (i.e. $\theta_c = 0$). Parameters $\theta'$ for proposed model $m'$ (which is the same as m except that it includes $\theta_c$) are obtained as

$$\theta'_k = \theta_k \qquad (k \neq c)$$

$$\theta'_c = u,$$

where $u$ is generated from a proposal density with appropriate support, e.g. uniform or normal. Alternatively, consider the case where $\theta_c$ is currently included in model $m$, while the proposed model $m'$ excludes $\theta_c$ (i.e. sets $\theta_c$ to zero). Then

$$\theta'_k = \theta_k \qquad (k \neq c)$$

$$u = \theta_c,$$

$$\theta'_c = 0.$$

The relations between current and candidate parameters are defined by identity functions, and the Jacobean $\partial(\theta')/\partial(\theta, u)$ accordingly equals 1 (King et al., 2010). The code is

```
n =53;    npar = 5;    T =10000; B=1000; B1 <- B+1
# Input data
y <- c(0,0,0,0,0,0,0,0,1,0,0,0,0,1,0,0,0,0,0,0,0,0,1,0,1,1,0,0,0,
0,0,0,1,1,1,0,0,0,0,0,0,1,1,1,1,1,1,1,1,1,1,1,1,1)
 X <- matrix(c(1,-0.73,0,0,0,
1,-0.58,0,0,0,
.....
1,-0.12,1,1,0,
1,0.23,1,1,1),nrow=53,byrow=T)
# arrays for posterior mean and sd, and retention rates
    postmn <-  poststd <- totret <- retrate <- array(0,npar)
# Normal priors on regression coefficients
    mu <- c(0,0.75,0.75,0.75,0.75); sig2 <- c(100,25,25,25,25);
    sig <- sqrt(sig2)
# Parameters for MH updates (Uniform random walk)
    kappa <- c(1,1,1,1,1)
# Initial parameter values
```

```
    theta <- c(0,0,0,0,0)
# RJ model probabilities
    RJprob <- c(1,0.5,0.5,0.5,0.5)
# Parameters for RJ proposals (Normal mean, variance)
    muRJ <- c(0,0,0,0,0);       sigRJ2 <- c(1,1,1,1,1)
    sigRJ <- sqrt(sigRJ2)
# sampled parameters and retention indicators
    sample <- samp2 <- Ret <- array(0, dim=c(T, npar));
# calculate log likelihood
calcLK <- function(n,y,theta,npar,X){ p <- array(0,n);   likhood <- 0
for (i in 1:n) { eta <- sum(theta[1:npar]*X[i,1:npar]);
                p[i] <- 1/(1+exp(-eta))
                    likhood <- likhood + y[i]*log(p[i])
                        +(1-y[i])*log(1-p[i])}
                    likhood}
# Calculate log(likelihood) for initial parameters
    LK <- calcLK(n, y, theta, npar, X)
# Start overall loop
    for (t in 1:T)
# start within model MH loop
{ for (i in 1:npar) {   oldtheta <- theta[i]
# Propose parameter using uniform random walk
            theta[i] <- runif(1, theta[i]-kappa[i], theta[i]
                                    +kappa[i])
# likelihood for proposed parameter
            newLK <- calcLK(n, y, theta, npar, X)
            num <- newLK + log(dnorm(theta[i],mu[i],sig[i]))
            den <- LK + log(dnorm(oldtheta,mu[i],sig[i]))
            A <- min(1,exp(num-den))
            u <- runif(1)
# Accept/reject within model M-H step
                if (u <= A) {   LK <- newLK              }
                else { theta[i] <- oldtheta}  }
# end within model loop
# start RJ loop
        r <- sample(2:5,1);          oldtheta <- theta[r];
        LK <- calcLK(n, y, theta, npar, X)
# Covariate not currently in model, propose value for parameter
        if (theta[r] == 0) { theta[r] <- rnorm(1, muRJ[r], sigRJ[r])}
# For covariate currently in model, try omitting
        else {   theta[r] <- 0}
# likelihood
    newLK <- calcLK(n, y, theta, npar, X)
    if (theta[r] != 0) {# Covariate potentially being added
        num <- newLK + log(dnorm(theta[r],mu[r],sig[r])) + log(RJprob[r])
        den <- LK + log(dnorm(theta[r],muRJ[r],sigRJ[r])) +
                                log(1-RJprob[r])}
    else {# Covariate potentially being removed
        num<- newLK+log(dnorm(oldtheta,muRJ[r],sigRJ[r]))
                                +log(1-RJprob[r])
        den <- LK+log(dnorm(oldtheta,mu[r],sig[r]))   + log(RJprob[r])}
# Accept/reject RJ step
        A <- min(1,exp(num-den));          u <- runif(1)
        if (u <= A) { LK <- newLK          }
                else { theta[r] <- oldtheta}
# end RJ loop
# Record parameter values and retention inidcators:
    for (i in 1:npar) { sample[t,i] <- theta[i];   samp2[t,i] <- theta[i]^2 }
```

```
       for (r in 1:npar) { if (theta[r] == 0) {Ret[t,r] <- 0}
                           else { Ret[t,r] <- 1}}
}
# End overall loop
# posterior means and sd, retention rates
for (i in 1:npar) { totret[i] <- sum(Ret[B1:T,i])
        postmn[i] <- sum(sample[B1:T,i])/totret[i]
        retrate[i] <- totret[i]/(T-B)
        poststd[i] <- sqrt((sum(samp2[B1:T,i])-totret[i]
                           *postmn[i]^2)/totret[i]) }
```

Note that posterior means for coefficients may be conditional on retention, or unconditional, with postmn[i] having divisors totret[i] or T-B respectively in the second line of the
final for loop. Model selection and marginal retention rates for individual parameters may
depend on the proposal densities used in the model selection stage.

For example, initially $N(0, 0.5)$ densities are used. From a run of $T = 10\,000$ iterations
with 1000 burn-in, posterior retention rates (retrate in the code) then exceed 0.95 for $x_1$ and
$x_2$, so that posterior odds of the predictor being necessary to the regression exceed 19 to 1.
However, retention rates for $x_3$ and $x_4$ are below 0.95, and for $x_4$ the rate is only 0.74. If more
diffuse $N(0, 1)$ proposal densities are used instead in the model selection stage, the profile
of posterior retention rates is similar to that using a $N(0, 0.5)$ proposal, though the retention
rate for $x_1$ falls to 0.94. As a final option, we take RJ proposal densities $u_k \sim N(m_k, 1)$ centred
approximately at the coefficient means $m_k$ from the initial analysis without predictor selection;
specifically $m = c(-1.5, 2.7, 2.3, 1.7, 0.9)$. This restores the marginal retention rate for $x_1$ to
above 0.95, while that for $x_4$ falls to 0.71.

Let $R_k = 1$ if $x_k$ is retained, $R_k = 0$ otherwise. There are $2^4$ possible models. Model
indicators $(M \in 1, \ldots, 16)$ may be calculated for each iteration as $M = 1 + 8I(R_1 = 1) + 4I$
$(R_2 = 1) + 2I(R_3 = 1) + I(R_4 = 1)$. Analysis in a spreadsheet of the final 9000 iterations under
the $u_k \sim N(m_k, 1)$ proposals shows the most frequently selected model to be $(1, x_1, x_2, x_3, x_4)$
with probability 0.57, while the model $(1, x_1, x_2, x_3)$ has posterior probability 0.26.

## 1.4    Monitoring MCMC chains and assessing convergence

An important practical issue involves assessment of convergence of the sampling process used
to estimate parameters, or more precisely update their densities. In contrast to convergence
of optimising algorithms (maximum likelihood or minimum least squares, say) convergence
here is used in the sense of convergence to a density rather than single point, namely the target
density $p(\theta|Y)$.

The worked examples above involved single chains, but it is preferable in achieving convergence to use two or more parallel chains to ensure a complete coverage of the sample space,
and lessen the chance that the sampling will become trapped in a relatively small region. Single long runs may, however, often be adequate for relatively straightforward problems, or as
a preliminary to obtain inputs to multiple chains.

A run with multiple chains requires overdispersed starting values (Cowles and Carlin,
1996), and these might be obtained from a preliminary single chain run; for example, one
might take the 1st and 99th percentiles of parameters from a trial run as initial values in a two
chain run (Bray, 2002), or the posterior means from a trial run combined with null starting
values. Another option might combine null parameters in one chain with parameters obtained
as a random draw[3] from a trial run in the other. Null starting values might be zeroes for

---

[3] Obtained, for example, by using the state space command in BUGS.

regression parameters, ones for precisions or variances, and identity matrices for precision or covariance matrices. Note that in BUGS not all parameters need necessarily be initialised, and parameters may instead be initialised by generating[4] from their priors.

Problems of MCMC convergence may reflect model formulation: highly parameterised models fitted to small datasets, adoption of certain forms of diffuse prior for variance parameters in hierarchical models (Browne and Draper, 2006), correlations between parameters, or general problems in model identifiability. Re-parameterisation to reduce correlation – such as centring predictor variables in regression – may improve convergence (Gelfand et al., 1995; Zuur et al., 2002). In non-linear regressions, a log transform of a parameter may be better identified than its original form. Running multiple chains will often highlight unsuspected issues in model identifiability. In some situations (e.g. models with multiple random effects, some of which are not identifiable), it may be acceptable to monitor identifiable parameters rather than impose additional constraints which may slow sampling (e.g. Besag et al., 1995, p. 15). For example, centred random effects may be identifiable (e.g. Rodrigues and Assuncao, 2008).

Techniques applied within MCMC itself to aid convergence include over-relaxation methods and state-space augmentation (Gilks and Roberts, 1996; Damien et al., 1999; Green, 2001), and block sampling of parameters (Roberts and Sahu, 1997). Over-relaxation involves generating multiple samples of each parameter and then choosing the one least correlated with the current value, so potentially reducing the tendency for sampling to become trapped in a highly correlated random walk.

### 1.4.1    Convergence diagnostics

Convergence for multiple chains may be assessed using Gelman–Rubin scale reduction factors, which are included in BUGS, whereas single chain diagnostics require use of the CODA or BOA packages in R (Plummer et al., 2006; Smith, 2007). The scale reduction factors compare variation in the sampled parameter values within and between chains. If parameter samples are taken from a complex or poorly identified model, then a wide divergence in the sample paths between different chains will be apparent (e.g. Gelman, 1996, Figure 8.1), and variability of sampled parameter values between chains will considerably exceed the variability within any one chain. Therefore define

$$V_j = \sum_{t=B+1}^{T} (\theta_j^{(t)} - \overline{\theta}_j)^2/(T - B - 1)$$

as the variability of the parameter samples $\theta_j^{(t)}$ within the $j$th chain ($j = 1, \ldots J$). This is assessed over T-B iterations after a burn in of $B$ iterations. An overall estimate of variability within chains is the average $V_W$ of the $V_j$. Let the average of the chain means $\overline{\theta}_j$ be denoted $\overline{\theta}_.$. Then the between chain variance is

$$V_B = \frac{T - B}{J - 1} \sum_{j=1}^{J} (\overline{\theta}_j - \overline{\theta}_.)^2.$$

The scale reduction factor (SRF) compares a pooled estimator of var($\theta$), given by

$$V_P = V_B/(T - B) + (T - B) V_W/(T - B - 1),$$

---

[4] This involves 'gen ints' in BUGS.

with the within sample estimate $V_W$. Specifically the SRF is

$$(V_P/V_W)^{0.5}$$

and values of the SRF, or Gelman–Rubin statistic, under 1.2 indicate convergence. Brooks and Gelman (1998) propose a multivariate version of the SRF.

Another multi-chain convergence statistic developed by Brooks and Gelman (1998) involves a ratio of parameter interval lengths. For each chain, the length of the $100(1 - \alpha)\%$ interval for parameter $\theta$ is obtained, namely the gap between $0.5\alpha$ and $(1 - 0.5\alpha)$ points from T-B samples. This provides $J$ within-chain interval lengths, with mean $I_W$. For the pooled output of $(T - B)J$ samples, the same $100(1 - \alpha)\%$ interval $I_P$ is also obtained. The ratio $I_P/I_W$ converges to 1 under convergent mixing over chains.

The analysis of sampled values from a single MCMC chain or parallel chains may be seen as an application of time series methods in regard to problems such as assessing stationarity in an autocorrelated sequence (Roberts, 1996). Thus the autocorrelation at lags 1, 2, and so on, may be assessed from the original series of sampled values $\theta^{(t)}$, at lag 1 from the samples $\theta^{(t+1)}, \theta^{(t+2)} \cdots$, or from more widely spaced sub-samples $k$ steps apart $\theta^{(t)}, \theta^{(t+k)}, \theta^{(t+2k)}$. High autocorrelation in samples means that more samples are needed to provide a given precision in the posterior summaries. The effective sample size ESS (obtainable from CODA) represents the equivalent number of independent iterations that the chain provides, and is calculated as

$$\text{ESS} = \frac{T}{1 + 2 \sum_{k=1}^{\infty} \rho_k}$$

where $\rho_k$ is the autocorrelation at lag $k$.

Geweke (1992) developed a $t$-test applicable to assessing convergence in runs of sampled parameter values, both in single and multiple chain situations. Let $\bar{\theta}_a$ be the posterior mean of sampled $\theta$ values from the first $n_a$ iterations in a chain (after burn-in) and $\bar{\theta}_b$ be the mean from the last $n_b$ draws. If there is a substantial run of intervening iterations then the two samples should be independent. Let $V_a$ and $V_b$ be the variances of these averages[5] . Then the statistic

$$Z = (\bar{\theta}_a - \bar{\theta}_b)/(V_a + V_b)^{0.5}$$

should be approximately $N(0, 1)$. In CODA, this test is obtained with fractions of samples in the first and second sub-samples specified, e.g. geweke.diag(th, frac1 = 0.1, frac2 = 0.5).

### 1.4.2    Model identifiability

As mentioned above, delayed convergence may reflect problems in model identifiability due to over-fitting or redundant parameters. Use of diffuse priors increases the chances of a poorly identified model, or even effective impropriety, especially in complex hierarchical models

---

[5] If by chance the successive samples $\theta_a^{(t)}, t = 1, \ldots n_a$ and $\theta_b^{(t)}, t = 1, \ldots, n_b$ were independent, then $V_a$ and $V_b$ would be obtained as the population variance of the $\theta^{(t)}$, namely $V(\theta)$, divided by $n_a$ and $n_b$. In practice dependence in the sampled values is likely, and $V_a$ and $V_b$ must be estimated by allowing for the autocorrelation. Thus

$$V_a = (1/n_a) \left[ \gamma_0 + \sum_{j=1}^{n_a - 1} \gamma_j \left( \frac{n_a - j}{n_a} \right) \right]$$

where $\gamma_j$ is the autocovariance at lag $j$. In practice only a few lags may be needed.

(Hobert and Casella, 1996; Gelfand and Sahu, 1999; Berger *et al.*, 2005), and elicitation of more informative priors may assist identification and convergence. Slow convergence will show in poor mixing with high autocorrelation in the successive sampled values of parameters, apparent graphically in trace plots that wander rather than rapidly fluctuating around a stable mean.

Examples where identifiability issues are raised include random effects in nested models, for instance

$$y_{ij} = \mu + \eta_i + u_{ij} \qquad i = 1, \ldots, n; j = 1, \ldots, m \tag{1.6}$$

where $\eta_i \sim N(0, \sigma_\eta^2), u_{ij} \sim N(0, \sigma_u^2)$. Poor mixing may occur because the mean of the $\eta_i$ and the global mean $\mu$ are confounded: a constant may be added to the $\eta_i$ and subtracted from $\mu$ without altering the likelihood (Gilks and Roberts, 1996). Transformation in (1.6) may be applied to produce identifiability (Vines *et al.*, 1996), using new parameters

$$v = \mu + \bar{\eta}; \alpha_i = \eta_i - \bar{\eta}$$

in the model

$$y_{ij} = v + \alpha_i + u_{ij},$$

$$\alpha_1 \sim N(0, (m-1)\sigma_\eta^2/m),$$

$$\alpha_j \sim N\left(-\sum_{k=1}^{j-1} \alpha_k, \frac{m-j}{m-j+1}\sigma_\eta^2\right), \qquad (1 < j < m),$$

$$\alpha_m = -\sum_{k=1}^{m-1} \alpha_k.$$

More complex examples occur in a spatial disease model with both iid and spatially structured random errors (Gelfand *et al.*, 1998), sometimes known as a spatial convolution model and considered in Example 1.4 below, and in a particular kind of multiple random effects model, the age-period-cohort model (Knorr-Held and Rainer, 2001). Identifiability issues may also occur in discrete mixture regressions (Chapter 4), and structural equation models (Chapter 9), due to label switching during the MCMC sampling (Jasra *et al.*, 2005; Dunson, 2010). Such instances of non-identifiability will show as essentially non-convergent parameter series between chains, whereas simple constraints on parameters may achieve identifiability. For example, if a structural equation model involved a latent construct such as alienation, and indicator loadings on this construct were not suitably constrained, then one chain might fluctuate around a loading of −0.8 on social integration (the obverse of alienation), and another chain fluctuate around a loading of 0.8 on alienation.

### Example 1.3   MCMC diagnostics in R

To illustrate the range of diagnostic tools available in R, consider sampled data for a normal linear regression with $n = 1000$ observations, $N(0, 1)$ residuals, and a single predictor. The coda and rjags libraries are used. The code calls the BUGS program linear_reg.bug which can be located in the working directory (the cat command can also be used). For this particular example, autocorrelations are extremely low and the ESS is essentially equivalent to the total number of samples, except for the precision parameter.

The BUGS code in linear_reg.bug (a text file) is

```
model {for (i in 1:N) {y[i] ~ dnorm(b[1] + b[2]* x[i], tau)}
       for (j in 1:2) {b[j] ~ dnorm(0, 0.0001)}
       tau ~ dgamma(1,0.001)}
```

The code for a two chain analysis of 10000 iterations with 1000 burn in is

```
library(coda); library(rjags)
# generate data
N <- 1000; x <- runif(N,-1,1); eps <- rnorm(N, 0, 1); y <- x + eps
setwd("C://R files")
inits <- list(b =c(0,0), tau = 1)
M   <- jags.model('linear_reg.bug', data = list('x' = x, 'y' = y,   'N' = N),
                  n.chains = 2,n.adapt = 1000)
params <- c('b', 'tau')
samples <- coda.samples(M,  params, n.iter=10000)
plot(samples)
burn.in <- 1000
# summaries
summary(window(samples, start = burn.in))
HPDinterval(samples, prob = 0.95)
# diagnostics and plots
autocorr.diag(samples)
effectiveSize(samples)
gelman.diag(samples)
gelman.plot(samples)
```

## 1.5   Model assessment

Having achieved convergence with a suitably identified model, a number of processes may be required to firmly establish the status of a model. These include model checks (also called model criticism), and sensitivity assessment of the relation of posterior inferences to prior assumptions. If these are satisfactory, one may move on to model choice (or possibly model averaging).

### 1.5.1   Sensitivity to priors

As mentioned above, with small samples, or with models where latent effects are to some extent identified by their prior, there may be sensitivity in posterior inferences to the assumed priors. There may also be sensitivity if an informative prior based on accumulated knowledge is adopted.

One strategy is to consider a limited range of alternative priors and assess changes in inferences; this is known as 'informal' sensitivity analysis (Gustafson, 1996). One might also consider more formal approaches to robustness based perhaps on non-parametric priors (such as the Dirichlet process prior) or via mixture ('contamination') priors. For instance, one might assume a two group mixture with larger probability $1 - p$ on the 'main' prior $\pi_1(\theta)$, and a smaller probability such as $p = 0.1$ on a contaminating density $\pi_2(\theta)$, which may be any density (Berger, 1990; Gustafson, 1996). One might consider the contaminating prior to be a flat reference prior, or one allowing for shifts in the main prior's assumed parameter values (Berger, 1990). For instance, if $\pi_1(\theta)$ is $N(0, 1)$, one might take $\pi_2(\theta) \sim N(m_2, v_2)$, where higher stage priors set $m_2 \sim U(-0.5, 0.5)$ and $v_2^{0.5} \sim U(0.5, 1.5)$.

In large datasets and with simple non-hierarchical models (e.g. regression analyses), inferences may be robust to changes in prior unless priors are heavily informative. By contrast, in hierarchical models there may be sensitivity in inferences regarding underlying latent unit effects (Chapter 2). For example, in small area mortality applications, meta-analysis or in league table comparisons, the goal may be the smoothing of unreliable rates based on small event counts or small sample sizes. Results may be sensitive to priors on variance parameters in such models, especially where different types of random effect are present (Gelfand *et al.*, 1998; Daniels, 1999).

A strategy of adopting just proper priors on variances (or precisions) may be advocated in terms of letting the data speak for themselves, such as a gamma $Ga(\epsilon, \epsilon)$ priors on precisions with $\epsilon = 0.001$ or $\epsilon = 0.0001$. However, this may cause slow convergence and relatively weak identifiability, and there may be sensitivity in inferences between analyses using different supposedly vague priors (Kelsall and Wakefield, 1999). One might introduce stronger priors favouring particular values more than others (e.g. a $Ga(1,0.1)$ prior on a precision), or data based priors taking account of observed variability. Mollié (1996) suggests such a strategy for the spatial convolution model. Alternatively the model might specify that random effect variances are interdependent in their prior; this is a form of extra information (see Example 1.4).

### 1.5.2   Model checks

Posterior predictive checks whether based on some form of cross-validation (including leave one out methods), or using all the data, have been discussed above. These may be complemented by outlier and influence analysis. Thus in a linear regression model with normal errors

$$y_i = \beta_1 + \beta_2 x_{1i} + \dots \beta_{p+1} x_{pi} + e_i$$

the posterior mean of $\widehat{e}_i = y_i - X_i\widehat{\beta}$ compared to its posterior standard deviation provides an indication of outlier status (Pettit and Smith, 1985; Chaloner, 1994). An overall assessment of normality might involve a posterior p-test (Meng, 1994) using an established omnibus normality criterion, or Q-Q plots of posterior estimates of residuals. Alternatively the CPO predictive quantity $CPO_i = p(y_i|y_{[i]})$ may be used as an outlier diagnostic, and as a basis for influence measures. Weiss and Cho (1998) consider possible divergence criteria in terms of the ratios

$$a_i = \frac{p(\theta|y_{[i]})}{p(\theta|y)} = \frac{CPO_i}{p(y_i|\theta)},$$

such as the $L_1$ norm, with the influence of case $i$ on the totality of model parameters then represented by $d(a_i) = 0.5|a_i - 1|$.

Residual analysis is more complicated for discrete responses or when there are departures from residual normality. Models which can replicate the normal linear model by introducing latent data lead to particular types of Bayesian residual (Albert and Chib, 1995). Thus in a binary probit or logit model, underlying the observed binary $y$ are latent continuous variables $z$, confined to negative or positive values according as $y$ is 0 or 1. An estimate of the residual density is then provided by monitoring $z_i - X_i\beta$, as in a normal errors model. To exemplify applications where latent variable representation is not possible, model criticism for beta-binomial regression is discussed by Pires and Diniz (2012). Thus with

$$y_i \sim Bin(n_i, \pi_i),$$
$$\pi_i \sim Be(\alpha_{1i}, \alpha_{2i}),$$
$$\rho_i = 1/(\alpha_{1i} + \alpha_{2i} + 1),$$

consider fitted values $\tilde{y}_i$ that correspond to the maximum of $CPO_i$. Then as well as deviance residuals (Spiegelhalter *et al.*, 2002), one may define residuals

$$r_i = \frac{y_i - \tilde{y}_i}{\{n_i \hat{\pi}_i (1 - \hat{\pi}_i)[1 + (n_i - 1)\hat{\rho}_i]\}^{0.5}}$$

where $\hat{\pi}_i$ and $\hat{\rho}_i$ are posterior mean estimates.

Residual analysis and model criticism for hierarchical models (e.g. Albert and Chib, 1997; Yan and Sedransk, 2010; Scheel *et al.*, 2011) may involve detecting unanticipated hierarchical structure, as against detecting outliers, or assessing conflict between prior and likelihood information at the node level.

## 1.5.3   Model choice

Model choice is considered in Chapter 2 and certain further aspects which are particularly relevant in regression modelling are discussed in Chapter 3. While marginal likelihood, and the Bayes factor based on comparing such likelihoods, defines the canonical model choice, in practice (e.g. for complex random effects models or models with diffuse priors), this method may be relatively difficult to implement. Tractable approaches based on the marginal likelihood principle include the importance sampling method (e.g. Gelfand and Dey, 1994; Lenk and Desarbo, 2000), and the method of Chib (1995) based on the marginal likelihood identity.

Methods such as cross validation by single case omission lead to a form of pseudo Bayes factor based on multiplying the estimated CPO over all cases (Gelfand, 1996, p. 150). Formal cross-validation when based on actual omission of each case in turn may be only practical with relatively small samples. Other sorts of partitioning of the data into training samples and hold-out (or validation) samples may be applied and are less computationally intensive.

In subsequent chapters the main methods of model choice used are those based on predictive criteria, comparing model predictions $y_{rep}$ with actual observations[6] (e.g. Gelfand and Ghosh, 1998), and the DIC criterion (Spiegelhalter *et al.*, 2002); see Chapter 2. These are admittedly not formal Bayesian choice criteria, but are relatively easy to apply over a wide range of models including non-conjugate and heavily parameterised models.

The marginal likelihood approach leads to posterior probabilities or weights on different models, which in turn are the basis for parameter estimates derived by model averaging (Wasserman, 2000). Model averaging has particular relevance for regression models, especially in multicollinear situations where competing specifications provide closely comparable explanations for the data, and so there is a basis for weighted averages of predictions and parameters over different models. Model averaging under predictor selection methods (e.g. George and McCulloch, 1993; Kuo and Mallick, 1998) is discussed in Chapter 3.

---

[6] A simple approach to predictive fit generalises the method of Laud and Ibrahim (1995). Let $y_i$ be the observations, $\phi$ be parameters, and $y_{rep,i}$ be replicates sampled from $p(y_{rep}|\phi)$. Suppose $v_i$ and $\zeta_i$ are the posterior mean and variance of $y_{rep,i}$, then one possible criterion for any $w > 0$ is

$$C = \sum_{i=1}^{n} \zeta_i + [w/(w+1)] \sum_{i=1}^{n} (v_i - y_i)^2$$

Typical values of $w$ at which to compare models might be $w = 0.1, w = 1$, and $w = 10$. Larger values of $w$ put more stress on the closeness of fit between $v_i$ and $y_i$, and so downweight the precision of predictions.

## Example 1.4    Self-harm in London Small Areas

This example demonstrates the use of BUGS in a spatial example, and also illustrates possible influence of prior specification on regression coefficients and random effects. Consider a small area health outcome, namely hospital admissions for self-harm $y_i$ in a three year period (ICD10 X60–X84, financial years 2004/05 to 2006/07) in 625 London small areas[7] (electoral wards). Focusing first on regression effects, there is overwhelming accumulated evidence that self-harm risk is higher in deprived areas. Having allowed for the impact of age differences via indirect standardisation (to provide expected admissions $E_i$), the following model is assumed

$$y_i \sim \text{Po}(\mu_i)$$

$$\mu_i = E_i \rho_i$$

$$\log(\rho_i) = \beta_1 + \beta_2 x_i,$$

where the deprivation score $x_i$ is in standard form. The only parameters, $\beta_1$ and $\beta_2$, are assigned diffuse $N(0, 1000)$ priors. The sum of observed and expected events is the same, and since $x$ is standardised, one might expect $\beta_1$ to be near zero. Two sets initial values are adopted, $\beta = (0,0)$ and $\beta = (0, 0.1)$, with the latter based on the mean of a trial (single chain) run. A two chain run then shows early convergence via Gelman–Rubin criteria (at under 1000 iterations). From iterations 1001–10000 pooled over the chains a mean and 95% credible interval for $\beta_2$ of 0.117 (0.103, 0.131) are obtained.

However, there may well be information which would provide more informative priors. Relative risks $\rho_i$ between areas for health outcomes reflect gradients in risk for individuals over attributes such as income, occupation, health behaviours, household tenure, ethnicity etc. These gradients typically show at most five-fold variation between social categories. Accumulated evidence, including evidence for London wards, suggests that extreme relative contrasts in standard morbidity ratios ($100 \times \rho_i$) between areas are unlikely to exceed 5 or 10 (e.g. a range from 50 to 250). Simulating with the known covariate $x_i$ and expectancies $E_i$, it is possible to obtain or elicit priors consistent with these prior beliefs. For instance, one might consider taking a $N(0, 1)$ prior on $\beta_1$ and a $N(0.25, 1.5)$ prior on $\beta_2$. The latter favours positive values but still has a large part of its density over negative values.

Values of $y_i$ are simulated (see Model 2 in the code for Example 1.4) with these priors; note that initial values are by definition generated from the priors (i.e. the option 'gen inits' is necessarily adopted) and since this is pure simulation there is no notion of convergence. Relative risks are summarised in the ratios $y_i/E_i$, and the 2.5% and 97.5% percentiles of posterior median relative risk are 0.62 and 1.88. The range in posterior median relative risks is from 0.58 to 3.71. So this informative prior specification appears broadly in line with accumulated evidence.

One might then see how far inference about $\beta_2$ is affected by adopting the $N(0.25, 1.5)$ prior instead of the $N(0, 1000)$ diffuse prior[8] when the observations are restored. In fact, the posterior mean and 95% credible interval from a two chain run is found to be the same as under the diffuse prior.

A different example of sensitivity analysis involves using a contamination prior on $\beta_2$. Thus suppose $\pi_1(\beta_2)$ is $N(0.25, 1.5)$ as above, but that for $\pi_2(\beta_2)$ a Student $t$ with 2 degrees of freedom but same mean zero and variance is adopted, and $p = 0.1$. Again the same credible interval for $\beta_2$ is obtained as before (Model 3 in Example 1.4). In a simple non-hierarchical

---

[7] The first is the City of London (as one combined ward), then wards are alphabetic within boroughs arranged alphabetically (Barking, Barnet, ... , Westminster). Spatial interaction is based on contiguity.

[8] The prior on the intercept is changed to $N(0, 1)$ also.

prior specification such as this, one might take the contaminating prior to be completely flat (dflat() in BUGS), and this is suggested as an exercise. In the current example, inferences on $\beta_2$ appear robust to alternative priors, and this is frequently the case with regression parameters in large samples – though with small datasets there may well be sensitivity.

An example where sensitivity in inferences is more likely is in estimation of underlying area effects when observations consist of small event counts or populations at risk (Manton *et al.*, 1987). In such circumstances, borrowing strength procedures using hierarchical random effects imply smoothing of the rate for any unit towards the average implied under the density of the effects, especially for units with small sample sizes. In spatial applications, two types of random variability have been suggested: the first is unstructured variation, under which smoothing is towards a global average, and the second is spatially structured variation whereby smoothing is towards the average in the neighbourhood of adjacent wards. Then assuming

$$\log(\rho_i) = \beta_1 + \alpha_i$$

the total latent area effect $\alpha_i$ consists of an unstructured or 'pure heterogeneity' effect $v_i$ and a spatial effect $\phi_i$. While the data holds information about which type of effect is more predominant, the prior on the variances $\sigma_v^2$ and $\sigma_\phi^2$ may also be important in identifying the relative roles of the two error components.

A popular prior used for specifying spatial effects, the conditional autoregressive prior (Besag *et al.*, 1991) raises an identifiability issue in that it specifies differences in risk between areas $i$ and $j$, namely $\phi_i - \phi_j$, but not the average level (i.e. the location) of the spatial risk (see Chapter 8). This prior can be specified in a conditional form, namely

$$\phi_i \sim N\left(\sum_{j \in A_i} \phi_j/M_i, \sigma_\phi^2/M_i\right),$$

where $M_i$ is the number of areas adjacent to area $i$, $A_i$ denotes that set of areas, and $\sigma_\phi^2$ is interpreted as a conditional rather than marginal variance. The latter may be estimated as the empirical variance of sampled $\phi_i$. To resolve the identifiability problem one may centre the sampled $\phi_i$ at each MCMC iteration and so provide a location; i.e. actually use in the model to predict $\log(\rho_i)$ the shifted effects $\phi_i' = \phi_i - \bar{\phi}$.

Following Sun *et al.* (1999), identifiability can also be gained by introducing a propriety parameter $\gamma$

$$\phi_i \sim N\left(\gamma \sum_{j \in A_i} \phi_j/M_i, \sigma_\phi^2/M_i\right) \tag{1.7}$$

taken to have prior $\gamma \sim U(0, 1)$. Issues still remain in specifying priors on $\sigma_v^2$ and $\sigma_\phi^2$ (or their inverses), and in identifying both these variances and the separate risks $v_i$ and $\phi_i$ in each area in the model

$$\log(\rho_i) = \beta_1 + v_i + \phi_i,$$

where $v_i \sim N(0, \sigma_v^2)$. A 'diffuse prior' strategy might be to adopt gamma priors $Ga(\epsilon_1, \epsilon_2)$ on the precisions $1/\sigma_v^2$ and $1/\sigma_\phi^2$, where $\epsilon_1 = \epsilon_2 = \epsilon$ and $\epsilon$ is a small constant such as $\epsilon = 0.001$, but possible problems in doing this are noted above. One might, however, set priors on $\epsilon_1$ and $\epsilon_2$ themselves rather than presetting them (Daniels and Kass, 1999), analogous to contamination priors in allowing for higher level uncertainty.

Identifiability might also be improved by instead linking the specification of $v_i$ and $\phi_i$ (see Model 4 in Example 1.4). For example, one might adopt a bivariate prior on these random effects as in Langford *et al.* (1999), and discussed in Chapter 8. Or one might still keep $v_i$ and $\phi_i$ as univariate errors, but allow their variances to be interdependent, for instance taking $\sigma_v^2 = c\sigma_\phi^2$ so that one variance is conditional on the other and a pre-selected value of $c$. Bernardinelli *et al.* (1995) recommend $c = 0.7$. A prior on $c$ might also be used, e.g. a gamma prior with mean 0.7. One might alternatively take a bivariate prior (e.g. bivariate normal) on the log transformed variances, $\log(\sigma_v^2)$ and $\log(\sigma_\phi^2)$. Another possibility is a shrinkage prior (Daniels, 1999), as in a $U(0, 1)$ prior on $\sigma_v^2/[\sigma_v^2 + \sigma_\phi^2]$.

Here we first consider independent $Ga(0.5, 0.0005)$ priors on $1/\sigma_v^2$ and $1/\sigma_\phi^2$ in a two chain run. One set of initial values on the precisions is provided by default unity values, and the other by posterior mean values under an initial single chain run. The problems possible with independent diffuse priors show in the relatively slow convergence of $\sigma_\phi$; not until 5000 iterations does the Gelman–Rubin statistic fall below 1.1. The standard deviations $\sigma_v$ and $\sigma_\phi$ of the random effects have posterior medians 0.45 and 0.038, with unstructured variation predominant.

In a second analysis the variances[9] are interrelated with $\sigma_v^2 = c\sigma_\phi^2$ and $c$ taken as $Ga(0.7, 1)$, drawing on the recommendation of Bernardinelli *et al.* (1995). This is relatively informative prior reflecting an expectation that a small area health outcome will probably show both types of variability. Further the prior on $1/\sigma_\phi^2$ allows for uncertainty in the parameters, i.e. instead of a default prior such as $1/\sigma_\phi^2 \sim Ga(1, 0.001)$, it is assumed that

$$1/\sigma_\phi^2 \sim Ga(a_1, a_2)$$

with

$$a_1 \sim Exp(1)$$

$$a_2 \sim Ga(1, 0.001).$$

The priors for $a_1$ and $a_2$ reflect the option sometimes used for a diffuse prior on precisions such as $1/\sigma_\phi^2$, namely a $Ga(1, v)$ prior (with $v$ preset at a small constant, such as $v = 0.001$).

This model achieves earlier convergence (before iteration 2000) in the random effect standard deviations. The last 7500 iterations of a two chain run of 10 000 iterations yield a posterior median for $c$ of 0.43, and for $\sigma_v$ and $\sigma_\phi$ of 0.25 and 0.38. The posterior medians of $a_1$ and $a_2$ are 1.68 and 0.35.

So sensitivity is apparent regarding variances of random effects in this example, despite the relatively large sample, though substantive inferences on area relative risks may be more robust. A suggested exercise is to experiment with other priors allowing interdependent variances or errors, e.g. a $U(0, 1)$ prior on $\sigma_v^2/[\sigma_v^2 + \sigma_\phi^2]$. A further exercise might involve summarising sensitivity on the inferences about relative risk, e.g. how many of the mean relative risks shift upward or downward by more than 2.5%, and how many by more than 5%, in moving from one random effects prior to another.

# References

Al-Awadhi, F., Hurn, M. and Jennison, C. (2004) Improving the acceptance rate of reversible jump MCMC proposals. *Statistics and Probability Letters*, **69**(2), 189–198.

Albert J. (2007) *Bayesian Computation with R*. Springer, New York, NY.

---

[9] In BUGS this interrelationship involves precisions.

Albert, J. and Chib, S. (1993) Bayesian analysis of binary and polychotomous response data. *Journal of the American Statistical Association*, **88**, 669–679.

Albert, J. and Chib, S. (1995) Bayesian residual analysis for binary response regression models. *Biometrika*, **82**, 747–956.

Albert, J. and Chib, S. (1997) Bayesian tests and model diagnostics in conditionally independent hierarchical models. *Journal of the American Statistical Association*, **92**(439), 916–925.

Alqallaf, F. and Gustafson, P. (2001) On cross-validation of Bayesian models. *Canadian Journal of Statistics*, **29**, 333–340.

Alston, C., Mengersen, K. and Pettitt, A. (eds) (2012) *Case Studies in Bayesian Statistical Modelling and Analysis*. Wiley, Chichester, UK.

Andrieu, C. and Moulines, É. (2006) On the ergodicity properties of some adaptive MCMC algorithms. *Annals of Applied Probability*, **16**(3), 1462–1505.

Barbieri, M. and Berger, J. (2004) Optimal predictive model selection. *Annals of Statistics*, **32**(3), 870–897.

Bayarri, M. and Berger, J. (2004) The interplay of Bayesian and frequentist analysis. *Statistical Science*, **19**(1), 58–80.

Bayarri, M. and Castellanos, M. (2007) Bayesian checking of the second levels of hierarchical models. *Statistical Science*, **22**, 322–343.

Berger, J. (1990) Robust Bayesian analysis: Sensitivity to the prior. *Journal of Statistical Planning and Inference*, **25**(3), 303–328.

Berger, J., Strawderman, W. and Tang, D. (2005) Posterior propriety and admissibility of hyperpriors in normal hierarchical models. *Annals of Statistics*, **33**, 606–646.

Bernardinelli, L., Clayton, D. and Montomoli C. (1995) Bayesian estimates of disease maps: how important are priors? *Statistics in Medicine*, **14**, 2411–2431.

Besag, J., Green, P., Higdon, D. and Mengersen, K. (1995) Bayesian computation and stochastic systems. *Statistical Science*, **10**, 3–41.

Besag, J., York, J. and Mollié, A. (1991) Bayesian image restoration, with two applications in spatial statistics. *Annals of the Institute of Statistical Mathematics*, **43**(1), 1–20.

Bray, I. (2002) Application of Markov chain Monte Carlo methods to projecting cancer incidence and mortality. *Journal of the Royal Statistics Society C*, **51**, 151–164.

Brooks, S. (1998) Markov chain Monte Carlo method and its application. *Journal of the Royal Statistical Society D*, **47**(1), 69–100.

Brooks, S. and Gelman, A. (1998) General methods for monitoring convergence of iterative simulations. *Journal of Computational and Graphical Statistics*, **7**, 434–456.

Brooks, S., Giudici, P. and Roberts, G. (2003) Efficient construction of reversible jump MCMC proposal distributions. *Journal of the Royal Statistics Society B*, **65**, 3–56.

Brooks, S., Gelman, A., Jones, G. and Meng, X.-L. (eds) (2011) *Handbook of Markov Chain Monte Carlo*. CRC, Boca Raton, FL.

Browne, W. and Draper, D. (2006) A comparison of Bayesian and likelihood-based methods for fitting multilevel models. *Bayesian Analysis*, **1**(3), 473–514.

Chaloner, K. (1994) Residual analysis and outliers in Bayesian hierarchical models. In P.R. Freeman (ed.), *Aspects of Uncertainty: A Tribute to D. V. Lindley*, pp. 149–157. Wiley, Chichester, UK.

Chib, S. (1995) Marginal likelihood from the Gibbs output. *Journal of the American Statistical Association*, **90**, 1313–1321.

Chib, S. (2013) Markov chain Monte Carlo Methods in Bayesian Theory and Applications, (eds) P. Damien, P. Dellaportas, N. Polson, D. Stephens. OUP.

Chib, S. and Greenberg, E. (1995) Understanding the Metropolis-Hastings algorithm. *The American Statistician*, **49**(4), 327–335.

Chib, S. and Jeliazkov, I. (2005) Accept–reject Metropolis–Hastings sampling and marginal likelihood estimation. *Statistica Neerlandica*, **59**(1), 30–44.

Clark, J. and Gelfand, A. (eds) (2006) *Hierarchical Modelling for the Environmental Sciences: Statistical Methods and Applications*. Oxford University Press, Oxford, UK.

Collett, D. (1991) *Modelling Binary Data*. Chapman & Hall, London.

Cowles, M. and Carlin, B. (1996) Markov chain Monte Carlo convergence diagnostics: a comparative review. *Journal of the American Statistical Association*, **91**(434), 883–904.

Damien, P., Wakefield, J. and Walker, S. (1999) Gibbs sampling for Bayesian non-conjugate and hierarchical models by using auxiliary variables. *Journal of the Royal Statistical Society B*, **61**(2), 331–344.

Damien, P., Dellaportas, P., Polson, N. and Stephens, D. (2013) *Bayesian Theory and Applications*. Oxford University Press, Oxford, UK.

Daniels, M. (1999) A prior for the variance in hierarchical models. *Canadian Journal of Statistics*, **27**(3), 567–578.

Daniels, M. and Kass, R. (1999) Nonconjugate Bayesian estimation of covariance matrices and its use in hierarchical models. *Journal of the American Statistical Association*, **94**, 1254–1263.

Daniels, M., Chatterjee, A. and Wang, C. (2012) Bayesian model selection for incomplete data using the posterior predictive distribution. *Biometrics*, **68**(4), 1055–1063.

Ding, V., Hubbard, R., Rutter, C. and Simon, G. (2013) Assessing the accuracy of profiling methods for identifying top providers: performance of mental health care providers. *Health Services and Outcomes Research Methodology*, **13**(1), 1–17.

Dunson, D. (2010) Flexible Bayes regression of epidemiologic data. In A. O'Hagan and M. West (eds), *The Oxford Handbook of Applied Bayesian Analysis*. Oxford University Press, Oxford, UK.

Fan, Y. and Sisson, S. (2011) Reversible jump Markov chain Monte Carlo. In S. Brooks, A. Gelman, G. Jones and X.-L. Meng (eds), *Handbook of Markov Chain Monte Carlo*. CRC, Boca Raton, FL.

Garthwaite, P., Kadane, J. and O'Hagan, A. (2005) Statistical methods for eliciting probability distributions. *Journal of the American Statistical Association*, **100**(470), 680–701.

Gelfand, A. (1996) Model determination using sampling-based methods. In W. Gilks, S. Richardson and D. Spiegelhalter (eds), *Markov Chain Monte Carlo in Practice*, pp.145–161, Chapman Hall, London.

Gelfand, A. and Dey, D. (1994) Bayesian model choice: Asymptotics and exact calculations. *Journal of the Royal Statistical Society B*, **56**(3), 501–514.

Gelfand, A. and Ghosh, S. (1998) Model choice: A minimum posterior predictive loss approach. *Biometrika*, **85**(1), 1–11.

Gelfand, A. and Sahu, S. (1999) Identifiability, improper priors, and Gibbs sampling for generalized linear models. *Journal of the American Statistical Association*, **94**, 247–253.

Gelfand, A. and Smith, A. (1990) Sampling-based approaches to calculating marginal densities. *Journal of the American Statistical Association*, **85**, 398–409.

Gelfand, A., Dey, D. and Chang, H. (1992) Model determination using predictive distributions with implementation via sampling-based methods. In J.M. Bernardo, J.O. Berger, A.P. Dawid and A.F.M. Smith (eds), *Bayesian Statistics 4*, pp. 147–68. Oxford University Press, Oxford, UK.

Gelfand, A., Sahu, S. and Carlin, B. (1995) Efficient parameterizations for normal linear mixed models. *Biometrika*, **82**, 479–488.

Gelfand, A., Ghosh, S., Knight, J. and Sirmans, C. (1998) Spatio-temporal modeling of residential sales markets. *Journal of Business and Economic Statistics*, **16**, 312–321.

Gelman, A. (1996) Inference and monitoring convergence. In W. Gilks, S. Richardson, and D. Spiegelhalter (eds), *Practical Markov Chain Monte Carlo*, pp. 131–143. Chapman and Hall, London.

George, E. and McCulloch, R. (1993) Variable selection via Gibbs sampling. *Journal of the American Statistical Association*, **88**(423), 881–889.

Geweke, J. (1992) Evaluating the accuracy of sampling-based approaches to calculating posterior moments. In J.M. Bernardo, J.O. Berger, A.P. Dawid and A.F.M. Smith (eds), *Bayesian Statistics 4*. Clarendon Press, Oxford, UK.

Geyer, C. (2011) Introduction to Markov Chain Monte Carlo. In S. Brooks, A. Gelman, G. Jones and X.-L. Meng (eds), *Handbook of Markov Chain Monte Carlo*, chapter 1. CRC, Boca Raton, FL.

Ghosh, M. and Rao, J. (1994) Small area estimation: an appraisal. *Statistical Science*, **9**, 55–76.

Gilks, W. and Roberts, C. (1996) Strategies for improving MCMC. In W. Gilks, S. Richardson and D. Spiegelhalter (eds), *Practical Markov Chain Monte Carlo*, pp. 89–114. Chapman and Hall, London, UK.

Gilks, W. and Wild, P. (1992) Adaptive rejection sampling for Gibbs sampling. *Applied Statistics*, **41**, 337–48.

Gilks, W., Roberts, G. and Sahu, S. (1998) Adaptive Markov chain Monte Carlo through regeneration. *Journal of the American Statistical Association*, **93**, 1045–1054.

Gill, J. and Walker, L. (2005) Elicited priors for bayesian model specifications in political science research. *Journal of Politics*, **67**(3), 841–872.

Green, P. (1995) Reversible jump MCMC computation and Bayesian model determination. *Biometrika*, **82**, 711–732.

Green, P. (2001) A primer on Markov chain Monte Carlo. In O. Barndorff-Nielsen, D. Cox and C. Kluppelberg (eds), *Complex Stochastic Systems*, chapter 1, pp 1–62. Chapman and Hall, London, UK.

Green, P. (2003) Trans-dimensional Markov Chain Monte Carlo. In P. Green, N. Hjort and S. Richardson (eds), Highly Structured Stochastic Systems, pp. 179–198. Oxford University Press, Oxford, UK.

Griffin, J. and Stephens, D. (2013) Advances in Markov chain Monte Carlo. In P. Damien, P. Dellaportas, N. Polson and D. Stephens (eds), *Bayesian Theory and Applications*. Oxford University Press, Oxford, UK.

Gustafson, P. (1996) Robustness considerations in Bayesian analysis. *Statistical Methods in Medical Research*, **5**, 357–373.

Hamelryck, T., Mardia, K. and Ferkinghoff-Borg J. (eds) (2012) *Bayesian Methods in Structural Bioinformatics*. Springer, New York, NY.

Harvey, A. (1993) *Time Series Models*, 2nd edn. Harvester-Wheatsheaf, Hemel Hempstead, UK.

Hjort, N., Dahl, F. and Steinbakk, G. (2006) Post-processing posterior predictive p values. *Journal of the American Statistical Association*, **101**, 1157–1174.

Hobert, J. and Casella, G. (1996) The effect of improper priors on Gibbs sampling in hierarchical linear mixed models. *Journal of the American Statistical Association*, **91**(436), 1461–1473.

Hozo, S., Djulbegovic, B. and Hozo, I. (2005) Estimating the mean and variance from the median, range, and the size of a sample. *BMC Medical Research Methodology*, **5**(1), 13.

Jasra, A., Holmes, C.C. and Stephens, D.A. (2005) Markov chain Monte Carlo methods and the label switching problem in Bayesian mixture modeling. *Statistical Science*, **20**, 50–67.

Kelsall, J. and Wakefield, J. (1999) Discussion on Bayesian models for spatially correlated disease and exposure data (by N.G. Best *et al.*). In J. Bernardo *et al.* (eds), *Bayesian Statistics 6: Proceedings of the Sixth Valencia International Meeting*. Clarendon Press, Oxford, UK.

King, R., Morgan, B., Gimenez, O. and Brooks, S. (2010) Bayesian analysis for population ecology. CRC Press.

Knorr-Held, L. and Rainer, E. (2001) Prognosis of lung cancer mortality in West Germany: a case study in Bayesian prediction. *Biostatistics*, **2**, 109–129.

Kuo, L. and Mallick, B. (1998) Variable selection for regression models. *Sankhyā: The Indian Journal of Statistics B*, **60**, 65–81.

Langford, I., Leyland, A., Rasbash, J. and Goldstein, H. (1999) Multilevel modelling of the geographical distributions of diseases. *Journal of the Royal Statistical Society C*, **48**, 253–268.

Laud, P. and Ibrahim, J. (1995) Predictive model selection. *Journal of Royal Statistical Society B*, **57**, 247–262.

Lenk, P. and Desarbo, W. (2000) Bayesian inference for finite mixtures of generalized linear models with random effects. *Psychometrika*, **65**(1), 93–119.

Lu, D., Ye, M. and Hill, M. (2012) Analysis of regression confidence intervals and Bayesian credible intervals for uncertainty quantification. *Water Resources Research*, **48**(9), DOI: 10.1029/2011WR011289.

Lunn, D., Best, N. and Whittaker, J. (2008) Generic reversible jump MCMC using graphical models. *Statistics and Computing*, **19**, 395–408.

Manton, K., Woodbury, M., Stallard, E., Riggan, W., Creason, J. and Pellom, A. (1989) Empirical Bayes procedures for stabilizing maps of US cancer mortality rates. *Journal of the American Statistical Association*, **84**, 637–650.

Marshall, E. and Spiegelhalter, D. (2007) Identifying outliers in Bayesian hierarchical models: a simulation-based approach. *Bayesian Analysis*, **2**, 409–444.

Martin, A. and Quinn, K. (2006) Applied Bayesian inference in R using MCMCpack. *R News*, **6**(1), 2–7.

Manton, K., Stallard, E., Woodbury, M, Riggan, W., Creason, J. and Mason, T. (1987) Statistically adjusted estimates of geographic mortality profiles. *Journal of the National Cancer Institute*, **78**, 805–815.

Meng, X.L. (1994) Posterior predictive p-values. *Annals of Statistics*, **22**(3), 1142–1160.

Mollié, A. (1996) Bayesian mapping of disease. In W. Gilks, S. Richardson and D. Spieglehalter (eds), *Markov Chain Monte Carlo in Practice*, pp 359–380. Chapman and Hall, London, UK.

Neal, R. (2003) Slice sampling. *Annals of Statistics*, **31**, 705–741.

Parent E. and Rivot E. (2012) *Introduction to Hierarchical Bayesian Modeling for Ecological Data.* CRC, Boca Raton, FL.

Pettit, L. and Smith, A. (1985) Outliers and influential observations in linear models. In J. Bernardo, M. DeGroot, D. Lindley and A. Smith (eds), *Bayesian Statistics 2*, pp. 473–494. North- Holland, Amsterdam.

Pires, R. and Diniz, C. (2012) Bayesian residual analysis for beta-binomial regression models. *AIP Conference Proceedings*, **1490**, 259–267.

Plummer, M., Best, N., Cowles, K. and Vines, K. (2006) CODA: convergence diagnosis and output analysis for MCMC. *R News*, **6**, 7–11.

Racz, M. and Sedransk, J. (2010) Bayesian and frequentist methods for provider profiling using risk-adjusted assessments of medical outcomes. *Journal of the American Statistical Association*, **105**(489), 48–58.

Ritter, C. and Tanner, M. (1992) Facilitating the Gibbs sampler: the Gibbs stopper and the griddy-Gibbs sampler. *Journal of the American Statistical Association*, **87**(419), 861–868.

Roberts, G. (1996) Markov chain concepts related to sampling algorithms. In W. Gilks, S. Richardson and D. Spiegelhalter (eds), *Markov Chain Monte Carlo in Practice*, pp. 45–58, Chapman and Hall, London, UK.

Roberts, G. and Sahu, S. (1997) Updating schemes, correlation structure, blocking and parameterization for the Gibbs sampler. *Journal of the Royal Statistical Society B*, **59**, 291–317.

Rodrigues, A. and Assuncao, R. (2008) Propriety of posterior in Bayesian space varying parameter models with normal data. *Statistics and Probability Letters*, **78**, 2408–2411.

Rouder, J. and Lu, J. (2005). An introduction to Bayesian hierarchical models with an application in the theory of signal detection. *Psychonomic Bulletin and Review*, **12**(4), 573–604.

Rue, H., Martino, S. and Chopin, N. (2009) Approximate Bayesian inference for latent gaussian models using integrated nested laplace approximations. *Journal of the Royal Statistical Society B*, **71**, 319–392.

Sahu, S. (2002) Bayesian estimation and model choice in item response models. *Journal of Statistical Computation and Simulation*, **72**(3), 217–232.

Scheel, I., Green, P. and Rougier, J. (2011) A graphical diagnostic for identifying influential model choices in Bayesian hierarchical models. *Scandinavian Journal of Statistics*, **38**(3), 529–550.

Sherlock, C., Fearnhead, P. and Roberts, G. (2010) The random walk metropolis: linking theory and practice through a case study. *Statistical Science*, **25**(2), 172–190.

Shiffrin, R., Lee, M., Kim, W. and Wagenmakers, E.J. (2008) A survey of model evaluation approaches with a tutorial on hierarchical Bayesian methods. *Cognitive Science*, **32**(8), 1248–1284.

Sisson, S. (2005) Transdimensional Markov chains: A decade of progress and future perspectives. *Journal of the American Statistical Association*, **100**(471), 1077–1089.

Smith, B. (2007) boa: an R package for MCMC output convergence assessment and posterior inference. *Journal of Statistical Software*, **21**(11), 1–37.

Spiegelhalter, D., Best, N., Gilks, W. and Inskip, H. (1996) Hepatitis B: a case study of Bayesian methods. In W. Gilks, S. Richardson and D. Spieglehalter (eds), *Markov Chain Monte Carlo in Practice*, pp 21–43. Chapman and Hall, London, UK.

Spiegelhalter, D., Best, N., Carlin, B. and Van Der Linde, A. (2002) Bayesian measures of model complexity and fit. *Journal of the Royal Statistical Society B*, **64**(4), 583–639.

Statisticat LLC (2013) LaplacesDemon: Complete Environment for Bayesian Inference. R package version 13.10.07. http://www.bayesian-inference.com/software.

Sun, D., Tsutakawa, R. and Speckman, P. (1999) Posterior distribution of hierarchical models using CAR(1) distributions. *Biometrika*, **86**, 341–350.

Sutton, A. and Abrams, K. (2001) Bayesian methods in meta-analysis and evidence synthesis. *Statistical Methods in Medical Research*, **10**, 277–303.

Vehtari, A. and Ojanen, J. (2012) A survey of Bayesian predictive methods for model assessment, selection and comparison. *Statistics Surveys*, **6**, 142–228.

Vines, S., Gilks, W. and Wild, P. (1996) Fitting Bayesian multiple random effects models. *Statistics and Computing*, **6**, 337–346.

Wasserman, L. (2000) Bayesian model selection and model averaging. *Journal of Mathematical Psychology*, **44**, 92–107.

Weiss, R. and Cho, M. (1998) Bayesian marginal influence assessment. *Journal of Statistical Planning and Inference*, **71**, 163–177.

Wikle, C. (2003) Hierarchical Bayesian models for predicting the spread of ecological processes. *Ecology*, **84**(6), 1382–1394.

Yan, G. and Sedransk, J. (2010) A note on Bayesian residuals as a hierarchical model diagnostic technique. *Statistical Papers*, **51**(1), 1–10.

Zuur, G., Gartwaite, P. and Fryer, R. (2002) Practical use of MCMC methods: lessons from a case study. *Biometrical Journal*, **44**, 433–455.

# 2

# Hierarchical models for related units

## 2.1 Introduction: Smoothing to the hyper population

A relatively simple Bayesian problem but one which has motivated much research is that of ensemble estimation, namely estimating the parameters of a common distribution thought to underlie a collection of indicators for similar types of units, sometimes called a hyper-population. Among possible examples are medical, sports, or educational: exam success rates in different schools, results from a set of randomised controlled trials, surgical mortality rates in a set of hospitals (Austin *et al.*, 2001; Austin, 2002), or a run of batting averages for baseball players. A histogram plot of such indicators will typically be ragged, but often appear to suggest an underlying smoother density. While point estimates (e.g. based on fixed effects maximum likelihood) are generally considered unbiased, they are subject to variance instability and may provide unreliable inferences from comparisons, and other properties (e.g. improved stability and precision) may be relevant (Greenland, 2000). While technically unbiased, fixed effects estimators are also subject to measurement or design errors.

So underlying the unit-specific estimates one may posit an underlying set of latent effects characterized by a common density. The hyperparameters of this density are themselves unknowns, and generally include an average effect size and variance, the latter representing heterogeneity in latent outcomes between units, as distinct from the sampling variance of each individual estimate (Pauler and Wakefield, 2000). Given the parameters of the common density one seeks to make conditional estimates of the underlying outcome rate in each unit of observation. Because of the conditioning on the higher stage densities, such estimation for sets of similar units is often known as hierarchical modelling (Bernardo, 1996; Madigan and Ridgeway, 2004), though other terms (e.g. shrinkage estimation) are sometimes used. For instance, in the first stage of the Poisson–gamma model considered below, the observed counts are conditionally independent given the unknown means that are taken to have generated them. At the second stage, these means are themselves determined by the gamma density parameters, while the density for the gamma parameters forms the third stage.

*Applied Bayesian Modelling*, Second Edition. Peter Congdon.
© 2014 John Wiley & Sons, Ltd. Published 2014 by John Wiley & Sons, Ltd.

Hierarchical modelling usually results in a smoothing or shrinkage of estimates for each unit towards the average outcome rate (e.g. Xie *et al.*, 2012), and such estimates typically have greater precision and better out of sample predictive performance (James and Stein, 1961). Specifically, Rao (1975) shows that with respect to a quadratic loss function, Bayes estimators outperform classical (e.g. least squares) estimators in problems of simultaneous inference regarding a set of related parameters. Hierarchical methods generally alleviate small sample problems (where some rates are imprecisely estimated), and, as compared to fixed effect procedures, provide a flexible framework to quantify variability and compare units. Various comparisons are possible, such as rankings of smoothed estimates (Goldstein and Spiegelhalter, 1996; Xie *et al.*, 2009), or comparisons of rates or relative risks against thresholds (Deely and Smith, 1998; Austin, 2002), and summary measures characterising the latent effects can also be obtained (e.g. inequality indices for smoothed small area mortality rates) (Congdon and Southall, 2005). A hierarchical Bayesian method allows for full posterior densities of ranks of institutions (or areas, etc.) to be obtained, and outliers identified (Deely and Smith, 1998; Coory *et al.*, 2009). Policy intervention or prioritization may be better based on ranks allowing for uncertainty in observed trends or relativities (Sauer *et al.*, 2005). These procedures may, however, imply a risk of bias as against unadjusted maximum likelihood or fixed effects estimates, a dilemma known as the bias–variance trade-off (e.g. Austin *et al.*, 2003; Austin, 2005).

Hierarchical procedures for 'borrowing strength' or 'pooling strength' rest on implicit assumptions: that the units are exchangeable, or similar enough to justify an assumption of a common density (Lindley and Smith, 1972; Bernardo, 1996), and that the hierarchical model chosen is an appropriate one, for example, that a single normal hyperpopulation is appropriate (Marshall and Spiegelhalter, 1998). It may be that units are better considered exchangeable within sub-groups of the data, for example, if responses from randomised trials form one sub-group of units, while case–control studies form another. More robust options for hierarchical models include outlier accommodation or discrete mixtures of standard hierarchical schemes, such as discrete mixtures of the normal–normal or Poisson–gamma models.

Assessing the fit and appropriateness of assumptions made in hierarchical models raises questions about model choice and checking. So this chapter begins by setting out some guidelines as to model comparison and assessment, which are applicable to this and later chapters. There are no set 'gold standard' model choice criteria, though some arguably come closer to embodying true Bayesian principles than others.

## 2.2     Approaches to model assessment: Penalised fit criteria, marginal likelihood and predictive methods

There is usually uncertainty about model parameterisation, with a risk of underfitting or overfitting, and a guiding principle involves achieving a balance between parsimony and goodness of fit. Adding more parameters may improve within-sample fit but at the expense of weaker identifiability or less precise out-of-sample predictions. Penalised fit criteria obtained from MCMC estimation, and formal Bayesian model comparison, both seek to address such aspects of model comparison. The Bayesian approach to model choice and its implementation via MCMC sampling methods has benefits in comparisons of non-nested models, when classical procedures may encounter problems with parameters on the boundary of the parameter space. Examples are provided by discrete mixture models involving different numbers of components, and comparing a beta-binomial model as against a discrete mixture of binomials (Morgan, 2000).

## 2.2.1   Penalised fit criteria

Penalised goodness of fit measures (Burnham and Anderson, 2004; Bozdogan, 2000; Akaike, 1973) are used to prevent overfitting, and involve an adjustment to the model log-likelihood or deviance to reflect the number of parameters in the model. Thus suppose $L$ denotes the likelihood and $D$ the deviance of a model involving $p$ parameters. The deviance may be simply defined as minus twice the log likelihood, $D = -2 \log L$, or as a scaled deviance:

$$D_s = -2 \log(L/L_E),$$

where $L_E$ is the saturated likelihood obtained by an exact fit of predicted to observed data. To allow for the number of parameters (or 'dimension' of the model) one may use criteria such as the Akaike information criterion (or AIC), typically with the deviance evaluated at the maximum likelihood estimate $\hat{\theta}_{ML}$ expressed either as $D(\hat{\theta}_{ML}) + 2p$, or $D_s(\hat{\theta}_{ML}) + 2p$.

Another criterion used generally as a penalised fit measure, though also justified as an asymptotic approximation to the Bayesian posterior probability of a model, is the Bayesian information criterion (BIC) (Schwarz, 1978). Depending on the simplifying assumptions made it may take different forms, but the most common version is, for sample of size $n$,

$$\text{BIC} = D(\hat{\theta}_{ML}) + p[\log_e(n)].$$

The difference in BIC (divided by 2) between two models is an approximation to the logarithm of the Bayes factor, considered below (Kass and Raftery, 1995).

The BIC is classed as a consistent criterion in that it chooses the correct model with probability converging to 1 under certain conditions (e.g. nested models including the true model), whereas a conservative criterion such as the AIC asymptotically selects over-parameterised models with a non-zero probability (e.g. Vandewalle et al., 2013). So larger models (with more parameters) are more heavily penalised under the BIC than the AIC (Kadane and Lazar, 2004; Ward, 2008; Min et al., 2010). The BIC approximation for model $k$ is derived by considering the posterior probability for the model $M_k$ as in (2.1) below, and by expanding minus twice the log of that quantity around the maximum likelihood estimate. Extensions of the BIC have been proposed (e.g. Konishi et al., 2004; Chen and Chen, 2008).

For a model of known dimension, the AIC or BIC may be calculated either using average deviances over an MCMC chain (e.g. averages over all iterations of deviances $D^{(t)}$ derived using the sampled parameters $\theta^{(t)}$ at each iteration), or using deviances evaluated at the posterior mean $\bar{\theta}$ (Dempster, 1974). AIC or BIC comparisons of this kind allow a preliminary sifting of models when formal Bayesian methods of model choice may be difficult to apply, and more comprehensive selection methods (or averaging) are reserved to a final stage of the analysis involving a few closely competing models. However, they may also be used for model choice when one model has a clear minimum average AIC or BIC (Leeb and Pötscher, 2006; Baldassarre et al., 2009).

Monitoring fit measures such as the deviance over an MCMC run has utility if one seeks penalised fit measures taking account of model dimension. A complication is that the number of parameters in complex random effects models (including the hierarchical models considered in this chapter) is not actually defined. The approach of Spiegelhalter et al. (2002) may be used to estimate the effective number of parameters, denoted $p_e$. For data $y$ and parameters $\theta$, $p_e$ is approximated by the difference $E[D|y, \theta] - D(\bar{\theta}|y)$ between the expected deviance $E[D|y, \theta]$, as measured by the posterior mean of sampled deviances $D^{(t)} = D(\theta^{(t)})$ at iterations $t = 1, \dots, T$ in a long MCMC run, and the deviance $D(\bar{\theta}|y)$, evaluated at the posterior mean of the parameters. Then one may define the deviance information criterion or DIC, a penalised fit measure analogous to the AIC, namely

$$\text{DIC} = D(\bar{\theta}|y) + 2p_e.$$

Alternatively a modified BIC

$$\text{BIC} = D(\bar{\theta}|y) + p_e \log(n)$$

may be used as this takes account of sample size. Note that $p_e$ might also be obtained by comparing an average likelihood with the likelihood at the posterior mean and then multiplying by 2. Related work on effective parameters when the average likelihoods of two models are compared appears in Aitkin (1991).

## 2.2.2    Formal model selection using marginal likelihoods

The formal Bayesian model assessment scheme involves marginal likelihoods, and while it follows a theoretically clear procedure may in practice be difficult to implement in complex models or large samples (Gelfand and Ghosh, 1998). The Bayes factor may also be sensitive to the information contained in diffuse priors and is not defined for improper priors (Kadane and Lazar, 2004). Suppose $K$ models denoted $M_k, k = 1, \ldots, K$, have prior probabilities $\phi_k = \Pr(M_k)$ of being true, with $\sum_{k=1}^{K} \phi_k = 1$. Let $\theta_k$ be the parameter set in model $k$, with prior $\pi(\theta_k)$. Then the posterior probabilities attaching to each model after observing data $y$ are

$$\Pr(M_k|y) = \frac{\phi_k \int p(y|\theta_k)\pi(\theta_k)d\theta_k}{\sum\limits_{j=1}^{K} \phi_j \int p(y|\theta_j)\pi(\theta_j)d\theta_j}, \tag{2.1}$$

where $p(y|\theta_k) = L(\theta_k|y)$ is the likelihood of the data under model $k$. The integrals in both the denominator and numerator of (2.1) are known as prior predictive densities or marginal likelihoods (Gelfand and Dey, 1994). They give the probability of the data conditional on a model as

$$p(y|M_k) = m_k(y) = \int p(y|\theta_k)\pi(\theta_k)d\theta_k. \tag{2.2}$$

The marginal likelihood also occurs in Bayes formula for updating the parameters $\theta_k$ of model $k$, namely

$$p(\theta_k|y) = p(y|\theta_k)\pi(\theta_k)/m_k(y),$$

where $p(\theta_k|y)$ denotes the posterior density of the parameters. This is equivalently expressible as the marginal likelihood identity

$$m_k(y) = p(y|\theta_k)\pi(\theta_k)/p(\theta_k|y). \tag{2.3}$$

Model assessment can in principle be reduced to a sequential set of choices between two competing models. The formal method for comparing two competing models in a Bayesian framework involves the Bayes factor $\text{BF}_{12}$, obtained equivalently as the ratio of marginal likelihoods for model $M_1$ compared to model $M_2$, or as the ratio of posterior odds to prior odds. For $K = 2$, the posterior probability for model $k$ is obtained as

$$\Pr(M_k|y) = \frac{p(y|M_k)\Pr(M_k)}{p(y|M_1)\Pr(M_1) + p(y|M_2)\Pr(M_2)},$$

and the posterior odds are then

$$\frac{\Pr(M_1 \mid y)}{\Pr(M_2 \mid y)} = \frac{p(y|M_1)\Pr(M_1)}{p(y|M_2)\Pr(M_2)}. \tag{2.4}$$

The Bayes factor is obtained as

$$BF_{12} = \frac{\Pr(M_1 \mid y)}{\Pr(M_2 \mid y)} \bigg/ \frac{\Pr(M_1)}{\Pr(M_2)} = \frac{p(y|M_1)}{p(y|M_2)}.$$

The main obstacle to routine implementation of such a selection process is evaluation of the integral $\int p(y|\theta_k)\pi(\theta_k)d\theta_k$. This can in principle be evaluated by sampling from the prior and averaging over the resulting likelihoods, and is sometimes available analytically, but more complex methods are usually needed, and in highly parameterised or non-conjugate models a fully satisfactory procedure has yet to be developed. Several approximations have been suggested some of which are described below.

Another issue with formal model selection concerns sensitivity to prior assumptions, and stability of the Bayes factor when flat or just proper non-informative priors are used on parameters (Liu and Aitkin, 2008). It can be demonstrated that such priors lead (when models are nested within each other) to simple models being preferred over more complex models – this is Lindley's paradox (Lindley, 1957, 1997), with related discussions in Gelfand and Dey (1994), DeSantis and Spezzaferri (1997), and Casella et al. (2009). By contrast, likelihood ratios used in classical testing tend to favour more complex models by default (Gelfand and Dey, 1994). Even under proper priors, with sufficiently large sample sizes the Bayes factor tends to attach too little weight to the correct model and too much to a less complex or null model. Hence some advocate a less formal view to Bayesian model selection based on predictive criteria other than the Bayes factor (see Section 2.2.4). These may lead to model checks analogous to classical $p$ tests, or to pseudo Bayes factors of various kinds.

### 2.2.3   Estimating model probabilities or marginal likelihoods in practice

MCMC simulation methods are typically applied to deriving posterior densities $p(\theta|Y)$, or sampling predictions $y_{new}$ in models considered singly. However, they have extended to include parameter estimation and model choice in the joint parameter and model space $\{\theta_k, M_k\}$ for $k = 1, \dots, K$ (Carlin and Chib, 1995). Thus at iteration $t$ there might be a switch between models (e.g. from $M_j$ to $M_k$) and updating only on the parameters in model $k$. For equal prior model probabilities, the best model is the one chosen most frequently, and the posterior odds follow from (2.4). The reversible jump algorithm of Green (1995) also provides a joint space estimation method.

However, following studies such as Chib (1995), Lenk and Desarbo (2000) and Gelfand and Dey (1994), the marginal likelihood of a single model may be approximated from the output of MCMC chains. The most simple apparent estimator of the marginal likelihood would apply the usual Monte Carlo methods for estimating integrals in equation (2.2). Thus for each of a large number of draws, $t = 1, \dots, T$ from the prior density of $\theta$, one may evaluate the likelihood $L^{(t)} = L(\theta^{(t)}|y)$ at each draw, and calculate the average. This may be feasible with an informative prior, but would require a considerable number of draws ($T$ perhaps in the millions).

Since (2.3) is true for any point this suggests another estimator for $m(y)$ based on an approximation for the posterior density $\hat{p}(\theta|Y)$, evaluated at a high density point such as the posterior mean $\bar{\theta}$ or mode $\theta_m$. At the posterior mean, one obtains a log marginal likelihood estimate as

$$\log[\hat{m}(y)] \approx \log[p(y|\bar{\theta})] + \log[\pi(\bar{\theta})] - \log[\hat{p}(\theta|Y)]. \tag{2.5}$$

Alternatively, following DiCiccio *et al.* (1997), Gelfand and Dey (1994, p. 511), and others importance sampling may be used. In general the integral of a function $h(u)$ may be written as

$$H = \int h(u)\mathrm{d}u = \int \{h(u)/g(u)\}g(u)\mathrm{d}u.$$

where $g(u)$ is the importance function. Suppose $u^{(1)}, u^{(2)}, \ldots, u^{(T)}$ are a series of draws from this function $g$ which approximates $h$, whereas $h$ itself which is difficult to sample from. An estimate of $H$ is then

$$T^{-1} \sum_{t=1}^{T} h(u^{(t)})/g(u^{(t)}).$$

As a particular example, the marginal likelihood may be expressed as

$$m(y) = \int p(y|\theta)\pi(\theta)\mathrm{d}\theta = \int [p(y|\theta)\pi(\theta)/g(\theta)]g(\theta)\mathrm{d}\theta$$

where $g$ is a normalised importance function for $p(y|\theta)\pi(\theta)$. The sampling estimate of $m(y)$ is then

$$\hat{m}(y) = T^{-1} \sum_{t=1}^{T} L(\theta^{(t)})\pi(\theta^{(t)})/g(\theta^{(t)}),$$

where $\{\theta^{(1)}, \theta^{(2)}, \ldots, \theta^{(T)}\}$ are draws from the importance function $g$, and $\pi(\theta^{(t)})$ are evaluations of the prior densities at $\theta^{(t)}$. In practice only an unnormalised density $g^*$ may be known and the normalisation constant is estimated as $T^{-1} \sum_{t=1}^{T} \pi(\theta^{(t)})/g^*(\theta^{(t)})$, with corresponding sampling estimate for $m(y)$,

$$\hat{m}(y) = \sum_{t=1}^{T} L(\theta^{(t)})w(\theta^{(t)})/ \sum_{t=1}^{T} w(\theta^{(t)}) \tag{2.6}$$

where $w(\theta^{(t)}) = \pi(\theta^{(t)})/g^*(\theta^{(t)})$. Following Geweke (1989) it is desirable that the tails of the importance function $g$ decay slower than those of the posterior density that the importance function is approximating. So if the posterior density is multivariate normal (for analytic reasons or by inspection of MCMC samples) then a multivariate Student $t$ with low degrees of freedom is most appropriate as an importance density.

A special case occurs if $g^* = L\pi$, leading to cancellation in (2.6), and to the harmonic mean of the likelihoods as an estimator for $m(y)$, namely

$$\hat{m}(y) = \frac{T}{\sum_{t=1}^{T} 1/L^{(t)}}.$$

For small samples this estimator may, however, be subject to instability (Chib, 1995). For an illustration of this criterion in disease mapping see Hsiao *et al.* (2000). Another estimator for the marginal likelihood based on importance sampling ideas is obtainable from the relation

$$[m(y)]^{-1} = \int \frac{g(\theta)}{L(\theta|y)\pi(\theta)}p(\theta|Y)\mathrm{d}\theta,$$

so that

$$m(y) = \int L(\theta|y)\pi(\theta)d\theta,$$

$$= \left\{ E\left[ \frac{g(\theta)}{L(\theta|y)\pi(\theta)} \right] \right\}^{-1},$$

where the expectation is with respect to the posterior distribution of $\theta$. The marginal likelihood may then be approximated by

$$\hat{m}(y) = 1/[T^{-1} \sum_{t=1}^{T} g^{(t)}/\{L^{(t)}\pi^{(t)}\}] \tag{2.7}$$

$$= T/ \left[ \sum_{t=1}^{T} g^{(t)}/\{L^{(t)}\pi^{(t)}\} \right].$$

It is generally recommended for $g$ to be a function (or product of separate functions) that approximates $p(\theta|Y)$ (e.g. Lenk and Desarbo, 2000). So in fact two phases of sampling are typically involved: an initial MCMC analysis to provide approximations $g$ to $p(\theta|Y)$ or its components, and a second run recording $g^{(t)}$, $L^{(t)}$ and $\pi^{(t)}$ at iterations $t = 1, \ldots T$, namely the values of the importance density, the likelihood and the prior as evaluated at the sampled values $\theta^{(t)}$, which are either from the posterior (after convergence), or from $g$ itself. The importance density and prior value calculations, $g^{(t)}$ and $\pi^{(t)}$, may well involve a product over relevant components for individual parameters.

For numeric reasons (i.e. underflow of likelihoods $L^{(t)}$ in larger samples), it may be more feasible to obtain estimates of $\log[m(y)]$ in (2.7), and then take exponentials to provide a Bayes factor. This involves monitoring

$$\delta^{(t)} = \log[g^{(t)}/\{L^{(t)}\pi^{(t)}\}] = \log(g^{(t)}) - \log(L^{(t)}) - \log(\pi^{(t)}), \tag{2.8a}$$

over MCMC iterations. Then a spreadsheet may be used to obtain

$$\Delta^{(t)} = \exp[\delta^{(t)}], \tag{2.8b}$$

and then minus the log of the average of the $\Delta^{(t)}$ calculated, so that

$$\log[\hat{m}(y)] = -\log(\overline{\Delta}).$$

If exponentiation in (2.8b) leads to numeric overflow a suitable constant (such as the average of the $\delta^{(t)}$) can be subtracted from the $\delta^{(t)}$, before they are exponentiated, and then also subtracted from $-\log(\overline{\Delta})$.

## 2.2.4    Approximating the posterior density

In (2.5) and (2.7) above, an estimate of the marginal likelihood involves a function $g$ that approximates the posterior $p(\theta|y)$ using MCMC output. One possible approximation entails taking moment estimates of the joint posterior density of all parameters, or a product of moment estimate approximations of posterior densities of individual parameters or subsets of parameters. Suppose $\theta$ is of dimension $q$ and the sample size is $n$. Then, as Gelfand and Dey (1994) state, a possible choice for $g$ to approximate the posterior would be a multivariate normal or Student $t$ with mean of length $q$ and covariance matrices of dimension $q \times q$ that are computed from the sampled $\theta_j^{(t)}$, $t = 1, \dots, T$; $j = 1, \dots, q$. The formal basis for this assumption of multivariate normality of the posterior density, possibly after selective parameter transformation, rests with the Bayesian version of the central limit theorem (Kim and Ibrahim, 2000).

In practice for complex models with large numbers of parameters, one might split the parameters into sets (Lenk and Desarbo, 2000), such as regression parameters, variances, dispersion matrices, mixture proportions, and so on. Suppose the first subset of parameters in a particular problem consists of regression parameters with sampled values $\{\beta_j^{(t)}, t = 1, \dots, T;$ $j = 1, \dots, q_1\}$. For these the posterior density might be approximated by taking $g_1(\beta)$ to be multivariate normal or multivariate $t$, with the mean and dispersion matrices defined by the posterior means and the $q_1 \times q_1$ dispersion matrix taken from a long MCMC run of $T$ iterations on the $q_1$ parameters. Geweke (1989) considers more refined methods such as split normal or $t$ densities for approximating skew posterior densities, as might occur in non-linear regression.

The next set, indexed $j = q_1 + 1, \dots, q_2$ might be the parameters of a precision matrix $\Upsilon = \Sigma^{-1}$ for interdependent errors. For a precision matrix $\Upsilon$ of order $r = q_2 - q_1$, with Wishart prior $W(Q_0, r_0)$, the importance density $g_2(\Upsilon)$ may be provided by a Wishart with $n + r_0$ degrees of freedom and scale matrix $Q = \hat{S}(n + r_0)$, where $\hat{S}$ is the posterior mean of $\Upsilon^{-1}$. The set indexed by $j = q_2 + 1, \dots, q_3$ might be variance parameters $\xi_j$ for independent errors. Since variances themselves often have skewed densities, the posterior of $\chi_j = \log(\xi_j)$ may better approximate normality. The parameters indexed $j = q_3 + 1, \dots, q_4$ might be components $\psi = (\psi_1, \psi_2, \dots, \psi_J)$ of a Dirichlet density of dimension $J = q_4 - q_3$. Suppose $J = 2$, as in Example 2.4 below, then there is one free parameter $\psi$ to consider with a prior beta density. If the posterior mean and variance of $\psi$ from a long MCMC run are $k_\psi$ and $V_\psi$, then these may be equated to the theoretical mean and variance, as in $M_\psi = a_p/H$, and $V_\psi = a_p b_p / H^2[H + 1]$, where $H = (a_p + b_p)$. Solving gives an approximation to the posterior density of $\psi$ as a beta density with sample size

$$H = [k_\psi(1 - k_\psi) - V_\psi]/V_\psi$$

and success probability $k_\psi$.

So for the MCMC samples $\theta^{(t)} = \{\beta^{(t)}, \Upsilon^{(t)}, \chi^{(t)}, \psi^{(t)}, \dots\}$, the values taken by the approximate posterior densities, namely $g_1^{(t)}(\beta), g_2^{(t)}(\Upsilon), g_3^{(t)}(\chi)$ and $g_4^{(t)}(\psi)$, and other stochastic quantities, are evaluated. Let the values taken by the product of these densities be denoted $g^{(t)}$. This provides the values of each parameter sample in the approximation to the posterior density $p(\theta|y)$ (Lenk and Desarbo, 2000, p. 117) and these are used to make the estimate $\hat{m}(y)$ in (2.5) or (2.7). An example of how one might obtain the components of $g$ using this approach, a beta-binomial mixture is considered below (Example 2.4).

Chib (1995) proposes a method for approximating the posterior in analyses when integrating constants of all full conditional densities are known as they are in standard conjugate models. Suppose the parameters fall into $b = 1, \dots, B$ blocks (e.g. $B = 2$ in linear univariate regression with one block being regression parameters and the other being the variance). Consider the posterior density as a series of conditional densities, with

$$p(\theta|y) = p(\theta_1|y)p(\theta_2|\theta_1, y)p(\theta_3|\theta_1, \theta_2, y) \dots p(\theta_B|\theta_{B-1}, \theta_{B-2}, \dots, \theta_1, y).$$

In particular

$$p(\theta^*|y) = p(\theta_1^*|y)p(\theta_2^*|\theta_1^*, y)p(\theta_3^*|\theta_1^*, \theta_2^*, y) \dots p(\theta_B^*|\theta_{B-1}^*, \theta_{B-2}^*, \dots, \theta_1^*, y), \qquad (2.9)$$

where $\theta^*$ is a high density point, such as the posterior mean, where the posterior density in the marginal likelihood identity (2.3) may be estimated.

Suppose a first run is used to provide $\theta^*$. Then the value of the first of these densities, namely $p(\theta_1^*|y)$ is analytically

$$p(\theta_1^*|y) = \int p(\theta_1^*|y, \theta_2, \theta_3, \ldots, \theta_B)p(\theta_2, \theta_3, \ldots, \theta_B|y)d\theta_2 d\theta_3, \ldots, d\theta_B,$$

and may be estimated in a subsequent MCMC run with all parameters free. If this run is of length T then the average of the full conditional density of $\theta_1 = \theta_1^*$ evaluated at the samples of the other parameters provides

$$\hat{p}(\theta_1^*|y) = T^{-1} \sum_{t=1}^{T} p(\theta_1^*|\theta_2^{(t)}, \theta_3^{(t)}, \ldots, \theta_B^{(t)}).$$

However, the second density on the right side of (2.9) conditions on $\theta_1$ fixed at $\theta_1^*$, and requires a secondary run in which only parameters in the $B - 1$ blocks apart from $\theta_1$ are free to vary ($\theta_1$ is fixed at $\theta_1^*$ and is not updated). The value of the full conditional $p(\theta_2^*|y, \theta_1, \theta_3, \ldots, \theta_B)$ is taken at that fixed value of $\theta_1$ but at the sampled values of other parameters, $\{\theta_b^{(t)}, b > 2\}$, i.e. $p(\theta_2^*|y, \theta_1^*, \theta_3^{(t)}, \ldots, \theta_B^{(t)})$. So

$$p(\theta_2^*|\theta_1^*, y) = T^{-1} \sum_{t=1}^{T} p(\theta_2^*|\theta_1^*, \theta_3^{(t)}, \ldots, \theta_B^{(t)}).$$

In the third density on the right hand side of (2.9), both $\theta_1^*$ and $\theta_2^*$ are known, and another secondary run is required where all parameter blocks except $\theta_1$ and $\theta_2$ vary freely, and so on. One may then substitute the logs of the likelihood, prior and estimated posterior at $\theta^*$ in (2.5). Chib (1995) considers the case where latent data $z$ are also part of the model, as with latent normal outcomes in a probit regression.

## 2.2.5    Model averaging from MCMC samples

Model averaged parameters $\theta_A$ under the formal Bayesian paradigm are obtained as

$$p(\theta_A|y) = \sum_{k=1}^{K} \Pr(M_k|y)p(\theta_k|y, m = k).$$

However, following Congdon (2007), model averaging may be based on MCMC samples from models sampled in parallel. For models $(1, \ldots, K)$ define the parameter set for all models $\theta = (\theta_1, \ldots, \theta_K)$. One may obtain model weights at each iteration, and estimates of posterior probabilities for each model, or of model averaged parameters by averaging over samples. The analysis samples from all models concurrently without switching between models. In BUGS, repeat copies of the same data are needed for each model.

From the model probability formula $p(m = j|y) = \int p(m = j, \theta|y)d\theta = \int p(m = j|y, \theta)p(\theta|y)d\theta$, weights for model $j$ at iteration $t$ are obtainable as

$$w_j^{(t)} = p(m = j|y, \theta^{(t)}) = \frac{p(m = j, y, \theta^{(t)})}{p(y, \theta^{(t)})} = \frac{p(y|m = j, \theta^{(t)})p(\theta^{(t)}|m = j)p(m = j)}{p(y, \theta^{(t)})}$$

where $\{\theta^{(t)} = (\theta_1^{(t)}, \ldots, \theta_2^{(t)}, \ldots, \theta_K^{(t)}), t = 1, T\}$ are parameter samples for each model. The term

$$p(\theta|m = j) = p(\theta_1|m = j)p(\theta_2|m = j) \ldots p(\theta_j|m = j), \ldots, p(\theta_K|m = j)$$

in the numerator involves both the model $j$ prior $p(\theta_j | m = j)$, and pseudo-priors $p(\theta_k | m = j$, $k \neq j)$. The choice for the pseudo prior is arbitrary and the simplification

$$p(\theta_k | m = j) = g_k = \hat{p}(\theta_k | m = k, y), \qquad \text{(for all } j)$$

is adopted, namely the estimated posterior density of $\theta_k$ given $y$ and $m = k$.

Hence

$$p(\theta | m = j) = p(\theta_j | m = j) \prod_{k \neq j} p(\theta_k | m = j) = p(\theta_j | m = j) g_1 g_2 \cdots g_{j-1} g_{j+1} \cdots g_K,$$

and model weights at iteration $t$ are then

$$w_j^{(t)} = \frac{p(y | m = j, \theta_j^{(t)}) p(\theta_j^{(t)} | m = j) \left[ \prod_{h \neq j} g_h^{(t)} \right] p(m = j)}{p(y, \theta^{(t)})},$$

where the denominator is

$$p(y, \theta^{(t)}) = \sum_{k=1}^{K} p(y, \theta^{(t)}, m = k) = \sum_{k=1}^{K} \{ p(y | \theta^{(t)}, m = k) p(\theta_k^{(t)} | m = k) \left[ \prod_{h \neq k} g_h^{(t)} \right] p(m = k) \}$$

$$= \sum_{k=1}^{K} p(y | \theta_k^{(t)}, m = k) p(\theta_k^{(t)} | m = k) \left[ \prod_{h \neq k} g_h^{(t)} \right] p(m = k).$$

Dividing through by the product of the $K$ pseudo priors $(g_1 g_2, \ldots, g_K)$ one has

$$w_j^{(t)} = \frac{p(y | m = j, \theta_j^{(t)}) p(\theta_j^{(t)} | m = j) \left[ \prod_{h \neq j} g_h^{(t)} \right] p(m = j)}{\sum_{k=1}^{K} p(y | \theta_k^{(t)}, m = k) p(\theta_k^{(t)} | m = k) \left[ \prod_{h \neq k} g_h^{(t)} \right] p(m = k)}$$

$$= \frac{\left\{ \dfrac{p(y | m = j, \theta_j^{(t)}) p(\theta_j^{(t)} | m = j) p(m = j)}{g_j^{(t)}} \right\}}{\sum_{k=1}^{K} \left\{ \dfrac{p(y | \theta_k^{(t)}, m = k) p(\theta_k^{(t)} | m = k) p(m = k)}{g_k^{(t)}} \right\}}.$$

For example, the weights for model 1 when $K = 2$ are

$$w_1^{(t)} = \frac{\dfrac{p(y | \theta_1^{(t)}, m = 1) p(\theta_1^{(t)} | m = 1) p(m = 1)}{g_1^{(t)}}}{\dfrac{p(y | \theta_1^{(t)}, m = 1) p(\theta_1^{(t)} | m = 1) p(m = 1)}{g_1^{(t)}} + \dfrac{p(y | \theta_2^{(t)}, m = 2) p(\theta_2^{(t)} | m = 2) p(m = 2)}{g_2^{(t)}}}.$$

In the simplified notation of preceding sections, this comparison becomes

$$w_1^{(t)} = \frac{\dfrac{L_1^{(t)} \pi_1^{(t)} \phi_1}{g_1^{(t)}}}{\dfrac{L_1^{(t)} \pi_1^{(t)} \phi_1}{g_1^{(t)}} + \dfrac{L_2^{(t)} \pi_2^{(t)} \phi_2}{g_2^{(t)}}}.$$

As a byproduct of this comparison one may obtain posterior means (i.e. arithmetic averages) of the ratios $L^{(t)}\pi^{(t)}/g^{(t)}$, which may be compared to marginal likelihood estimates based on the harmonic averages of $L^{(t)}\pi^{(t)}/g^{(t)}$, as in (2.7).

Model averaging is based on the profile of weights $w_k^{(t)}$ obtained for all models $k$ each iteration $t$. For practical purposes it is preferable to work in the log scale, defining

$$q_k^{(t)} = \log\{p(y|\theta_k^{(t)}, m = k)p(\theta_k^{(t)}|m = k)\left[\prod_{h\neq k} g_h\right]p(m = k)\}$$

$$= \log(L_k^{(t)}) + \log(\pi_k^{(t)}) + \log(\phi_k) - \log(g_k^{(t)}),$$

and deviations $\Delta q_k^{(t)} = q_k^{(t)} - \max_k(q_k^{(t)})$, with $w_k^{(t)}$ then obtained by exponentiating:

$$w_k^{(t)} = \frac{\exp(\Delta q_k^{(t)})}{\sum_k \exp(\Delta q_k^{(t)})}.$$

Approximations to the log marginal likelihood may be obtained from the posterior means of what may be denoted composite log-likelihoods, namely

$$\eta_k^{(t)} = -\delta_k^{(t)} = \log(L_k^{(t)}) + \log(\pi_k^{(t)}) - \log(g_k^{(t)}),$$

where $\delta_k^{(t)}$ are as in (2.8a). Application of this procedure can be illustrated for two cases where marginal likelihoods and Bayes factors are available analytically.

### Example 2.1   Fixed and unknown mean

First, consider the likelihood $y_i \sim N(\theta, 1)$, for $i = 1, \dots, n$, and models $M_1$: $\theta = 0$ and $M_2$: $\theta \neq 0$. The sufficient statistic is $\bar{y}$ where $\bar{y} \sim N(\theta, 1/n)$. For $M_1$ one has $p(\bar{y}|M_1) = \sqrt{\frac{n}{2\pi}}\exp(-n\bar{y}^2/2)$. For $M_2$ assume $p(\theta|m = 2) = 1/(2c)$, that is $\theta \sim U(-c, c)$. Then (Lunn $et\ al.$, 2013, p. 171) one has

$$p(\bar{y}|M_2) \approx 1/(2c)$$

and Bayes factor $B_{12} = p(\bar{y}|M_1)/p(\bar{y}|M_2) = 2c\sqrt{\frac{n}{2\pi}}\exp(-n\bar{y}^2/2)$.

Assuming the value $\theta = 0.5$, and $n = 20$, BUGS is used to sample $y$ values for $y_i \sim N(\theta, 1)$. One simply compiles the code

```
model {for (i in 1:20) {y[i] ~dnorm(0.5,1)}}
```

and generates initial values. Then using info/node info, the sampled $y$ are obtained as $y_{samp} =$ (0.166, −1.815, 0.899, 0.419, 1.390, 1.345, −1.305, −0.165, 0.492, 0.148, −1.250, 0.619, 0.364, 1.258, 1.530, 1.420, −1.211, 1.078, −0.382, 1.271), with average $\bar{y}_{samp} = 0.3136$.

For model 2 applied to these data, the value $c = 2$ is assumed, and the model applied in BUGS to obtain an estimate for the density $p(\theta|m = 2)$. With data $y_{samp}$ and code

```
model {theta ~dunif(-2,2)
for (i in 1:n) {ysamp[i] ~dnorm(theta,1)}}
```

the posterior density $p(\theta|m = 2)$ is estimated as

$$\theta \sim N(0.313, 0.223^2).$$

From the formula $B_{12} = 2c\sqrt{\frac{n}{2\pi}}\exp(-n\bar{y}^2/2)$, and with $y = y_{samp}$, one has $B_{12} = 2.669$. Alternatively, using the model weights procedure above, and with equal prior model probabilities of 0.5, the following code may be implemented:

```
model {# log-lkd for model 1
for (i in 1:n) {y1[i] ~dnorm(0,1)
LL1[i] <- -0.5*log(2*3.14159)-0.5*pow(y1[i],2)}
# log-lkd for model 2
for (i in 1:n) {y2[i] ~dnorm(theta,1)
LL2[i] <- -0.5*log(2*3.14159)-0.5*pow(y2[i]-theta,2)}
# prior and log prior ordinates for model 2
theta ~dunif(-2,2); prior2[1] <- log(1/4)
# log pseudo-prior ordinate for model 2
g2[1] <- -0.5*log(2*3.14159)-log(s.theta)-0.5*pow(theta-mu.theta,2)/
                                              (s.theta*s.theta)
# Combine loglikelihoods, log-priors, log pseudo-priors, and log
                                     prior model probs
TL[1] <- sum(LL1[])+log(0.5)
TL[2] <- sum(LL2[])+prior2[1]-g2[1]+log(0.5)
# Scale against largest combined likelihood
maxL<- ranked(CL[],2);
for (j in 1:2) {SL[j] <- CL[j]-maxL; expSL[j] <- exp(SL[j])}
# model weights
w[1] <- expSL[1]/sum(expSL[]); w[2] <- expSL[2]/sum(expSL[])}
```

The datasets y1[1:20] and y2[1:20] both contain $y_{samp}$ as above, while mu.theta $= 0.313$ and s.theta $= 0.223$. The last 99 000 of a two chain sequence of 100 000 iterations provides posterior mean estimates $w_1 = 0.72752$, $w_2 = 0.27248$, and so an estimated Bayes factor of $BF_{12} = 2.669$, the same as the analytic estimate.

### Example 2.2    Regression with simulated data

This example compares posterior model weights and marginal likelihoods obtained by the parallel sampling method with analytic Bayes factors and marginal likelihoods (Denison et al., 2002, p. 20; Koop et al., 2007, p. 295). The application is to normal linear regression using the conjugate normal inverse-gamma prior. Simulated standard normal predictors $x_1$ and $x_2$ are obtained for $n = 100$ cases, and values for a response obtained as

$$y_i = \beta_0 + \beta_1 x_{1i} + e_i,$$

where $e_i \sim N(0, 0.2)$, $\beta_0 = 1$, and $\beta_1 = 0.5$. Two models are compared: the true model ($M_1$) with just $x_1$ as a predictor, and a model ($M_2$) with both $x_1$ and $x_2$ as predictors. A gamma Ga(3,0.4) prior is assumed on $1/\sigma^2$, and $N(0, 4\sigma^2)$ priors assumed for regression coefficients. The following Matlab code includes the sampled data and produces analytic marginal likelihood estimates of $-58.12$ and $-60.96$ for $M_1$ and $M_2$ respectively. Thus the Bayes factor $B_{12}$ of 17.06 favours $M_1$, with posterior model probabilities of 0.945 on $M_1$, and 0.055 on $M_2$.

```
load regyx1x2.txt; y=regyx1x2(:,1); x1=regyx1x2(:,2); x2=regyx1x2(:,3);
n=100;
% model 1
X = [ones(n,1) x1]; p = 2;
% prior hyperparameters;
a = 3;b = 0.4;
```

```
mu_beta = zeros(p,1); V_beta = 4*eye(p); invV_beta = inv(V_beta);
% Marg Lkd Calculation
D = (det(V_beta))*det( (invV_beta + X'*X));
E = eye(n) - X*inv(invV_beta + X'*X)*X';
C = (D^(-1/2))* (b^a)*gamma(n/2+a) /(gamma(a)*(2*pi)^(n/2));
ML1 = C*(b + 0.5*y'*E*y)^(-( n/2 + a));
disp('Log Marginal Likelihood'); log(ML1)
% model 2
X = [ones(n,1) x1 x2]; p = 3;
mu_beta = zeros(p,1); V_beta = 4*eye(p); invV_beta = inv(V_beta);
% Marg Lkd Calculation
D = (det(V_beta))*det( (invV_beta + X'*X));
E = eye(n) - X*inv(invV_beta + X'*X)*X';
C = (D^(-1/2))* (b^a)*gamma(n/2+a) /(gamma(a)*(2*pi)^(n/2));
ML2 = C*(b + 0.5*y'*E*y)^(-( n/2 + a));
disp('Log Marginal Likelihood'); log(ML2)
BF12 = ML1/ML2
pM1=ML1/(ML1+ML2)
pM2=ML2/(ML1+ML2)
```

To apply the parallel sampling procedure of Congdon (2007) to comparing $M_1$ and $M_2$, initial runs of BUGS are carried out on each model separately to obtain posterior density estimates $g_k = \hat{p}(\theta_k|m = k, y)$ of the parameters. Because of the mode of data generation and the simplicity of the models, univariate normal densities are sufficient here to approximate the posterior densities of the $\beta$ (regression) coefficients, but in real data applications a multivariate normal (or Student $t$) is more likely to be needed.

These posterior approximations are then used in a final parallel MCMC analysis comparing the two models. This provides posterior model weights of 0.9426 on $M_1$, and 0.0574 on $M_2$, similar to the analytic estimates. The posterior means of the composite log-likelihoods,

$$\eta_k^{(t)} = \log(L_k^{(t)}) + \log(\pi_k^{(t)}) - \log(g_k^{(t)})$$

namely eta[1:2] in the BUGS code[1], are also close to the analytic marginal likelihoods. From the second half of a single chain run of 100 000 iterations, the posterior mean composite log-likelihoods are obtained as $-58.11$ and $-60.94$.

To illustrate model averaging, averaged predictions of $y_{14}$ are obtained under the formal paradigm as 0.635 (sd=0.386), based on posterior normal densities $N(0.6381, 0.165)$ and $N(0.6209, 0.174)$ under $M_1$ and $M_2$ respectively. Under the model weights parallel sampling approach, the posterior mean prediction for $y_{14}$ is obtained as 0.6355 (sd=0.384).

## 2.2.6   Predictive criteria for model checking and selection: Cross-validation

Another approach to model choice and checking is based on the principle of predictive cross-validation (e.g. Shao, 1993; Arlot and Celisse, 2010). Predictions for a subset $y_r$ of the data are made from a posterior updated only using the complement of $y_r$, denoted $y_{[r]}$. A common choice is when one case (say case $i$) is omitted at a time, with estimation of the model based only on $y_{[i]}$, namely the remaining $n - 1$ cases excluding $y_i$. With $\tilde{y}_i$ denoting the predicted response, this leads to the leave-one out posterior predictive distribution (Stern and Cressie, 2000, p. 2388)

$$p(\tilde{y}_i|y_{[i]}) = \int p(\tilde{y}_i|\theta)p(\theta|y_{[i]})d\theta,$$

which when evaluated at the observed $y_i$ provides the conditional predictive ordinate (CPO). A related criterion is the cross-validation probability integral transform or PIT (Dawid, 1984; Czado *et al.*, 2009)

$$\text{PIT}_i = \int \Pr(\tilde{y}_i \le y_i | \theta) p(\theta | y_{[i]}) d\theta = \Pr(\tilde{y}_i \le y_i | y_{[i]}).$$

Observations with low CPO or PIT values are not being reproduced successfully by the model. This principle can be extended to multilevel applications such as item response modelling; thus for subjects $i = 1, \dots, n$ with ability $\tau_i$, the posterior predictive density of response $y_{ij}$ by subject $i$ to item $j \in 1, \dots, J$ is based on the subject profile for items $k \ne j$, namely

$$p(\tilde{y}_{ij} | y_{i[j]}) = \int p(\tilde{y}_{ij} | \tau_i) p(\tau_i | y_{i[j]}) d\tau_i.$$

For particular combinations of density and prior, the leave-one out posterior predictive density may have known form (e.g. van der Linden and Guo, 2008).

Cross-validation procedures assist in model checking via discrepancy measures comparing actual data with cross-validatory predictions. An example is the cross-validation residual (Johnson and Albert, 1999, p. 100)

$$d_{1i} = y_i - E(\tilde{y}_i | y_{[i]}).$$

If $\sigma_{i,CV}^2 = \text{var}[\tilde{y}_i | y_{[i]}]$, the standardised cross-validation residual is

$$d_{2i} = d_{1i} / \sigma_{i,CV}.$$

Under approximate posterior normality, 95% of the $d_{2i}$ should be within $(-2,2)$, and systematic patterns (e.g. from plots of $d_{2i}$ against $y_i$) indicate model inadequacy. Another check at each iteration, providing the PIT when aggregated over MCMC samples, is simply

$$d_{3i} = I(\tilde{y}_i \le y_i),$$

namely whether the cross-validation prediction equals or is less than the actual observation $y_i$ (Stern and Cressie, 2000, p. 2393). The expectation is $\Pr(\tilde{y}_i \le y_i | y_{[i]})$ and in an adequate model these are uniformly distributed with average around 0.5. High or low values for $\Pr(\tilde{y}_i \le y_i | y_{[i]})$ (e.g. over 0.95 or under 0.05) indicate underprediction and overprediction respectively. For binary (or multiple category) data, a more appropriate check (Johnson and Albert, 1999) is

$$d_{3i} = I(\tilde{y}_i = y_i).$$

For count data (e.g. with a Poisson or negative binomial likelihood) the appropriate check (Marshall and Spiegelhalter, 2007) is

$$d_{3i} = I(\tilde{y}_i < y_i) + 0.5I(\tilde{y}_i = y_i).$$

Another check is whether $\tilde{y}_i$ is contained in a small interval $(y_i - \epsilon, y_i + \epsilon)$ around the observed value. The function

$$d_{4i} = I(y_i - \epsilon \le \tilde{y}_i \le y_i + \epsilon)/2\epsilon$$

has expectation $p(\tilde{y}_i | y_{[i]})$, namely the CPO, when $\epsilon$ tends to zero.

Under cross-validation with training data $y_{[r]}$, an important feature is that even if the prior $\pi$ is improper, the predictive density

$$p(y_r|y_{[r]}) = m(y)/m(y_{[r]}) = \int p(y_r|\theta, y_{[r]})p(\theta|y_{[r]})d\theta$$

is proper because the posterior based on using only $y_{[r]}$ in estimating $\theta$, namely $p(\theta|y_{[r]})$, is proper. Hence sensitivity to the prior is diminished. Geisser and Eddy (1979) suggest the product

$$\hat{m}_{PS}(y) = \prod_{i=1}^{n} p(y_i|y_{[i]})$$

of leave-one-out predictive densities as an estimate for a pseudo marginal likelihood, analogous to the usual marginal likelihood. The ratio of two such quantities under models $M_1$ and $M_2$ provides a pseudo Bayes factor (sometimes abbreviated as PsBF):

$$PsBF = \prod_{i=1}^{n} \{p(y_i|y_{[i]}, M_1)/p(y_i|y_{[i]}, M_2)\}.$$

The psML and PsBF may be obtained using a Monte Carlo estimator of the CPO: this involves monitoring the inverse likelihoods for each case (without actually omitting $y_i$ from the analysis), and obtaining their posterior averages. The sum over subjects of the logged inverses of these posterior averages produces an estimator of log pseudo marginal likelihood, or LPML (e.g. Chen et al., 2000; Dey et al., 1997). The CPO estimator is

$$\hat{p}(y_i|y_{[i]}) = \left[ T^{-1} \sum_{t=1}^{T} \frac{1}{L_i(\theta^{(t)})} \right]^{-1}.$$

Cross validation via case-influence assessment compares the complete data posterior $p(\theta|y)$ to the observation excluded posterior $p(\theta|y_{[i]})$ via divergence criteria such as the Kullback–Leibler divergence

$$K(y, y_{[i]}) = \int p(\theta|y) \log \frac{p(\theta|y)}{p(\theta|y_{[i]})} d\theta.$$

$K(y, y_{[i]})$ measures the effect of excluding the $i$th observation on the joint posterior distribution of $\theta$. Following Cho et al. (2009) and Pires and Diniz (2012) one may estimate $K(y, y_{[i]})$ as

$$K(y, y_{[i]}) = -\log(CPO_i) + E_\theta[\log\{p(y|\theta)\}|y],$$

where the second term is the posterior mean of the log-likelihood for case $i$. A calibration for $K(y, y_{[i]})$ is

$$p_i^* = 0.5\left[ 1 + \sqrt{1 - \exp(-2K(y, y_{[i]}))} \right]$$

with values between 0.5 and 1, and high values (e.g. 0.8 or 0.9) indicating that the $i^{th}$ observation is influential on the posterior.

To illustrate how CPO and Kullback–Leibler diagnostics can be obtained via BUGS, consider the Gesell adaptive score data. For a linear regression of score $y$ on age $x$, the CPO statistics are estimated (e.g. in a spreadsheet) as the inverse of the posterior mean G[i] in the following code:

```
model { for (i in 1:21) {y[i] ~dnorm(mu[i],tau); e[i] < - y[i]-mu[i]
LL[i] <- -0.92+0.5*log(tau)-0.5*tau*e[i]*e[i]
G[i] <- 1/exp(LL[i])
mu[i] <- beta[1] + beta[2]*x[i]}
for (j in 1:2) {beta[j] ~dflat()}
tau ~dgamma(1,0.001)}.
```

The Kullback–Leibler criteria are estimated as $-\log(CPO_i)$ plus the posterior mean of LL[i]. Table 2.1 shows these outputs using the last 90 000 of a single chain run of 100 000 iterations, which may be compared with Table 2 in Chaloner and Brant (1988).

Cross-validation also underlies the intrinsic Bayes factor proposed by Berger and Perrichi (1996). This involves defining a small subset of the observed data, $y_M$ as a minimal training sample. The posterior $p(\theta|y_M)$ for $\theta$ derived from such a training sample supplies a proper prior for analysing the remaining data $y_{[M]}$. For instance, with a logit regression with $p$ predictors, the training samples are of size $p + 1$. In practice, a large number of training samples may be needed, since for large sample sizes there are many such possible subsets (Kadane and Lazar, 2004).

Summary statistics using cross-validated replicates have been proposed as model choice criteria. For $n$ observations and $T$ samples from an MCMC run, Czado et al. (2009) mention

**Table 2.1**  Model diagnostics. Adaptive score data.

| Obs | Posterior mean G | CPO | $-\log(CPO)$ | Posterior mean LL | Kullback–Leibler diagnostic | Kullback–Leibler calibration |
|---|---|---|---|---|---|---|
| 1 | 28.5 | 0.035 | 3.35 | −3.34 | 0.014 | 0.58 |
| 2 | 48.0 | 0.021 | 3.87 | −3.79 | 0.083 | 0.70 |
| 3 | 93.0 | 0.011 | 4.53 | −4.43 | 0.100 | 0.71 |
| 4 | 40.6 | 0.025 | 3.70 | −3.68 | 0.029 | 0.62 |
| 5 | 40.8 | 0.024 | 3.71 | −3.69 | 0.020 | 0.60 |
| 6 | 28.3 | 0.035 | 3.34 | −3.33 | 0.014 | 0.58 |
| 7 | 29.7 | 0.034 | 3.39 | −3.38 | 0.014 | 0.58 |
| 8 | 28.9 | 0.035 | 3.36 | −3.35 | 0.014 | 0.58 |
| 9 | 29.8 | 0.034 | 3.39 | −3.38 | 0.016 | 0.59 |
| 10 | 35.0 | 0.029 | 3.55 | −3.53 | 0.022 | 0.60 |
| 11 | 51.9 | 0.019 | 3.95 | −3.89 | 0.059 | 0.67 |
| 12 | 30.2 | 0.033 | 3.41 | −3.39 | 0.016 | 0.59 |
| 13 | 93.0 | 0.011 | 4.53 | −4.43 | 0.100 | 0.71 |
| 14 | 67.1 | 0.015 | 4.21 | −4.15 | 0.058 | 0.67 |
| 15 | 30.9 | 0.032 | 3.43 | −3.42 | 0.015 | 0.59 |
| 16 | 28.4 | 0.035 | 3.35 | −3.33 | 0.013 | 0.58 |
| 17 | 39.7 | 0.025 | 3.68 | −3.66 | 0.021 | 0.60 |
| 18 | 65.3 | 0.015 | 4.18 | −3.76 | 0.419 | 0.88 |
| 19 | 6862.0 | 0.000 | 8.83 | −7.50 | 1.337 | 0.98 |
| 20 | 52.4 | 0.019 | 3.96 | −3.92 | 0.038 | 0.64 |
| 21 | 28.4 | 0.035 | 3.35 | −3.33 | 0.013 | 0.58 |

the mean ranked probability score

$$RPS = \frac{1}{n} \sum_{i=1}^{n} \left( \frac{1}{T} \sum_{t=1}^{T} |\tilde{y}_i - y_i| - \frac{1}{T} \sum_{t=1}^{T/2} |\tilde{y}_{it} - \tilde{y}_{i,t+T/2}| \right),$$

and the Dawid-Sebastiani (1999) score

$$DSS = \frac{1}{n} \sum_{i=1}^{n} \left[ \left( \frac{y_i - \tilde{\mu}_i}{\tilde{\sigma}_i} \right)^2 + 2\log(\tilde{\sigma}_i) \right],$$

where $\tilde{\mu}_i$ and $\tilde{\sigma}_i$ are posterior means and standard deviations of the sampled replicates $\tilde{y}_{it}$. Lower scores indicate better fitting models.

## 2.2.7    Predictive checks and model choice using complete data replicate sampling

Cross-validatory procedures become more difficult to apply with large samples, and predictive assessment without omitting data can also used both to check model performance and/or provide model choice criteria. Thus predictions may be made by sampling $y_{new,i}^{(t)}$ for each observation at each iteration in an MCMC chain, providing the posterior predictive distribution

$$p(y_{new,i}|y) = \int p(y_{new,i}|\theta)p(\theta|y)d\theta.$$

For instance, under a normal model with mean $\mu_i$ for case $i$ at iteration $t$, and variance $\xi^{(t)}$, such a sample would be obtained by sampling replicate data as additional unknowns:

$$y_{new,i}^{(t)} \sim N(\mu_j^{(t)}, \xi^{(t)}),$$

and from the converged sampling one may obtain posterior summaries on the $y_{new}$ or functions involving them. One may also define the full data posterior predictive $p$-value (Stern and Cressie, 2000, p. 2395; Marshall and Spiegelhalter, 2007, p. 424)

$$Pr(y_{new,i} \le y_i|y) = \int Pr(y_{new,i} \le y_i|\theta)p(\theta|y)d\theta.$$

Tail values for $Pr(y_{new,i} \le y_i|y)$ (e.g. over 0.95 or under 0.05) indicate poorly predicted cases. Alternatively the distributional profile of all $p$-values may be considered. For example, Yan and Sedransk (2007) assess two level hierarchical models (levels 1 and 2 having indices $i$ and $j$) using such $p$-values: if the distribution of $y_{new,ij}|y$ is the same as the distribution of $y_{ij}$, the values $p_{ij} = Pr(y_{new,ij} \le y_{ij}|y)$ should approximate a uniformly distributed random sample.

The associated predictive checks applied in an MCMC run are as defined in Section 2.2.6. Thus at iteration $t$ in an MCMC run, and with $\theta^{(t)}$ denoting parameters at all stages in a hierarchical model, the check leading to the posterior predictive p-value is

$$d_i^{(t)} = I(y_{new,i}^{(t)} \le y_i|\theta^{(t)}),$$

namely whether the prediction equals or is less than the actual observation $y_i$. For binary (or multiple category) data a more appropriate check (Johnson and Albert, 1999) is $I(y_{new,i}^{(t)} =$

$y_i|\theta^{(t)}$), while for count data the relevant check (Berry and Armitage, 1995; Marshall and Spiegelhalter, 2007) is

$$I(y^{(t)}_{new,i} < y_i|\theta^{(t)}) + 0.5I(y^{(t)}_{new,i} = y_i|\theta^{(t)}).$$

Posterior predictive checks may also use omnibus criteria (Meng, 1994; Crespi and Boscardin, 2009). Let $D(y; \theta)$ be a discrepancy criterion for the observed data (e.g. a chi-square statistic); similarly let the criterion based on replicate data be denoted $D(y_{new}; \theta)$. Then a reference distribution $p_R$ for the chosen criterion can be obtained from the joint distribution of $y_{new}$ and $\theta$, namely

$$p_R(y_{new}, \theta) = p(y_{new}|\theta)p(\theta|y),$$

and the actual value set against this reference distribution. The posterior predictive $p$-value based on criterion $D$ is obtained as

$$p_b(y) = p_R[D(y_{new}; \theta) \geq D(y; \theta)|y].$$

In practice, $D(y^{(t)}_{new}, \theta^{(t)})$ and $D(y, \theta^{(t)})$ are obtained at each iteration in an MCMC run and the proportion of iterations where $D(y^{(t)}_{new}, \theta^{(t)})$ exceeds $D(y, \theta^{(t)})$ calculated (see Example 2.4 for an illustration). Values near 0 or 1 indicate lack of fit, while mid-range values (between 0.2 and 0.8) indicate a satisfactory model. A predictive check procedure is also described by Gelfand (1996, p. 153) and involves obtaining 50%, 95% (etc.) intervals for replicates $y_{new,i}$, and then counting how many of the actual data points are located in these intervals.

Whether applied for assessing fit to particular cases, or to the data overall, full data posterior predictive criteria tend to be conservative because the data are used twice: first to update $\theta$, and second in deriving the discrepancy measure (Hjort et al., 2006; Steinbakk and Storvik, 2009; 1998). Robins et al. (2000) show that the asymptotic distributions of posterior predictive $p$-values are more concentrated around 0.5 than under a uniform distribution, and propose methods to adjust the $p$-values to ensure their distributions are asymptotically uniform.

Marshall and Spiegelhalter (2007) suggest a mixed posterior predictive check for hierarchical models with random effects $R$, such as case specific effects $R = (r_1, \dots, r_n)$, at the second stage, and third stage hyperparameters $\psi$ (see Section 2.3). The mixed predictive prior distribution has form $p(y_{new}|R_{new}, \psi)$, and with checks $d^{(t)}_i$ conditioning on a newly sampled set of random effects:

$$d^{(t)}_i = I(y^{(t)}_{new,i} \leq y_i|r^{(t)}_{new,i}, \psi^{(t)}).$$

Such a check reduces dependence on the actual data, as $y$ influences replicates $R_{new}$ only indirectly through $\psi^{(t)}$ (Riebler and Held, 2010). So one would sample new random effects from their hyperdistribution, and then replicate data:

$$r^{(t)}_{new,i} \sim \pi(r|\psi^{(t)})$$

$$y^{(t)}_{new,i} \sim p(y_{new}|r^{(t)}_{new,i}, \psi^{(t)}).$$

Mixed predictive $p$-values are therefore expected to be less conservative than posterior predictive $p$-values (see Example 2.3 for an illustration). As well as indicating poorly fit cases, the histogram pattern of the mixed predictive $p$-values can be used for other forms of check: a non-uniform histogram pattern in hierarchical applications may, for example, indicate non-exchangeability (Hein et al., 2006, p. 72).

Posterior predictive sampling provides overall measures of fit that may assist in model selection (e.g. Gelfand and Ghosh, 1998; Chen *et al.*, 2000, Czado *et al.*, 2009). Laud and Ibrahim (1995) argue that model selection criteria such as the AIC and BIC rely on asymptotic considerations, whereas the predictive density for a hypothetical replication $y_{new}$ of the trial or observation process leads to a criterion free of asymptotic definitions. For model $k \in (1, \dots, K)$ possible models, with parameters $\theta_k$, the predictive density is

$$p(y_{new,k}|y) = \int p(y_{new,k}|\theta_k)p(\theta_k|y)d\theta_k.$$

Laud and Ibrahim (1995) consider the measure

$$C_k^2 = \sum_{i=1}^{n} \left[ \{\mu_{new,k,i} - y_i\}^2 + \sigma_{new,k,i}^2 \right]$$

where $\mu_{new,k,i}$ and $\sigma_{new,k,i}^2$ are posterior means and variances of the sampled replicates $y_{new,k,i,t}$ over MCMC iterations $t$. This criterion combines the fit $F = \sum_{i=1}^{n} \{\mu_{new,k,i} - y_i\}^2$ of predicted to actual data, and the variability $V = \sum_{i=1}^{n} \text{var}(y_{new,k,i})$ of predictions, with the latter acting as a penalty on complexity. For underfitted models, $V$ will be relatively large and diminish as fit improves. But overfitted models (that add parameters with little reduction in $F$) will tend to see $V$ rising again. Cross-validatory fit criteria mentioned in Section 2.2.6 have full data equivalents; for example, the Dawid-Sebastiani score using full data replicates is

$$\text{DSS} = \frac{1}{n} \sum_{i=1}^{n} \left[ \left( \frac{y_i - \mu_{new,i}}{\sigma_{new,i}} \right)^2 + 2\log(\sigma_{new,i}) \right].$$

Gelfand and Ghosh (1998) generalise predictive loss measures to deviance forms appropriate to discrete outcomes and allow different weighting on the fit component $\{\mu_{new,k,i} - y_i\}^2$. Thus for continuous data and any $w > 0$

$$C_{wk}^2 = \sum_{i=1}^{n} [\sigma_{new,k,i}^2 + \frac{w}{w+1} \{\mu_{new,k,i} - y_i\}^2]. \tag{2.10}$$

Larger values of $w$ put more stress on the match between $E(y_{new,i})$ and $y_i$, and so downweight precision of predictions. For discrete data, let $\delta_{new,i}$ be the posterior average of the deviance term based on sampled replicates $y_{new,i,t}$. So for Poisson distributed count data $\delta_{new,i}$ is the mean of sampled values of $D(y_{new,i,t}) = y_{new,i,t} \log(y_{new,i,t}) - y_{new,i,t}$. Also let $\mu_i$ be the modelled Poisson mean, and define $M_i = (\mu_i + wy_i)/(1 + w)$, then the Poisson version of weighted predictive criterion is

$$2 \sum_{i=1}^{n} [\delta_{new,i} - D(\mu_i)] + 2(w+1) \sum_{i=1}^{n} \left[ \frac{D(\mu_i) + wD(y_i)}{1+w} - D(M_i) \right]$$

$$= 2 \sum_{i=1}^{n} [\delta_{new,i} - D(\mu_i)] + 2 \sum_{i=1}^{n} [D(\mu_i) + wD(y_i) - (w+1)D(M_i)]$$

with the first component acting as the penalty. To avoid numeric problems, the deviance form may be adapted to preclude taking logarithms of zero, namely $D(z) = (z + c) \log(z + c) - (z + c)$, where $c$ is a small positive constant.

## 2.3    Ensemble estimates: Poisson–gamma and Beta-binomial hierarchical models

We now return to the application theme of this chapter, in terms of models for smoothing a set of parameters for similar units or groups in a situation which does not involve regression for groups or members within groups. Much of the initial impetus to development of Bayesian and Empirical Bayesian methods came from this problem, namely simultaneous inference about a set of parameters for similar units of observation (schools, clinical trials, etc.) (Rao, 1975). Outcomes (e.g. average exam grades, mortality rates) over similar units (schools, hospitals) are expected to be related to each other, and to reflect (after accounting for measurement or sampling errors) true latent parameters drawn from a common density. Apart from BUGS, Bayesian estimation of simple hierarchical models of various types can be carried out using R-packages such as bspmma (Burr, 2012) and LearnBayes (Albert, 2009).

In some cases the notion of exchangeability may be modified: thus, one might consider hospital mortality rates to be exchangeable within one group of teaching hospitals and within another group of non-teaching hospitals, but not across all hospitals in both groups combined. For example, in an examination of UK cardiac surgery deaths, the performance of 12 centres was argued to be more comparable within two broad operative procedure types: 'closed' procedures involving no use of heart bypass during anaesthesia, and 'open' procedures where the heart is stopped and heart bypass needed (Spiegelhalter, 1999).

The data may take the form of aggregate observations $y_i$ from the units, e.g. means for a metric variable or numbers of successes for a binomial variable, or be disaggregated to observations $y_{ij}$ for subjects $j$ within each group or unit of observation $i$. The data are seen as generated by a compound or hierarchical process, where hyperparameters are at stage 3, parameters $\lambda_i$ relevant to units $i$ are sampled from a prior density at stage 2, and then at stage 1 the observations are sampled from a conditional distribution given the unit parameters.

A related theme but with a different emphasis has been in generalising the standard densities to allow for heterogeneity between sample units. Thus the standard densities (e.g. binomial, Poisson, normal) are modified to take account of heterogeneity in outcomes between units which is greater than postulated under that density. This heterogeneity is variously known as over-dispersion, extra-variation or (in the case of symmetric data on continuous scales) as heavy tailed data. Williams (1982) discusses the example of toxicological studies where proportions of induced abnormality between litters of experimental animals vary because of unknown genetic or environmental factors. Similarly in studies of illness, there is likely to be variation in frailty or proneness $\lambda$.

Under either perspective consider the first stage sampling density $p(y|\lambda, \omega)$, for a set of $n$ observations, $\{y_i, i = 1, \ldots, n\}$, continuous or discrete, conditional on unit level latent parameters $\{\lambda_1, \ldots, \lambda_n\}$, and additional parameters $\omega$ involved in the data likelihood. Often a single population wide value of $\lambda$ (i.e. $\lambda_j = \lambda$ for all $j$) will be inappropriate and we seek to model population heterogeneity. This typically involves either (A) distinct parameters $\{\lambda_1, \ldots, \lambda_n\}$ for each unit $i = 1, \ldots, n$, or (B) parameters $\{\Lambda_1, \ldots, \Lambda_J\}$ constant within $J$ sub-populations, with $\lambda_i = \Lambda_j$ if unit $i$ is allocated to sub-population $j$. The latter approach implies discrete mixtures (e.g. Melnykov and Maitra, 2010; Kaplon, 2010), while the first approach most commonly involves a parametric model, drawing the random effects $\lambda_i$ from a hyperdensity, with form

$$\lambda \sim \pi(\lambda|\psi).$$

where $\psi$ are hyperparameters. They will be assigned their own prior $\pi(\psi)$, which will involve further unknowns.

For example, consider a Poisson model $y \sim \mathrm{Po}(\lambda)$, where $y$ is the number of nonfatal ill-nesses or accidents in a fixed period (e.g. a year), and $\lambda$ is a measure of illness or accident proneness. Instead of assuming all individuals have the same proneness, we might well con-sider allowing $\lambda$ to vary over individuals according to a density $\pi(\lambda|\psi)$, for instance a gamma or log-normal density to reflect the positive skewness in proneness. Since $\lambda$ is necessarily positive, we then obtain the marginal distribution of the number of illnesses or accidents as

$$\mathrm{Pr}(y = k) = \int \int [\lambda^k \exp(-\lambda)/k!]\pi(\lambda|\psi)\pi(\psi)\mathrm{d}\lambda\mathrm{d}\psi,$$

where the range of the integration over $\lambda$ is restricted to positive values, and that for $\psi$ depends on the form of the parameters $\psi$. In this case

$$E(y) = E(\lambda),$$

and

$$\mathrm{var}(y) = E(\lambda) + \mathrm{var}(\lambda), \qquad (2.11)$$

so that $\mathrm{var}(\lambda) = 0$ corresponds to the simple Poisson. It is apparent from (2.11) that the mixed Poisson will always show greater variability than the simple Poisson. This formulation gener-alises to the Poisson process, where counts occur in a given time $t$ or over a given exposure $E$. Thus $y \sim \mathrm{Po}(\lambda t)$ over period of length $t$, as in occurrences of apnea and hypopnea per hours $t$ of sleep (Li $et$ $al.$, 2004), or $y \sim \mathrm{Po}(\lambda E)$ where $y$ might be deaths in areas and $E$ the area populations, or an expected deaths figure.

The model choice questions include assessing whether heterogeneity exists and if so, establishing the best approach to modelling it. Thus under a discrete mixture approach a major question is choosing the number of sub-populations, including whether one sub-population only (i.e. homogeneity) is the best option. Under a parametric approach we may test whether there is in fact heterogeneity, i.e. whether a model with $\mathrm{var}(\lambda)$ exceeding zero improves on a model with constant $\lambda$ over all subjects, and if so, what density might be adopted to describe it. One may also check whether particular observations are at odds with the hierachical model assumed.

## 2.3.1    Hierarchical mixtures for poisson and binomial data

Consider, for example, the question of Poisson heterogeneity or extravariation in counts $y_i$ for units $i$ with varying exposed to risk totals (or other forms of offset) such that $E_i$ events are the denominator or the expected total. An example of this is in small area mortality and disease studies where $y_i$ deaths are observed as against $E_i$ deaths expected on the basis of the global death rate average or more complex methods of demographic standardisation. Then a homogeneous model would assume

$$y_i \sim \mathrm{Po}(\Lambda E_i)$$

with $\Lambda$ a constant relative risk across all areas, while a heterogeneous model would take

$$y_i \sim \mathrm{Po}(\lambda_i E_i),$$

$$\lambda_i \sim \pi(\lambda_i|\psi),$$

$$\psi \sim \pi(\psi|c_1, c_2, \dots)$$

with $\pi(\lambda|\psi)$ and $\pi(\psi|c_1, c_2, \dots)$ the second and third stage priors respectively, and $\{c_1, c_2, \dots\}$ define the prior(s) on $\psi$.

The conjugate option is defined by a gamma prior $\lambda_i \sim \text{Ga}(a, b)$, with $\psi = (a, b)$, for the varying relative risks $\lambda_i$, so that $E(\lambda) = a/b$, and $\text{var}(\lambda) = a/b^2 = E(\lambda)/b$. The third stage might then be specified as

$$a \sim E(1),$$

$$b \sim \text{Ga}(1, 0.001).$$

These are relatively flat prior densities consistent with $a$ and $b$ being positive parameters. The posterior density of each $\lambda_i$ has the form

$$p(\lambda_i|y) \propto [\lambda_i^{a-1} \exp(-b\lambda_i)](\lambda_i E_i)^{y_i} \exp(-\lambda_i E_i)$$

$$\propto \lambda_i^{a-1+y_i} \exp(-\lambda_i[b + E_i])$$

which is the kernel of a gamma density, $\text{Ga}(y_i + a, E_i + b)$. A moment estimator (Böhning, 2000) for $\tau^2 = \text{var}(\lambda)$ is provided by

$$\hat{\tau}^2 = \frac{1}{n} \left[ \sum_{i=1}^{n} \frac{(y_i - \hat{\Lambda}E_i)^2}{E_i^2} - \hat{\Lambda} \sum_{i=1}^{n} \frac{1}{E_i} \right], \tag{2.12}$$

where $\hat{\Lambda}$ is a fixed effects estimate of the average relative risk, and may be used in setting up the priors for $a$ and $b$. The alternative to a hierarchical model is to take $a$ and $b$ as known, with a diffuse setting such as $a = b = \varepsilon$, where $\varepsilon$ is small (e.g. $\varepsilon = 0.0001$), effectively providing a 'fixed effects' analysis, with no borrowing of strength. The latter procedure is subject to variance instability for small $y_i$ and/or offsets such as $E_i$, giving rise to apparently extreme rates based on small numbers.

Poisson heterogeneity may also be modelled flexibly in transformed scale of $\lambda_i$, such as $\log(\lambda_i)$. This transformation extends over the real line, so heterogeneity would be summarised by zero mean normal random effects $u_i \sim N(0, \sigma^2)$, Student $t$ effects $u_i \sim t(0, \sigma^2, v)$, or other options (skew-normal, mixtures of normals, etc.)

$$\log(\lambda_i) = \mu + u_i,$$

or

$$\log(\lambda_i) = \log(E_i) + \mu + u_i,$$

if there are offsets or exposures $E_i$. For $u$ normal, the mean and variance of $\lambda$ are obtained as $E(\lambda) = \exp\left(\mu + \frac{\sigma^2}{2}\right)$, and $\text{var}(\lambda) = (e^{\sigma^2} - 1)\exp(2\mu + \sigma^2) = (e^{\sigma^2} - 1)[E(\lambda)]^2$. MCMC estimation of the Poisson-lognormal is straightforward whereas classical estimation is complicated (Izsak, 2007) by the absence of a closed form result for the integral

$$P(y_i|\mu, \sigma^2) = \frac{1}{y_i! \sigma \sqrt{2\pi}} \int_0^\infty e^{-\lambda_i} \lambda_i^{y_i-1} \exp\left[ -\frac{1}{2\sigma^2}(\log(\lambda_i) - \mu)^2 \right] d\lambda_i.$$

A Poisson-lognormal scheme is generally preferred when $\lambda_i$ is being modelled using regression or in multilevel models with various sources of random extra-variability (e.g. see Chapter 5). Modification of the implicit proportionality assumption regarding exposures $E_i$ is also faciliated under the Poisson log-normal. Christiansen and Morris (1995) set out a procedure – treating $\log(E_i)$ as a covariate – for checking exchangeability for data on transplant mortality counts, and detect two groups of hospitals according to total transplants performed.

For binomial data, suppose the observations consist of event counts $y_i$ and populations at risk $n_i$, with

$$y_i \sim \text{Bin}(n_i, p_i), \qquad i = 1, \dots , N.$$

Rather than assume $p_i = p$, suppose the parameters for units $i$ are drawn from a beta density

$$p_i \sim \text{Beta}(\alpha, \beta),$$

with $\alpha$ and $\beta$ considered as prior successes and failures. Under a hierarchical model, the hyper-parameters $\{\alpha, \beta\}$ are themselves assigned a prior, $\pi(\alpha, \beta)$, at the third stage. The beta prior can be re-parameterised as

$$p_i \sim \text{Beta}(\kappa H, (1 - \kappa)H),$$

with $\kappa = \alpha/H$, and $H = \alpha + \beta$ providing a prior sample size, or degree of confidence in the prior estimated success rate $\kappa$ (Greenland, 2000; Young-Xu and Chan, 2008). The alternative to a hierarchical model is to take $\alpha$ and or $\beta$ to be known, with diffuse settings such as $\alpha = \beta = 0.001$ approximating a 'fixed effects' analysis with no effective borrowing of strength and risk of variance instability, but providing unbiased estimates. The precision of the fixed effect estimate depends on the size of denominator $n_i$, and comparisons may spuriously suggest differences in the success rates between units.

The joint beta-binomial posterior density of $\{\alpha, \beta, p_i\}$ is proportional to

$$\pi(\alpha, \beta) \left[ \frac{\Gamma(\alpha + \beta)}{\Gamma(\alpha)\Gamma(\beta)} \right]^n \prod_i [p_i^{\alpha-1}(1 - p_i)^{\beta-1}] \prod_i [p_i^{y_i}(1 - p_i)^{n_i-y_i}],$$

where the constant of proportionality contains the binomial terms $\begin{bmatrix} n_i \\ y_i \end{bmatrix}$. The posterior density of the $p_i$ parameters can be seen to consist of beta densities with parameters $\alpha + y_i$ and $\beta + n_i - y_i$. The posterior mean under repeated sampling is

$$E(p_i|y) = \frac{\alpha + y_i}{\alpha + \beta + n_i} = \frac{\kappa H + y_i}{H + n_i} = \left( \frac{H}{n_i + H} \right) \kappa + \left( \frac{n_i}{n_i + H} \right) \hat{p}_i,$$

namely a weighted compromise between the prior population mean and fixed effects or point estimates $\hat{p}_i = y_i/n_i$.

An alternative approach to binomial heterogeneity is to include a zero mean (e.g. normal or student t) random effect in the model for a transform of $p_i$ to the real line, such as probit($p_i$) or logit($p_i$). With a logit link and normal random effects, one has the logistic-normal mixture (Aitchison and Shen, 1980; Austin, 2009),

$$\log \left( \frac{p_i}{1 - p_i} \right) = \theta + u_i,$$

$$u_i \sim N(0, \tau^2),$$

where $\theta$ is the mean outcome rate on the logit scale. This formulation extends more conveniently to multilevel situations (e.g. Marshall and Spiegelhalter, 1998), or to regression models. One may obtain the density function for $p_i = \exp(\theta + u_i)/[1 + \exp(\theta + u_i)]$ as

$$p(p_i|\theta, \tau^2) = \frac{1}{\tau\sqrt{2\pi}p_i(1 - p_i)} \exp\left[ -\frac{1}{2\tau^2} \left\{ \log \left( \frac{p_i}{1 - p_i} \right) - \theta \right\}^2 \right]$$

showing that $\tau^2$ determines the shape of the density (Duchateau and Jansen, 2005).

## Example 2.3     Hepatitis B in Berlin regions

As an illustration of Poisson outcomes subject to possible overdispersion, consider data presented by Böhning (2000) on observed and expected cases of Hepatitis B in 23 Berlin city regions, denoted $\{y_i, E_i\}, i = 1, \dots, 23$. Note that the standard is not internal, and so $\sum_i E_i = 361.2$ differs slightly from $\sum_i y_i = 368$. One may assess heterogeneity by considering a single parameter model

$$y_i \sim \text{Po}(\Lambda E_i),$$

and evaluating the resulting chi-square statistic, $\sum_i [(y_i - \hat{\Lambda} E_i)^2 / E_i]$. The overall mean relative risk in this case is expected to be approximately $368/361.2$, and a Bayesian analysis provides posterior mean $\hat{\Lambda} = 1.019$ and average chi square statistic 195, indicating excess dispersion (Knorr-Held and Rainer, 2001). The above moment estimator (2.12) for regional variability in hepatitis rates, $\hat{\tau}^2$, is 0.594.

A fixed effects model might be adopted to allow for such variations. Here the parameters $\lambda_i$ are drawn independently of each other (typically from flat gamma priors) without reference to an overall density. This leads to posterior estimates close to maximum likelihood relative incidence estimates for the *ith* region, namely $\hat{R}_i = y_i / E_i$. Alternatively a conjugate hierarchical model may be adopted involving a Gamma prior $\text{Ga}(a, b)$ for heterogeneous relative risks $\lambda_i$, with the parameters $a$ and $b$ themselves assigned relatively flat prior densities confined to positive values (e.g. Gamma, exponential). So one has

$$y_i \sim \text{Po}(\lambda_i E_i),$$

$$\lambda_i \sim \text{Ga}(a, b),$$

$$a \sim \text{Ga}(c_1, c_2), b \sim \text{Ga}(d_1, d_2),$$

where $c_1, c_2, d_1$ and $d_2$ are known. Here the setting $c_k = d_k = 0.001$ for $k = 1, 2$ is used. In BUGS, initial values for $a$ and $b$ should be provided (say as $a = a_0, b = b_0$ with values such as $a_0 = b_0 = 1$) and initial values for $\lambda_i$ may be generated from the gamma prior $\text{Ga}(a_0, b_0)$ using the BUGS command 'gen inits'. One could of course provide initial values for the $\lambda_i$, but in problems with large sample sizes this may involve long vectors, and reasonable initial values are obtained by the 'gen inits' option provided $a$ and $b$ are not set at implicitly 'diffuse' values; for example, using initial values $a = b = 0.001$, and then applying 'gen inits' might lead to aberrant initial values for some $\lambda_i$.

Running three chains for 20 000 iterations, convergence is apparent early (at under 1000 iterations) in terms of BGR statistics. While there is a some sampling autocorrelation in the parameters $a$ and $b$ (around 0.20 at lag 10 for both), the posterior summaries on these parameters are altered little by sub-sampling every 10th iterate, or by extending the sampling a further 10 000 iterations.

In terms of fit and estimates with this model, the posterior mean of the chi-square statistic comparing $y_i$ and $\mu_i = \lambda_i E_i$ is now 23, so extra-variation in relation to available degrees of freedom is accounted for. Posterior mean estimates of $a$ and $b$ are 2.06 and 2.1, while the posterior mean of $\text{var}(\lambda_i)$ (which may be monitored directly, rather than being calculated using the posterior means of $a$ and $b$) is 0.574. Comparison (Table 2.2) of the unsmoothed and smoothed incidence ratios, $\hat{R}_i$, and the $\lambda_i$, shows shrinkage towards the global mean greatest for regions 16, 17 and 19, each having the smallest total (just two) of observed cases. Smoothing is slightly less for area 23, also with 2 cases, but higher expected cases (based on a larger population at risk than in areas 16, 17 and 19), and so more evidence for a genuinely low underlying incidence rate.

**Table 2.2**  Model means and predictive checks, Hepatitis B in Berlin regions.

| Region | y | E | R | $\lambda$ | Complete data posterior predictive probability | Mixed predictive posterior probability | Cross-validation posterior predictive probability | posterior median prediction under cross-validation |
|---|---|---|---|---|---|---|---|---|
| 1 | 29 | 10.71 | 2.71 | 2.43 | 0.685 | 0.955 | 0.970 | 8 |
| 2 | 26 | 17.99 | 1.45 | 1.40 | 0.576 | 0.778 | 0.784 | 14 |
| 3 | 54 | 18.17 | 2.97 | 2.77 | 0.664 | 0.972 | 0.985 | 14 |
| 4 | 30 | 19.21 | 1.56 | 1.51 | 0.586 | 0.812 | 0.819 | 15 |
| 5 | 16 | 21.96 | 0.73 | 0.75 | 0.501 | 0.446 | 0.442 | 18 |
| 6 | 15 | 14.63 | 1.03 | 1.02 | 0.536 | 0.608 | 0.615 | 12 |
| 7 | 6 | 9.62 | 0.62 | 0.69 | 0.481 | 0.400 | 0.384 | 8 |
| 8 | 35 | 17.27 | 2.03 | 1.91 | 0.619 | 0.894 | 0.915 | 13 |
| 9 | 17 | 18.82 | 0.90 | 0.91 | 0.522 | 0.546 | 0.550 | 15 |
| 10 | 7 | 18.27 | 0.38 | 0.44 | 0.437 | 0.218 | 0.210 | 15 |
| 11 | 43 | 32.18 | 1.34 | 1.32 | 0.549 | 0.742 | 0.748 | 26 |
| 12 | 17 | 24.59 | 0.69 | 0.71 | 0.496 | 0.423 | 0.424 | 20 |
| 13 | 15 | 8.40 | 1.79 | 1.63 | 0.636 | 0.850 | 0.859 | 7 |
| 14 | 11 | 15.64 | 0.70 | 0.74 | 0.497 | 0.435 | 0.432 | 13 |
| 15 | 11 | 11.83 | 0.93 | 0.94 | 0.534 | 0.564 | 0.563 | 9 |
| 16 | 2 | 9.95 | 0.20 | 0.34 | 0.342 | 0.119 | 0.104 | 8 |
| 17 | 2 | 10.83 | 0.18 | 0.31 | 0.339 | 0.105 | 0.091 | 9 |
| 18 | 9 | 18.34 | 0.49 | 0.54 | 0.459 | 0.288 | 0.283 | 15 |
| 19 | 2 | 5.18 | 0.39 | 0.55 | 0.412 | 0.261 | 0.252 | 4 |
| 20 | 3 | 10.95 | 0.27 | 0.38 | 0.383 | 0.155 | 0.148 | 9 |
| 21 | 11 | 20.01 | 0.55 | 0.59 | 0.472 | 0.335 | 0.325 | 17 |
| 22 | 5 | 13.83 | 0.36 | 0.44 | 0.415 | 0.208 | 0.199 | 12 |
| 23 | 2 | 12.79 | 0.16 | 0.27 | 0.329 | 0.082 | 0.069 | 11 |

Suppose one seeks to assess whether the hierarchical model improves over the homogenous Poisson model using a penalized fit measure. A scaled deviance is used with $\mu_i = \lambda_i E_i$, and

$$D_s(y, \mu) = 2 \sum_i [y_i \log(y_i/\mu_i) - (y_i - \mu_i)].$$

On fitting the homogeneous Poisson model an average deviance of 178.2 is obtained, or a *DIC* of 179.2; following Spiegelhalter *et al.* (2002), the *DIC* is obtained as either (a) the deviance at the posterior mean $D(\bar{\theta})$ plus $2p$ or (b) the mean deviance plus $p$. Comparing $\bar{D}$ and $D(\bar{\theta})$ under the hierarchical model suggests an effective number of parameters of 18.4, since the average deviance is 23.0, but the deviance at the posterior mean $\bar{\theta}$ (defined here by the posterior averages of the $\lambda_i$, since $a$ and $b$ do not enter into the first-stage likelihood for the observations) is 4.6. (Note that $D(\bar{\theta})$ is obtained from the BUGS code 'Deviance at posterior mean evaluation' without any updating but simply using the Info/Node Info sequence to find the deviance value corresponding to the inputs.) The *DIC* under the gamma mixture model is therefore 41.4, a clear gain in fit over the homogeneous Poisson model.

Even though fit is considerably improved under the hierarchical model, predictive checks can be applied to see whether there are still model discrepancies. One may use checks based on all the data, mixed predictive checks, or cross-validatory checks using the leave one out principle. The complete data posterior predictive probabilities $\Pr.comp_i$, appropriate for count data, are

$$\Pr(y_{new,i} < y_i|y) + 0.5\,\Pr(y_{new,i} = y_i|y),$$

and vary from 0.329 to 0.685, with little indication of model failure (see Table 2.2).

The mixed predictive checks $\Pr.mx_i$ are based on the sampling sequence

$$\lambda_{new,i}^{(t)} \sim \text{Ga}(a^{(t)}, b^{(t)})$$

$$y_{new,i}^{(t)} \sim \text{Po}(\lambda_{new,i}^{(t)} E_i),$$

$$d_i^{(t)} = I(y_{new,i}^{(t)} < y_i) + 0.5 I(y_{new,i}^{(t)} = y_i).$$

The cross-validation posterior predictive probabilities are:

$$\Pr.cv_i = \Pr(\tilde{y}_i^{(t)} < y_i|y_{[i]}) + 0.5\,\Pr(\tilde{y}_i^{(t)} = y_i|y_{[i]}),$$

where $\tilde{y}_i^{(t)}$ is sampled from a likelihood model excluding area $i$.

Unlike the complete data checks, both these checks indicate model failure, with predictive probabilities $\Pr.cv_i$ exceeding 0.95 for areas 1 and 3. This indicates underprediction in that $y_{new}$ is virtually always less than $y$. By contrast, $\Pr.cv_i$ for area 23 is 0.069, indicating overprediction. This can also be seen by comparing $y_i$ with the posterior medians for $\tilde{y}_i$ (last column of Table 2.2). It can be seen from Table 2.2 that the mixed replicate and full cross-validatory procedures yield similar diagnostics (cf. Green *et al.*, 2009). These discrepancies indicate that a single underlying hyper-population may be insufficient to represent the heterogeneity in the data (particularly that emanating from extreme relative risks), and one might consider instead a discrete mixture of Poisson–gamma models (e.g. Austin, 2009).

### Example 2.4 Hot hand in baseball

This example considers data on shooting percentages in baseball, as obtained by Vinnie Johnson over the 1985-89 seasons, and used by Kass and Raftery (1995) to illustrate different approximations for Bayes factors. The question of interest is whether the probability of successfully shooting goals $p$ is constant over games, as in simple binomial sampling (model $M_1$), so that

$$y_i \sim \text{Bin}(n_i, p),$$

where $n_i$ are attempts. Alternatively under $M_2$ the hypothesis is that Vinnie Johnson has a 'hot hand' – that is, he is significantly better in some games than would be apparent from his overall average. The latter pattern implies that $p$ is not constant over games, and instead there might be extra-binomial variation:

$$y_i \sim \text{Bin}(n_i, p_i) \tag{2.13a}$$

$$p_i \sim \text{Be}(\alpha, \beta). \tag{2.13b}$$

Here the models are compared via marginal likelihood approximations based on importance sampling and by estimating individual CPO values. To illustrate sensitivity of Bayes factors to prior settings, weakly informative priors use historic evidence on baseball batting averages (http://en.wikipedia.org/wiki/List_of_Major_League_Baseball_batting _champions). Overly diffuse priors may tend to favour the null model $M_1$.

There are other substantive features of Johnson's play that might be consistent with a hot hand, such as runs of several games with success rates $y_i/n_i$ larger than expected under the simple binomial. A posterior predictive check approach might be applied to assess this. This entails using different test statistics applied to observed and replicate data, $y$ and $y_{new}$, and preferably statistics that are sensible in the context of application. For example, Berkhof *et al.* (2000) compare the maximum success rate $\{y_{new,i}/n_i\}$ in the replicate data samples with the observed maximum.

In the binomial model $M_1$, the prior $p \sim Be(0.8, 1.2)$ is adopted in line with historic data which show batting champion averages around $0.35 - 0.4$. One may estimate the beta posterior density $Be(a_p, b_p)$ of $p$ using moment estimates of the parameters. Thus if $m_p$ is the posterior mean of $p$, and $V_p$ its posterior variance, then $H_p = a_p + b_p$ is estimated as $[m_p(1 - m_p) - V_p]/V_p$, with $a_p = m_p H_p$, and $b_p = (1 - m_p)H_p$. Thus with $m_p = 0.457$ and $V_p^{0.5} = 0.007173$, a posterior beta density with 'sample size' $H_p = 4822$, $a_p = 2204$ and $b_p = 2618$, is obtained. The log marginal likelihood estimator based on (2.7), namely sampled values of

$$\delta^{(t)} = \log(g^{(t)}) - [\log(L^{(t)}) + \log(\pi^{(t)})],$$

is then obtained as $-729.1$. Alternatively combining logs of the CPO estimates for each case, namely $\hat{p}(y_i|y_{[i]}) = \left[T^{-1} \sum_{t=1}^{T} \frac{1}{L_i(\theta^{(t)})}\right]^{-1}$, the estimated LPML of the binomial model is obtained as $-726.0$.

For the beta-binomial model $M_2$, ways of reparameterising the beta mixture parameters include $\alpha = v/\omega$, $\beta = (1 - v)/\omega$ (Kass and Raftery, 1995). Here the parameterisation with a prior sample size $H$, and expected success probability, $\kappa$ given $H$, is adopted:

$$p_i \sim Be(\kappa H, (1 - \kappa)H),$$

$$H \sim \exp(0.5),$$

$$\kappa|H \sim Be(0.4H, 0.6H).$$

in BUGS, initial values are set for $H$ and $\kappa$, with those for $p_i$ generated using 'gen inits'. The last 9500 iterations of a 10 000 three chain run are used to provide approximate posterior beta densities of the $p_i$ and $\kappa$ using the moment estimation procedure. The posterior density for $\kappa$ is approximated by a $Be(1637, 1986)$ density. For $H$, the posterior mean $m_H = 36.1$ is used to provide an approximate posterior exponential density with parameter 0.028.

In a subsequent run, the log marginal likelihood estimator for $M_2$ based on equation (2.7) is obtained as $-749.0$, giving a Bayes factor in favour of the simple binomial, $BF_{12} = \exp(19.9)$. The LPML under model 2 is $-739.6$, so the pseudo BF is $\exp(13.6)$. Unpenalised likelihood comparison by contrast shows that the beta-binomial has a higher posterior mean likelihood than the binomial ($-725.6$ for $M_1$ vs. $-698.9$ for $M_2$). The worsening marginal likelihood implies that the improved first stage likelihood obtained by the beta-binomial is not sufficient to offset the extra parameters it involves. The DIC reinforces this conclusion, providing values of 1452 for $M_1$ and 1494 for $M_2$.

Features of the game pattern such as highest and lowest success rates, or runs of 'cold' or 'hot' games (runs of games with consistent below or above average scoring) may or may not be consistent with global model assessments. Thus consider a predictive check under the simple binomial for the maximum shooting success rate, remembering that the observed maximum among the fixed effects estimates $y_i/n_i$ is 0.9. Setting $p_{new,i} = y_{new,i}/n_i$, the criterion

$$Pr(\max\{p_{new,i}\} > \max\{y_i/n_i\})$$

is found to be about 0.896. This is obtained using the node PPC in the code for model 1, and compares to 0.89 cited by Berkhof *et al.* (2000, p. 345). This significance level is approaching the thresholds which might in fact cast doubt on the simple binomial. Applying a discrete mixture (see Exercise 2.4) suggests some heterogeneity in the data.

## 2.4 Hierarchical smoothing methods for continuous data

For metric data, a typical problem involves originally nested data, for example at two levels, with $i = 1, \ldots , n_j$ replicated observations $y_{ij}$ within groups $j = 1, \ldots , J$. Often the observed data may be summarised in various ways (e.g. as odds ratios, or group averages) $y_j$ aggregating over individual observations, though with details provided on the variability within groups (or on confidence/credible intervals for the averages). Assuming the observed means are derived from similar observation settings or similar types of unit, they may be regarded as draws from an underlying common density for exchangeable, unobserved, true means $\theta_j$.

This assumption leads to a hierarchical model or meta-analysis with a first stage specifying an assumed density for the observations $p(y_j|\theta_j)$, a second stage specifying the density in the hyper-population $\pi(\theta_j|\psi)$, and a third stage specifying densities for hyperparameters $\pi(\psi)$ assumed to generate the second stage density. In the simplest normal hierarchical model, with normal likelihood for the $y_j$, the underlying normally distributed unit means $\theta_j$ differ by group but the global mean $\mu$ and variance $\tau^2$ are assumed constant over groups. Inferences may be required on the ranks of the underlying means (Goldstein and Spiegelhalter, 1996; Deely and Smith, 1998), or on outliers (Farrell *et al.*, 2010), though in analysis of variance situations the goal may be to assess whether the underlying group means are equal (i.e. whether a hierarchical model is needed) namely whether $\tau^2 = 0$ (Berger and Deely, 1988; Daniels, 1999). Suppose the observed summary statistics are assumed normally distributed,

$$y_j|\theta_j, \sigma_j^2 \sim N(\theta_j, \sigma_j^2),$$

where the sampling variances $\sigma_j^2$ are known. For example, for observations $y_j$ provided by logged odds ratios resulting from $2 \times 2$ tables of case versus exposure status, variances $\sigma_j^2$ are obtained by the usual formula in such applications (Woodward, 1999). Suppose the second stage prior for the $\theta$ is also assumed as normal, namely

$$\theta_j|\mu, \tau^2 \sim N(\mu, \tau^2).$$

The conditional posterior for the unit latent means is then obtained as

$$\theta_j|y_j, \mu, \sigma_j^2, \tau^2 \sim N([1 - T_j]y_j + T_j\mu, \tau^2 T_j),$$

$$T_j = \frac{\sigma_j^2}{\sigma_j^2 + \tau^2} = \frac{1/\tau^2}{1/\sigma_j^2 + 1/\tau^2}.$$

So the updated latent mean is a precision weighted average of the observation and the second stage mean, with the weight on the latter increasing with the precision of the $\theta_j$. Marginally one has

$$y_j|\tau^2 \sim N(\mu, \sigma_j^2 + \tau^2),$$

so that $\delta_j = \frac{y_j - \mu}{(\sigma_j^2 + \tau^2)^{0.5}}$, or the posterior means of $\delta_j$, can be used to assess normality assumptions (e.g. via QQ plots) (Dempster and Ryan, 1985).

In meta-analytic applications, subject matter may guide hyperparameter settings under a normal random effects $\theta_j \sim N(\mu, \tau^2)$ prior on latent log odds ratios, logit proportions, or log relative risks. Under this prior, 90% of $\theta_j$ values are expected to be in the interval $\mu \pm 1.645\tau$. Suppose the $\theta_j$ are ranked with $\theta_{[0.05]}$ denoting the 5th percentile, and $\theta_{[0.95]}$ the 95th percentile. Under the normal prior, the prior gap between these percentiles is $\theta_{[0.95]} - \theta_{[0.05]} = 3.29\tau$, or equivalently the quantile ratio $\exp(\theta_{[0.95]})/\exp(\theta_{[0.05]})$ in the original scale (odds ratios, relative risks, etc.) is $\exp(3.29\tau)$. So taking $\tau = 0.5$ amounts to a prior belief that the 95th percentile of the odds ratios (etc.) is $\exp(3.29 \times 0.5) = 5.18$ times the 5th percentile. A higher value such as $\tau = 1$ corresponds to a prior expectation of a 27-fold difference between the 95th and 5th percentiles, and may be considered implausible in some applications.

## 2.4.1    Priors on hyperparameters

A number of choices are available for priors on the third stage hyperparameters $\{\mu, \tau^2\}$. A flat prior on $\mu$ may be justified in terms of the likely strength of accumulated information in the observations about the level of the data. However, inferences about the $\theta_j$ may be sensitive to the prior on the higher level variance or precision (see Example 2.5). An improper prior on $\tau^2$, such as $\pi(\log \tau) \propto c$ may lead to impropriety in the posterior (Morris and Normand, 1992). A diffuse gamma prior on the second stage precision, namely $\tau^{-2} \sim Ga(\varepsilon, \varepsilon)$ with $\varepsilon$ small (e.g. $\varepsilon = 0.001$) puts the larger part of the prior mass on $\tau^2$ away from zero, and implies a marginal Student's $t$, prior for $\theta_j$ with 2 degrees of freedom (with tails heavier than a Cauchy). A moment estimate for $\tau^2$ when the stage 1 variances are known is

$$\hat{\tau}^2 = \frac{1}{J-1} \sum_{j=1}^{J} (y_j - \bar{y})^2 - \frac{1}{J} \sum_{j=1}^{J} \sigma_j^2,$$

and one could choose an inverse gamma prior centred at this value, but with downweighting. Another option is the uniform shrinkage prior, namely a uniform prior on the shrinkage weight $T_j = \frac{\sigma_j^2}{\sigma_j^2 + \tau^2}$, namely

$$T_j \sim U(0, 1),$$

which leads to a proper prior (Daniels, 1999; Natarajan and Kass, 2000). DuMouchel and Normand (2000) suggest the prior

$$\pi(\tau^2) = \frac{s_0}{2\tau(s_0 + \tau)^2}$$

where $s_0^2 = J / \sum_{j=1}^{J} \sigma_j^{-2}$ is the harmonic mean of the sampling variances, as it places a major part of the mass for $\tau^2$ at zero.

A similar property holds for the half normal prior proposed by Pauler and Wakefield (2000),

$$\tau^2 \sim N(0, V) \qquad I(\tau^2 > 0).$$

Following on the above discussion of prior beliefs regarding quantile ratios, and with $y_i$ representing $\log(OR_i)$ or $\log(RR_i)$ in study $i$, one might reasonably assume $V = 0.25$, namely

$$\tau \sim N(0, 0.25) \quad I(0, )$$

as values of $\tau > 1$ then have low probability.

Gustafson *et al.* (2006) suggest the conditional prior

$$\pi(\tau^2|\sigma^2) = \frac{1}{\sigma^2} \frac{a}{(1+\tau^2/\sigma^2)^{a+1}}$$

which is conservative (in the sense of avoiding over-estimation of $\tau^2$) for larger $a$, e.g. $a > 5$. The setting $a = 1$ is equivalent to the uniform shrinkage prior. This prior seeks to guard against substantial over-estimation of the random effect variability, as wrongly concluding there is significant variability may lead to costly interventions. However, in some applications, a perceived drawback of Bayesian hierarchical models is the possibility of overshrinkage, namely that $\tau^2$ is understated (Marshall and Spiegelhalter, 1998; Ohlssen *et al.*, 2007), whereby extreme occurrences are pulled towards the grand mean of the observations; an illustration is the prediction of football results (Baio and Blangiardo, 2010).

A modification of uniform shrinkage can be obtained by taking $R = T^\rho$ or $R_j = T_j^\rho$ (with $\rho > 0$ an additional unknown) to be uniform (see Example 2.5). This can be used to guard against both over and undershrinkage. Values of $\rho$ greater than 1 imply smaller $\tau^2$ and more smoothing at level 2 (i.e. a conservative smoothing), while values $0 < \rho < 1$ imply larger $\tau^2$ and less smoothing at level 2 (anti-conservative smoothing, preventing overshrinkage).

A uniform prior on $\tau$ with a large upper bound $B$, $\tau \sim U(0, B)$, is suggested by Gelman (2006). A more informative option is the half-Cauchy for $\tau$, obtained as a mixture of a positive normal variate

$$d_1 \sim N(0, D^2) \quad I(0,),$$

and a $\chi^2$ variate

$$d_2 \sim \chi_1^2$$

with $\tau = d_1/\sqrt{d_2}$. The scale $D$ of $d_1$ is equivalent to the median of the half-Cauchy, and preset based on prior evidence. In BUGS the required sequence is

```
invD2 <- D*D
d1 ~ dnorm(0,invD2) I(0,)
d2 ~ dchisqr(1)
tau <- d1/sqrt(d2).
```

## 2.4.2    Relaxing normality assumptions

Inferences may also be influenced by departures from normality such as skewness or discordant observations. Inappropriate use of the normal–normal model may result in overshrinkage in the $\theta_j$ (Marshall and Spiegelhalter, 1998), or distortion in the estimate of the overall latent effect size $\mu$, and the variance of the effects, $\tau^2$. The presence of exceptional observations may indicate a robust (heavy-tailed) alternative to the normal, such as the Student t. In a hierarchical model with observations $y_j$ with known variances $\sigma_j^2$, one would generally take these observations to be normal at stage 1, with the second stage density acting as the unifying population density. That unifying density would then include features that depart from normality, such as skewness or heavy tails (e.g. Choy and Smith, 1997; Xie *et al.*, 2007). Departures from symmetry or unimodality may also be modelled by discrete mixtures (see Section 2.5).

As mentioned in Chapter 1, the Student $t$ density is obtainable as a scale mixture of the normal, derived by mixing a $N(0, \sigma^2)$ variable $z_j$ with a positive random variable $\lambda_j$, so that

$$y_j = \mu + [w(\lambda_j)]^{0.5} z_j,$$

where $w(\lambda)$ is a weight function. Taking $w(\lambda) = 1/\lambda$ where $\lambda$ is a gamma variable leads to a Student-$t$ heavy-tailed alternative to the normal (Paddock *et al.*, 2004). Assuming a gamma scale mixture is applied at stage 2, one has

$$y_j \sim N(\theta_j, \sigma_j^2)$$

$$\theta_j \sim N(\mu, \tau^2/\lambda_j) \qquad j = 1, \dots, J$$

$$\lambda_j \sim Ga(v/2, v/2).$$

where $v$ is the degrees of freedom. The scaling parameters $\lambda_j$ are lowest for observations which are discrepant (i.e are potential outliers) from the main set, and outlier status may be assessed with the check $\Pr(\lambda_j < 1|y)$. If $v$ is taken an unknown, then a flat improper prior is not appropriate (Geweke, 1993), as the Lindley paradox indicates that this favours a hypothesis of normality. An exponential prior on $v$ is one option. Small values of $v$ (under 10) indicate that normality of the data is doubtful, while values in excess of 50 are essentially equivalent to normality. Heavy tailed departures from non-normality may also be modelled by exponential power distributions, such as the double exponential.

Departures from normality might also involve excess skewness, indicating skewed versions of the normal or Student t be applied (Fernandez and Steel, 1998; Branco and Dey, 2001; Counsell *et al.*, 2011). Following Azzalini and Capitanio (2003), for a symmetric density $g$ and cumulative distribution function $H$ (from a symmetric density $h$), then

$$p(y) = 2g(y)H(\lambda y),$$

is skew for non-zero $\lambda$. $\lambda$ is a skewness parameter, with $\lambda < 0$ for negative skew, and $\lambda > 0$ for positive skew. Taking $g = \phi$, and $H = \Phi$ (i.e. the standard normal pdf and cdf respectively) leads to the skew-normal distribution $y \sim SN(\mu, \tau^2, \lambda)$, which may be represented conditionally as

$$\theta_j \sim N(\mu + \lambda w_j, \tau^2),$$

where the $w_j$ are positive truncated standard normal variables. In BUGS one would specify

```
th[j] ~ dnorm(m[j],invtau2)
m[j]  <- mu+lambda*w[j]
w[j] ~ dnorm(0,1) I(0,).
```

with invtau2 representing $1/\tau^2$, or use the nonstandard likelihood option (e.g. the Poisson trick) to represent the density directly. Thus for log second stage densities L2 and suitable positive K,

```
L2[j] <- -log(tau*2.507) -0.5*pow((th[j] - mu)/tau, 2) +
                    log(phi(lambda*(th[j] -mu)/tau))
h[j] <- 0
h[j] ~ dpois(phi[j])
phi[j]<- - L2[j]+K.
```

An alternative skew-normal derivation underlies the variance gamma model (Fung and Seneta, 2010) developed for modelling skewed time series, with conditional representation

$$\theta_j \sim N(\mu + \lambda \delta_j, \tau^2 \delta_j),$$

where $\delta_j$ are positive, specifically gamma, variables. A variant of the skew $t$ model (Demarta and McNeil, 2005) has a similar conditional representation, namely

$$\theta_j \sim N(\mu + \lambda/\delta_j, \tau^2/\delta_j),$$

where again the $\delta_j$ are gamma, with $\delta_j \sim Ga(\frac{v}{2}, \frac{v}{2})$ and $v$ is the degrees of freedom parameter. In BUGS one would specify

```
th[j] ~dnorm(m[j],invtau[j])
invtau[j] <- delta[j]*invtau2
m[j] <- mu+lambda/delta[j]
delta[j] ~dgamma(nu.2,nu.2)
```

where nu.2 represents $v/2$. An alternative skew-$t$ model (Branco and Dey, 2001) specifies

$$\theta_j \sim N(\mu + \lambda w_j/\delta_j, \tau^2/\delta_j),$$

with $\delta_j \sim Ga(\frac{v}{2}, \frac{v}{2})$, and $w_j \sim N(0, 1)\, I(0, )$.

### 2.4.3   Multivariate borrowing of strength

Multivariate applications to borrow strength (Arends *et al.*, 2003; Kirkham *et al.*, 2012) may be used to strengthen inferences for a relatively rare outcome using information provided by a more frequent correlated outcome. Another application is in clinical studies with treatment and control groups, where the outcome (e.g. mortality, infection, recurrence) in the control group may be taken to capture baseline risk, and treatment effects may be correlated with baseline risk. Sensitivity and specificity of different diagnostic procedures (or a single test applied in different circumstances) may also be treated as correlated responses (Chu and Cole, 2006), with each study providing numbers of true and false positives and negatives, and goals including estimation of the underlying true sensitivity (true positives among diseased subjects) and specificity (true negatives among non-diseased subjects).

Consider the treatment-baseline effects scenario, and suppose $y_{i1}$ of $N_{i1}$ treated patients in a study $i$ show a particular response, as compared to $y_{i2}$ among $N_{i2}$ control patients. One approach focuses on the empirical log odds $z_{i1} = \log[y_{i1}/(N_{i1} - y_{i1})]$, and $z_{i2} = \log[y_{i2}/(N_{i2} - y_{i2})]$, and adopts a bivariate normal observation model for $(z_{i1}, z_{i2})$ at the first stage (van Houwelingen *et al.*, 2002), with outcomes assumed conditionally independent given latent treatment and baseline risks, $\theta_{i1}$ and $\theta_{i2}$ respectively. At stage 2 the latent risks are also bivariate normal but with correlation allowed. Thus

$$\begin{pmatrix} z_{i1} \\ z_{i2} \end{pmatrix} \sim N\left( \begin{pmatrix} \theta_{i1} \\ \theta_{i2} \end{pmatrix}, \begin{pmatrix} \sigma_{i1}^2 & 0 \\ 0 & \sigma_{i2}^2 \end{pmatrix} \right),$$

$$\begin{pmatrix} \theta_{i1} \\ \theta_{i2} \end{pmatrix} \sim N\left( \begin{pmatrix} \mu_1 \\ \mu_2 \end{pmatrix}, V \right),$$

where $V = \begin{pmatrix} \tau_1^2 & \rho\tau_1\tau_2 \\ \rho\tau_1\tau_2 & \tau_2^2 \end{pmatrix}$, with diagonal terms representing variability in the true treatment and control event rates, and with $\Delta = \mu_1 - \mu_2$ defining the underlying treatment effect. The differences $d_i = \theta_{i1} - \theta_{i2}$ represent smoothed estimates of the log odds ratios that may be compared with calculated empirical log-odds ratios (e.g. Turner *et al.*, 2000; van Houwelingen *et al.*, 2002, p. 593).

A binomial or Poisson likelihood may alternatively be adopted at stage 1, with appropriate link to bivariate normal latent effects. For example,

$$y_{i1} \sim \text{Bin}(N_{i1}, \pi_{i1}); \qquad y_{i2} \sim \text{Bin}(N_{i2}, \pi_{i2});$$

$$\text{logit}(\pi_{i1}) = \theta_{i1},$$

$$\text{logit}(\pi_{i2}) = \theta_{i2},$$

$$\begin{pmatrix} \theta_{i1} \\ \theta_{i2} \end{pmatrix} \sim N\left( \begin{pmatrix} \mu_1 \\ \mu_2 \end{pmatrix}, V \right).$$

### Example 2.5   Teacher expectancy and pupil IQ

To illustrate sensitivity to the second stage variance in a univariate hierarchical normal–normal model, consider data on experimental effects on children's IQ according to teacher expectations. Raudenbush (1984) reported on 19 studies with experimental groups who were told by teachers that there were high expectations regarding their ability, and a control group for whom no such encouragement was provided. Along with effect sizes the summary statistics from each study provided known sampling variances $\sigma_j^2$. The observed effect sizes vary widely, and some are based on small samples, so their precision $h_j = 1/\sigma_j^2$ is low. The variation $\tau^2$ in the underlying effect parameters $\theta_j$ is likely to be smaller than the variation between the observed effect sizes. So under the standard normal–normal model,

$$y_j \sim N(\theta_j, \sigma_j^2),$$

$$\theta_j \sim N(\mu, \tau^2).$$

A uniform shrinkage prior is taken, with

$$T = s_0^2/(s_0^2 + \tau^2),$$

$$\pi(T) = U(0, 1),$$

where $s_0^2 = J/\sum_{j=1}^{J} \sigma_j^{-2}$. This prior is compared with a uniform prior on $\tau$, and with a uniform prior on $R = T^\rho$, where $\rho$ is an unknown between 0 and 1.

Under the uniform shrinkage option, the last 9000 iterations from a two chain run of 10 000 iterations provide a posterior mean for $\tau^2$ of 0.023, with the corresponding posterior mean for $\tau = sd(\theta)$ of 0.136. Hence there is wide variability in the latent effect sizes, even if the average effect size $\mu = 0.083$ (with 95% credible interval from $-0.017$ to 0.200) is not quite conclusive regarding the teacher expectancy effect. The latent effect sizes $\theta_j$ range from $-0.03$ to 0.24 (Table 2.3), a considerable shrinkage compared to the raw effects. Shrinkage is greater for studies with less precisely estimated effects: the correlation between $|y_j - \theta_j|$ (one possible shrinkage indicator) and $h_j$ is $-0.51$.

A final point of note is that a mixed predictive check, using indicators at the $t^{th}$ iteration $d_j^{(t)} = I(y_{new,j}^{(t)} \leq y_j | \theta_{new,j}^{(t)}, \psi^{(t)})$, where $\psi = (\tau^2, \mu)$. This check suggests a limitation in the normal–normal model: two studies (4 and 10) with large positive raw effects $y_j$ have posterior means $d_j$ that exceed 0.99. This means replicates from the model almost always underpredict the actual observations.

Adopting a uniform prior $\tau \sim U(0, 100)$ (see Section 2.4.1) produces slightly less shrinkage, with a posterior mean for $\tau^2$ of 0.030, and for $\tau$ of 0.148. The posterior mean for $\theta_{10}$ is now 0.265, as opposed to 0.244 under the uniform prior on $T = s_0/(s_0 + \tau^2)$. The

**Table 2.3**  Teacher expectancy and pupil IQ.

| Study | Effect size estimate | Observed precision | Underlying effect | Mixed predictive check | Weeks in prior contact |
|-------|------|------|------|------|------|
| 1 | 0.03 | 62.5 | 0.05 | 0.395 | 2 |
| 2 | 0.12 | 45.5 | 0.10 | 0.576 | 3 |
| 3 | −0.14 | 35.7 | 0.00 | 0.160 | 3 |
| 4 | 1.18 | 7.2 | 0.22 | 0.996 | 0 |
| 5 | 0.26 | 7.4 | 0.11 | 0.669 | 0 |
| 6 | −0.06 | 90.9 | 0.00 | 0.204 | 3 |
| 7 | −0.02 | 90.9 | 0.02 | 0.276 | 3 |
| 8 | −0.32 | 20.8 | −0.03 | 0.066 | 3 |
| 9 | 0.27 | 37.0 | 0.15 | 0.805 | 0 |
| 10 | 0.8 | 15.9 | 0.24 | 0.992 | 1 |
| 11 | 0.54 | 11.0 | 0.16 | 0.914 | 0 |
| 12 | 0.18 | 20.0 | 0.11 | 0.640 | 0 |
| 13 | −0.02 | 11.9 | 0.06 | 0.383 | 1 |
| 14 | 0.23 | 11.9 | 0.11 | 0.673 | 2 |
| 15 | −0.18 | 40.0 | −0.02 | 0.116 | 3 |
| 16 | −0.06 | 35.7 | 0.03 | 0.261 | 3 |
| 17 | 0.3 | 52.6 | 0.18 | 0.858 | 1 |
| 18 | 0.07 | 111.1 | 0.07 | 0.483 | 2 |
| 19 | −0.07 | 33.3 | 0.03 | 0.252 | 3 |

Dawid–Sebastiani score for this option is based on complete data replicates for $y$, and their posterior means and variances, namely

$$\text{DSS} = \frac{1}{n} \sum_{i=1}^{n} \left[ \left( \frac{y_i - \mu_{new,i}}{\sigma_{new,i}} \right)^2 + 2 \log(\sigma_{new,i}) \right].$$

This score is lowered to −2.36 (i.e. a better complete data predictive fit), as compared to −2.24 under the uniform prior on $T$.

As noted above, a uniform prior on $R = T^{\rho}$ (i.e. $T = s_0^2/(s_0^2 + \tau^2) = R^{1/\rho}$) with values $0 < \rho < 1$ implies larger $\tau^2$. So this leads to less smoothing than under a uniform prior on $T$, and can be seen as 'anti-conservative' smoothing. Here the prior $\rho \sim U(0, 1)$ is adopted, giving an estimate for $\tau$ of 0.144, and posterior mean for $\theta_{10}$ of 0.260. The parameter $\rho$ has mean 0.63, and 95% interval (0.14, 0.98). The Dawid–Sebastiani score for this option is −2.28. Although complete data predictive fit improves slightly under a uniform prior on $\tau$, and under a uniform prior on $R = T^{\rho}$ ($0 < \rho < 1$), the mixed predictive checks for studies 4 and 10 remain problematic.

**Example 2.6    Bivariate meta-analysis for treatment of respiratory infections**

Consider data on respiratory tract infections $(y_{i1}, y_{i2})$ in treated and control groups of respective size $(N_{i1}, N_{i2})$ in 22 trials for selective decontamination of the digestive tract (Turner *et al.*, 2000). A favourable treatment effect is evinced in lower infection rates and hence negative log-odds ratios comparing treatment and control groups. The empirical log odds ratios are all negative but vary widely, from $-0.07$ (observed standard deviation, 0.53) to $-3.62$ (sd 1.48). A binomial is adopted at stage 1, with logit link to bivariate normal log-odds effects. Thus,

$$y_{i1} \sim \text{Bin}(N_{i1}, \pi_{i1}); \qquad y_{i2} \sim \text{Bin}(N_{i2}, \pi_{i2});$$

$$\text{logit}(\pi_{i1}) = \theta_{i1},$$

$$\text{logit}(\pi_{i2}) = \theta_{i2},$$

$$\begin{pmatrix} \theta_{i1} \\ \theta_{i2} \end{pmatrix} \sim N \left( \begin{pmatrix} \mu_1 \\ \mu_2 \end{pmatrix}, V \right).$$

A Wishart prior $V^{-1} \sim \text{Wishart}(R, k)$ may be adopted for the precision matrix, having expectation $kR^{-1}$, so $H = \frac{1}{k}R$ provides a prior elicitation of the unknown covariance matrix $V$. In meta-analysis applications, as discussed above, a prior expectation is often that values $\tau_j > 1$ on latent treatment or baseline effects (in logit or log scales) are unlikely, while $\tau_j = 0.5$ or $\tau_j = 0.75$ might be more supported. So a reasonable setting for prior variances for $\theta_{i1}$ and $\theta_{i2}$ is taken as $0.75^2 = 0.56$. These provide diagonal terms $(h_{11}, h_{22})$ in $H$, so with $k = 2$, the two diagonal terms in $R$ are 1.12. Off-diagonal terms in $R$ are taken as 0.

The second half of a two chain run of 20 000 iterations provides estimates for $\Delta = \mu_1 - \mu_2$, the overall treatment effect of $-1.37$ $(-1.84, -0.93)$, and also supports the adoption of a bivariate approach, in that $\rho$ is estimated as 0.76 (0.49, 0.92). Smoothed study-specific estimates of the log odds ratio range from $-0.39$ (s.d. 0.28) to $-2.66$ (s.d. 0.77).

A second analysis replicates the prior scheme used in R-INLA (see http://www.r-inla.org /examples/volume-ii/code-for-model-bivariatemetaanalysis), with log-gamma priors on the log-precisions in $V^{-1}$, and an effectively uniform prior on $\rho$. The code is

```
model { for (i in 1:22) {y1[i] ~dbin(p[i,1],N1[i]);
                         y2[i] ~ dbin(p[i,2],N2[i])
logit(p[i,1]) <- th[i,1]; logit(p[i,2]) <- th[i,2]
# smoothed log-odds-ratios
d[i] <- th[i,1]-th[i,2]
th[i,1:2] ~dmnorm(mu[1:2],V.inv[,]) }
V.inv[1:2,1:2] <- inverse(V[,])
V[1,2] <- rho/sqrt(inv.tau2[1]*inv.tau2[2]); V[2,1] <- V[1,2]
rho ~dunif(-1,1)
for (j in 1:2) {mu[j] ~dflat(); inv.tau2[j] ~ dgamma(0.25,0.025);
                               V[j,j] <- 1/inv.tau2[j]}
# overall treatment effect
Del <- mu[1]-mu[2]}
```

Posterior estimates are similar to those under the Wishart prior except that the variances of the treatment and baseline effects are slightly smaller, with the diagonal terms of V estimated as 0.84 (s.d. 0.36) and 1.80 (s.d. 0.70).

Implementation of a bivariate meta-analysis directly in R-INLA requires two lines of data for each study, with treatment and control arms identified by binary variables. Hence a file with relevant column headings and data for study 1 with $y_1 = 7$, $N_1 = 47$, $y_2 = 25$, $N_2 = 54$ is represented as

| id | infec | tpat | trt | ctrol |
|----|-------|------|-----|-------|
| 1  | 7     | 47   | 1   | 0     |
| 2  | 25    | 54   | 0   | 1     |

..................

The command sequence for the 22 studies (44 data lines) is then

```
library(INLA)
# columns headed id,infec,tpat,trt,ctrol
D <- read.table("resptract.txt", header = T)
F = infec ~trt+ctrol - 1+f(id,model ="2diid",
    param=c(0.25,0.025,0.25,0.025,0,0.2),n=44)
M <- inla(F, family = "binomial", Ntrials = tpat, data = D)
```

which provides a higher estimate for $\rho$ of 0.80 (0.53, 0.94) and higher variance terms in V.

## 2.5     Discrete mixtures and dirichlet processes

The priors considered above for the underlying hyper-population mixing density have a specific parametric form. However, to avoid being tied to particular parametric forms of prior, non-parametric options have been proposed, such as finite mixtures of parametric densities (Laird, 1982) or mixtures obtained using Dirichlet process priors. The goals in choosing semi- or non-parametric mixing include improved fit and robustness of inferences by approximating more closely the true density of the sample (West *et al.*, 1994; Da Silva, 2007, 2009). This is especially so for observations that are multimodal or asymmetric in form. Here a hierarchical model assuming a single underlying hyperpopulation (e.g. an underlying normal density) may result in overshrinkage (Marshall and Spiegelhalter, 1998). One option is to adopt finite mixtures of the stage 1 density (Clayton and Kaldor, 1987; Frühwirth-Schnatter, 2006). However, to allow hierarchical borrowing of strength within sub-populations, the finite mixture approach may be applied to conjugate mixtures (e.g. discrete mixtures of two or more poisson–gamma, or normal–normal models), or to non-conjugate mixtures (e.g. discrete mixtures of poisson-lognormal). The mixture will generally involve a small but unknown number of sub-populations, as determined by an underlying latent set of group membership variables for each observation. There are then likely to be issues of discrimination between models involving different numbers of sub-populations, especially in smaller samples or if data likelihoods are relatively flat; see Böhning (2000).

### 2.5.1     Finite mixture models

Discrete mixture models have wide flexibility in representing heterogeneous data, when a choice of parametric form for the heterogeneity is unclear, or when inferences are sensitive to particular choices of parametric mixture. Assume a mixture of $G$ subpopulations or groups, and let $L_i$ denote a latent group membership indicator for unit $i$. $L_i$ is a categorical variable that can take any value between 1 and $G$. The latent membership data may also be expressed by multinomial indicators $Z_{ig} = 1$ if $L_i = g$ and $Z_{ig} = 0$ otherwise. Let $\pi_g$ be the prior probability of belonging to sub-population $g$, with $\sum_g \pi_g = 1$. Under the conjugate scheme, the prior multinomial probability $\pi = (\pi_1, \ldots, \pi_G)$ for the latent membership indicators is drawn from a Dirichlet density with weights $\alpha_g$, $\pi \sim D(\alpha_1, \ldots, \alpha_G)$. The latter may be preset, or additional unknowns, assigned positive (e.g. gamma) priors.

For a finite mixture at stage 1 (i.e. the data likelihood stage), e.g. of normal sub-populations for continuous data, or Poisson subpopulations for count data, let the parameters of group $g$ be denoted $\omega_g$. Conditional on the latent group indicators $L_i$ one has

$$y_i | L_i = g \sim p(y_i | \omega_g),$$

with the marginal density

$$p(y_i) = \sum_{g=1}^{G} \pi_g p(y_i | \omega_g)$$

So conditioning on $L_i$, a finite mixture of normal densities would take

$$y_i | L_i = g \sim N(y_i | \mu_g, \sigma_g^2),$$

$$L_i | \pi \sim \text{Categoric}(\pi),$$

$$\pi \sim D(\alpha_1, \dots, \alpha_G).$$

The equivalent marginal density is

$$p(y_i) = \sum_{g=1}^{G} \pi_g N(y_i | \mu_g, \sigma_g^2),$$

with one possible prior scheme being

$$\pi(\mu_g | \lambda, V) = N(\lambda, V),$$

$$\pi(h_g = 1/\sigma_g^2 | a, b) = \text{Ga}(a, b).$$

Because of the labelling issue under MCMC sampling, a constraint may additionally be applied (e.g. to the $\mu_g$) for unique identification (see below).

The posterior for $\pi$ is provided by a Dirichlet with elements $K_g + \alpha_g$ where $K_g$ is the number of sample members assigned to the $g^{th}$ group, and the posterior membership probabilities governing the allocation of units $i$ to groups $g$ are:

$$P_{ig} = \Pr(Z_{ig} = 1 | y) = \pi_j p(y_i | \omega_g) / \sum_{g=1}^{G} \pi_g p(y_i | \omega_g).$$

If $\mu_g$ denotes the updated group mean, the smoothed mean $\tilde{\mu}_i$ for the $i^{th}$ subject may then be estimated as a weighted sum of group means

$$\tilde{\mu}_i = \sum_g P_{ig} \mu_g.$$

Such means often show shrinkage towards the global mean $\bar{\mu} = \pi_1 \mu_1 + \pi_2 \mu_2 + \dots \pi_G \mu_G$ (e.g. in disease mapping applications) even if the units have high posterior probabilities of belonging to groups with means above or below the global mean.

For a finite mixture of hierarchical priors, with the discrete mixture at stage 2, one has:

$$y_i \sim p(y_i | \lambda_i),$$

$$\lambda_i | L_i = g \sim \pi(\lambda_i | \psi_g),$$

where separate hyperpriors $\psi_g \sim \pi_g(\psi_g | c_1, c_2, \dots)$ are assumed for each group at stage 3. For example, a finite mixture of normal–normal models could adopt a uniform shrinkage prior at stage 3

$$y_i \sim N(\theta_i, \sigma^2),$$

$$\theta_i | L_i = g \sim N(\mu_g, \tau_g^2),$$

$$\mu_g \sim N(\lambda_g, V_g),$$

$$h = 1/\sigma^2 \sim Ga(a, b)$$

$$T_g = \frac{\sigma^2}{\sigma^2 + \tau_g^2} \sim U(0, 1),$$

$$L_i | \pi \sim Categoric(\pi),$$

$$\pi \sim D(\alpha_1, \ldots, \alpha_G).$$

A particular form for the Dirichlet weights (Neal, 2000) which clarifies the infinite mixture (Dirichlet process) generalisation is $\alpha_g = \alpha/G$, namely

$$\pi \sim D(\alpha/G, \ldots, \alpha/G),$$

with $\alpha$ preset or unknown. In this case Gibbs sampling for the classification indices uses the full conditional density

$$p(L_i = g | L_{[i]}, \alpha, \omega) = \frac{K_{[i]g} + \alpha/G}{n - 1 + \alpha},$$

where $L_{[i]}$ contains all classification indices except $L_i$, and $K_{[i]g}$ is the number of observations (or latent means), except for $y_i$ ($\theta_i$), that belong to group $g$.

The major issues in identifying finite mixture models using data likelihoods $p(y_i | \omega_g)$, or second stage densities $\pi(\lambda_i | \psi_g)$, are the general question of identifiability in the face of possibly flat likelihoods, especially for values of $G$ around the optimal value (Böhning, 2000), and the specification of appropriate priors that are objective but also effective in estimation. Thus Wasserman (2000) cites the hindrance in mixture modelling arising from the fact that improper priors yield improper posteriors. More generally, vague priors even if proper, may lead to poorly identified posterior solutions, especially for small samples. Various approaches to prior specification in mixture modelling have been proposed and often informative priors based on subject matter knowledge may be employed.

There are also issues of 'label switching' in MCMC estimation of mixture models. If sampling takes place from an unconstrained prior with $G$ groups then the parameter space has $G!$ subspaces corresponding to different ways of labelling the states. In an MCMC run on an unconstrained prior there may be jumps between these subspaces. Constraints are generally imposed to ensure that components do not 'flip over' during estimation. One may specify that one mixture probability is always greater than another, or that means are ordered, $\mu_1 > \mu_2 > \ldots > \mu_G$, or that variances are ordered. It remains problematic whether such constraints distort the final estimate. Depending on the problem one sort of constraint may be more appropriate to a particular data set: constraining means may not be effective if subgroups with different means are not well identified, but groups with different variances are (Frühwirth-Schattner, 2001). A maximum likelihood analysis may justify one type of constraint as against another.

There is also the possibility of empty groups (e.g. only $G - 1$ groups of cases are chosen at a particular MCMC iteration when a $G$ group model is being fitted); to avoid this, a data dependent prior may be adopted avoid a null groups occurring (Wasserman, 2000). Chen *et al.* (2001) seek to overcome boundary and identifiability problems by introducing an extra penalty component to the likelihood that amounts to introducing $GC$ extra 'observations', $C$ with mean $\mu_1$, $C$ with mean $\mu_2$, and so on. As an example, let there be $G = 2$ sub-groups with probabilities $\pi$ and $(1 - \pi)$ differing only in their means $\mu_1$ and $\mu_2$, with $\mu_1 < \mu_2$. The number $2C$ of extra observations is related to the precision on the prior for the $\mu_j$'s. Thus if $\{\mu_1, \mu_2\}$ are assigned uniform $U(-R, R)$ priors, then Chen *et al.* take $C = \log(R)$. For a normal prior on the $\mu_j$ one might take $C = \log(3\sigma)$ where $\sigma^2$ is the prior variance of the $\mu_j$. The likelihood is penalised as follows:

$$\sum_i \{\log[(1-\pi)p(y_i|\mu_2) + \pi p(y_i|\mu_1)]\} + C\log[4\pi(1-\pi)].$$

Finite mixture models can be represented as a regression on a latent binary or multinomial indicator (Tarpey and Petkova, 2010). Finite mixture models implicitly assume that distinct latent groups exist in the population, but this may not always be substantively plausible; examples are tendency to experience a placebo effect or unobserved influences on area health risks, which are likely to vary continuously. Instead of a discrete latent predictor as assumed for finite mixtures, one may posit a continuous latent predictor. Thus for $y$ normal one has

$$y_i \sim N(\beta_0 + \beta_1 x_i, \sigma^2),$$

$$x_i \sim Be(\alpha_1, \alpha_2),$$

with $x_i$ as the latent predictor. Assuming $x$ is binary leads to a discrete mixture (with $G = 2$), but generalizing to a continuous beta density on $(0, 1)$ can produce bimodal densities and may provide a plausible alternative to discrete sub-groups.

## 2.5.2   Dirichlet process priors

Dirichlet Process Priors (abbreviated as DPP priors) offer another approach avoiding parametric assumptions and unlike a mixture of parametric densities, are less impeded by uncertainty about the appropriate number of sub-groups (Dey *et al.*, 1999; Teh *et al.*, 2006; Teh, 2010). The DPP method deals with possible clustering in the data without trying to specify the number of clusters, except perhaps a maximum conceivable number. Let $\{y_i, i = 1, \ldots, n\}$ be drawn from a distribution with unknown parameters $\{\theta_i, \varphi_i\}$, where a Dirichlet process prior is adopted for the $\theta_i$, but a conventional parametric prior for the $\varphi_i$. The Dirichlet process involves a random distribution $G$ centred on a baseline prior $G_0$ from which candidate values for $\theta_i$ are drawn, and precision parameter $\alpha$ describing the concentration of the mass around $G_0$. For instance in Example 2.3 most of the incidence ratios $\lambda_i$ are likely to be between 0.25 and 4 in a hierarchical model, and a suitable baseline prior $G_0$ on $\theta_i = \log(\lambda_i)$ might be a N(0,0.5) density.

One has

$$y_i|\theta_i, \varphi_i \sim p(y|\theta_i, \varphi_i),$$

$$\theta_i|G \sim G$$

$$G|\alpha, G_0 \sim DP(\alpha, G_0),$$

where $G_0$ is a specified density, which may itself contain unknown parameters. Suppose the $\theta_i$ are unknown means (or transformed means, such as log relative risks or logit probabilities) for each case, and that clustering in these values is expected. Then for cases within cluster $m$, the same value $\theta_m \sim G_0$ would be appropriate for them. One may consider the process as a mixture with an infinite number of point masses (Teh, 2010)

$$G = \sum_{m=1}^{\infty} \pi_m \delta_{\theta_m},$$

though a straightforward adaptation (Ohlssen *et al.*, 2007) expresses the Dirichlet Process to be a mixture of continuous densities

$$G = \sum_{m=1}^{\infty} \pi_m f(|\theta_m).$$

In practice, the Dirichlet process may be approximated by truncating the mixture at $M$ components with $\sum_{m=1}^{M} \pi_m = 1$ (with $M$ large, usually $M \leq n$). A set of $M$ potential values of $\theta$, $\{\theta_m^*, m = 1, \ldots, M\}$, is drawn from $G_0$. The most appropriate value $\theta_m^*$ for case $i$ is then selected using multinomial sampling with $M$ groups, and Dirichlet prior on the group probabilities having prior weights $\{\alpha_m = \alpha/M, m = 1, \ldots, M\}$. So the cluster indicator for case $i$ is chosen according to

$$L_i | \pi \sim \text{Categorical}(\pi),$$

$$\pi \sim D(\alpha/M, \ldots, \alpha/M),$$

with larger $\alpha$ resulting in more smoothing and a larger number of clusters (Ishwaran and James, 2002). Then case $i$ is assigned the parameter $\theta_{L_i}$. The actual number of clusters $M^*$ may be considerably less than $M$, and one may monitor $M^*$ over MCMC iterations.

   The parameter $\alpha$ may be preset, to provide faster sampling (typical values are $\alpha = 1$ or $\alpha = 5$). For $\alpha$ known, Ohlssen *et al.* (2007) suggest $M$ be obtained as

$$M \simeq 1 - \alpha \log(\varepsilon),$$

where $E(\alpha_M) \simeq \varepsilon$ is the probability assigned to the final mass point. However, $\alpha$ may itself be assigned a prior. For example, the R package dpmixsim (http://cran.r-project.org/web/packages/dpmixsim/index.html) adopts a default $\alpha \sim \text{Ga}(2, 2)$; see also Ishwaran and James (2002). The conditional density for the classification indices $L_i$ can be obtained by taking $M \to \infty$ in the conditional density under the finite mixture case above, namely

$$p(L_i = m | L_{[i]}, \alpha) = \frac{K_{[i]m}}{n - 1 + \alpha},$$

$$p(L_i \neq L_j, i \neq j | L_{[i]}, \alpha) = \frac{\alpha}{n - 1 + \alpha},$$

where $K_{[i]m}$ is the number of observations, except for $y_i$, that belong to group $m$.

   The prior $\pi \sim D(\alpha/M, \ldots, \alpha/M)$ for the group probabilities ($M$ large) can be used to approximate the Dirichlet process (Ishwaran and Zarepour, 2001). Alternatively, the random mixture weights $\pi_m$ under the truncated mixture may be constructed by drawing a sequence $r_1, r_2, \ldots, r_{M-1}$ of $\text{Be}(1, \alpha)$ random variables (Sethuraman, 1994; Ishwaran and James, 2001). Thus sample

$$r_j \sim \text{Be}(1, \alpha) \qquad m = 1, \ldots, M - 1,$$

and, defining $r_M = 1$, set

$$\pi_1 = r_1,$$

$$\pi_2 = r_2(1 - r_1),$$

$$\pi_3 = r_3(1 - r_2)(1 - r_1),$$

$$\ldots$$

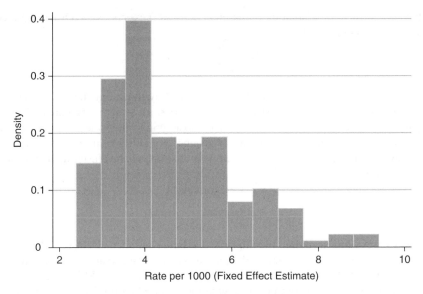

**Figure 2.1**    Abortion rates.

and so on. This is known as a stick-breaking prior since at each stage what is left of a stick of unit length is broken and the length of the broken portion assigned to the current value $\pi_m$. The stick-breaking prior can be written $\pi \sim \text{GEM}(\alpha)$, where the letters stand for Griffiths, Engen, and McCloskey (Teh, 2010). Other sampling strategies for the $r_m$ are discussed by Ishwaran and James (2001); one option takes the random weights as $r_j \sim \text{Be}(a_j, b_j)$ where $a_j = 1 - \alpha$, $b_j = j\alpha$, and $0 < \alpha < 1$.

**Example 2.7    English abortion rates**

Abortion rates during 2006-08 for women aged under 16 show wide variation over 151 English health agencies (Primary Care Trusts). The data are abortions $y_i$ and populations $T_i$ of women aged 13-15. The England wide rate $1000y./T.$ is 4.2 per 1000, with moment estimated rates for individual agencies, $1000y_i/T_i$, ranging from 2.4 to 9.4 per 1000. A plot (Figure 2.1) of these rates suggests positive skewness, and the first option considered involves a stage 2 skew normal in the logits of the rates:

$$y_i \sim \text{Bin}(T_i, r_i),$$

$$\text{logit}(r_i) = \theta_i,$$

$$\theta_j \sim N(\mu + \lambda w_j, \tau^2),$$

with the $w_j$ being positive $N(0, 1)$ variables, and with additional priors $\pi(\lambda) = N(0, 10)$, and $\pi(1/\tau^2) = \text{Ga}(1, 0.001)$. Note that $\mu$ is confounded with $\lambda w_j$, and the identified mean level of the $\theta_j$ is obtained by monitoring $\mu^{*(t)} = \mu^{(t)} + \lambda^{(t)}\overline{w}^{(t)}$ over iterations t.

Inferences are based on the final 4000 iterations from a 2 chain run of 5000, with initial values in BUGS only set for $\mu$, $\lambda$ and $1/\tau^2$, and with the $\theta_j$ generated from the priors using 'gen inits'. The skew parameter $\lambda$ is significant, with mean 0.48 and 95% interval (0.35, 0.57), with $\mu^*$ having mean $-5.43$. For predictive assessment, a mixed predictive check is used,

**Table 2.4**  Mixture Model, Abortion rates.

| Parameter | Mean | Standard deviation | 2.5% | 97.5% |
|---|---|---|---|---|
| $\mu_1$ | −5.7 | 0.06 | −5.8 | −5.5 |
| $\mu_2$ | −5.2 | 0.09 | −5.4 | −5.0 |
| $\pi_1$ | 0.44 | 0.15 | 0.19 | 0.76 |
| $\pi_2$ | 0.56 | 0.15 | 0.24 | 0.81 |

based on sampling replicate $\theta_j$, namely $d_j^{(t)} = I(y_{new,j}^{(t)} \le y_j | \theta_{new,j}^{(t)}, \psi^{(t)})$, where $\psi = (\tau^2, \lambda, w_j)$. There are two areas with $\Pr(y_{new,j} \le y_j | y)$ under 0.05, indicating that the model over-predicts as compared to the actual rates for these areas. Additionally, complete data replicate samples $y_{new,i} \sim \text{Bin}(T_i, r_i)$ are taken, and using them the DSS criterion is evaluated as 4.958.

A second model has a gamma scale mixture at stage 2 (i.e. a Student $t$) with unknown degrees of freedom $v$, namely

$$\theta_j \sim N(\mu, \tau^2/\gamma_j),$$

$$\gamma_j \sim \text{Ga}\left(\frac{v}{2}, \frac{v}{2}\right).$$

The prior on $v$ is indirect, namely

$$v = 1/\kappa,$$

$$\kappa \sim U(0.01, 0.5),$$

with the posterior mean for $v$ of 24.6 (from a two chain run of 5000 iterations) indicating a mild departure from normality. There are again two areas with $\Pr(y_{new,j} \le y_j | y)$ under 0.05, but none with $\Pr(y_{new,j} \le y_j | y)$ over 0.95. The DSS criterion also improves, albeit slightly, to 4.939.

Skewness and multimodality can often be dealt with by discrete mixture approaches. Finally considered therefore is a two group mixture of normal–normal models, namely

$$y_i \sim \text{Bin}(T_i, r_i),$$

$$\text{logit}(r_i) = \theta_i,$$

$$\theta_i | L_i \sim N(\mu_{L_i}, \tau_{L_i}^2),$$

$$L_i \sim \text{Categoric}(\pi_1, \pi_2),$$

with $\mu_1$ assigned a N(−5.5,100) prior, and with the constraint $\mu_2 = \mu_1 + \delta$, where $\delta \sim U(0, 1)$. The precisions $1/\tau_g^2$ are assigned Ga(1,0.001) priors, and $\pi = (\pi_1, \pi_2)$ is assigned a Dirichlet prior $\pi \sim D(5, 5)$. The prior on $\delta$ is data-based (informative) and designed to prevent the second component becoming empty. Convergence is apparent relatively soon, and the second half of a 20 000 iteration two chain run provides mixture parameters as in Table 2.4. There is a minority subpopulation with a lower second stage mean $\mu_g$. However, there are now 7 counties with $\Pr(y_{new,j} \le y_j | y)$ over 0.975, and three with $\Pr(y_{new,j} \le y_j | y)$ under 0.025. Nevertheless the complete data based DSS criterion (Section 2.2.7) is 4.955, a slight improvement on the single group model.

**Example 2.8    Discrete mixture model for Hepatitis B in Berlin regions**

Example 2.3 drew attention to possible limitations in a Poisson–gamma mixture for the Berlin Hepatitis B counts. Here various discrete mixture options are considered. A baseline model is a simple mixture of Poisson densities, and on the basis of maximum likelihood analysis, $G = 3$ groups are assumed for illustrative purposes. The fmm package in stata shows a major gain in likelihood in moving from $G = 2$ to $G = 3$ but only small improvements on further disaggregation. On substantive grounds also, one often expects a group of areas with low risk, another with typical risk, and another with high risk.

The model includes an identifiability constraint on the group means $\lambda_g$, namely

$$y_i \sim \text{Po}(E_i \lambda_{L_i}),$$

$$L_i \sim \text{Categoric}(\pi_1, \pi_2, \pi_3),$$

$$(\pi_1, \pi_2, \pi_3) \sim D(1, 1, 1),$$

$$\lambda_1 \sim \text{Ga}(1, 0.01),$$

$$\lambda_g = \lambda_{g-1} + \delta_g, \qquad g = 2, 3$$

$$\delta_g \sim \text{Ga}(1, 0.01).$$

The Dirichlet prior on $\pi$ is implemented using $G = 3$ separate gamma priors with index 1 and shape 1. A mixed predictive assessment is not possible for this model, and instead the Poisson weighted predictive criterion of Gelfand and Ghosh (1998) is used, with $F$ and $V$ denoting the fit and penalty components respectively, $w = 1$ in (2.10), and $T = F + V$. The group means have posterior means (95% intervals) $\lambda_1 = 0.53$ (0.40, 0.65), $\lambda_2 = 1.41$ (1.03, 1.81), and $\lambda_3 = 5.02$ (2.06, 3.90), with $\pi$ estimated as (0.59, 0.27, 0.14). Although a well-defined solution is obtained, this finite Poisson mixture shows a deterioration in complete data predictive performance as against the Poisson–gamma hierarchy of Example 2.3, with $T$ rising from 46.7 to 51.9.

A more plausuble scenario may be a discrete mixture over two or more conjugate or non-conjugate hierarchical priors, such as the Poisson–gamma or Poisson log-normal. To illustrate this approach, a discrete mixture of Poisson–gamma hierarchies is fitted, with $G = 3$. This involves the gamma parameterisation $\lambda \sim \text{Ga}(\Lambda b, b)$, with $E(\lambda) = \Lambda$. The hierarchical sequence, including a constraint on the group means $\Lambda_g$, is then

$$y_i \sim \text{Po}(E_i \lambda_{L_i}),$$

$$L_i \sim \text{Categoric}(\pi_1, \pi_2, \pi_3),$$

$$\lambda_i | L_i = g \sim \text{Ga}(\Lambda_g b_g, b_g),$$

$$\Lambda_1 \sim \text{Ga}(1, 0.01),$$

$$\Lambda_g = \Lambda_{g-1} + \delta_g, \delta_g \sim U(0, 5), \qquad g = 2, 3$$

$$b_g \sim \text{Ga}(1, 0.01),$$

$$(\pi_1, \pi_2, \pi_3) \sim D(1, 1, 1).$$

In BUGS, initial values in a two chain run (of 10 000 iterations) are provided for $\Lambda, b$, and $\delta$, with other unknowns generated from their priors using 'gen inits'. The full data predictive $T$ is obtained as 48.0, worse than the single group Poisson–gamma of Example 2.3. On the other hand, mixed predictive checks improve, with posterior predictive probabilities Pr.$mx_i$

now having maximum and minimum values of 0.78 (area 3) and 0.07 (area 23). The group means are $\Lambda = (0.54, 1.43, 2.82)$, with $\pi$ estimated as $(0.55, 0.29, 0.16)$.

A third option adopts a Dirichlet process prior with the maximum clusters $M$ taken as 10 (corresponding to $\alpha = 2, \varepsilon = 0.01$), and $G_0$ taken as a single gamma density with unknown parameters:

$$y_i \sim \text{Po}(E_i \lambda_{L_i}),$$

$$\lambda_m | G \sim G, \qquad m = 1, \ldots, M,$$

$$G \sim DP(\alpha, G_0),$$

$$G_0 = \text{Ga}(b\Lambda, b), \Lambda \sim E(1), b \sim \text{Ga}(1, 0.01),$$

$$L_i \sim \text{Categoric}(\pi_1, \ldots, \pi_M),$$

$$\pi \sim \text{GEM}(\alpha),$$

$$\alpha \sim \text{Ga}(2, 2).$$

Mixed predictive checks can be obtained by the sequence

$$y_{new,i} \sim \text{Po}(E_i \lambda_{L_i,new}),$$

$$\lambda_{m,new} | G \sim G, \qquad m = 1, \ldots, M.$$

Initial values are provided for $\alpha$, with other unknowns generated from the priors. Using a two chain run of 10 000 iterations, posterior means for $\alpha$ and $M^*$ are 1.7 and 6.1 respectively. The full data predictive $T$ is obtained as 49.0, while mixed predictive probabilities Pr.$mx_i$ have maximum and minimum values of 0.95 (area 3) and 0.07 (area 23). The realised random effects $\lambda_{L_i}$ can be monitored, and show (Figure 2.2) most shrinkage for areas which have low observed relative risks, $y_i/E_i$, based on small $E_i$. The shrinkage applied to area 3 may be regarded as overshrinkage (hence the adverse mixed predictive check), despite the Dirichlet process approach.

Analogous to the three group finite mixture of Poisson–gamma densities, one can, however, extend the Dirichlet process to be a mixture over $M$ gamma densities (Ohlssen *et al.*, 2007). Thus

$$y_i \sim \text{Po}(E_i \lambda_i),$$

$$\lambda_i | G \sim \text{Ga}(b_{L_i}\Lambda_{L_i}, b_{L_i}),$$

$$G \sim DP(\alpha, G_0),$$

$$G_0 = \text{Ga}(b_m\Lambda_m, b_m), m = 1, \ldots, M,$$

$$\Lambda_m \sim E(1) \qquad I(0.1, )$$

$$b_m \sim \text{Ga}(1, 0.1),$$

$$L_i \sim \text{Categoric}(\pi_1, \ldots, \pi_M),$$

$$\pi \sim \text{GEM}(\alpha),$$

$$\alpha \sim \text{Ga}(2, 2).$$

The lower limit for $\Lambda_m$ is adopted to avoid numeric problems when low gamma scale values are selected. The posterior means for $\alpha$ and $M^*$ are now 1.2 and 4.1 respectively. The full data

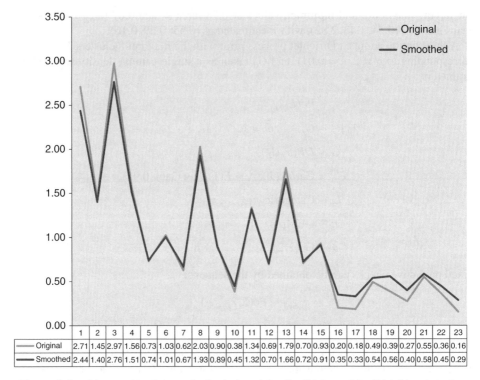

| | 1 | 2 | 3 | 4 | 5 | 6 | 7 | 8 | 9 | 10 | 11 | 12 | 13 | 14 | 15 | 16 | 17 | 18 | 19 | 20 | 21 | 22 | 23 |
|---|---|---|---|---|---|---|---|---|---|---|---|---|---|---|---|---|---|---|---|---|---|---|---|
| Original | 2.71 | 1.45 | 2.97 | 1.56 | 0.73 | 1.03 | 0.62 | 2.03 | 0.90 | 0.38 | 1.34 | 0.69 | 1.79 | 0.70 | 0.93 | 0.20 | 0.18 | 0.49 | 0.39 | 0.27 | 0.55 | 0.36 | 0.16 |
| Smoothed | 2.44 | 1.40 | 2.76 | 1.51 | 0.74 | 1.01 | 0.67 | 1.93 | 0.89 | 0.45 | 1.32 | 0.70 | 1.66 | 0.72 | 0.91 | 0.35 | 0.33 | 0.54 | 0.56 | 0.40 | 0.58 | 0.45 | 0.29 |

**Figure 2.2**    Observed and smoothed relative risks, Berlin hepatitis, Dirichlet process.

predictive $T$ is obtained as 45.2, with mixed predictive probabilities Pr.$mx_i$ ranging from 0.14 (area 23) to 0.87 (area 3).

A different underlying rationale for heterogeneity is provided by regression on a latent continuous predictor, which may be taken as beta or logistic-normal (Tarpey and Petkova, 2010). Thus

$$y_i \sim \text{Po}(E_i \lambda_i),$$

$$\log(\lambda_i) = \beta_0 + \beta_1 x_i,$$

$$x_i \sim \text{Be}(\alpha_1, \alpha_2).$$

The multiplicative term $\beta_1 x_i$ is subject to an identifiability issue, and to ensure that higher values of $x$ correspond to higher hepatitis B risk, it is assumed that $\beta_1 > 0$. Specifically $\alpha_j \sim \text{Ga}(1, 0.01)$, and $\beta_1 \sim N(0, 10\,000)I(0, )$. A two chain run of 100 000 iterations with inferences from the 2nd half, produces posterior means $\beta = (-6.7, 31.4)$ and $\alpha = (50.2, 146.3)$. The full data predictive $T$ is 45.4, similar to that for the Dirichlet process mixture of gamma densities.

## 2.6    General additive and histogram smoothing priors

Many types of smoothing problems involve a series of observations $\{y_1, \dots, y_n\}$ at ordered observation or design points (e.g. consecutive ages, income groups, etc.), where the points may be equally or unequally spaced. While this is a feature of time series, it is worth emphasizing the scope for more general applications. For example, mortality rates for areas often show a gradient according to the area deprivation category, such as deprivation decile, and a relevant smoothing model would take this ordering into account. Similarly heteroscedasticity often varies systematically with the response (e.g. variance increasing or decreasing with the level of the data), and instead of assuming level 2 heterogeneity $\tau^2$ as constant, a suitably transformed $\tau_t^2$ may be modelled using a smoothness prior. In all these instances the ordering of the data units (i.e. their labelling) becomes relevant; the units are now subject to structuring under the prior (leading to 'structured' priors), rather than a prior for exchangeable units which is indifferent to a relabelling of units.

Among the approaches which have figured in the Bayesian literature, we consider here two relatively simple structured prior methodologies which are suitable for a wide range of problems. One is based on smoothness priors derived from differencing an underlying series of true values (Kitagawa and Gersch, 1996). The other, histogram smoothing, is applicable to frequency plots, including those deriving from originally metric data aggregated to equal length intervals – an example involving weight gains in pigs is considered by Leonard and Hsu (1999).

## 2.6.1  Smoothness priors

For the smoothness prior approach, the analogue of the normal–normal prior of Section 2.4 is provided by the case where $y_t$ are metric with normal errors, and observed at equally spaced design points. Extensions to unequally spaced observation intervals are also considered below, as are extensions to grouped observations – multiple observations at a single design point $t$. The model for metric observations is

$$y_t = \theta_t + e_t,$$

where $e_t \sim N(0, \sigma^2)$, or $e_t \sim N(0, \sigma_t^2)$, and where the $\theta_t$ may be taken as the underlying smooth series, free of the measurement error in the observed series. The case with $\sigma_t^2$ known corresponds to the known stage 1 variance observation scheme in Section 2.4. For discrete outcomes, the focus of smoothing is typically on the latent mean counts, risks or probabilities. Thus if the $y_t$ were counts assumed to be Poisson distributed, one might have

$$y_t \sim Po(\mu_t); \qquad \log(\mu_t) = \theta_t$$

while if they were binomial with populations $n_t$, then the model might be

$$y_t \sim Bin(\pi_t, n_t); \qquad logit(\pi_t) = \theta_t$$

The smoothing model explicitly uses the ordering in the design. A widely used smoothing model assumes normal or Student distributed random walks in the first, second or higher differences of the $\theta_t$. For example, a normal random walk (RW) in the first difference $\Delta\theta_t$ is equivalent to the smoothness prior

$$\theta_t \sim N(\theta_{t-1}, \tau^2),$$

and a random walk in the second difference $\Delta^2\theta_t$ leads to the prior

$$\theta_t \sim N(2\theta_{t-1} - \theta_{t-2}, \tau^2).$$

These models have to be set in motion by assumptions about the initial parameter(s): this involves separate priors on the initial parameters, such as $\theta_1$ in the case of a first order RW,

$\theta_1$ and $\theta_2$ in the case of a second order RW, and so on. A common practice is to ascribe vague priors with large variances to these parameters. A rationale for this initial conditions prior is provided by the measurement error model for continuous normal data. Here the ratio $\lambda = \sigma^2/\tau^2$ is a smoothing parameter, which for a second order random walk, $RW(2)$, appears in the penalised least squares criterion

$$\sum_{t=1}^{n} e_t^2 + \lambda \sum_{t=3}^{n} \Delta^2 \theta_t.$$

With flat initial priors on the initial $\theta_t$ in the series the posterior modes for $\theta_t$ are equivalently those that minimise this criterion. Thus one might set initialising priors

$$\theta_j \sim N_r(0, kI), \qquad j = 1, \dots, r$$

where $r$ is the order of differencing in the smoothness prior and $k$ is large.

It may be noted that one is not confined to asymmetric priors: if the design index $t$ were something like income group or deprivation category then it might make sense for $\theta_t$ to depend on adjacent points at either side. Thus a generalisation of the $1^{st}$ order random walk prior might involve dependence on both $\theta_{t-1}$ and $\theta_{t+1}$. For example

$$\theta_t = 0.5\theta_{t-1} + 0.5\theta_{t+1} + u_t \qquad 2 \le t \le n-1$$
$$\theta_1 = \theta_2 + u_1$$
$$\theta_n = \theta_{n-1} + u_n.$$

Further issues concern specialised priors in the event of unequal spacing or tied observations. For unequal spacing, it is necessary to weight each preceding point differently, and to change the precision such that wider spaced points are less tied to their predecessor than closer spaced points. Thus suppose the observations were at points $t_1, t_2, \dots, t_n$ with $\delta_1 = t_2 - t_1, \delta_2 = t_3 - t_2, \dots, \delta_{n-1} = t_n - t_{n-1}$. The RW1 prior becomes

$$\theta_t \sim N(\theta_{t-1}, \delta_t \tau^2),$$

so that wider gaps $\delta_t$ translates into larger variances. Similarly the RW2 prior becomes

$$\theta_t \sim N(v_t, \delta_t \tau^2),$$

where $v_t = \theta_{t-1}(1 + \delta_t/\delta_{t-1}) - \theta_{t-2}(\delta_t/\delta_{t-1})$, which reduces to the equal space form if $\delta_t = \delta_{t-1}$. For multiple observations at one design point in the series, the stage 2 prior is for the $G$ distinct values of the series index. For instance suppose the observations $y_t$ were 1,3,4,7,11,15 ($n = 6$) at observation points 1,1,2,2,3,4 (i.e. $G = 4$). Then if $g_t$ denotes the grouping index to which observation $t$ belongs, and a first order random walk in the underlying series is assumed, one has

$$y_t \sim N(\theta[g_t], \sigma^2) \qquad t = 1, \dots, n$$
$$\theta_k \sim N(\theta_{k-1}, \tau^2). \qquad k = 2, \dots, G$$

## 2.6.2   Histogram smoothing

Suppose observations (possibly of an originally continuous variable) are summarised by frequency counts, with $y_j$ being the number of observations between cut points $K_{j-1}$ and $K_j$.

Examples are counts $y_j$ of events (e.g. migrations, deaths) arranged in terms of a monotonic classifier, at levels $j = 1, \dots, J$, such as age band or income level, frequently with denominators $N_j$. Often sampling variability will mean the observed frequencies $y_j$ or crude rates $y_j/N_j$ in the $j^{th}$ interval will be ragged in form when substantive considerations would imply smoothness. Random walk smoothness priors, as in Section 2.6.1, can be applied to this problem, and to extensions such as scatterplot smoothing (Eilers and Goeman, 2004).

An alternative methodology is set out by Leonard and Hsu (1999), who consider the pig weight gain data of Snedecor and Cochran (1989) where weight gains of 522 pigs are represented in histogram based on $J = 21$ intervals of equal width. Let prior beliefs about smoothness in the histogram be represented by an underlying density $h(u)$, and let $\pi_j$ be the probability that an observation is in the $j^{th}$ interval

$$\pi_j = \int_{K_{j-1}}^{K_j} h(u)du.$$

The observed frequencies are assumed to be multinomial with parameters $\{\pi_1, \pi_2, \dots, \pi_J\}$, and to imperfectly reflect the underlying smooth density. A prior structure on the $\pi_j$ that includes smoothness considerations is appropriate. Specifically express $\pi_j$ via a multiple logit model

$$\pi_j = \exp(\phi_j)/\sum_k \exp(\phi_k),$$

where the $\phi_k$ are taken to be multivariate normal with means $\{a_1, a_2, \dots, a_J\}$ and $J \times J$ covariance matrix $V$. A noninformative prior on the $\pi_j$ would take them to be equal in size, i.e. $\pi_j = 1/J$, which is equivalent to

$$a_j = -\log(J).$$

In the case of frequencies arranged by age, income, etc., there are likely to total population or exposure totals $N_j$, and one may then set

$$a_j = \log(N_j/\sum_k N_k).$$

The covariance matrix $V$ is structured to reflect dependence between neighbouring frequencies on the histogram classifier. Thus one option is $1^{st}$ order dependence with correlation $\rho$, as discussed further in Chapter 5, with

$$V_{ij} = \rho^{|i-j|}\sigma^2.$$

Then $\rho = 0$ leads to exchangeability in the histogram probabilities, that is to a joint distribution that is unaffected by permutation of the suffixes $1, 2, \dots, J$. This is contrary to expectation in many situations where greater similarity is anticipated between $\pi_j$ and $\pi_{j+1}$ than between $\pi_j$ and $\pi_{j+k}$, $k > 1$. Note that, with $\kappa = 1/(\sigma^2 - \sigma^2\rho^2)$, the elements of $T = V^{-1}$ are given by

$$T_{11} = T_{JJ} = \kappa,$$

$$T_{jj} = \kappa(1 + \rho^2) \qquad j = 2, \dots, J-1$$

$$T_{j,j+1} = T_{j+1,j} = -\kappa\rho \qquad j = 1, \dots, J-1$$

$$T_{jk} = 0 \qquad \text{elsewhere.}$$

### Example 2.9    TB cases in Alaska

Consider annual TB incident cases in Alaska between 1988 and 2010, with data on cases and populations $\{d_t, N_t\}$. Alaskan native populations are known to have excess TB risk, and for this reason Alaska has one of the highest TB incidence rates among US states. Following Breslow (1984), for large $d_t$, one may consider $y_t = \log(d_t/N_t)$ to be normal with means $\theta_t$ and variance $\sigma^2 + 1/d_t$. One may assume that the underlying rates $\theta_t$ in the log scale change gradually through time, and a random walk $\Delta^r\theta_t \sim N(0, \tau^2)$ in the $r^{th}$ differences expresses this belief. A two stage model including an RW1 latent series has the form

$$y_t \sim N(\theta_t, \sigma^2 + 1/d_t),$$

$$\theta_t \sim N(\theta_{t-1}, \tau^2),$$

$$\theta_1 \sim N(0, 1000),$$

$$1/\sigma^2 \sim Ga(1, 0.001),$$

$$T = \frac{\sigma^2}{\sigma^2 + \tau^2},$$

$$T|\sigma^2 \sim U(0, 1)$$

Figure 2.3 shows the original and smoothed incidence series of rates per hundred thousand, with the latter obtained as $100\,000e^{\theta_t}$. The smoothed rates vary between 7 and 13.7 per 100 000, whereas the observed rates vary between 5.3 and 17.2. The parameter $\theta_1$ here acts

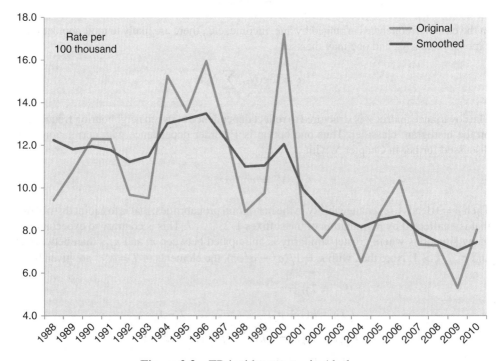

**Figure 2.3**    TB incidence rates in Alaska.

to define the level of the series (i.e. acts as a form of fixed effect), and so mixed predictive checks take

$$y_{new,t} \sim N(\theta_{new,t}, \sigma^2 + 1/d_t),$$

$$\theta_{new,1} = \theta_1,$$

$$\theta_{new,t} \sim N(\theta_{t-1}, \tau^2) \qquad t > 1.$$

Posterior mixed predictive $p$-values range between 0.12 for 2009 and 0.97 for the year 2000, when the observed rate per $100\,000$ ($10\,0000 d_t/N_t$) peaked at 17.2.

**Example 2.10    Hospitalisations by age**

As an example comparing histogram and state space smoothing, consider hospitalisations $y_j$ during a year for mental illness at age $j$ over two London boroughs. These are classified by $J = 70$ single years of age from ages 15 to 84, and may be seen as binomial outcomes in relation to age specific population estimates $N_j$. A histogram smoothing model is applied with $\phi_k$ taken as multivariate normal with means $\{a_1, a_2, \ldots, a_J\}$ and $J \times J$ covariance matrix with elements $V_{ij} = \rho^{|i-j|}\sigma^2$, and with

$$a_j = \log(N_j / \sum_k N_k),$$

where $N_j$ is the population at risk at age $j$. A $U(0, 1)$ prior on $\rho$ is assumed and gamma $Ga(1, 0.001)$ prior on $\tau = 1/\sigma^2$. A three chain run shows convergence of Gelman–Rubin criteria for these two parameters at around iteration 1500, and the remaining iterations of a run of 5000 provide the summary. The smoothing of originally ragged frequencies reflects the operation of the prior in adjacent categories of the histogram. Thus the observed schedule shows 19 hospital cases at age 21, 6 at age 22 and 25 at age 23 ($j = 7, \ldots, 9$). The smoothed version shows much less discontinuity with 16.5, 13.6, and 18.7 as the estimated frequencies of hospitalisation at these ages. A posterior mean estimate of $\rho = 0.97$ is obtained with 95% interval from 0.92 to 0.995, confirming the high correlation in the histogram probabilities. A two stage binomial-logit normal smoothing model is then applied. A first order random walk prior in the logits of the latent probabilities is adopted, namely

$$y_t \sim Bin(N_t, \pi_t), \qquad logit(\pi_t) = \theta_t,$$

$$\theta_t \sim N(\theta_{t-1}, \tau^2), \qquad t > 1,$$

$$\theta_1 \sim N(0, 1000),$$

$$1/\tau^2 \sim Ga(1, 0.001).$$

From a three chain run (5000 iterations with 500 burn in) a posterior precision averaging 23 is obtained. Values for the smoothed frequencies are close to those obtained under the histogram smoothing approach. The random walk and histogram smooth model are expected to be broadly similar for values of $\rho$ close to 1.

# Exercises

2.1.   Lindley's paradox may be illustrated by taking less informative priors in Example 2.1. Thus for $M_2$, assume $\theta \sim U(-10, 10)$ instead of $\theta \sim U(-2, 2)$, and apply the model weights procedure to obtain posterior mean weights $w_1 = 0.9303$, and $w_2 = 0.0697$. The estimated and analytical Bayes factors are both now 13.35, or higher evidence favouring the simpler model.

**2.2.** In Example 2.3, estimate a log-normal mixture model for the incidence counts $y_i \sim$ $Po(\lambda_i)$, namely

$$\log(\lambda_i) = \log(E_i) + \lambda_0 + u_i,$$

where the $u$ are normal with unknown variance, and obtain the complete data and mixed predictive $p$-values for each Berlin area.

**2.3.** Part of the reason for the worse marginal likelihood of the beta-binomial in Example 2.4 (Vinnie Johnson baseball averages) may be the relatively low sample sizes for each play. The posterior $p_i$ are estimated relatively imprecisely (with wide credible intervals that include the global probability), and so the extent of heterogeneity may be masked. Repeat the beta-binomial and marginal likelihood estimation, but with both $y_i$ and $n_i$ doubled for all $i$, (e.g. the first observation $y_1 = 3$, $n_1 = 10$ becomes $y_1 = 6$, $n_1 = 20$), so that the adjusted data are more 'informative' regarding heterogeneity, and assess whether the beta-binomial becomes a more suitable option.

**2.4.** Apply a three-group discrete mixture model to the baseball average data. A two group mixture, with code as below, gives an LPML (for the original data) of $-726.5$, with plots of $p_{new}$ suggesting some bimodality – though the posterior predictive criterion regarding maximum $p_{new}$ is still approaching 0.9.

```
model { for (i in 1:380) {y[i] ~dbin(p[G[i]],n[i]);
# latent groups
G[i] ~dcat(pi[1:2])
ynew[i] ~dbin(p[G[i]],n[i]); pnew[i] <- ynew[i]/n[i]
LL[i] <- logfact(n[i])-logfact(y[i])-logfact(n[i]-y[i])
        +y[i]*log(p[G[i]])+(n[i]-y[i])*log(1-p[G[i]])
# LPML is sum of negatives of logged posterior means of invLK
invLK[i] <- 1/exp(LL[i])}
# mixture probabilities
pi[1:2] ~ddirch(alph[1:2])
for (j in 1:2) {alph[j] <- 1}
# constrained prior on group hitting rates
logit(p[1]) <- del[1]
del[1] ~dnorm(0,1)
logit(p[2]) <- del[1]+del[2]
del[2] ~dnorm(0,1) I(0,)
# posterior predictive criterion
PPC <- step(ranked(pnew[],380)-0.9)}
```

**2.5.** In Example 2.5, apply a scale mixture alternative at stage 2 to the normal–normal model, namely

$$y_j \sim N(\theta_j, \sigma_j^2)$$

$$\theta_j \sim N(\mu, \tau^2/\lambda_j) \qquad j = 1, \ldots, J$$

$$\lambda_j \sim Ga(v/2, v/2)$$

with $v$ an unknown. It is suggested to take $v = 1/\kappa$ where $\kappa \sim U(0.02, 0.5)$. Assess the effect of this strategy on the parameter $\mu$, the average latent mean for the teacher expectation effect. Also compare the shrinkage for studies 4 and 10 to that obtained under the normal–normal model. A code for this extension is

```
model {for (j in 1:J) {y[j] ~dnorm(th[j],h[j])
y.new[j] ~dnorm(th[j],h[j])
```

```
th[j] ~dnorm(mu,invTau2[j])
invTau2[j] <- 1/Tau2[j]; Tau2[j] <- tau2/lam[j]
lam[j] ~dgamma(nu.2,nu.2)
# replicates and checks
th.new[j] ~dnorm(mu,invTau2[j])
y.new.mx[j] ~dnorm(th.new[j],h[j])
d.mx[j] <- step(y[j]-y.new.mx[j])
lam.lo[j] <- step(1-lam[j])}
# priors
kap ~dunif(0.01,0.5); nu <- 1/kap; nu.2 <- nu/2
mu ~dflat(); tau <- sqrt(tau2)
s0 <- J/sum(h[]); T ~dunif(0,1); tau2 <- s0*(1-T)/T}
```

Note that even this adaptation leaves high mixed predictive $p$-values for studies 4 and 10, and indicates an alternative stage 2 model (e.g. one allowing positive skew in the $\theta_j$).

**2.6.** Many evaluations of Poisson mixture models consider aggregated data, for example numbers of consumers making 0,1,2, etc. purchases. Brockett *et al.* (1996) present such data for purchases of 'salty snacks', which are re-analysed by Kaplon (2010). Thus consider total subjects in a consumer panel making 0,1,2, ... ,64,65 purchases, namely data in the form

| Purchases | Number of Consumers |
|-----------|---------------------|
| 0 | 352 |
| 1 | 183 |
| 2 | 150 |
| .... | |
| 64 | 19 |
| 65 | 18. |

To analyse such data in BUGS they can be disaggregated to unit level, as demonstrated in the following code which fits a single mean Poisson model. There are n=3852 subjects in all. The input data are in the form

| t[] | n1[] | n2[] |
|-----|------|------|
| 0 | 1 | 352 |
| 1 | 353 | 535 |
| 2 | 536 | 685 |
| .... | | |
| 65 | 3835 | 3852 |

with the code to fit a single mean Poisson model (and obtain predictions) then being

```
model {for (j in 1:66) { for (i in n1[j]:n2[j]) {y[i] <- t[j]}}
for (i in 1:3852) {y[i] ~dpois(mu)
yrep[i] ~dpois(mu)
# classify replicates (0,1,...65 purchases
for (j in 1:66) {ind.rep[i,j] <- equals(yrep[i],j-1)}}
mu ~dgamma(1,0.001).
```

Add code to obtain the predicted numbers of subjects making 0,1,2, ... , 65 purchases, say $\{f_{rep,j}, j = 1,66\}$ and use a predictive $p$-test to compare these with the observed numbers $\{f_{obs,j}, j = 1,66\}$. Compare Poisson–gamma and Poisson-lognormal mixtures to establish which has the lower number of extreme predictive $p$-tests from the 66 (either under 0.05 or over 0.95). As mentioned above the appropriate indicator for count data is $I(f_{rep,j}^{(t)} < f_{obs,j}|\theta^{(t)}) + 0.5I(f_{rep,j}^{(t)} = f_{obs,j}|\theta^{(t)})$. Also note that the BUGS step(a-b) comparison is equivalent to $a \le b$ rather than $a < b$.

**2.7.** A request in the UK parliament (http://www.theyworkforyou.com/wrans/?id=2011-03 -07c.44095.h) related to 2009 mortality rates (per 100 000 population) in Wales according to income decile of neighbourhood of residence at time of death (decile 1 is least deprived, and decile 10 is most deprived). The data provided include 95% intervals:

| Decile | Rate | 2.5% | 97.5% |
|--------|------|------|-------|
| 1 | 444 | 426 | 462 |
| 2 | 469 | 451 | 487 |
| 3 | 507 | 489 | 525 |
| 4 | 545 | 526 | 564 |
| 5 | 545 | 526 | 564 |
| 6 | 588 | 568 | 608 |
| 7 | 627 | 605 | 648 |
| 8 | 682 | 659 | 705 |
| 9 | 742 | 717 | 767 |
| 10 | 807 | 780 | 834 |

Obtain the known precisions $h_t = 1/\sigma_t^2$ $(t = 1, \dots, 10)$ at stage 1, with the rates then $y_t \sim N(\theta_t, 1/h_t)$. At stage 2 apply a RW1 prior for the latent series $\theta_t$ with variance $\tau^2$. Use a uniform shrinkage prior on $\tau^2$ involving the harmonic mean variance $s_0^2 = 10/\sum_{t=1}^{10} h_t$. Obtain mixed posterior p-values by the sequence

$$y_{new,t} \sim N(\theta_{new,t}, 1/h_t),$$

$$\theta_{new,1} = \theta_1,$$

$$\theta_{new,t} \sim N(\theta_{t-1}, \tau^2) \qquad t > 1.$$

and identify the decile with the highest mixed posterior p-value (also the decile with the greatest shrinkage).

**2.8.** Apply the normal-latent beta model discussed in Section 2.5 to data on short-term changes in depression ratings (Tarpey and Petkova, 2010) (see Exercise 2.8.odc). Such changes in depression are unlikely to be due to the pharmacological intervention, and can be plausibly explained by a latent continuous placebo effect, namely

$$y_i \sim N(\beta_0 + \beta_1 x_i, \sigma^2),$$

$$x_i \sim Be(\alpha_1, \alpha_2),$$

with $x_i$ as the latent predictor. For identifiability, it is recommended to assume $\beta_1 > 0$.

# Notes

1. The model weights code for comparing linear regression models in Example 2.2 is

```
model {# log-lkd for model 1
for (i in 1:n) {y1[i] <- y[i]; y1[i] ~dnorm(mu1[i],tau1)
```

```
y1new[i] ~dnorm(mu1[i],tau1)
mu1[i] <- beta1[1]+beta1[2]*x1[i]
LL1[i] <- 0.5*log(tau1/(2*3.14159))-0.5*tau1*pow(y1[i]-mu1[i],2)}
for (j in 1:2) {beta1[j] ~dnorm(0,tau.beta1[j])
v.beta1[j] <- 4*sig2.1; tau.beta1[j] <- 1/v.beta1[j];
                    pr.s.beta1[j] <- 2*sqrt(sig2.1)}
tau1 ~dgamma(3,0.4); sig2.1 <- 1/tau1
# log-lkd for model 2
for (i in 1:n) {y2[i] <- y[i]; y2[i] ~dnorm(mu2[i],tau2)
y2new[i] ~dnorm(mu2[i],tau2)
mu2[i] <- beta2[1]+beta2[2]*x1[i]+beta2[3]*x2[i]
LL2[i] <- 0.5*log(tau2/(2*3.14159))-0.5*tau2*pow(y2[i]-mu2[i],2)}
for (j in 1:3) {beta2[j] ~dnorm(0,tau.beta2[j])
v.beta2[j] <- 4*sig2.2; tau.beta2[j] <- 1/v.beta2[j];
                    pr.s.beta2[j] <- 2*sqrt(sig2.2)}
tau2 ~dgamma(3,0.4); sig2.2 <- 1/tau2
# log prior ordinates for model 1
prior.ord1[1] <- -0.5*log(2*3.1416)-log(pr.s.beta1[1])
-0.5*pow(beta1[1],2)/(pr.s.beta1[1]*pr.s.beta1[1])
prior.ord1[2] <- -0.5*log(2*3.1416)-log(pr.s.beta1[2])
-0.5*pow(beta1[2],2)/(pr.s.beta1[2]*pr.s.beta1[2])
prior.ord1[3] <- 3*log(0.4)-loggam(3)+2*log(tau1)-0.4*tau1
# log pseudo-prior ordinates for model 1
g.ord1[1] <- -0.5*log(2*3.14159)-log(s.beta1[1])
-0.5*pow(beta1[1]-mu.beta1[1],2)/(s.beta1[1]*s.beta1[1])
g.ord1[2] <- -0.5*log(2*3.14159)-log(s.beta1[2])
-0.5*pow(beta1[2]-mu.beta1[2],2)/(s.beta1[2]*s.beta1[2])
g.ord1[3] <- a.tau1*log(b.tau1)-loggam(a.tau1)+(a.tau1-1)
                              *log(tau1)-b.tau1*tau1
# log prior ordinates for model 2
prior.ord2[1] <- -0.5*log(2*3.1416)-log(pr.s.beta2[1])
-0.5*pow(beta2[1],2)/(pr.s.beta2[1]*pr.s.beta2[1])
prior.ord2[2] <- -0.5*log(2*3.1416)-log(pr.s.beta2[2])
-0.5*pow(beta2[2],2)/(pr.s.beta2[2]*pr.s.beta2[2])
prior.ord2[3] <- -0.5*log(2*3.1416)-log(pr.s.beta2[3])
-0.5*pow(beta2[3],2)/(pr.s.beta2[3]*pr.s.beta2[3])
prior.ord2[4] <- 3*log(0.4)-loggam(3)+2*log(tau2)-0.4*tau2
# log pseudo-prior ordinates for model 2
g.ord2[1] <- -0.5*log(2*3.14159)-log(s.beta2[1])
-0.5*pow(beta2[1]-mu.beta2[1],2)/(s.beta2[1]*s.beta2[1])
g.ord2[2] <- -0.5*log(2*3.14159)-log(s.beta2[2])
-0.5*pow(beta2[2]-mu.beta2[2],2)/(s.beta2[2]*s.beta2[2])
g.ord2[3] <- -0.5*log(2*3.14159)-log(s.beta2[3])
-0.5*pow(beta2[3]-mu.beta2[3],2)/(s.beta2[3]*s.beta2[3])
g.ord2[4] <- a.tau2*log(b.tau2)-loggam(a.tau2)+(a.tau2-1)
                              *log(tau2)-b.tau2*tau2
# Combining loglikelihoods, log-priors, log importance samples,
                                        log model probs
TL[1] <- sum(LL1[])+sum(prior.ord1[1:3])-sum(g.ord1[1:3])+log(0.5)
TL[2] <- sum(LL2[])+sum(prior.ord2[1:4])-sum(g.ord2[1:4])+log(0.5)
# Composite log-likelihoods
eta[1] <- sum(LL1[])+sum(prior.ord1[1:3])-sum(g.ord1[1:3])
eta[2] <- sum(LL2[])+sum(prior.ord2[1:4])-sum(g.ord2[1:4])
# Scaling against the largest likelihood
maxL <- ranked(TL[],2);
for (j in 1:2) {SL[j] <- TL[j]-maxL; expSL[j] <- exp(SL[j])}
# model weights
w[1] <- expSL[1]/sum(expSL[]); w[2] <- expSL[2]/sum(expSL[])
```

```
# model averaged prediction
ynew14 <- w[1]*y1new[14]+w[2]*y2new[14]}
```

The first dataset (including parameters of estimated posterior densities) is

```
list(n=100, s.beta1=c(0.04029,0.04026),mu.beta1=c(0.9463,0.4499),
s.beta2=c(0.0404,0.0402,0.0466),mu.beta2=c(0.9469,0.4503,-0.0077),
a.tau1=53,b.tau1=8.5,a.tau2=52.4,b.tau2=8.36).
```

The predictor and covariate values are as in Table 2.5.

**Table 2.5**   Simulated linear regression data.

| Case | y | x1 | x2 | Case | y | x1 | x2 | Case | y | x1 | x2 |
|---|---|---|---|---|---|---|---|---|---|---|---|
| 1 | 1.211 | 0.759 | −0.433 | 35 | 0.988 | 0.586 | 1.290 | 69 | 1.145 | 0.532 | 0.040 |
| 2 | 0.118 | −0.583 | −1.666 | 36 | 0.036 | −1.034 | 0.669 | 70 | 1.198 | 1.442 | 0.677 |
| 3 | 1.154 | 1.015 | 0.125 | 37 | 0.811 | 0.039 | 1.191 | 71 | 1.971 | 2.072 | 0.569 |
| 4 | 0.915 | 0.590 | 0.288 | 38 | 0.992 | 0.645 | −1.203 | 72 | 0.389 | −2.226 | −0.256 |
| 5 | 0.027 | −1.051 | −1.147 | 39 | 1.283 | 0.865 | −0.020 | 73 | 0.007 | −0.625 | −0.378 |
| 6 | 0.427 | 0.405 | 1.191 | 40 | 1.181 | 0.575 | −0.157 | 74 | −0.263 | −2.596 | −0.296 |
| 7 | 1.099 | 0.252 | 1.189 | 41 | 1.393 | −0.864 | −1.604 | 75 | 0.455 | −0.941 | −1.475 |
| 8 | 0.588 | −1.081 | −0.038 | 42 | 0.830 | −0.333 | 0.257 | 76 | 0.619 | −1.115 | −0.234 |
| 9 | 1.922 | 1.506 | 0.327 | 43 | 0.776 | 0.644 | −1.057 | 77 | 0.881 | −0.219 | 0.118 |
| 10 | 0.998 | −0.241 | 0.175 | 44 | 2.069 | 0.841 | 1.415 | 78 | 0.370 | −0.206 | 0.315 |
| 11 | 0.620 | −0.393 | −0.187 | 45 | 0.100 | −1.520 | −0.805 | 79 | 1.078 | −0.901 | 1.444 |
| 12 | 0.613 | −1.063 | 0.726 | 46 | 0.188 | −0.411 | 0.529 | 80 | 1.278 | 1.067 | −0.351 |
| 13 | 2.412 | 2.256 | −0.588 | 47 | 0.494 | 0.062 | 0.219 | 81 | 1.271 | 0.928 | 0.623 |
| 14 | 0.241 | −0.687 | 2.183 | 48 | 1.126 | −0.725 | −0.922 | 82 | 1.284 | 0.352 | 0.799 |
| 15 | 1.617 | −0.048 | −0.136 | 49 | 0.855 | 0.059 | −2.171 | 83 | 1.176 | 0.740 | 0.941 |
| 16 | 1.415 | 0.432 | 0.114 | 50 | 0.994 | 0.300 | −0.059 | 84 | 0.742 | −0.001 | −0.992 |
| 17 | 1.241 | 0.902 | 1.067 | 51 | 1.650 | 0.009 | −1.011 | 85 | 1.274 | 0.675 | 0.212 |
| 18 | 1.965 | 1.765 | 0.059 | 52 | 0.808 | −0.747 | 0.615 | 86 | 2.371 | 1.801 | 0.238 |
| 19 | 1.537 | 0.685 | −0.096 | 53 | 0.350 | −0.190 | 0.508 | 87 | 1.024 | 0.074 | −1.008 |
| 20 | 0.401 | −1.359 | −0.832 | 54 | 0.723 | 1.295 | 1.692 | 88 | −0.033 | −1.135 | −0.742 |
| 21 | 0.712 | −0.044 | 0.294 | 55 | 1.136 | −1.361 | 0.591 | 89 | 0.891 | −0.822 | 1.082 |
| 22 | 1.807 | 0.712 | −1.336 | 56 | 1.947 | 2.020 | −0.644 | 90 | 0.216 | −1.220 | −0.132 |
| 23 | 1.711 | −0.751 | 0.714 | 57 | 0.104 | −0.359 | 0.380 | 91 | 0.788 | −0.362 | 0.390 |
| 24 | 0.725 | −0.956 | 1.624 | 58 | 1.941 | 1.002 | −1.009 | 92 | 0.476 | −0.462 | 0.088 |
| 25 | 1.819 | 0.712 | −0.692 | 59 | 1.559 | 0.895 | −0.020 | 93 | 1.518 | 0.236 | −0.636 |
| 26 | 0.271 | −0.132 | 0.858 | 60 | 0.655 | −0.533 | −0.048 | 94 | 1.094 | −1.147 | −0.560 |
| 27 | 0.808 | −0.477 | 1.254 | 61 | 0.376 | −1.662 | 0.000 | 95 | 0.675 | −0.832 | 0.444 |
| 28 | 0.430 | −0.367 | −1.594 | 62 | 1.063 | 1.064 | −0.318 | 96 | −0.318 | −1.991 | −0.950 |
| 29 | 0.101 | −1.420 | −1.441 | 63 | −0.495 | −2.654 | 1.095 | 97 | 0.791 | −1.322 | 0.781 |
| 30 | 0.773 | 0.426 | 0.571 | 64 | 1.044 | 1.003 | −1.874 | 98 | 1.182 | −0.506 | 0.569 |
| 31 | 1.166 | −0.472 | −0.400 | 65 | 1.488 | 0.864 | 0.428 | 99 | 0.138 | 0.127 | −0.822 |
| 32 | 1.538 | 0.769 | 0.690 | 66 | 1.280 | 0.378 | 0.896 | 100 | 0.635 | −0.758 | −0.266 |
| 33 | 0.427 | −1.063 | 0.816 | 67 | 0.796 | 0.417 | 0.731 | | | | |
| 34 | 0.105 | −1.129 | 0.712 | 68 | 1.452 | 0.169 | 0.578 | | | | |

# References

Aitchison, J. and Shen, S. (1980) Logistic-normal distributions: Some properties and uses. *Biometrika*, **67**, 261–272.

Aitkin, M. (1991) Posterior Bayes factors. *Journal of the Royal Statistical Society B*, **53**, 111–114.

Akaike, H. (1973). Information theory and an extension of the maximum likelihood principle. In *Second International Symposium on Information Theory*, pp. 267–281. Akademinai Kiado.

Albert, J. (2009) *Bayesian Computation with R*, 2nd edn. Springer, New York, NY.

Arends, L., Vokjo, Z. and Stijnen, T. (2003) Combining multiple outcome measures in a meta-analysis: An application. *Statistics in Medicine*, **22**, 1335–1353.

Arlot, S. and Celisse, A. (2010) A survey of cross-validation procedures for model selection. *Statistics Survey*, **4**, 40–79.

Austin, P. (2002) A comparison of Bayesian methods for profiling hospital performance. *Medical Decision Making*, **22**(2), 163–72.

Austin, P. (2005) The reliability and validity of Bayesian measures for hospital profiling: a Monte Carlo assessment. *Journal of Statistical Planning and Inference*, **128**, 109–122.

Austin, P. (2009) Are (the log-odds of) hospital mortality rates normally distributed? Implications for studying variations in outcomes of medical care. *Journal of Evaluation and Clinical Practice*, **15**, 514–23.

Austin, P., Naylor, C. and Tu, J. (2001) A comparison of a Bayesian vs. a frequentist method for profiling hospital performance. *Journal of Evaluation and Clinical Practice*, **7**, 35–45.

Austin, P., Alter, D. and Tu, J. (2003) The use of fixed- and random-effects models for classifying hospitals as mortality outliers: a Monte Carlo assessment. *Medical Decision Making*, **23**, 526–539.

Azzalini, A. and Capitanio, A. (2003) Distributions generated by perturbation of symmetry with emphasis on a multivariate skew t-distribution. *Journal of the Royal Statistical Society B*, **65**(2), 367–389.

Baio, G. and Blangiardo, M. (2010) Bayesian hierarchical model for the prediction of football results. *Journal of Applied Statistics*, **37**, 253–264.

Baldassarre, G., Laio, F. and Montanari, A. (2009) Design flood estimation using model selection criteria. *Physics and Chemistry of the Earth*, **34**, 606–611.

Berger, J. and Deeley, J. (1988) A Bayesian approach to ranking selection of related means with alternatives to analysis-of-variance methodology. *Journal of the American Statistical Association*, **83**, 364–373.

Berger, J. and Perrichi, L. (1996) The intrinsic Bayes factor for model selection and prediction. *Journal of the American Statistical Association*, **91**(433), 109–122.

Berkhof, J., van Mechelen, I. and Hoijtink, H. (2000) Posterior predictive checks: principles and discussion. *Computational Statistics*, **15**(3), 337–354.

Bernardo, J.M. (1996) The concept of exchangeability and its applications. *Far East Journal of Mathematical Sciences*, **4**, 111–121.

Berridge, D. (2003) Logistic-normal mixture models applied to data on the development of children's reasoning. In L. Smith (ed.), *Reasoning by Mathematical Induction in Children's Arithmetic*, pp. 141–149. Pergamon, New York, NY.

Berry, G. and Armitage, P. (1995). Mid-p confidence intervals: a brief review. *Statistician*, **44**, 417–423.

Böhning, D. (2000) *Computer Assisted Analysis of Mixtures and Applications*. Chapman & Hall, London, UK.

Bozdogan, H. (2000) Akaike's Information Criterion and recent developments in information complexity. *Journal of Mathematical Psychology*, **44**(1), 62–91.

Branco, M. and Dey, D. (2001) A general class of multivariate skew elliptical distributions. *Journal of Multivariate Analysis*, **79**, 99–113.

Breslow, N.E. (1984) Extra-Poisson variation in log-linear models. *Applied Statistics*, **33**, 38–44.

Brockett, P., Golden, L. and Panjer, H. (1996) Flexible purchase frequency modeling. *Journal of Marketing Research*, **33**, 94–107.

Brooks, S. and Gelman, A. (1998) General methods for monitoring convergence of iterative simulations. *Journal of Computational and Graphical Statistics*, **7**, 434–455.

Burnham, K. and Anderson, D. (2004) Multimodel inference: understanding *AIC* and *BIC* in model selection. *Sociological Methods and Research*, **33**, 261–305.

Burr, D. (2012) bspmma: An R package for Bayesian semiparametric models for meta-analysis. *Journal of Statistical Software*, **50**(4), 1–23.

Carlin, B. and Chib, S. (1995) Bayesian model choice via Markov chain Monte Carlo methods. *Journal of the Royal Statistical Society B*, **57**(3), 473–484.

Casella, G., Girón, J., Martínez, L. and Moreno, E. (2009) Consistency of Bayesian procedures for variable selection. *Annals of Statistics*, **37**, 1207–1122.

Chaloner, K. and Brant, R. (1988) A Bayesian approach to outlier detection and residual analysis. *Biometrika*, **75**(4), 651–659.

Chen, J. and Chen, Z. (2008) Extended Bayesian information criteria for model selection with large model spaces. *Biometrika*, **95**, 759–771.

Chen, M., Shao, Q. and Ibrahim, J. (2000) *Monte Carlo Methods in Bayesian Computation*. Springer, New York, NY.

Chen, H., Chen, J. and Kalbfleisch, J. (2001) A modified likelihood ratio test for homogeneity in finite mixture models. *Journal of the Royal Statistical Society B*, **63**(1), 19–30.

Chib, S. (1995) Marginal likelihood from the Gibbs output. *Journal of the American Statistical Association*, **90**, 1313–1321.

Cho, H., Ibrahim, J.G., Sinha, D. and Zhu, H. (2009) Bayesian case influence diagnostics for survival models. *Biometrics*, **65**(1), 116–124.

Choy, S. and Smith, A. (1997) Hierarchical models with scale mixtures of normal distribution. *Test*, **6**, 205–211.

Christiansen, C. and Morris, C. (1995) Fitting and checking a two-level Poisson model: modeling patient mortality rates in heart transplant patients. In D. Berry and D. Stangl (eds), *Bayesian Biostatistics*. Marcel Dekker, New York, NY.

Chu, H. and Cole, S. (2006) Bivariate meta-analysis of sensitivity and specificity with sparse data: a generalized linear mixed model approach. *Journal of Clinical Epidemiology*, **59**(12), 1331–1332.

Clayton, D. and Kaldor, J. (1987) Empirical Bayes estimates of age-standardised relative risks for use in disease mapping. *Biometrics*, **43**, 671–681.

Congdon, P. (2007) Model weights for model choice and averaging. *Statistical Methodology*, **4**(2), 143–157.

Congdon, P. and Southall, H. (2005) Trends in inequality in infant mortality in the north of England, 1921–1973, and their association with urban and social structure. *Journal of the Royal Statistical Society A*, **168**(4), 679–700.

Coory, M., Wills, R. and Barnett, A. (2009) Bayesian versus frequentist statistical inference for investigating a one-off cancer cluster reported to a health department. *BMC Medical Research Methodology*, **9**, 30.

Counsell, N., Cortina-Borja, M., Lehtonen, A. and Stein, A. (2011) Modelling psychiatric measures using skew-Normal distributions. *European Psychiatry*, **26**, 112–114.

Crespi, C. and Boscardin, W. (2009) Bayesian model checking for multivariate outcome data. *Computational Statistics and Data Analysis*, **53**(11), 3765–3772.

Czado, C., Gneiting, T. and Held, L, (2009) Predictive model assessment for count data. *Biometrics*, **65**, 1254–1261.

Daniels, M. (1999) A prior for the variance in hierarchical models. *Canadian Journal of Statistics*, **27**, 567–578.

Da Silva, A. (2007) A Dirichlet process mixture model for brain MRI tissue classification. *Medical Image Analysis*, **11**, 169–182.

Da Silva, A. (2009) Bayesian mixture models of variable dimension for image segmentation. *Computational Methods Programs Biomedicine*, **94**, 1–14.

Dawid, A. (1984) Statistical theory: the prequential approach. *Journal of the Royal Statistical Society A*, **147**, 278–292.

Dawid, A. and Sebastiani, P. (1999) Coherent dispersion criteria for optimal experimental design. *Annals of Statistics*, **27**, 57–81.

Deely, J. and Smith, A. (1998) Quantitative refinements for comparisons of institutional performance. *Journal of the Royal Statistical Society A*, **161**(1), 5–12.

Demarta, S. and McNeil, A.J. (2005). The t copula and related copulas. *International Statistical Review*, **73**(1), 111–129.

Dempster, A. (1974) The direct use of likelihood for significance testing, In *Proceedings of Conference on Foundational Questions in Statistical Inference*, pp. 335–352. Department of Theoretical Statistics, University of Aarhus.

Dempster, A. and Ryan, L. (1985) Weighted normal plots. *Journal of the American Statistical Association*, **80**, 845–850.

Denison, D., Holmes, C., Mallick, B. and Smith, A. (2002) *Bayesian Methods for Nonlinear Classification and Regression*. Wiley, Chichester, UK.

DeSantis, F. and Spezzaferri, F. (1997) Alternative Bayes factors for model selection. *Canadian Journal of Statistics*, **25**, 503–515.

Dey, D.K., Chen, M.H. and Chang, H. (1997) Bayesian approach for nonlinear random effects models. *Biometrics*, **53**, 1239–1252.

Dey, D., Muller, P. and Sinha, D. (1999) *Practical Nonparametric and Semiparametric Bayesian Statistics. Lecture Notes in Statistics 133*. Springer, New York, NY.

DiCiccio, T., Kass, R., Raftery, A. and Wasserman, L. (1997) Computing Bayes factors by combining simulation and asymptotic approximations. *Journal of the American Statistical Association*, **92**, 903–915.

Duchateau, L. and Jansen, P. (2005) Understanding heterogeneity in generalized mixed and frailty models. *The American Statistician*, **59**, 143–146.

DuMouchel, W. (1990) Bayesian meta-analysis. In D. Berry (ed.), *Statistical Methodology in the Pharmaceutical Sciences*. Marcel Dekker, New York, NY.

DuMouchel, W. and Normand, S.L. (2000) Computer-modeling and graphical strategies for meta-analysis. In D. Stangl and D. Berry (eds), *Meta-Analysis in Medicine and Health Policy*, Ch. 6, pp. 108–154. Marcel Dekker, New York, NY.

Eilers, P. and Goeman, J. (2004) Enhancing scatterplots with smoothed densities. *Bioinformatics*, **20**, 623–628.

Farrell, P., Groshen, S., MacGibbon, B. and Tomberlin, T. (2010) Outlier detection for a hierarchical Bayes model in a study of hospital variation in surgical procedures. *Statistical Methods in Medicine Research*, **19**, 601–619.

Fernandez, C. and Steel, M. (1998) On Bayesian modelling of fat tails and skewness. *Journal of the American Statistical Association*, **93**, 359–367.

Fruhwirth-Schattner, S. (2001) Markov Chain Monte Carlo estimation of classical and dynamic switching and mixture models. *Journal of the American Statistical Association*, **96**, 194–209.

Frühwirth-Schnatter, S. (2006) *Finite Mixture and Markov Switching Models*. Springer, New York, NY.

Frühwirth-Schnatter, S. and Pyne, S. (2010) Bayesian inference for finite mixtures of univariate and multivariate skew-normal and skew-t distributions. *Biostatistics*, **11**, 317–336.

Fung, T. and Seneta, E. (2010) Modelling and estimation for bivariate financial returns. *International Statistical Review*, **78**(1), 117–133.

Geisser, S. and Eddy, W. (1979) A predictive approach to model selection. *Journal of the American Statistical Association*, **74**, 153–160.

Gelfand, A. (1996) Model determination using sampling based methods. In W. Gilks, S. Richardson and D. Spieglehalter (eds), *Markov Chain Monte Carlo in Practice*. Chapman and Hall/CRC, Boca Raton, FL.

Gelfand, A. and Dey, D. (1994) Bayesian model choice: asymptotics and exact calculations. *Journal of the Royal Statistical Society B*, **56**(3), 501–514.

Gelfand, A. and Ghosh, S. (1998) Model choice: A minimum posterior predictive loss approach. *Biometrika*, **85**(1), 1–11.

Gelman, A. (2006) Prior distributions for variance parameters in hierarchical models. *Bayesian Analysis*, **1**, 515–533.

Geweke, J. (1989) Bayesian inference in econometric models using Monte Carlo integration. *Econometrica*, **57**, 1317–1339.

Geweke, J. (1993) Bayesian treatment of the independent Student-t linear model. *Journal of Applied Econometrics*, **8**, S19–S40.

Goldstein, H. and Spiegelhalter, D. (1996) League tables and their limitations: statistical issues in comparisons of institutional performance (with discussion). *Journal of the Royal Statistical Society A*, **159**, 385–443.

Green, P. (1995) Reversible jump Markov Chain Monte Carlo computation and Bayesian model determination. *Biometrika*, **82**(4), 711–732.

Green, M.J., Medley, G. and Browne, W.J. (2009) Use of posterior predictive assessments to evaluate model fit in multilevel logistic regression. *Veterinary Research*, **40**, 30. DOI: 10.1051/vetres /2009013.

Greenland, S. (2000) Principles of multilevel modelling. *International Journal of Epidemiology*, **29**(1), 158–167.

Gustafson, P., Hossain, S. and MacNab, Y. (2006) Conservative prior distributions for variance parameters in hierarchical models. *Canadian Journal of Statistics*, **34**, 377–390.

Hein, A., Lewin, A. and Richardson, S. (2006) Bayesian hierarchical models for inference in microarray data. In K. Do, P. Muller and M. Vanucci (eds), *Bayesian Inference for Gene Expression and Proteomics*, pp. 53–70. Cambridge University Press, Cambridge, UK.

Hjort, N., Dahl, F. and Steinbakk, G. (2006) Post-processing posterior predictive p values. *Journal of the American Statistical Association*, **101**(475), 1157–1174.

Hsiao, C., Tzeng, J. and Wang, C. (2000) Comparing the performance of two indices for spatial model selection: application to two mortality data sets. *Statistics in Medicine*, **19**, 1915–1930.

Ishwaran, H. and James, L. (2001) Gibbs sampling methods for stick-breaking priors. *Journal of the American Statistical Association*, **96**, 161–173.

Ishwaran, H. and James, L. (2002) Approximate Dirichlet process computing in finite normal mixtures: smoothing and prior information. *Journal of Computational and Graphical Statistics*, **11**, 508–532.

Ishwaran, H. and Zarepour, M. (2001) Markov chain Monte Carlo in approximate Dirichlet and beta two-parameter process hierarchical models. *Biometrika*, **87**, 371–339.

Izsak, R. (2007) Maximum likelihood fitting of the Poisson lognormal distribution. *Environmental and Ecological Statistics*, **15**, 143–156.

James, W. and Stein, C. (1961) Estimation with quadratic loss. In *Proceedings of the Fourth Berkeley Symposium on Mathematical Statistics and Probability*, **1**, 361–379.

Johnson, V. and Albert, J. (1999) *Ordinal Data Modeling*. Springer, New York, NY.

Kadane, J. and Lazar, N. (2004) Methods and criteria for model selection. *Journal of the American Statistical Association*, **99**(465), 279–290.

Kaplon, R. (2010) Heterogeneity in models of purchase frequency. a comparison of Poisson-gamma mixtures with finite poisson mixtures. *Operations Research and Decisions*, **21**, 3–4.

Kashiwagi, N. and Yanagimoto, T. (1992) Smoothing serial count data through a state-space model. *Biometrics*, **48**, 1187–1194.

Kass, R. and Raftery, A. (1995) Bayes factors. *Journal of the American Statistical Association*, **90**, 773–795.

Kim, S. and Ibrahim, J. (2000) Default Bayes factors for generalized linear models. *Journal of Statistical Planning and Inference*, **87**(2), 301–315.

Kirkham, J., Riley, R. and Williamson, P. (2012) A multivariate meta-analysis approach for reducing the impact of outcome reporting bias in systematic reviews. *Statistics in Medicine*, **31**, 2179–2195.

Kitagawa, G. and Gersch, W. (1996) *Smoothness Priors Analysis of Time Series. Lecture Notes in Statistics 116*. Springer, New York, NY.

Knorr-Held, L. and Rainer, E. (2001) Projections of lung cancer mortality in West Germany: a case study in Bayesian prediction. *Biostatistics*, **2**(1), 109–129.

Konishi, S., Ando, T. and Imoto, S. (2004) Bayesian information criteria and smoothing parameter selection in radial basis function networks. *Biometrika*, **91**, 27–43.

Koop, G., Poirier, D. and Tobias, J. (2007) *Bayesian Econometric Methods*. Cambridge University Press, Cambridge.

Laird, N. (1982) Empirical Bayes estimates using the nonparametric maximum likelihood estimate for the prior. *Journal of Statistical and Computational Simulation*, **15**, 211–220.

Laud, P. and Ibrahim, J. (1995) Predictive model selection. *Journal of the Royal Statistical Society B*, **57**(1), 247–262.

Leeb, H. and Pötscher, B. (2006) Model selection and inference: facts and fiction. *Econometric Theory*, **21**, 21–59.

Lenk, P. and Desarbo, W. (2000) Bayesian inference for finite mixtures of generalized linear models with random effects. *Psychometrika*, **65**, 93–119.

Leonard, T. (1973) A Bayesian method for histograms. *Biometrika*, **60**, 297–308.

Leonard, T. (1980) The roles of inductive modelling and coherence in Bayesian statistics. In J. Bernardo, M. DeGroot, D. Lindley and A. Smith (eds), *Bayesian Statistics I*, pp. 537–555. University Press, Valencia.

Leonard, T. and Hsu, J. (1999) *Bayesian Methods: An Analysis for Statisticians and Interdisciplinary Researchers*. Cambridge University Press, Cambridge, UK.

Li, L., Palta, M. and Shao, J. (2004) A measurement error model with a Poisson distributed surrogate. *Statistics in Medicine*, **23**, 2527–2536.

Lindley, D. (1957) A statistical paradox. *Biometrika*, **44**, 187–192.

Lindley, D. (1997) Some comments on Bayes factors. *Journal of Statistical Planning and Inference*, **61**(1), 181–189.

Lindley, D.V. and Smith, A.F. (1972) Bayes estimates for the linear model. *Journal of the Royal Statistical Society B*, **34**, 1–41.

Liu, C. and Aitkin, M. (2008) Bayes factors: prior sensitivity and model generalizability. *Journal of Math Psychology*, **52**, 362–375.

Lunn, D., Jackson, C., Spiegelhalter, D., Best, N. and Thomas, A. (2012) *The BUGS Book: A Practical Introduction to Bayesian Analysis*. CRC, Boca Raton, FL.

Madigan, D. and Ridgeway, G. (2004) Bayesian data analysis. In N. Ye (ed.), *Handbook of Data Mining*, pp. 103–132. Lawrence Erlbaum, Mahwah, New Jersey.

Marshall, E. and Spiegelhalter, D. (1998) Comparing institutional performance using Markov chain Monte Carlo methods. In B. Everitt and G. Dunn (eds), *Recent Advances in the Statistical Analysis of Medical Data*, pp. 229–250. Arnold, London, UK.

Marshall, E. and Spiegelhalter, D. (2007) Identifying outliers in Bayesian hierarchical models: a simulation-based approach. *Bayesian Analysis*, **2**(2), 409–444.

McCullagh, P. and Nelder, J. (1989) *Generalized Linear Models*. Chapman & Hall/CRC, Boca Raton, FL.

Melnykov, V. and Maitra, R. (2010) Finite mixture models and model-based clustering. *Statistics Survey*, **4**, 80–116.

Meng, X.-L. (1994) Posterior predictive p-values. *Annals in Statistics*, **22**, 1142–1160.

Miller, G. (2007) *Statistical Modelling* of Poisson/log normal data. *Radiat Prot Dosimetry*, **124**, 155–163.

Min, A., Holzmann, H. and Czado, C. (2010) Model selection strategies for identifying most relevant covariates in homoscedastic linear models. *Computational Statistics and Data Analysis*, **54**, 3194–3211.

Morgan, B. (2000) *Applied Stochastic Modelling*. Arnold, London, UK.

Morris, C. and Normand, S. (1992) Hierarchical models for combining information and meta-analyses. In J. Bernardo, J. Berger, A. Dawid and A. Smith (eds), *Bayesian Statistics 4*, pp. 321–344. Oxford University Press, Oxford, UK.

Natarajan, R. and Kass, R. (2000) Reference Bayesian methods for generalized linear mixed models. *Journal of the American Statistical Association*, **95**, 227–237.

Neal, R. (2000) Markov chain sampling methods for dirichlet process mixture models. *Journal of Computational and Graphical Statistics*, **9**, 249–265.

Ohlssen, D., Sharples, L. and Spiegelhalter, D. (2007) Flexible random-effects models using Bayesian semi-parametric models: applications to institutional comparisons. *Statistics in Medicine*, **26**, 2088–2112.

Paddock, S., Wynn, B., Carter, G. and Buntin, M. (2004) Identifying and accommodating statistical outliers when setting prospective payment rates for inpatient rehabilitation facilities. *Health Service Reseach*, **39**, 1859–1879.

Pauler, D. and Wakefield, J. (2000) Modeling and implementation issues in Bayesian meta-analysis. In D. Stangl and D. Berry (eds), *Meta-Analysis in Medicine and Health Policy*, pp. 205–230. Marcel Dekker, New York, NY.

Pires, R. and Diniz, C. (2012) Correlated binomial regression models. *Computational Statistics and Data Analysis*, **56**(8), 2513–2525.

Rao, C. (1975) Simultaneous estimation of parameters in different linear models and applications to biometric problems. *Biometrics*, **31**, 545–554.

Raudenbush, S. (1984) Magnitude of teacher expectancy effects on pupil IQ as a function of the credibility of expectancy induction: A synthesis of findings from 18 experiments. *Journal of Educational Psychology*, **76**(1), 85–97.

Richardson, S. and Green, P. (1997) On *Bayesian Analysis* of mixtures with an unknown number of components. *Journal of the Royal Statistical Society B*, **59**, 731–758.

Riebler, A. and Held, L. (2010) The analysis of heterogeneous time trends in multivariate age–period–cohort models. *Biostatistics*, **11**(1), 57–69.

Robins, J.M., van der Vaart, A. and Ventura, V. (2000) Asymptotic distribution of P values in composite null models. *Journal of the American Statistical Association*, **95**(452), 1143–1156.

Sauer, J., Link, W. and Royle, J. (2005) Hierarchical models and Bayesian analysis of bird survey information. In R. John and R. Terrell (eds), *Bird Conservation, Implementation and Integration in the Americas: Proceedings 3rd International Partners in Flight Conference 2002*, pp. 762–770. U.S. Department of Agriculture, Forest Service, Pacific Southwest Research Station.

Schwarz, G. (1978) Estimating the dimension of a model. *Annals of Statistics*, **6**, 461–464.

Sethuraman, J. (1994) A constructive definition of Dirichlet priors. *Statistica Sinica*, **4**(2), 639–650

Shao, J. (1993) Linear model selection by cross-validation. *Journal of the American Statistical Association*, **88**(422), 486–494.

Snedecor, G. and Cochran, W. (1989) *Statistical Methods*, 8th edn. Iowa State University Press, Ames, IA.

Spiegelhalter, D. (1999) An initial synthesis of statistical sources concerning the nature and outcomes of paediatric cardiac surgical services at Bristol relative to other specialist centres from 1984 to 1995, Bristol Royal Infirmary Inquiry. http://www.bristol-inquiry.org.uk/brisdsanalysisfinal.htm# Background Papers.

Spiegelhalter, D. (2005) Handling over-dispersion of performance indicators. *Qual Saf Health Care*, **14**, 347–351.

Spiegelhalter, D., Best, N. and Carlin, B. (1998) Bayesian deviance, the effective number of parameters, and the comparison of arbitrarily complex models, *Research Report 98-009*, Division of Biostatistics, University of Minnesota.

Spiegelhalter, D., Aylin, P., Best, N., Evans, S. and Murray, G. (2002) Commissioned analysis of surgical performance using routine data: lessons from the Bristol Inquiry. *Journal of the Royal Statistical Society A*, **165**, 191–231.

Steinbakk, G. and Storvik, G. (2009) Posterior Predictive p-values in Bayesian Hierarchical Models. *Scandinavian Journal of Statistics*, **36**(2), 320–336.

Stephens, M. (2000) *Bayesian Analysis* of mixture models with an unknown number of components – An alternative to reversible jump methods. *Annals of Statistics*, **28**(1), 40–74.

Stern, H. and Cressie, N. (2000) Posterior predictive model checks for disease mapping models. *Statistics in Medicine*, **19**(17–18), 2377–2397.

Tarpey, T. and Petkova, E. (2010) Latent regression analysis. *Statistical Modelling*, **10**(2), 133–158.

Teh, Y. (2010) Dirichlet processes. *In Encyclopedia of Machine Learning*, Eds C. Sammut and G. Webb. Springer, New York, NY.

Teh, Y., Jordan, M., Beal, M. and Blei, D. (2006) Hierarchical Dirichlet processes. *Journal of the American Statistical Association*, **101**, 1566–1581.

Turner, R., Omar, R., Yang, M., Goldstein, H. and Thompson, S. (2000) A multilevel model framework for meta-analysis of clinical trials with binary outcomes. *Statistics in Medicine*, **19**(24), 3417–3432.

van der Linden, W. and Guo, F. (2008) Bayesian procedures for identifying aberrant response-time patterns in adaptive testing. *Psychometrika*, **73**(3), 365–384.

Vandewalle, V., Biernacki, C., Celeux, G. and Govaert, G. (2013) A predictive deviance criterion for selecting a generative model in semi-supervised classification, *Computational Statistics and Data Analysis*, **64**, 220–236.

van Houwelingen, H., Arends, L. and Stijnen, T. (2002) Advanced methods in meta-analysis: multivariate approach and meta-regression. *Statistics in Medicine*, **21**(4), 589–624.

Ward, E. (2008) A review and comparison of four commonly used Bayesian and maximum likelihood model selection tools. *Ecological Modelling*, **211**, 1–10.

Wasserman, L. (2000) Asymptotic inference for mixture models using data-dependent priors. *Journal of the Royal Statistical Society B*, **62**(1), 159–180.

West, M., Müller, P. and Escobar, M. (1994) Hierarchical priors and mixture models, with application in regression and density estimation. In P. Freeman and A. Smith (eds), *Aspects of Uncertainty: A Tribute to D. V. Lindley*. Wiley, Chichester, UK.

Williams, D.A. (1982) Extra-binomial variation in logistic linear models. *Applied Statistics*, **31**, 144–14.

Woodward, M. (1999) *Epidemiology. Study Design and Data Analysis*. Chapman and Hall/CRC, London, UK.

Xie, D., Raghunathan, T. and Lepkowski, J. (2007) Estimation of the proportion of overweight individuals in small areas – a robust extension of the Fay-Herriot model. *Statistics in Medicine*, **26**, 2699–2715.

Xie, M., Singh, K. and Zhang, C.-H. (2009) Confidence intervals for population ranks in the presence of ties and near ties. *Journal of the American Statistical Association*, **104**(486), 775–788.

Xie, X., Kou, S. and Brown, L. (2012) SURE estimates for a heteroscedastic hierarchical model. *Journal of the American Statistical Association*, **107**, 1465–1479.

Yan, G. and Sedransk, J. (2007) Bayesian diagnostic techniques for detecting hierarchical structure. *Bayesian Analysis*, **2**(4), 735–760.

Young-Xu, Y. and Chan, K. (2008) Pooling overdispersed binomial data to estimate event rate. *BMC Medical Research Methodology*, **8**, 58.

# 3

# Regression techniques

## 3.1 Introduction: Bayesian regression

Methods for Bayesian estimation of the normal linear regression model, whether with univariate or multivariate outcome, are well established. With an inverse gamma prior on the residual variance in univariate regression, and conjugate normal prior on the regression coefficients (conditional on the residual variance), analytic formulae for the posterior densities of these coefficients and other relevant quantities (e.g. predicted responses for new predictor values) are available. These permit direct estimation with no need for repeated sampling. However, the normal linear regression model is restricted to continuous responses and makes assumptions regarding the error structure, the form of relationship, and the appropriate form of predictors that are not necessarily met in practice. Parameter estimation under alternative assumptions or responses such as heteroscedastic linear regression (Peña *et al.*, 2009), generalised linear models (e.g. Gerwinn *et al.*, 2010), non-linear or varying coefficient relationships (e.g. Blum and François, 2010), and non-conjugate priors (e.g. Fang and Dawid, 2002) are typically facilitated by a sampling based approach to estimation. Similar advantages from iterative sampling apply in assessing the density of model parameters, or structural quantities defined by functions of parameters and data. The Bayesian approach may also be used to benefit with regression model selection, in terms of priors adapted to screening out marginally important or irrelevant predictors, or in comparisons between non-nested models. Bayesian computation for regression may be implemented using BUGS, R-INLA, MCMCpack (http://cran.r-project.org/web/packages/MCMCpack/), bayesm (http://cran.r-project.org/web/packages/bayesm/), monomvn and ARM (http://cran.r-project.org/web/packages/arm/).

The development below is selective among the wide range of modelling issues which have been explored from a Bayes perspective, but intended to illustrate potential benefits of the Bayes approach. The first section below reviews normal linear regression from a Bayesian perspective, including model checks and fitting. Section 3.3 considers the broader generalized linear model perspective including binomial, binary and Poisson responses. Section 3.4 considers latent variable approaches in binary regression. Section 3.5 then moves to consider regression model choice, focussing especially on different methods for predictor selection. Section 3.6 considers Bayesian regression as applied to multiple category outcomes, which includes ordinal and nested category response data.

*Applied Bayesian Modelling*, Second Edition. Peter Congdon.
© 2014 John Wiley & Sons, Ltd. Published 2014 by John Wiley & Sons, Ltd.

## 3.2   Normal linear regression

The normal linear regression model

$$y_i = X_i\beta + \varepsilon_i,$$

$$\varepsilon \sim N(0, \sigma^2 I),$$

describes the relation between a response $y_i$ and a vector of predictor or explanatory variables $X_i = (x_{1i}, x_{2i}, \dots, x_{Pi})$ over $i = 1, \dots, n$ observations (where usually $x_{1i} = 1$). The errors $\varepsilon = (\varepsilon_1, \dots, \varepsilon_n)$ are assumed to be mutually uncorrelated, with $E(\varepsilon_i \varepsilon_j | X_i, X_j) = 0$ (all $i \neq j$), and independent of the predictors with $E(\varepsilon_i x_{ki}) = 0$ for all $i$ and $k$. This model assumes a metric outcome though the technique can be applied to latent variables underlying binary or category responses (see Section 3.4), and is applicable to various forms of predictor: continuous, binary, and categorical variables. Such predictors can be taken as known, that is, observed without measurement error, or as random, with density $\pi(X_i | \theta_x)$ where $\theta_x$ does not include $\{\beta, \sigma^2\}$.

Assuming the $x$ variables are known without error, and defining the error precision $h = 1/\sigma^2$, one may write the likelihood for each observation as

$$p(y_i | \beta, h) = \left(\frac{h}{2\pi}\right)^{0.5} \exp\left[-\frac{h}{2}(y_i - X_i\beta)^2\right],$$

and under the independent errors assumption, the total likelihood is a product over observations, namely

$$p(y | \beta, h) = \left(\frac{1}{2\pi}\right)^{n/2} h^{n/2} \exp\left[-\frac{h}{2}\sum_{i=1}^{n}(y_i - X_i\beta)^2\right].$$

Denoting the classical linear regression estimator as $\hat{\beta} = (X'X)^{-1}X'y$, one may write the sum of squared errors term in the likelihood as

$$\sum_{i=1}^{n}(y_i - X_i\beta)^2 = (y - X\hat{\beta})(y - X\hat{\beta}) + (\beta - \hat{\beta})(X'X)(\beta - \hat{\beta}),$$

and setting $T_L = (y - X\hat{\beta})(y - X\hat{\beta})$, and $R(\beta) = (\beta - \hat{\beta})(X'X)(\beta - \hat{\beta})$, the likelihood can be stated as

$$p(y | \beta, h) = \left(\frac{1}{2\pi}\right)^{n/2} h^{n/2} \exp\left[-\frac{h}{2}[T_L + R(\beta)]\right].$$

A range of prior schemes are possible. Consider the conditional dependence prior scheme $\pi(\beta, h) = \pi(h)\pi(\beta | h)$, with the prior on the regression coefficient conditioning on the precision,

$$h \sim \text{Ga}\left(\frac{n_0}{2}, \frac{n_0 T_0}{2}\right), \qquad \beta | h \sim N\left(c_0, \frac{1}{h}C_0\right),$$

which may be designated as the normal-gamma $NG(n_0, T_0, c_0, C_0)$ prior. This prior can be represented

$$\pi(\beta, h) \propto h^{P/2} \exp\left[-\frac{h}{2}(\beta - c_0)'C_0^{-1}(\beta - c_0)\right] h^{(n_0/2-1)} \exp\left[-\frac{h}{2}n_0 T_0\right]$$

$$= h^{(P+n_0)/2-1} \exp\left[-\frac{h}{2}\left\{(\beta - c_0)'C_0^{-1}(\beta - c_0) + n_0 T_0\right\}\right].$$

Combining likelihood and prior, the posterior can be represented as

$$p(\beta, h|y) \propto p(y|\beta, h)\pi(\beta, h)$$

$$\propto h^{\frac{n+P+n_0}{2}-1} \exp\left[-\frac{h}{2}\left[n_0 T_0 + T_L + (\beta - c_0)' C_0^{-1}(\beta - c_0) + R(\beta)\right]\right]$$

$$\propto h^{(P+n_1)/2-1} \exp\left[-\frac{h}{2}\left\{(\beta - c_1)' C_1^{-1}(\beta - c_1) + n_1 T_1\right\}\right],$$

where $n_1 = n_0 + n$,

$$C_1^{-1} = C_0^{-1} + X'X,$$

$$c_1 = C_1(C_0^{-1}c_0 + X'y),$$

$$n_1 T_1 = n_0 T_0 + (y - Xc_1)'y + (c_0 - c_1)' C_0^{-1}c_0.$$

Hence under this prior, one obtains full conjugacy, since the posterior is also normal gamma, namely $NG(n_1, T_1, c_1, C_1)$. Particular forms of this prior may be particularly relevant in selecting predictors, such as the g-prior (Liang *et al.*, 2008) obtained with $n_0 = T_0 = 0$ and $C_0 = gX'X$.

This conjugacy property does not extend to the independent prior scheme $\pi(\beta, h) = \pi(h)\pi(\beta)$, namely

$$h \sim \text{Ga}\left(\frac{n_0}{2}, \frac{n_0 T_0}{2}\right), \qquad \beta \sim N(c_0, C_0),$$

though the conditionals $p(h|y, \beta)$ and $p(\beta|y, h)$ are respectively gamma and normal, a property known as conditional conjugacy. Such conditional conjugacy allows simple Gibbs sampling from the posterior. Specifically,

$$p(h|y, \beta) = \text{Ga}\left(\frac{n + n_0}{2}, \frac{n_1 T_1}{2}\right),$$

where now $T_1 = R(\beta) + T_L + n_0 T_0$. Also one has

$$p(\beta|y, h) = N(c_1, C_1)$$

where $C_1$ is as above, but now

$$c_1 = C_1(C_0^{-1}c_0 + hX'y).$$

## 3.2.1    Linear regression model checking

As mentioned above, the suitability of the linear regression model rests on several assumptions: linearity and homogeneity of predictor effects, error normality, and constant variance. In practice departures such as outlier points, nonlinear effects of predictors, groupwise heteroscedasticity, heavy tailed or skewed errors will suggest modified models. Model checks to assess the assumptions include residual analysis (e.g. Zellner and Moulton, 1985; Chaloner and Brant, 1988), analysis of influential cases (e.g. Kass *et al.*, 1989; Zhu *et al.*, 2011), and comparison with the data of predictions or replicate samples

$$y_{rep,i} \sim N(X_i\beta, \sigma^2),$$

from the linear regression. Such samples from the posterior predictive density

$$p(y_{rep}|y) = \int p(y_{rep}|\theta)p(\theta|y)d\theta,$$

enable various posterior predictive checks against observations (Rubin, 1984). These include forms of probability test, providing a probability estimate that the observations arose by chance, given the model assumptions.

One check (Zellner and Moulton, 1985; Gelfand, 1996, p. 202) is simply whether observations $y_i$ are contained within 95% or 99% predictive intervals, which can be obtained by repeated sampling of $y_{rep,i}$. In a model that reproduces the data satisfactorily, at least $(1 - \alpha)\%$ of the observations should be contained within $(1 - \alpha)\%$ predictive intervals based on replicate data. Extreme posterior predictive probabilities

$$\Pr(y_{rep,i} \geq y_i|y)$$

of falling in extreme tails (e.g. values under 0.025 or over 0.975) indicate poorly fitted cases; these can be obtained using binary indicators $I(y_{rep,i}^{(t)} \geq y_i)$ at each MCMC iteration.

Posterior predictive probability tests are obtained using discrepancy statistics $D(y, \theta)$ that can be evaluated for both data and replicates, and for MCMC implementation involve sampling $D(y_{rep}^{(t)}, \theta^{(t)})$ and $D(y, \theta^{(t)})$ at each iteration (Meng, 1994). Then the probability

$$p_{PP} = \Pr[D(y_{rep}, \theta) > D(y, \theta)]$$

may be estimated by the proportion of iterations where $D(y_{rep}^{(t)}, \theta^{(t)})$ exceeds $D(y^{(t)}, \theta^{(t)})$, namely

$$\hat{p}_{PP} = \frac{1}{T}\sum_{t=1}^{T} I[D(y_{rep}^{(t)}, \theta^{(t)}) > D(y, \theta^{(t)})].$$

Extreme $p_{PP}$ values indicate that the observations are unlikely under the model assumed. For a linear regression, one may also derive predictive residuals $\varepsilon_{rep} = y_{rep} - X\beta$, and use $p$-test comparisons of discrepancy measures $D(\varepsilon_{rep}, \theta)$ against the data based versions $D(\varepsilon, \theta)$. The discrepancy statistic for the observed data may sometimes be known, e.g. the maximum or interquartile range $Q_3 - Q_1$ of the data values, or the median $M_d$. For example, one can obtain Bowley's coefficient of skewness $(Q_3 - 2M_d + Q_1)/(Q_3 - Q_1)$ from the data, and compare with the the same coefficient evaluated using $y_{rep}$.

Residual checking for normal linear regression may include comparison of omnibus measures (e.g. of skew and kurtosis) against their expected values under error normality, quantile-quantile plots of posterior means of realised residuals $\varepsilon_i = y_i - X_i\beta$, or posterior probabilities that standardised residuals $\varepsilon_i^s = \varepsilon_i/\sigma$, or absolute standardised residuals, exceed a relevant significance threshold, $\Pr(|\varepsilon_i^s||y) > k$. To assess skew and kurtosis in the residuals, one may use sampled values of $\beta^{(t)}$ and $\sigma^{(t)}$ to calculate, at each iteration, the following checks

$$J_{sk}^{(t)} = I\left[\frac{1}{n}\sum_{i=1}^{n}(y_i - X_i\beta^{(t)})^3/\sigma^{3(t)} > 0\right] = I\left[\frac{1}{n}\sum_{i=1}^{n}\varepsilon_i^{3(t)}/\sigma^{3(t)} > 0\right],$$

$$J_{kt}^{(t)} = I\left[\frac{1}{n}\sum_{i=1}^{n}(y_i - X_i\beta^{(t)})^4/\sigma^{4(t)} > 3\right] = I\left[\frac{1}{n}\sum_{i=1}^{n}\varepsilon_i^{4(t)}/\sigma^{4(t)} > 3\right],$$

with probability estimates $\frac{1}{T}\sum_{t=1}^{T} J_{sk}^{(t)}$ and $\frac{1}{T}\sum_{t=1}^{T} J_{kt}^{(t)}$ to assess whether skew or kurtosis are elevated relative to that expected under error normality.

Predictive check assessments of skew and kurtosis can also be made; for example, a $p$-test could compare skewness using predictive errors $D_{rep}^{sk} = \frac{1}{n}\sum_{i=1}^{n} \varepsilon_{rep,i}^{3(t)}/\sigma_{rep}^{3(t)}$ against $D_{data}^{sk} = \frac{1}{n}\sum_{i=1}^{n} \varepsilon_{i}^{3(t)}/\sigma^{3(t)}$. Other normality checks include comparing empirical CDF proportions

$$\hat{F}_i = \frac{1}{n}\sum_{i \neq k}^{n} I(\varepsilon_k^s \leq \varepsilon_i^s)$$

for standardised residuals against the expected proportions from the cumulative normal distribution function, $F_i^{\exp} = \Phi(\varepsilon_i^s)$. The resulting chi-square discrepancy

$$\chi^2 = \sum \frac{(\hat{F}_i - F_i^{\exp})}{F_i^{\exp}(1 - F_i^{\exp})}$$

can be compared with the same chi-square discrepancy obtained using $\varepsilon_{rep,i}^s = \varepsilon_{rep,i}/\sigma$. Heteroscedasticity may be assessed using one of the range of available tests (e.g. White's test) applied to $\varepsilon_i$ and $\varepsilon_{rep,i}$ (see Example 3.1).

Discrepancies based on replicates sampled from full data models may be subject to masking: a potential outlier contributes to the model parameters and so deflates any discrepancy. Masking is avoided by using cross-validatory (CV) approaches, which for linear regression applications focus on the predictive density

$$p(y_r|y_{[r]}) = \int p(y_r|\beta, \sigma^2, X_r)p(\beta, \sigma^2|y_{[r]}, X_{[r]})d\beta d\sigma^2,$$

where $y_r$ is a subset of cases, and only the remaining observations $y_{[r]}$ are used to update parameters. Cross-validation methods play a role in identifying outliers and influential cases, and also in assessing fit. In particular, consider the predictive density based on single case omission, or conditional predictive ordinate (CPO), namely

$$p(y_{rep,i}|y_{[i]}) = \int p(y_{rep,i}|\sigma^2, \beta, X_i)p(\beta, \sigma^2|y_{[i]})d\beta,$$

where $y_{[i]}$ denotes the response data omitting case $i$ (known predictor data for case $i$ are still used). As discussed in Chapter 2, conditional predictive ordinates are estimable using Monte Carlo methods (e.g. Dey *et al.*, 1997, equation 2.6), and a predictive or pseudo marginal likelihood is obtained as the product $q(y) = \prod_i p(y_{rep,i}|y_{[-i]})$. The ratio of $q_1(y)$ and $q_2(y)$ for models $M_1$ and $M_2$ then provides a predictive or pseudo Bayes factor (Dey *et al.*, 1997; Mukhopadhyay *et al.*, 2005).

### Example 3.1 Permanent income and consumption

Zellner and Moulton (1985) evaluate the association between permanent consumption $c$ and income $x$ over $n = 26$ countries under differing combinations of response form and prior. The analysis here uses the response $y = \log[r/(1-r)]$ where $r = c/x$ and predictor $\log(x)$ (called the logit transformation model or LTM). Also adopted are flat priors (uniform along the real line) for both $\beta$ and $\log(\sigma)$. From the last 9000 of a two chain run of 10 000 iterations, the coefficient on $\log(x)$ is obtained as $-0.19$ with 95% interval $(-0.42,0.04)$. To assess possible outlier status, one may monitor $I(\varepsilon_{rep,i}^{(t)} > 0) = I(y_{rep,i}^{(t)} > y_i)$ and $I(|\varepsilon_i^s| > 1.96)$. Posterior means of these indicators highlight country 14 (Malta) as having lower observed $z$

than predicted by the model, with $\Pr(abs(\varepsilon_{14}^s)|z > 1.96) = 0.93$, and $\Pr(z_{rep,i}^{(t)} > z_i|z) = 0.99$. $\Pr(z_{rep,i}^{(t)} > z_i|z)$ is also relatively high, namely 0.96, for country 18 (Japan).

However, other diagnostics do not suggest pronounced departures from the linear regression assumptions[1]. The probability estimate $\frac{1}{T}\sum_{t=1}^{T} J_{sk}^{(t)}$ for error skewness is rather low at 0.14, and a plot of the posterior means of $\varepsilon_i$ confirms some negative skew. However, the p-test of chi square statistics, comparing empirical and expected cumulative error distributions for observations and replicates, is 0.29, indicating no model failure. White's test involves regression of $\varepsilon_i^2$ and $\varepsilon_{rep,i}^2$ against $\log(x_i)$ and $[\log(x_i)]^2$, and comparing the values of $nR^2$ obtained from this regression against a $\chi_2^2$ density. A test using the indicator $I(nR^{2(t)} > 5.99)$, where 5.99 is the 0.95 point of the cumulative $\chi_2^2$ distribution, shows no evidence of heteroscedasticity, with $\Pr(nR^2 > 5.99|y) = 0.0055$. A posterior predictive p-test also does not suggest a departure from the homoscedasticity assumption, with the relevant p value being 0.38.

## 3.3 Simple generalized linear models: Binomial, binary and Poisson regression

The normal linear model is a special case of the generalised linear model, which includes density forms where an unconstrained regression prediction may not be suitable (e.g. for necessarily positive variables), and where the variance and mean may be interrelated. Under this generalisation, the observations are drawn from a density belonging to the exponential family, namely

$$p(y|\theta_i, \phi_i) = \exp\left[\frac{y_i\theta_i - b(\theta_i)}{\phi_i} + c(y_i, \phi_i)\right],$$

where $\theta_i$ determines the location of the distribution, and $\phi_i = \phi/w_i$ are dispersions that can include weights $w_i$. For $y$ following a density within the exponential family, one has $E(y_i) = \mu_i = b'(\theta_i)$, and $var(y_i) = b''(\theta_i)\phi_i = \phi_i var(\mu_i)$. This family includes both continuous densities (normal, log-normal, inverse-Gaussian, exponential, gamma, Weibull, etc.), and discrete distributions (binomial, Bernoulli, Poisson, geometric, etc.). The mean $\mu_i$ is related to a regression term $\eta$ via a link function $g$,

$$g(\mu_i) = \eta_i = X_i\beta.$$

There may be benefits in using the canonical form of the link function, when $\eta$ is the same as the location parameter $\theta$ of the distribution. The inverse link function $h(\eta_i)$ maps the value of the linear predictor to the conditional mean $\mu_i$.

### 3.3.1 Binary and binomial regression

Binary and binomial regression are widely applied forms of generalised linear model, especially for analysing survey data and assessing risks in epidemiology. Under binary regression, $y_i \sim \text{Bern}(\pi_i)$ where $\pi_i = \Pr(y_i = 1|X_i)$ is the success probability to be predicted using regressors $X_i$. Binomial regression involves grouped data, with $n_i$ subjects in the $i^{th}$ group and $y_i$ subjects with response 1, and predicts the probability of success given the risk profile $X_i$ of the $i^{th}$ group. The regression involves a distribution function $F = h$, with $\pi_i = F(X_i\beta)$ and link function $g = F^{-1}$ relating the success probability to the regression term, namely $g(\pi_i) = \eta_i = X_i\beta$. Commonly adopted options are the logistic, normal and extreme value, respectively:

$$F(X_i\beta) = \exp(X_i\beta)/[1 + \exp(X_i\beta)],$$

$$F(X_i\beta) = \Phi(X_i\beta),$$

$$F(X_i\beta) = 1 - \exp[-e^{X_i\beta}].$$

The corresponding links are the logit, probit and complementary log-log, respectively

$$\log\left(\frac{\pi_i}{1 - \pi_i}\right) = X_i\beta,$$

$$\pi_i = \Phi^{-1}(X_i\beta),$$

$$\log(-\log(1 - \pi_i)) = X_i\beta.$$

with the logit being the canonical link function.

By contrast to normal linear regression, under a normal prior $\beta \sim N(c_0, C_0)$ for the regression coefficients, the posterior density $p(\beta|y)$ under binomial and binary regression is not a standard density, though log-concavity of the full conditionals allows Gibbs sampling via adaptive rejection (Dellaportas and Smith, 1993). Specifically, assuming a logit link and $i = 1, \ldots, G$ groups, the binomial case gives

$$p(\beta|y) \propto \exp[-\frac{1}{2}(\beta - c_0)' C_0^{-1}(\beta - c_0) + \sum_{i=1}^{G} [y_i\eta_i + n_i\log(1 - \pi_i)\}].$$

While a diffuse prior on $\beta$ is often convenient in linear regression, numeric stability and prior choice in binomial or binary regression may be assisted by scaling predictors (Raftery, 1996). Gelman *et al.* (2008) advocate scaling nonbinary predictors in logistic regression to have mean 0 and standard deviation 0.5, and then place Cauchy priors centred at 0 and with scale 2.5 on regression coefficients.

An additional consideration in setting priors is that binomial and binary regressions are often used where there is accumulated evidence on direction and size of coefficients, and this can be incorporated in evidence-based priors. For example, prior evidence on risk patterns may be included by eliciting the success probability associated with combinations of predictor values. Under the conditional mean priors approach of Bedrick *et al.* (1997), priors on $\beta$ are indirectly elicited using notional success probabilities. For $R$ predictors (including an intercept) the typical scenario (one providing proper priors on all regression coefficients) is to select $R$ predictor combinations $\tilde{x}_r$ in the observed or observable range of predictors. These combinations include contrasting values on predictors. One then specifies beta densities

$$\tilde{\pi}_r \sim \text{Beta}(\tilde{n}_r\tilde{\kappa}_r, \tilde{n}_r(1 - \tilde{\kappa}_r)),$$

for success probabilities $\{\tilde{\pi}_1, \ldots, \tilde{\pi}_R\}$ at these values, where $\tilde{n}_r$ is interpretable as a prior sample size or measure of strength of evidence, and $\tilde{\kappa}_r$ is the elicited mean for $\tilde{\pi}_r$ when $x = \tilde{x}_r$. One replaces the conventional (e.g. normal) prior on $\beta$ with $R$ beta priors on $\tilde{\pi}_r$, leading to an induced prior on $\beta$,

$$\pi(\beta) \propto \prod_r \{F(\tilde{x}_r\beta)\}^{\tilde{n}_r\tilde{\kappa}_r - 1}\{1 - F(\tilde{x}_r\beta)\}^{\tilde{n}_r(1-\tilde{\kappa}_r)-1} \left|\frac{\partial F(\tilde{x}_r\beta)}{\partial \beta}\right|.$$

For a logit link regression, the Jacobean is proportional to $F(\tilde{x}_r\beta)[1 - F(\tilde{x}_r\beta)]$, and the induced prior takes the form

$$\pi(\beta) \propto \prod_r \{F(\tilde{x}_r\beta)\}^{\tilde{n}_r\tilde{\kappa}_r}\{1 - F(\tilde{x}_r\beta)\}^{\tilde{n}_r(1-\tilde{\kappa}_r)},$$

and so is conjugate with the binomial likelihood. If only $R' < R$ combinations are chosen, then priors on some $\beta$ coefficients will be improper (Bredrick *et al.*, 1996, p. 1455).

In practice, the induced prior on $\beta$ may be obtained by solving the expressions for the link functions $g(\tilde{\pi}_r)$ for $\beta$. Suppose $R = 2$, with covariate vector $(1, x_i)$, and that $\tilde{x}_1 = (1, x_{LQ})$ and $\tilde{x}_2 = (1, x_{UQ})$ are defined at the the lower and upper quartiles of $x_i$. Then the CMP prior is

$$\tilde{\pi}_1 \sim \text{Beta}(a_1, b_1), \tilde{\pi}_2 \sim \text{Beta}(a_2, b_2),$$

where $(a_r, b_r)$ are chosen in line with the expected success rates at $x_{LQ}$ and $x_{UQ}$. For a logit link one has

$$\text{logit}(\tilde{\pi}_1) = \beta_1 + \beta_2 x_{LQ},$$
$$\text{logit}(\tilde{\pi}_2) = \beta_1 + \beta_2 x_{UQ},$$

from which $\beta_2^{(t)}$ and $\beta_1^{(t)}$ can be obtained for sampled $\tilde{\pi}_r^{(t)}$. At each MCMC iteration, one obtains the induced regression coefficients as

$$\beta_2 = [\text{logit}(\tilde{\pi}_2) - \text{logit}(\tilde{\pi}_1)]/(x_{UQ} - x_{LQ}),$$
$$\beta_1 = \text{logit}(\tilde{\pi}_1) - \beta_2 x_{LQ}.$$

A number of outputs from binary and binomial regression are of interest beyond the regression coefficients. The overall predictive or discriminatory value of a binary regression (e.g. in medical diagnosis) may be summarised by selecting a risk score threshold $\eta^T$, or corresponding probability $\pi^T = g^{-1}(\eta^T)$, and classifying cases above the threshold as having a positive classification or diagnosis, with the remainder classed as negative. These classifications can be compared with the true diagnoses $y = 1$ (positive) and $y = 0$ (negative) respectively. Prediction rates for true or false classification can be combined with cross-validation (Gelman *et al.*, 2008), where part of the data is used as a training sample.

Let $n_1$ and $n_0$ respectively denote the number of subjects with $y = 1$ and $y = 0$. The total of subjects (denoted $n_{11}$) correctly classified as positive will be subjects with $y_i = 1$ and $\eta_i > \eta^T$. This total will vary between MCMC iterations and the sensitivity of the selected threshold will be measured as $\bar{n}_{11}/n_1$, where $\bar{n}_{11}$ is the posterior mean. Subjects with $y = 0$, but with $\eta_i > \eta^T$, are 'false positives'. The number of subjects (denoted $n_{00}$) correctly classified as negative will be subjects with $y_i = 0$ and $\eta_i \leq \eta^T$, and the specificity of the selected threshold will be measured as $\bar{n}_{00}/n_0$. Subjects with $y = 1$ but $\eta_i \leq \eta^T$ are 'false negatives'.

Predictive assessments for binary or binomial regression involve sampling replicates $y_{rep,i} \sim \text{Bern}(\pi_i)$ or $y_{rep,i} \sim \text{Bin}(n_i, \pi_i)$, and comparing these with the observations, either directly or via posterior predictive tests using discrepancy measures. For binary data, the probabilities $\Pr(y_{rep,i} = y_i|y)$ that replicate binary data values equal the observed values replace the probabilities $\Pr(y_{rep,i} \geq y_i|y)$ appropriate in linear regression for metric responses. Low probabilities $\Pr(y_{rep,i} = y_i|y)$ indicate poorly fitted cases. Checks for binary data can also be performed using the latent outcome approach, where metric response checks are possible (Section 3.4). For binomial response data, the relevant probabilities are $\Pr(y_{rep,i} > y_i|y) + 0.5 \Pr(y_{rep,i} = y_i|y)$, and both low and high probabilities indicate poor fit. Predictive checks for binomial data may extend to posterior predictive p-tests for overdispersion, using tests such as those in Dean (1992) or Baksh *et al.* (2011), involving discrepancy functions evaluated both for actual and replicate data.

To deal with the masking problem in using complete data replicates for model checking, one may consider the binary and binomial cross-validatory (CV) predictive density based on single case omission, obtained (Johnson and Albert, 1999) as

$$p(y_{rep,i}|y_{[i]}) = \int p(y_{rep}|X_i, \beta)p(\beta|y_{[i]})d\beta,$$

where $y_{[i]}$ denotes the response data omitting case $i$. Obtaining the CV predictive density from first principles (actually fitting the model to $n$ separate samples formed by omitting $y_i$ in turn) is computationally intensive, albeit feasible for small datasets, as illustrated in Example 3.2. Example 3.2 also illustrates how the CV density can be closely approximated by Monte Carlo estimates of conditional predictive ordinates.

### 3.3.2 Poisson regression

Poisson regression is a natural choice when the response is a small count, though also applicable for large counts when these are observed in relation to an offset $t_i$, as in $y_i \sim \text{Po}(t_i\rho_i)$ and the underlying rate $\rho_i$ is low. Poisson regression has a relative risk interpretation: the effect of a predictor is multiplicative on the rate, leading to a changes in relative risk as the predictor varies. The canonical link function is the natural logarithm, namely

$$y_i \sim \text{Po}(\rho_i),$$

$$\log(\rho_i) = X_i\beta,$$

with a unit increase in predictor $x_{ji}$ then associated with a multiplicative increase $\exp(\beta_j)$ in $\rho_i$. When there is an offset, the rate $\rho_i$ will be corrected for extent or time of exposure. For example, Bedrick *et al.* (1996) consider number of fish caught as Poisson with mean $\rho_i t_i$ where $t_i$ is hours spent fishing, and $\rho_i$ is then average number of fish caught per hour. In epidemiological applications, the exposure is often expected disease events $E_i$, and with an internal standard (Schoenbach and Rosamond, 2000, Ch. 6), the sum of actual and expected events are equal, $\sum_i y_i = \sum_i E_i$. Then the Poisson mean will be $\rho_i E_i$, where $\rho_i$ is a measure of relative risk (with average 1) in different units $i$, such as patient categories or areas.

Just as priors in binomial regression may be expressed via notional success rates, prior beliefs on Poisson regression effects may be easier to express in terms of relative risk. Thus in an epidemiological application, one may have a gamma prior on the relative risk $\tilde{\rho}_r$ at selected values $\tilde{x}_r$ in the range of $x$. In the gamma prior

$$\text{Gamma}(y|a, b) \propto y^{a-1} \exp[-by]$$

with mean $a/b$ and variance $a/b^2$ the second parameter can be given a prior sample size interpretation. So the CMP priors have the form $\tilde{\rho}_r \sim \text{Gamma}(\tilde{n}_r\tilde{\kappa}_r, \tilde{n}_r)$ where $\tilde{\kappa}_r$ is the elicited mean relative risk at $\tilde{x}_r$.

One replaces the conventional (e.g. normal) prior on $\beta$ with $R$ gamma priors on $\tilde{\rho}_r$, leading to an induced prior on $\beta$. With inverse link $h$, one has

$$\pi(\beta) \propto \prod_r \{h(\tilde{x}_r\beta)\}^{\tilde{n}_r\tilde{\kappa}_r-1} \exp[-\tilde{n}_r h(\tilde{x}_r\beta)] \left|\frac{\partial h(\tilde{x}_r\beta)}{\partial \beta}\right|.$$

With $g$ as the log link, and $h$ an exponential, the Jacobean is proportional to $h(\tilde{x}_r\beta)$ and

$$\pi(\beta) \propto \prod_r \{h(\tilde{x}_r\beta)\}^{\tilde{n}_r\tilde{\kappa}_r} \exp[-\tilde{n}_r h(\tilde{x}_r\beta)],$$

and so $\pi(\beta)$ is conjugate with the Poisson likelihood.

For routine MCMC implementation the induced priors are obtained by solving the $g(\tilde{\rho}_r)$ for $\beta$. Suppose $R = 2$, with predictor vector $(1, x_i)$ where $x_i$ is a standardised metric variable, and that $\tilde{x}_1 = (1, x_L) = (1, -1)$ and $\tilde{x}_2 = (1, x_U) = (1, 1)$. The CMP prior is expressed as

$$\tilde{\rho}_1 \sim \text{Gamma}(a_1, b_1),$$

$$\tilde{\rho}_2 \sim \text{Gamma}(a_2, b_2),$$

$$\log(\tilde{\rho}_1) = \beta_1 + \beta_2 x_L = \beta_1 - \beta_2,$$

$$\log(\tilde{\rho}_2) = \beta_1 + \beta_2 x_U = \beta_1 + \beta_2.$$

At each MCMC iteration, one obtains the induced regression coefficients as

$$\beta_2 = [\log(\tilde{\rho}_2) - \log(\tilde{\rho}_1)]/2,$$

$$\beta_1 = [\log(\tilde{\rho}_2) + \log(\tilde{\rho}_1)]/2.$$

Consider a positive risk factor for a disease outcome, such as area percent agricultural workers $x$ (standardised) for area cancer incidence totals $y$, and that $\tilde{\rho}_1 \sim \text{Gamma}(1.5, 1)$ at $x = 1$, and $\tilde{\rho}_2 \sim \text{Gamma}(0.5, 1)$ at $x = -1$ (i.e. $\tilde{n}_r = 1, r = 1, 2$). In terms of BUGS, one has

```
model { for (i in 1:N) { y[i] ~dpois(mu[i])
log(mu[i]) <- beta[1]+ beta[2]*x[i]}
# conventional prior
# for (j in 1:2) {beta[j] ~dnorm(0,0.001)}}
# CMP prior
rho.AD.1 ~dgamma(1.5,1); rho.AD.2 ~dgamma(0.5,1);
beta[1] <- (log(rho.AD.1)+log(rho.AD.2))/2
beta[2] <- (log(rho.AD.2)-log(rho.AD.1))/2}
```

Predictive assessments for Poisson regression involve sampling replicates $y_{rep,i} \sim \text{Po}(\rho_i)$ or $y_{rep,i} \sim \text{Po}(t_i \rho_i)$ and comparing these with the observations, either directly or via posterior predictive tests using discrepancy measures. The relevant probabilities are $\Pr(y_{rep,i} > y_i|y) + 0.5\Pr(y_{rep,i} = y_i|y)$, and both low and high probabilities indicate poor fit. Overdispersion is commonly present in count data, leading on to more specialized regression methods (see Chapter 4). Predictive checks for overdispersion in Poisson regression may extend to posterior predictive $p$-tests using criteria such as those in Dean and Lawless (1989) as discrepancy functions, evaluated both for actual and replicate data. For example, one possibility involves comparing $D = 0.5 \sum_{i=1}^{n}\{(y_i - \rho_i)^2 - y_i\}$ with its counterpart

$$D_{rep} = 0.5 \sum_{i=1}^{n}\{(y_{rep,i} - \rho_i)^2 - y_{rep,i}\}$$

based on replicate data (see Exercise 3.3).

## Example 3.2    Student attainment data

To illustrate assessment of a binary logistic regression, consider pass-fail outcomes $y_i$ for $n = 30$ students in relation to centred SAT-Math (SAT-M) scores (Johnson and Albert, 1999, p. 77). To assess any poorly fitted cases, predictive probabilities of correct classification $\Pr(y_{rep,i} = y_i|y)$, and Monte Carlo estimates of conditional predictive ordinates $p(y_{rep,i}|y_{[i]})$ are obtained. To demonstrate assessment of classification accuracy, a set of $D = 11$ thresholds $\eta_d^T$ are set, from $\eta_1^T = -1$ through to $\eta_{11}^T = 1$ at intervals of 0.2, and aggregates $n_{11d}$ and

$n_{00d}$ (totals correctly classified as positive and as negative respectively) obtained at each threshold. A separate code, with explicit omission of each data point, and 30 separate model estimations, is used to obtain cross-validatory predictive probabilities $\Pr(y_{rep,i} = y_i|y_{[i]})$ from first principles.

Using the last 9000 iterations from two chain runs of 10 000 iterations[2], Monte Carlo estimates of the conditional predictive ordinates $p(y_{rep,i}|y_{[i]})$, are found to be virtually identical with $\Pr(y_{rep,i} = y_i|y_{[i]})$ obtained from first principles, with a correlation over all cases of 0.99996. The CPO estimates are obtained as the exponential of the negative logarithms of the posterior mean of the inverse likelihoods, namely G[i] as in the code fragment:

```
model { for (i in 1:n) {y[i] ~dbern(p[i])
LL[i] <- y[i]*log(p[i])+(1-y[i])*log(1-p[i])
G[i] <- 1/exp(LL[i]).
```

For the first five subjects, the CPO estimates are (0.6579,0.5476,0.5525,0.0270,0.9234), as compared to (0.6635,0.5494,0.5504,0.0257,0.9239) from first principles. These two cross-validatory measures of residual status provide lower probabilities than the complete data predictive checks $\Pr(y_{rep,i} = y_i|y)$. However, total subjects with low predictive probabilities (e.g. under 0.05 or under 0.10) are similar for the cross-validatory criteria as against complete data criteria.

As to establishing a classification threshold, for higher values of $\eta_d^T$, sensitivity declines while specificity increases, as expected. A combined measure of predictive accuracy is simply the total of the sensitivity and specificity (Böhning et al., 2008) and on this basis, the highest level of accuracy (namely 1.195) is for $\eta_4^T = -0.4$.

## 3.4  Augmented data regression

MCMC sampling of binary, ordinal and multinomial regression models may be simplified by considering latent data $z$ (e.g. utilities, frailties) underlying the observed $y$ (Albert and Chib, 1993). Introducing augmented data may also assist in predictive assessment and predictor selection. For example, the probit regression for binary data specifies $\Pr(y_i = 1|X_i, \beta) = \Phi(X_i\beta)$ where $\Phi$ is the cumulative distribution function for a standard normal. With a prior $\pi(\beta) \sim N_P(b_0, V_0)$ on the regression parameters, the posterior density has the form

$$p(\beta|y, X) \propto p(y|X, \beta)\pi(\beta) = N_P(b_0, V_0)\prod_{i=1}^{n}\{\Phi(X_i\beta)\}^{y_i}\{1 - \Phi(X_i\beta)\}^{1-y_i}.$$

This does not reduce to a convenient density form, so that sampling $\beta$ involves Metropolis–Hastings methods. However, consider the equivalent linear regression

$$z_i = X_i\beta + u_i,$$

where $y_i = I(z_i > 0)$ and $u_i \sim N(0, 1)$. Then

$$\Pr(y_i = 1|X_i, \beta) = \Pr(z_i > 0|X_i, \beta) = \Pr(u_i > -X_i\beta) = 1 - \Phi(-X_i\beta) = \Phi(X_i\beta).$$

This reformulation leads to simpler Gibbs sampling, namely alternate samples from $p(\beta^{(t)}|Z^{(t)}, y)$, and truncated normal samples from $p(Z^{(t)}|\beta^{(t)}, y)$, with the form of truncation depending whether $y$ is 1 or 0. The complete conditional for $\beta$ corresponds to that for normal linear regression with $Z$ regarded as responses, namely

$$p(\beta|Z,y) \sim N(b_n, V_n),$$

where $V_n = (X'X + V_0^{-1})^{-1}, d_n = X'Z + V_0^{-1}b_0$, and $b_n = d_n V_n$. For $y_i = 1$, drawing from $p(Z^{(t)}|\beta^{(t)}, y)$ involves sampling positive $Z_i$ in a normal regression with mean $X_i\beta$ and variance 1. If $y_i = 0$, $z_i$ is sampled from the same normal density but constrained to be negative. So

$$z_i \sim N(X_i\beta, 1) \qquad I(0, \infty) \qquad if \qquad y_i = 1$$
$$z_i \sim N(X_i\beta, 1) \qquad I(-\infty, 0) \qquad if \qquad y_i = 0.$$

Slow mixing may result from posterior correlation between $Z$ and $\beta$. An alternative MCMC estimation strategy for augmented data probit regression uses the composition sampler of Holmes and Held (2006). This involves sampling $\beta$ and $Z$ jointly, with $\beta$ sampled conditional on $Z$, but $Z$ sampled directly from its marginal distribution (integrating over $\beta$).

Alternative forms of sampling $Z$ can be adopted for other links. The logit link is achieved via

$$z_i \sim logistic(X_i\beta, 1), \qquad I(0, \infty) \qquad y_i = 1$$
$$z_i \sim logistic(X_i\beta, 1), \qquad I(-\infty, 0) \qquad y_i = 0$$

where the logistic density logistic$(\mu, \tau)$ with mean $\mu$ and scale parameter $\tau$ is

$$p(x|\tau, \mu) = \tau \exp(\tau[x - \mu])/\{1 + \exp(\tau[x - \mu])\}^2$$

with variance $\kappa^2/\tau^2$ where $\kappa^2 = \pi^2/3$. Note that the standard logistic density with mean 0 and variance 1 has the form

$$p(x) = \kappa \exp(\kappa x)/\{1 + \exp(\kappa x)\}^2.$$

Alternatively the logit link may be approximated by sampling $z_i$ from a Student $t$ with 8 degrees of freedom (Albert and Chib, 1993). This can be implemented using constrained normal sampling, but with a scale mixture element, as in:

$$z_i \sim N(X_i\beta, 1/\lambda_i) \qquad I(0, \infty) \qquad y_i = 1$$
$$z_i \sim N(X_i\beta, 1/\lambda_i) \qquad I(-\infty, 0) \qquad y_i = 0$$
$$\lambda_i \sim Ga(\nu/2, \nu/2)$$

with $\nu = 8$. Another possible augmentation strategy for logit regression involves a mixture of normal error distributions,

$$z_i = X_i\beta + m_{G_i} + u_i,$$

with $y_i = I(z_i > 0)$ and $u_i \sim N(0, V_{G_i})$. The normal mixture of errors approximates the type I extreme value density, and in Frühwirth-Schnatter and Frühwirth (2007), $G_i \sim Mult(\pi[1, \ldots, 10])$ where $\pi, m[1, \ldots, 10]$ and $V[1, \ldots, 10]$ are known constants.

In BUGS one may use the dbern.aux function to implement augmented binary regression (Lunn et al., 2006, 2009) (see Exercise 3.4). For example to fit a logistic model the code would be

```
model {for (i in 1:n) {y[i] ~dbern.aux(z[i]); z[i] ~dlogis(eta[i], 1)
eta[i] <- beta[1]+beta[2]*x[i,1]+beta[3]*x[i,2]+...+beta[p]*x[i,p]}
for (j in 1:p) {beta[j] ~dnorm(0,0.001)}}.
```

Alternatively, one can define lower/upper sampling limits $\{A[i],B[i]\}$ for each case according to whether $y = 1$ or 0, with limits set at extreme density points with effectively zero support beyond them; for example, $A[i] = -10$, and $B[i] = 0$ when $y = 0$, and $A[i] = 0$, $B[i] = 10$ when $y = 1$. Then for logistic regression

```
model {for (i in 1:n) {z[i] ~dlogis(eta[i], 1) I(A[i],B[i])
eta[i] <- beta[1]+beta[2]*x[i,1]+beta[3]*x[i,2]+...+beta[p]*x[i,p]}
for (j in 1:p) {beta[j] ~dnorm(0,0.001)}}.
```

A potential benefit of the augmented data approach for model checking is that the residuals $z_i - X_i\beta$ are nominally a random sample from the underlying cumulative distribution, such as $\Phi$ for the probit (Johnson and Albert, 1999), so providing a mechanism to assess poorly fitted cases. Thus for probit regression, the realised residuals

$$\varepsilon_i = z_i - X_i\beta,$$

are approximately $N(0,1)$ if the model is appropriate, whereas a posterior distribution of $\varepsilon_i$ significantly different from $N(0,1)$ implies conflict with the observed $y_i$. So one may compare the posterior probability $\{\Pr(|\varepsilon_i||y) > 1.96\}$ with its prior value under $\Phi$, namely 0.05. For the augmented data logit, one monitors

$$\Pr(|\varepsilon_i|/\kappa^{0.5}) > 2.02,$$

where $\kappa = \pi/\sqrt{3}$, and $2.02\kappa^{0.5}$ is the 0.975 point of the cumulative standard logit, while for the scale mixture approximation to the logit, one monitors

$$\Pr(|\varepsilon_i|\lambda_i^{\,0.5}) > 1.96.$$

The regression coefficients from binary regression models may be difficult to interpret as they measure the change in the latent $Z$ associated with a change in a predictor. An alternative is provided by marginal effects (or marginal probability effects), or partial effects of each predictor on the probability $\Pr(y_i = 1)$, namely

$$ME_{ji} = \frac{\partial \Pr(y_i = 1)}{\partial x_{ji}} = \frac{\partial F(\eta_i)}{\partial x_{ji}}$$

where $F$ is the cumulative density function (e.g. normal, logistic) defining $\Pr(y = 1)$. For probit regression, where $F = \Phi$, with $x_{ji}$ included linearly in $\eta_i = X_i\beta$, one has

$$ME_{ji} = \beta_j\phi(\eta_i),$$

where $\phi$ is the standard normal density. For logit regression the marginal effect is $ME_{ji} = \beta_j\frac{\exp(\eta_i)}{[1+\exp(\eta_i)]^2}$. Of interest for interpreting predictor impacts are marginal effects at reference values of $x_j$ such as $x_{ji} = 1$ when $x$ is binary, or at the mean of $x_j$ when $x_j$ is continuous. Letting $\eta_s$ denote the regression term at these selected values, then the probit marginal effect is $ME_j = \beta_j\phi(\eta_s)$. The densities of these effects can be estimated from MCMC sampling, avoiding the need for delta approximations (Anderson and Newell, 2003).

## 3.5   Predictor subset choice

Model uncertainty in regression analysis may involve different aspects of model specification, such as the error structure of the residuals (e.g. normal vs Student $t$ in linear regression), whether transformations should be applied to predictors and/or response, and whether there are non-linear regression effects. However, an over-riding consideration in regression concerns choice of the best subset of predictors. While approaches based on marginal likelihood comparison, penalised fit criteria (e.g. BIC or DIC), or using predictive (cross-validatory) criteria have been successfully applied to regression model selection, distinctive selection indicator methods have been proposed for Bayesian predictor choice. One of the major motivations for predictor selection occurs if there is collinearity between predictors, $\{x_1, x_2, \ldots, x_P\}$. Taken as a single predictor, the coefficient $\beta_j$ for predictor $x_j$ may have a clearly defined posterior density in line with subject matter knowledge (e.g. a 95% credible interval confined to positive values, assuming $x_j$ was expected to have a positive effect on $y$). However, with several predictors operating together, coefficients on particular $x_j$ may be reduced to 'insignificance' (in terms of a 95% interval neither clearly positive nor negative), or even take signs opposite to substantively based expectations.

Choice among $P$ predictors to reduce such parameter instability leads to regression selection schemes for including or excluding each predictor, involving prior probabilities $r_j = \Pr(\gamma_j = 1)$ for binary inclusion indicators $\gamma_j$. An alternative are priors including a penalty (e.g. an $L_1$ norm) designed to shrink unnecessary regression effects towards zero, which do not necessarily include inclusion indicators (Lykou and Ntzoufras, 2013). Under binary inclusion, the most common choice is

$$\gamma_j \sim \text{Bern}(0.5),$$

implying $2^P$ equally probable models, and that about half the variables are to be retained a priori (O'Hara and Sillanpää, 2009, p. 88). Posterior marginal retention probabilities $\Pr(\gamma_j = 1|y)$ are estimated by the proportion of MCMC iterations when $\gamma_j = 1$, while posterior model probabilities are based on the sampled frequency of different combinations of retained predictors. A Bayes factor for marginal retention can be obtained by comparing the posterior retention odds $\Pr(\gamma_j = 1|y)/\Pr(\gamma_j = 0|y)$ with the prior odds $\Pr(\gamma_j = 1)/\Pr(\gamma_j = 0)$. Interest generally lies with predictors having posterior retention probabilities exceeding their prior probability. For example, assuming $\Pr(\gamma_j = 1) = 0.5$, Barbieri and Berger (2004) define the median probability model as that defined by predictors with posterior inclusion probabilities exceeding 0.5 (equivalent to the posterior median for $\gamma_j$ being 1) (see also Farcomeni, 2010).

George and McCulloch (1993) propose a stochastic search variable selection scheme (SVSS), whereby $\beta_j$ has a vague prior centred at zero (or some other value) when $\gamma_j = 1$, but when $\gamma_j = 0$ is selected, the prior is centred at zero with high precision (i.e. $\beta_j$ is zero for all practical purposes). So for $\gamma_j = 1$, one might set a relatively large prior variance $V_j$ for the $j^{th}$ coefficient, but multiply this by a small constant $0 < c < 1$ (so the implied impact of $x_j$ on $y$ is limited) when $\gamma_j = 0$. Ntzoufras (2002) recommends the value $c = 0.001$, while George and McCulloch (1997, p. 344) recommend $c > 0.0001$. This leads to a mixture prior with $\pi(\beta_j, \gamma_j) = \pi(\beta_j|\gamma_j)\pi(\gamma_j)$, where

$$\gamma_j \sim \text{Bern}(r_j),$$
$$\pi(\beta_j|\gamma_j) = (1 - \gamma_j)N(0, c\tau_j^2) + \gamma_j N(0, \tau_j^2).$$

The Bernoulli parameters $r_j$ can be preset, for example, to a relatively low value such as 0.25 (favouring parsimonious models), or to 0.5, the 'indifference' value. George and McCulloch (1997) recommend establishing thresholds of practical significance, $\Delta_j = \Delta y/\Delta x_j$ where $\Delta y$ indicates a small change in $y$ corresponding to a relatively large shift in $x_j$. If $\beta_j < |\Delta_j|$ then the linear effect of $x_j$ on $y$ is negligible. One would then choose the prior variance $\tau_j^2$ and $c_j$ so that

$$\Delta_j^2 = \log\left(\frac{1}{c_j}\right)\left(\frac{1}{c_j\tau_j^2} - \frac{1}{\tau_j^2}\right).$$

Alternatively independent priors $\pi(\beta_j, \gamma_j) = \pi(\beta_j)\pi(\gamma_j)$ for binary retention indicators $\gamma_j \sim \text{Bern}(r_j)$, and coefficients $\beta_j$, may be adopted (Kuo and Mallick, 1998). Thus the $\beta_j$ are typically taken as normal

$$\beta_j \sim N(0, V_j)$$

with $V_j$ large enough to ensure full coverage of the parameter space, but with priors not highly diffuse as this tends to favour overly parsimonious models (an example of Lindley's paradox). Inferences on regression coefficients are based on monitoring the 'realised' coefficients

$$\delta_j = \gamma_j\beta_j.$$

Let $\delta_j^*$ denote $\delta = (\delta_1, \ldots, \delta_P)$ assuming $\gamma_j = 1$, and so with $j^{th}$ entry $\beta_j$, and also denote $\varepsilon_i^* = y_i - X_i\delta_j^*$. Similarly let $\delta_j^{**}$ denote $\delta$ but with $j^{th}$ entry 0, with $\varepsilon_i^{**} = y_i - X_i\delta_j^{**}$. Defining $c_j = r_j \exp\left[-0.5 \sum_{i=1}^{n} (\varepsilon_i^*)^2/\sigma^2\right]$ and $d_j = (1 - r_j)\exp\left[-0.5 \sum_{i=1}^{n} (\varepsilon_i^{**})^2/\sigma^2\right]$, the conditional posterior of $\gamma_j$ is Bernoulli with updated retention probability

$$r_j^u = c_j/(c_j + d_j).$$

For normal linear regression with both $y$ and the $x$ variables standardised, the $\beta_j$ are standardised regression coefficients, and prior expectations about plausible values for $\beta_j$ are more easily obtained. For example, standardised regression coefficients exceeding unity (a change of one standard deviation in a predictor produces an equivalent change in $y$) may be unusual. Kuo and Mallick suggest a prior variance for standardised regression coefficients $\beta_j$ as

$$V_j = k^2, \quad k \in [0.5, 4].$$

If the response is not standardised but the $x$ variables are, then this setting becomes

$$V_j = s_y^2 k^2, k \in [0.5, 4],$$

where $s_y^2$ is the variance of the response data.

Dellaportas et al. (2002) develop a Gibbs variable sampling (GVS) method designed to avoid sampling $\beta_j$ from too vague a prior when $\gamma_j = 0$. The conditional prior on $\beta_j$ given $\gamma_j$ is

$$P(\beta_j|\gamma_j) = \gamma_j N(0, V_j) + (1 - \gamma_j)N(b_{0j}, B_{0j})$$

where $V_j$ is set large to allow unrestricted parameter search, but $\{b_{0j}, B_{0j}\}$ are pseudo prior values (usually recommended to approximate the posterior mode) obtained from a pilot run or frequentist analysis.

The full RJMCMC method for regression selection (e.g. Ntzoufras *et al.*, 1999) involves a Metropolis–Hastings algorithm that may switch between models as well as regression parameters. Consider models $M_k = (k, \beta_k)$ and $M_{k'} = (k', \beta_{k'})$ with differing dimensions $k$ and $k'$. The algorithm jumps between such models to obtain samples from the joint model space distribution $p(k, \beta_k | y)$, and is reversible in the sense of ensuring detailed balance in an irreducible aperiodic chain that converges to the correct target density. Assume the current location of the Markov chain is $M_k = (k, \beta_k)$, and define the total model density as

$$h_k = p(y|k, \beta_k)\pi(\beta_k|k)\pi(k),$$

with ratio of new and existing model densities as

$$H(k, k') = h(k')/h(k).$$

Then one possible RJMCMC algorithm update proceeds with the steps:

1. propose switching to model $M_{k'}$ with probability $J(k, k')$;

2. sample $b$ from a proposal density $q(b|\beta_k, k, k')$;

3. set $(\beta_{k'}, b') = g_{k,k'}(\beta_k, b)$, where $g_{k,k'}()$ is a bijection between $(b, \beta_k)$ and $(b', \beta_{k'})$, and where $(b, b')$ are defined to match the dimensions of both vectors;

4. accept the new model $(k', \beta_{k'})$ with probability $\min(1, H(k, k')Q(k, k'))$ where

$$Q(k, k') = \frac{J(k', k)}{J(k, k')} \frac{q(b'|\beta_{k'}, k', k)}{q(b|\beta_k, k, k')} \frac{\partial g_{k,k'}(\beta_k, b)}{\partial(\beta_k, b)}.$$

Repeated application of the algorithm will generate a sample of model indicators $k^{(t)} \in (1, \ldots, K)$, and the posterior probability of model $k^*$, $\Pr(k^*|y)$ is estimated by $\frac{1}{T}\sum I(k^{(t)} = k^*)$.

In WinBUGS, RJMCMC for normal linear regression can be implemented using the jump.lin.pred and jump.model.id commands (Lunn *et al.*, 2006, 2009). Assuming $P$ predictors apart from the intercept, the following code includes a prior on the number of predictors (P.ret) to be included, on the residual precision (tau), and on the regression coefficient precision, tau.beta (with implicit prior means of zero on the coefficients):

```
model { for (i in 1:n) { y[i] ~dnorm(eta[i], tau) }
eta[1:n] <- jump.lin.pred(X[1:n, 1:P], P.ret, tau.beta)
# Under Inference/samples, enter model.id
# Then select jump/summarize and type model.id in "id node"
model.id <- jump.model.id(eta[1:n])
tau.beta <- tau/k; k ~dexp(1000)
tau ~dgamma(1, 0.001); P.ret ~dbin(0.5, P)}
```

Constrained selection priors can be used when there are hierarchical elements in inclusion. For example, a power (squared, cubed, etc.) in a predictor $x_j$ would not usually be included if $x_j$ itself were not included. A systematic approach to such situations, based on an extended SVSS method (Farcomeni, 2010), focuses on covariate groups defined by parent covariates (e.g. $x_j$), and by subordinate covariates (e.g. $x_j^2$) only included if one or more of the parent covariates are included. Thus in the linear model

$$y = \beta_0 + \beta_1 x_1 + \beta_2 x_1^2 + \beta_3 x_2 + \beta_4 x_2^2 + \beta_5 x_1 x_2,$$

there are two hierarchical predictor groups $\{x_1, x_1^2\}$, and $\{x_2, x_2^2\}$, and an additional constraint crossing groups, such that $x_1 x_2$ is included only when both $x_1$ and $x_2$ is included. Let $\omega_j \sim$ Bern$(\pi_\omega)$ be base binary indicators. Then the ultimate retention indicators $\gamma_j$ for subordinate covariates are products of their own and parent base indicators. For the preceding example

$$\eta_i = \beta_0 + \gamma_1 \beta_1 x_{1i} + \gamma_2 \beta_2 x_{1i}^2 + \gamma_3 \beta_3 x_{2i} + \gamma_4 \beta_4 x_{2i}^2 + \gamma_5 \beta_5 x_{1i} x_{2i},$$

$$\gamma_1 = \omega_1; \gamma_2 = \omega_2 \omega_1; \gamma_3 = \omega_3; \gamma_4 = \omega_4 \omega_3; \gamma_5 = \omega_5 \omega_1 \omega_3;$$

$$\omega_j \sim \text{Bern}(\pi_\omega),$$

$$P(\beta_j | \gamma_j) = (1 - \gamma_j) N(0, c\tau_j^2) + \gamma_j N(0, \tau_j^2).$$

There may be benefits in focusing on prior model probabilities (called a model choice focus in Example 3.3), rather than on prior probabilities of retention for individual predictors under an all possible subsets approach. In certain applications it may be preferable to focus on a small number of plausible models (Heumann and Grenke, 2010). For example, with $P = 4$ predictors, the focus of interest might not be in comparing all 16 possible models but only two, say $(x_1, x_2, x_4)$ and $(x_1, x_2, x_3, x_4)$. So one sets $\gamma_{mj} = 1$ if model $m$ includes predictor $j$, and $\gamma_{mj} = 0$ otherwise; in the two models just mentioned, $\gamma_{11} = 1, \gamma_{12} = 1, \gamma_{13} = 0, \gamma_{14} = 1$ and $\gamma_{2j} = 1$ for all $j$. The regression term $\eta_i$ is then conditional on a sampled model indicator $M \sim \text{mult}(\pi_1, \ldots, \pi_K)$ where $K$ is the number of possible models, so that

$$(\eta_i | M = m) = \beta_0 + \gamma_{m1} \beta_1 x_{1i} + \cdots + \gamma_{mP} \beta_P x_{Pi},$$

where the $\beta_j$ could follow a $g$-prior or Lasso prior (see Sections 3.5.1 and 3.5.2). There is flexibility over the choice of model prior. For example, one may set equal prior model probabilities

$$\pi_k = \text{Pr}(M_k) = 1/K,$$

or choose to downweight models of larger dimension, as in

$$\pi_k \propto 1/(1 + \kappa)^{P_k}$$

where $\kappa > 0$, and $P_k$ is the dimension of $M_k$.

The above selection techniques extend to general linear models. For example, suppose a Poisson density with mean $\mu_i$ is assumed for count outcome $y_i$, so that if all predictors (assumed standardised) are included in a log link, one has

$$\log(\mu_i) = \xi_i = \beta_0 + \beta_1 x_{1i} + \beta_2 x_{2i} + \cdots + \beta_P x_{Pi}.$$

Under predictor selection, binary variables $\gamma_j$ are introduced relating to retention of predictors, namely

$$\log(\mu_i) = \xi_i = \beta_0 + \gamma_1 \beta_1 x_{1i} + \gamma_2 \beta_2 x_{2i} + \cdots + \gamma_P \beta_P x_{Pi}.$$

To provide suitable priors for $\beta_j$ under selection, one might take $s_\xi^2 = var(\xi)$ as analogous to the variance of the response under normal linear regression. Adapting Kuo and Mallick (1998), the prior variance of $\beta_j$, would then be

$$V_j = s_\xi^2 k^2, \quad k \in [0.5, 4].$$

### 3.5.1  The *g*-prior approach

Results of regression model selection depend on the priors adopted for regression coefficients, with diffuse (non-informative) priors generally not recommended. Information on the predictors (though not responses) is made use of under the g-prior approach, originally proposed for linear regression with a metric outcome $y$, usually assuming normal errors, and known predictors (Fernandez *et al.*, 2001; Liang *et al.*, 2008). With $X_\gamma$ denoting the $n \times P_\gamma$ predictor matrix under model $M_\gamma$ (defined by retaining the subset of $P_\gamma$ predictors with $\gamma_j = 1$), the g-prior[3] specifies

$$\beta_\gamma | g, \tau, \gamma \sim N\left(0, \frac{g}{\tau}[X'_\gamma X_\gamma]^{-1}\right),$$

where $\tau = 1/\sigma^2$ is the residual precision, and $g > 0$ can be interpreted as an inverse prior sample size (Bové and Held, 2011). Larger $g$ values generally result in selection of less complex models. This is because the precision matrix for the regression coefficients has the form $\tau X'_\gamma X_\gamma / g$ so that larger values of $g$ lead to greater downweighting of the information $\tau X'_\gamma X_\gamma$. Denote the posterior mean for the coefficient of determination as $R^2$. With a flat prior on $\beta_0$, and the prior

$$\pi(\tau) = 1/\tau,$$

on the precision, one may obtain a Bayes factor

$$\text{BF} = (1 + g)^{(n-P_\gamma-1)/2}[1 + g(1 - R^2)]^{-(n-1)/2},$$

relative to the null (intercept only) model with $\mu_i = \beta_0$ (Liang *et al.*, 2008, Section 2.1), showing directly that larger values of $g$ are a penalty on model dimension.

The setting or prior adopted for $g$ is therefore important for model choice. The value of $g$ may be preset: the setting $g = n$ defines the unit information prior (Kass and Wasserman, 1995), while Fernandez *et al.* (2001) suggest $g = \max(n, P^2)$. However, Liang *et al.* (2008) suggest that $g$ be assigned a prior in order to avoid the so-called information paradox obtained when $g$ is fixed, since the Bayes factor then tends to a constant value as $R^2 \to 1$. Among the options are

$$\frac{g}{1 + g} \sim \text{Be}\left(1, \frac{a}{2} - 1\right)$$

where values $2 < a \leq 4$ are proposed as reasonable. A positive valued prior (e.g. inverse gamma) on $1 + g$ can also be assumed (Cui and George, 2008, p. 891). The g-prior approach can be combined with predictor (and model) selection with $\beta_j = 0$ when $\gamma_j = 0$ (i.e. in models excluding $x_j$) (Fernandez *et al.*, 2001, Section 2).

The g-prior approach may be adapted to Poisson or binomial regression when there is overdispersion modelled using random effects. For example, a logistic model for overdispersed binomial data, and a g-prior on the regression coefficients, would take the form

$$y_i \sim \text{Bin}(n_i, \pi_i)$$

$$\text{logit}(\pi_i) = \beta_0 + \gamma_1\beta_1 x_{1i} + \ldots + \gamma_P\beta_P x_{Pi} + u_i,$$

$$u_i \sim N(0, 1/\tau)$$

$$\beta_\gamma | \tau, g, \gamma \sim N\left(0, \frac{g}{\tau}[X'_\gamma X_\gamma]^{-1}\right).$$

Similarly for binary responses, a logit or probit regression implemented by augmented data sampling has an implicit error variance of 1 (Yang and Song, 2010). This leads to the $g$-prior

$$\beta_\gamma | g, \gamma \sim N(0, g[X'_\gamma X_\gamma]^{-1}).$$

A general scheme for hyper-$g$ priors for generalized linear models (Bové and Held, 2011), adapts the usual $g$-prior in terms of a constant $c$ and a diagonal matrix $W = diag(w)$ of the weights $w_i$ that define the exponential prior. Thus

$$\beta_\gamma | g, \tau, \gamma \sim N\left(0, \frac{gc}{\tau}[X'_\gamma W X_\gamma]^{-1}\right),$$

where $c = 1$ except for binomial/Bernoulli regression, where $c$ depends on the link assumed. Choice of the prior for $g$ is left open, though options

$$\pi(g) = IG(0.5, 0.5n);$$

$$\pi(g) = \frac{1}{n}\left(1 + \frac{g}{n}\right)^{-2};$$

$$\pi(g) = IG(0.001, 0.001)$$

are compared in an application (Bové and Held, 2011, Section 4).

Variants on the $g$-prior scheme for normal linear regression are to standardise the predictors (i.e. take account of the information on the diagonal of $X'X$), but scale the residual variance $\sigma^2$ by a factor $\lambda^2$ in a conditional prior $\pi(\beta_j | \sigma^2)$ for $\beta_j$. Thus

$$\beta_j | \sigma^2 \sim N(0, \lambda^2 \sigma^2),$$

with Meyer and Wilkinson (1998) recommending values of $\lambda$ between 0.5 and 3, with retention indicators $\gamma_j \sim Bern(r_j)$, and $r_j$ either preset or taking a beta prior. This prior is adopted by Woodward and Walley (2009) for variable selection involving categorical predictors (sometimes called categorical factors). Thus for a predictor $C_i$ with $K$ possible categories, one specifies an exchangeable prior on the coefficients for each category, namely

$$\eta_i = \beta_0 + \delta_1 I(C_i = 1) + \delta_2 I(C_i = 2) + \ldots + \delta_K I(C_i = K),$$

$$\delta_k = \gamma_k \beta_k,$$

$$\gamma_k \sim Bern(r_k),$$

$$\beta_k \sim N(0, 2\lambda^2 \sigma^2),$$

with the multiplier 2 in the preceding following from equivalent representation of the $K = 2$ case by dummy variables taking values $-1$ and $+1$. This scheme implies that particular categories of the predictor could be included while others are excluded, that is $C_i$ is partially included. An alternative strategy (e.g. Wang and George, 2007) is that a categorical variable is either completely included or completely excluded: if the retention indicator for any category of $C$ is zero, then $C$ is entirely excluded as a predictor, as under constrained variable selection (Farcomini, 2010).

### 3.5.2    Hierarchical lasso prior methods

Hierarchical prior methods for regression coefficients can be applied to modify the above described selection schemes. Such priors may take unconditional or conditional priors on the $\beta_j$, as in

$$y = X\beta + \varepsilon, \qquad \varepsilon \sim N(0, \sigma^2), \qquad \beta_j \sim N(0, \tau^2),$$

where $\tau^2$ is an additional unknown (i.e. Bayesian ridge regression), or

$$\beta_j|\sigma^2 \sim N(0, \lambda^2\sigma^2),$$

where $\lambda^2$ is an additional unknown. These priors may or may not be combined with retention indicators and associated priors. The posterior in Bayesian ridge regression is

$$p(\beta|y) \propto \exp\left[-\frac{1}{2\sigma^2}(y - X\beta)'(y - X\beta)\right] \exp\left(-\frac{1}{2\tau^2}\beta'\beta\right)$$

and maximising this is equivalent to the minimising the penalised criterion

$$(y - X\beta)'(y - X\beta) + \lambda\beta'\beta,$$

with noise to signal ratio $\lambda = \sigma^2/\tau^2$.

A different penalty is implied by the Lasso method for linear regression models (Park and Casella, 2008; Tibshirani, 2011; Lykou and Ntzoufras, 2013). For $y$ values centred to have mean zero, and standardised $x$ variables, estimates of $\beta$ are obtained by imposing an $L_1$ norm: let $T(\beta) = \sum_{j=1}^{P} |\beta_j|$, then $\sum (y_i - \mu_i)^2$ is minimised subject to $T(\beta) \le t_\beta$. Lower values of $t_\beta$ lead to greater shrinkage, while above a certain threshold for $t_\beta$, there is essentially equivalence to the usual linear regression estimates (Efron *et al.*, 2004).

Lasso estimates can be interpreted as the posterior mode under a conditional Laplace (double exponential) prior distribution for the $\beta_j$. The conditional prior (Park and Casella, 2008) with parameter $\lambda/\sigma$, has the form

$$\pi(\beta_j|\lambda, \sigma^2) = \frac{\lambda}{2\sigma} \exp[-\lambda|\beta_j|/\sigma].$$

so that the prior variance of $\beta_j$ is $2\sigma^2/\lambda^2$. The parameter $\lambda$ can be preset or taken as an extra unknown, with smaller values of $\lambda$ implying more parsimonious models (as the variance of $\beta_j$ is raised). This prior may be combined with selection indicators (Lykou and Ntzoufras, 2013)

$$\mu_i = \beta_0 + \gamma_1\beta_1 x_{1i} + \ldots + \gamma_P\beta_P x_{Pi} + u_i,$$

where $\gamma_j$ are binary variables. Alternatively one might estimate the effective probability of exclusion, for example by the proportion of iterations where $\beta_j \le b_0$ where $b_0$ is a small number.

The Laplace density can be represented as a scale mixture (Yuan and Lin, 2005; Yi and Xu, 2008), namely a zero-mean normal prior with an independent exponentially distributed variance. For the conditional Laplace prior one has

$$\beta_j \sim N(0, \sigma^2\tau_j),$$

$$\tau^2 \sim E\left(\frac{\lambda^2}{2}\right),$$

with $\lambda^2/2$ itself assigned a gamma prior such as $\frac{\lambda^2}{2} \sim Ga(0.1, 0.1)$ (Yi and Xu, 2008). More generally one may specify (Bae and Mallick, 2004)

$$\beta_j \sim N(0, \lambda_j),$$

with the exponential prior

$$\lambda_j \sim E(\kappa_j),$$

implying a Laplace prior for $\beta$. Alternatively inverse gamma priors can be used for $\lambda_j$. The hierarchical version has the benefit that each $\lambda_j$ can be assigned different $\kappa_j$, or one may assume $\kappa_j = \kappa$ and assign a preset value to $\kappa$.

## Example 3.3    Elementary school attainment

These data relate to the academic performance scores of $n = 400$ elementary schools sampled from the California Department of Education's API 2000 dataset. There are $P = 9$ potential predictors:

$x_1$: %English language learners

$x_2$: %pupils receiving free meals

$x_3$: year-round school (binary)

$x_4$: %pupils in 1st year in school (pupil turnover)

$x_5$: average class size in kindergarten through 3rd grade

$x_6$: average class size in kindergarten, 4th-6th grades

$x_7$: %of teachers with full credentials

$x_8$: %of teachers with emergency credentials

$x_9$: total number of students.

There are isolated missing values for $x_4, x_5$ and $x_6$ which are sampled assuming missingness at random. The predictors are standardised using the observed means and standard deviations, as the impact on the nominally unknown population means and standard deviations of the isolated missing values is very small. The performance scores are not standardised.

For an initial model without predictor selection, convergence is apparent before iteration 1000 and inferences are based on the second half of a two chain run of 5000 iterations. Checks on a normal linear regression including all predictors show no evidence of skewness in the residuals, or of kurtosis inconsistent with normality. These are assessed by monitoring $Sk = \frac{(\sum_i e_i^3/n)}{(\sigma^2)^{1.5}}$ and $Kt = \frac{(\sum_i e_i^4/n)}{(\sigma^2)^2}$ respectively. For example, the posterior 95% interval for $Kt$ is (2.27, 3.95). A Q-Q plot of the posterior means of $e_i = y_i - x_i\beta$ also shows no evidence of skewness in the tails. However, the effects of predictors $x_5, x_7, x_8$ and $x_9$ are inconclusive, in that the 95% credible intervals include zero (Table 3.1).

The Kuo–Mallick prior is applied with prior inclusion probabilities $\Pr(\gamma_j = 1) = 0.5$, and $V_j = s_y^2 k^2$, with $k = 0.5$ and $s_y = 142$. Hence the prior precisions on $\beta_j$ are set at $1/V_j = 1/71^2 \approx 0.0002$. This analysis (with inferences from the second half of a two chain run of 10 000 iterations in BUGS) produces five predictors with posterior inclusion probabilities $\Pr(\gamma_j = 1|y)$ exceeding 0.5 (namely $x_1, x_2, x_3, x_4, x_8$). The highest such probabilities are for $x_1$ and $x_2$, namely $\Pr(\gamma_1 = 1|y) = \Pr(\gamma_2 = 1|y) = 1$, though $x_3, x_4, x_6, x_7$, and $x_8$ also have posterior

**Table 3.1** Elementary school perfomance, parameter summary, regression without selection.

|            | Mean   | Standard deviation | 2.5%    | Median | 97.5%  |
|------------|--------|--------------------|---------|--------|--------|
| $\alpha$   | 648    | 3                  | 642     | 648    | 653    |
| $\beta_1$  | −23.37 | 5.01               | −33.14  | −23.32 | −13.30 |
| $\beta_2$  | −91.71 | 5.24               | −101.80 | −91.75 | −81.45 |
| $\beta_3$  | −8.22  | 3.76               | −15.46  | −8.24  | −0.85  |
| $\beta_4$  | −10.16 | 3.20               | −16.48  | −10.15 | −3.78  |
| $\beta_5$  | 2.12   | 3.10               | −4.09   | 2.17   | 8.06   |
| $\beta_6$  | 7.71   | 3.01               | 1.89    | 7.71   | 13.68  |
| $\beta_7$  | 8.87   | 6.58               | −3.27   | 8.78   | 21.85  |
| $\beta_8$  | −8.71  | 6.62               | −20.96  | −8.73  | 4.42   |
| $\beta_9$  | −2.18  | 3.71               | −9.53   | −2.14  | 5.05   |
| $\sigma^2$ | 3200   | 230                | 2779    | 3194   | 3672   |
| $K$        | 3.03   | 0.44               | 2.27    | 3.00   | 3.95   |
| $S$        | 0.09   | 0.16               | −0.22   | 0.09   | 0.40   |

retention probabilities exceeding 0.5. Only $x_1$ and $x_2$ have 95% credible intervals (assessed from the realised coefficients $\delta_j = \gamma_j \beta_j$) excluding zero, namely (−33,−12) with mean −23 for $x_1$, and (−107,−85) with mean −96 for $x_2$.

For the SSVS prior, the prior coefficient variances $V_j = s_y^2 k^2$, with $k = 0.5$ and $s_y = 142$ are retained for the case $\gamma_j = 1$. For $\gamma_j = 0$, the multiplier $c = 0.01$ is adopted so that $P(\beta_j | \gamma_j) = (1 − \gamma_j)N(0, 0.01V_j) + \gamma_j N(0, V_j)$. This option (with inferences from the second half of a two chain run of 10 000 iterations) produces five predictors with posterior inclusion probabilities $\Pr(\gamma_j = 1|y)$ exceeding 0.5 (namely $x_1, x_2, x_3, x_4, x_8$), with the highest such probabilities again for $x_1$ and $x_2$, namely $\Pr(\gamma_1 = 1|y) = \Pr(\gamma_2 = 1|y) = 1$.

The GVS prior uses posterior means and variances from regression without selection to provide pseudo-prior values $\{b_{0j}, B_{0j}\}$. As might be expected this produces a slightly less parsimonious section, in that seven predictors $\{x_1, x_2, x_3, x_4, x_6, x_7, x_8\}$ have posterior inclusion probabilities $\Pr(\gamma_j = 1|y)$ exceeding 0.5. The highest such probabilities are for $x_1$ and $x_2$, namely $\Pr(\gamma_1 = 1|y) = \Pr(\gamma_2 = 1|y) = 1$.

Two options for the shrinkage approach are applied. Both also include selection indicators with prior probabilities $\Pr(\gamma_j = 1) = 0.5$. The first takes Laplace priors on $\beta_j$ with parameter $\lambda/\sigma$ where $\lambda$ is itself assumed to be an unknown, with prior $\pi(\lambda) = E(1)$. A two chain run of 5000 iterations (inferences from the last 4500) gives the median probability model $\{x_1, x_2, x_3, x_4, x_6, x_8\}$ with the highest marginal inclusion probabilities (of 1) for $x_1$ and $x_2$. Taking $\pi(\lambda) = E(20)$ produces a more parsimonious median probability model $\{x_1, x_2, x_8\}$.

The second shrinkage model has the hierarchical prior scheme

$$\beta_j | \lambda_j \sim N(0, \lambda_j),$$

$$\lambda_j | \kappa \sim E(\kappa),$$

$$\kappa \sim E(a_\kappa),$$

where $a_\kappa$ is specified at alternative values of 0.1, 1, and 10. Under the first two values for $a_\kappa$, the median probability model is $\{1, 2, 3, 4, 6, 7, 8\}$ so that only two predictors are of doubtful relevance, but with $a_\kappa = 10$, the median probability model is $\{1,2,3,4,8\}$. Under all models $Pr(\gamma_1|y) = Pr(\gamma_2|y) = 1$.

For RJMCMC, two options[3] on the number of retained predictors $P_{ret}$ are specified. A N(0,10000) prior is assumed for the predictors, which will tend to favour sparse models, and Ga(1,0.001) prior on the residual precision. Despite these settings, with $P_{ret} \sim Bin(P,0.5)$, the posterior mean for $P_{ret}$ is 4.8, with marginal retention probabilities exceeding 0.5 for five predictors: $x_1,x_2,x_3,x_4$, and $x_8$. Posterior model probabilities are far from decisive, with the highest probability of 0.13 on $\{x_1,x_2,x_3,x_4,x_6,x_7\}$. With a more parsimonious $P_{ret} \sim Bin(P,0.25)$, the posterior mean for $P_{ret}$ is reduced to 3.6, with marginal retention probabilities exceeding 0.5 for only three predictors, $x_1,x_2$, and $x_8$. The highest posterior model probabilities are 0.39 on $\{x_1,x_2,x_8\}$ and 0.13 on $\{x_1,x_2,x_3,x_8\}$.

The Chib and Laplace approximations to the analytic marginal likelihood for these two models can be obtained using MCMCpack. A N(0,10000) prior is again assumed for the predictors, with a more diffuse prior for the intercept. The code is

```
library(MCMCpack); setwd("C://R files")
# read data with headings
D=read.table(file='elemsch.txt',header=T);
# Model 1
m1C <- MCMCregress(y~x1+x2+x8, burnin=1000,b0=c(0, 0, 0, 0),data=D,
B0=c(0.000001,0.0001, 0.0001, 0.0001), c0=2, d0=0.002,
marginal.likelihood="Chib95", mcmc=100000, verbose=10000)
# Model 2
m2C <- MCMCregress(y~x1+x2+x3+x8, burnin=1000,b0=c(0, 0, 0, 0, 0), data=D,
B0=c(0.000001,0.0001, 0.0001, 0.0001, 0.0001), c0=2, d0=0.002,
     marginal.likelihood="Chib95", mcmc=100000, verbose=10000)
# Bayes factors and model probs
BF <- BayesFactor(m1C, m2C); print(BF); m.probs
          <- PostProbMod(BF); print(m.probs).
```

Marginal likelihoods, and Bayes factors, are obviously affected by the priors assumed. With the relatively diffuse priors adopted there is little to choose between the models with $Pr(M_1|y) = 0.323$ and $Pr(M_2|y) = 0.677$. More informative priors (e.g. with higher precision on the beta coefficients) will raise $Pr(M_2|y)$.

The g-prior

$$\beta_j \sim N\left(0, \frac{g}{\tau}[X'X]^{-1}\right) \qquad j = 1, \ldots, P,$$

is applied[4] to the two models with high posterior probability according to the RJMCMC analysis, namely $(x_1,x_2,x_8)$ (model 1) and $(x_1,x_2,x_3,x_8)$ (model 2). Analysis without selection and using $g = n$ shows, for model 1, a log (BF) $= \frac{(n-P-1)}{2} \log(1+g) - \frac{(n-1)}{2} \log[1+g(1-R^2)] = 347.25$ as compared to the null model (resulting from $R^2 = 0.8344$). For model 2, $R^2 = 0.8366$ and log (BF) $= 346.89$. The implied posterior model probabilities are $p(M_1|y) = 0.59$ and $p(M_2|y) = 0.41$.

A subsequent analysis[5] combines a model choice focus with a g-prior on regression coefficients, with the preset option $g = n$ initially retained. There are $K = 2$ alternatives: $(x_1,x_2,x_8)$ (model 1) and $(x_1,x_2,x_3,x_8)$ (model 2), with equal prior model probabilities assumed, namely $\pi_k = Pr(M_k) = 1/K = 0.5$. After relabelling predictors $(x_1,x_2,x_3,x_8)$ to $(x_1,x_2,x_3,x_4)$, one has $\gamma_{11} = 1, \gamma_{12} = 1, \gamma_{13} = 0, \gamma_{14} = 1$ and $\gamma_{2j} = 1$ for all $j$. Then the model indicator is multinomial:

$$M \sim mult(0.5, 0.5),$$

$$(\eta_i | M = m) = \beta_0 + \gamma_{m1}\beta_1 x_{1i} + \ldots + \gamma_{mP}\beta_P x_{Pi},$$

$$\beta_j \sim N(0, \frac{g}{\tau}[X'X]^{-1}).$$

A two chain run in WinBUGS14 gives (from the last 9000 of a two chain 10 000 iteration run) posterior probabilities 0.55 on $(x_1, x_2, x_8)$ and 0.45 on $(x_1, x_2, x_3, x_8)$. A second estimation takes

$$g/(1+g) \sim Be(1, 0.5)$$

and provides a posterior mean for $g$ of 770 and probabilities $p(M_1, M_2 | y) = (0.58, 0.42)$. Finally, a third estimation retains $g = n$, but lets model probabilities penalize larger models, via

$$\pi_k \propto 1/(1+\kappa)^{P_k}$$

with a prior $\kappa \sim E(1)$. This results in a posterior mean $\kappa = 0.99$ and $p(M_1, M_2 | y) = (0.61, 0.39)$.

**Example 3.4    Nodal involvement**

To illustrate predictor selection and model choice with augmented data binary regression, this example considers binary outcomes on prostatic cancer nodal involvement. There are four possible predictors: $x_1 = \log(\text{serum acid phosphate})$, $x_2 = \text{result of X-ray}$ (1 = +ve, 0 = -ve), $x_3 = \text{size of tumour}$ (1 = large, 0 = small) and $x_4 = \text{pathological grade of tumour}$ (1 = more serious, 0 = less serious) (Collett, 1991). These data illustrate a frequent problem in model fitting: simple unpenalised fit measures such as the deviance may show a slight gain in fit as extra predictors (or other parameter sets) are added, but penalised fit measures, or marginal likelihood type measures, show that any improvement is offset by extra complexity. Cross validation approaches may also favour the less complex model in such situations.

As mentioned above, alternative links for binary regression may be used, and one may consider generalised links (see chapter 4). Here a probit link is adopted, so that

$$\pi_i = \Pr(y_i = 1 | x_i) = \Phi(\beta_0 + X_i\beta).$$

The model may be fitted directly using the Bernoulli likelihood, or by introducing latent normally distributed responses,

$$z_i \sim N(X_i\beta, 1) \qquad I(0, \infty) \qquad if \qquad y_i = 1$$
$$z_i \sim N(X_i\beta, 1) \qquad I(-\infty, 0) \qquad if \qquad y_i = 0.$$

In WinBUGS14 one may use the dbern.aux function (available under the WinBUGS jump add-on) for such an augmented data regression. Thus a probit regression, with priors as in Chib (1995), is implemented by the code

```
model {for (i in 1:n) {y[i] ~dbern.aux(Z[i]); Z[i] ~dnorm(eta[i], 1)
eta[i] <- alpha + sum(reg[i,])
for (j in 1:p) {reg[i,j] <- beta[j]*x[i,j]}}
for (j in 1:p) {beta[j] ~dnorm(0.75,0.04)}}
```

From a baseline analysis with this code, one may obtain estimates of $p(\theta|y)$ (where $\theta = \beta$) for two models to be subsequently compared, namely $M_1$ with only $x_1 - x_3$ included, as against $M_2$ including all four predictors.

Chib (1995) shows that model $M_1$ has a worse log-likelihood than $M_2$ ($-24.43$ as against $-23.77$), but a better marginal likelihood ($-34.55$ vs $-36.23$), with Bayes factor $B_{12} = \exp(1.68) = 5.37$. This can be obtained from MCMCpack by the sequence

```
library(MCMCpack); setwd("C://R files")
# read data with headings
D=read.table(file="nodal.txt",header=T);
# Model 1
m1C <- MCMCprobit(y~x1+x2+x3, burnin=1000,b0=0.75,data=D,
B0=0.04, marginal.likelihood="Chib95", mcmc=100000, verbose=10000)
m2C <- MCMCprobit(y~x1+x2+x3+x4, burnin=1000,b0=0.75,data=D,
B0=0.04, marginal.likelihood="Chib95", mcmc=100000, verbose=10000)
# Bayes factors and model probs
BF <- BayesFactor(m1C, m2C); print(BF);
m.probs <- PostProbMod(BF); print(m.probs)
```

By conventional criteria on interpreting Bayes factors (Kass and Raftery, 1995) a Bayes factor of this size counts as 'positive evidence' for the reduced model, though not conclusive. It may be noted that R-INLA gives the same marginal likelihood estimates. The relevant code is

```
require(INLA);
D <- read.table("nodal.txt",header=T); n <- 53
M1 = inla(y ~1 + x1+x2+x3+x4, data = D, family = "binomial",
                             Ntrials = rep(1, n),
control.family = list(link = "probit"), control.predictor = list(compute=T),
control.fixed=list(mean=c(0.75,0.75,0.75,0.75),mean.intercept=0.75,
prec.intercept=0.04,prec=c(0.04,0.04,0.04,0.04)))
M2 = inla(y ~1 + x1+x2+x3, data = D, family = "binomial",
                             Ntrials = rep(1, n),
control.family = list(link = "probit"),
   control.predictor = list(compute=T),
control.fixed=list(mean=c(0.75,0.75,0.75),mean.intercept=0.75,
prec.intercept=0.04,prec=c(0.04,0.04,0.04)))
```

Here alternative approaches to marginal likelihood approximation and model choice are first considered. An importance sample estimator of the marginal likelihood is provided by

$$m_k(y) = 1/\left[T^{-1}\sum_t g_k^{(t)} / \left\{L_k^{(t)}\pi_k^{(t)}\right\}\right] = T/\left[\sum_t g_k^{(t)} / \left\{L_k^{(t)}\pi_k^{(t)}\right\}\right],$$

where $L_k^{(t)}$ and $\pi_k^{(t)}$ are the likelihood and prior ordinates for samples $\theta_k^{(t)}$ of $\theta_k = (\beta_{0k}, \beta_{1k}, \ldots, \beta_{P_k k})$ from model $k$ (with $P_1 = 3$, $P_2 = 4$ predictors), and $g_k^{(t)}$ is the value of an importance function approximating the posterior density $p(\theta_k|y)$. This function is provided by a multivariate normal approximation. The importance sample estimates of the marginal likelihoods, are $-35.21$ (model 1), and $-36.88$ (model 2), and so $B_{12} = 5.31$. These are obtained as described in Chapter 2 by monitoring

$$\delta_k^{(t)} = \log[g_k^{(t)}/\{L_k^{(t)}\pi_k^{(t)}\}] = \log(g_k^{(t)}) - \log(L_k^{(t)}) - \log(\pi_k^{(t)}), \qquad k = 1, 2.$$

The method of Congdon (2007) can also be applied (program C) with model weights on $M_1$ and $M_2$ obtained at each iteration as

$$w_1^{(t)} = \frac{\dfrac{L_1^{(t)}\pi_1^{(t)}\phi_1}{g_1^{(t)}}}{\dfrac{L_1^{(t)}\pi_1^{(t)}\phi_1}{g_1^{(t)}} + \dfrac{L_2^{(t)}\pi_2^{(t)}\phi_2}{g_2^{(t)}}}; \quad w_2^{(t)} = \frac{\dfrac{L_2^{(t)}\pi_2^{(t)}\phi_2}{g_2^{(t)}}}{\dfrac{L_1^{(t)}\pi_1^{(t)}\phi_1}{g_1^{(t)}} + \dfrac{L_2^{(t)}\pi_2^{(t)}\phi_2}{g_2^{(t)}}};$$

where $\phi_1 = \phi_2 = 0.5$ are prior model probabilities. This method[6] also provides marginal likelihood estimates through monitoring

$$\zeta_k^{(t)} = -\delta_k^{(t)} = \log(L_k^{(t)}) + \log(\pi_k^{(t)}) - \log(g_k^{(t)}), \qquad k = 1, 2.$$

Posterior means of $\zeta_k$ are accordingly obtained as $-35.2$ and $-36.87$ respectively, giving $B_{12} = 5.31$. Posterior means of $w_1$ and $w_2$ are 0.8392 and 0.1608 giving a slightly lower estimate $B_{12} = 5.22$.

A benefit of the model weights method is in model averaging during the MCMC sequence (rather than subsequently), and accordingly case level probabilities of nodal involvement under each model, and averaged probabilities, are obtained at each iteration. For example, for subjects 15 and 17, the two models provide contrasting probabilities of nodal involvement (around twice as high under $M_2$). Thus for subject 15, the posterior mean model averaged probability $\pi_{15}$ is 0.056, compared to means of 0.048 and 0.095 under $M_1$ and $M_2$ respectively.

Estimates of the log marginal likelihood already discussed are virtually identical to those obtained from the application of the marginal likelihood identity. In general one may estimate $\log[m(y)]$ as

$$\log\{m[y]\} = \log\{p(y|\theta_{HD})\} + \log\{\pi(\theta_{HD})\} - \log\{\pi(\theta_{HD}|y)\}$$

at a high density point $\theta_{HD}$. Here (program D in the Example 3.4 code) an estimate for $\log[m(y)]$ is obtained using a multivariate normal approximation to the posterior $p(\theta|y)$, and taking $\theta_{HD}$ as the maximum likelihood estimate. The resulting estimates for $\log[m(y)]$ are $-35.14$ and $-36.82$ for models 1 and 2, giving $\text{BF}_{12} = 5.37$.

In probit regression, an alternative approximation to $\pi(\theta_{HD}|y)$ is available. Since augmented data $z$ is part of the model definition, and there is just one other parameter block (i.e. the regression coefficients), one may estimate $\pi(\theta_{HD}|y)$ using draws $z^{(t)}$ from a sampling run where all parameters are updated, and the probability of $\theta_{HD}$ (obtained from elsewhere, possibly via an earlier estimation or by maximum likelihood) is evaluated against the full conditional density for $\theta$ given $z^{(t)}$. The latter density is derived by considering $z$ as dependent variable in a normal errors model (Chib, 1995). If the prior for $\theta = (\beta_0, \beta_1, \dots, \beta_P)$ is $\theta \sim N_{P+1}(\theta_0, T_0^{-1})$, then the relevant full conditional is

$$p(\theta|z, y) = N_{P+1}(\theta_z, T^{-1}),$$

where $T = T_0 + X'X$, $X$ is $n \times (P+1)$ including a constant, and

$$\theta_z = T^{-1}[T_0\theta_0 + X'z],$$

with $z$ being the vector of sampled latent responses. Program E applies this approach (with $\theta_{HD}$ again as the maximum likelihood estimate) giving a marginal likelihood estimate (from the second half of single chains of 10,000 iterations) of $-33.81$ for model 1, and $-35.25$ for model 2, and a Bayes factor $\text{BF}_{12} = 4.22$.

Predictor selection methods such as RJMCMC and the Kuo–Mallick method show there are broader questions regarding appropriate model selection for these data. With $\eta_i = X_i\beta$, and with $x$ variables in standardized form, the standard deviation $s_\eta$ of the regression terms $\eta_i$ when all predictors are included is found from the baseline analysis to be 1.18. Two contrasting prior variance options in a normal prior for $\beta_j$, are then adopted, namely

$$V_{1j} = s_\eta^2 k^2, \quad k = 0.5,$$

and

$$V_{2j} = s_\eta^2 k^2, \quad k = 4,$$

following the recommendations of Kuo and Mallick (1998) on bounds for $k$. The median probability model under the more informative (lower variance) first prior is found to be $M_1$ as above, namely $\{x_1, x_2, x_3\}$. However, the individual marginal retention probabilties for these predictors, namely 0.73, 0.94 and 0.80 show that other models are being selected at particular iterations. The two most frequently selected models (from the second half of two chain sequence of 10 000 iterations) are $M_1 = \{x_1, x_2, x_3\}$ with probability 0.34, and $M_2 = \{x_1, x_2, x_3, x_4\}$ with probability 0.24, with $BF_{12} = 1.39$. Under the more diffuse prior variance option, one has $V_{2j} = s_\eta^2 k^2 = 22.3$, leading to a prior precision for $\beta_j$ of 0.045, close to that assumed by Chib (1995). The median probability model under this option is found to be $\{x_2, x_3\}$, with the marginal retention probabilty for $x_1$ now only 0.35. The models with the highest posterior probabilities are $\{x_2\}$ with probability 0.30, and $\{x_2, x_3\}$ with probability 0.23.

Model choice under RJMCMC is affected both by the assumed prior precision $V_j$ on the $\beta$ coefficients, and by the prior on the number of predictors $P_{ret}$ to be retained in the regression, namely

$$P_{ret} \sim \text{Bin}(P, \pi_{ret}).$$

Here it is assumed that $\pi_{ret}$ is itself unknown, namely $\pi_{ret} \sim \text{Be}(1, 1)$, so that all values of $P_{ret}$ are equally likely a priori.

The code for the RJMCMC analysis is as follows (with p = 4 and p1 = 5)

```
model { for (i in 1:n) {y[i] ~dbern.aux(Z[i]); Z[i] ~ dnorm(eta[i], 1)}
for (j in 1:p) {Retain[j] <- 1-equals(beta.jump[j], beta.jump[p1])
Marker[j,j] <- 1; Marker[p1,j] <- 0
for (k in 1:j-1) { Marker[j,k] <- 0; Marker[k,j] <- 0}}
eta[1:n] <- jump.lin.pred.int(x[1:n,1:p],p.ret,beta.prc,alph.mn,alph.prc)
ModSel <- jump.model.id(eta[1:n])
beta.jump[1:p1] <- jump.lin.pred.int.pred(eta[1:n], Marker[1:p1, 1:p])
# regrn parameters
alpha <- beta.jump[p1];
# realised coefficients
for (j in 1:p) { beta[j] <- beta.jump[j] - beta.jump[p1]}
# prior on number of retained predictors
p.ret ~dbin(pi.ret, p); pi.ret ~dbeta(1,1)
# assumed precision of beta coefficients
beta.prc <- 0.04; alph.prc <- 0.01; alph.mn <- 0}
```

Under the relatively diffuse prior $V_j = 25$ (precision of 0.04 on the beta coefficients) adopted by Chib (1995), the median probability model (from the second half of a two chain sequence of 10 000 iterations) is $\{x_1, x_2, x_3\}$. However, the respective marginal retention probabilities (0.66, 0.87, 0.66) on the first three predictors show that a variety of other models are chosen.

In fact, nine models need to be considered to ensure the cumulative probability $\sum_{J}^{j=1} \Pr(M_j|y)$ of the most favoured $J$ models exceeds 0.9. The probabilities on the five most frequently selected models are $\Pr(x_1, x_2, x_3|y) = 0.276$; $\Pr(x_1, x_2, x_3, x_4|y) = 0.175$; $\Pr(x_2|y) = 0.124$; $\Pr(x_2, x_3|y) = 0.113$, and $\Pr(x_1, x_2|y) = 0.076$. The Bayes factor favouring $M_1$, namely $\{x_1, x_2, x_3\}$ over $M_2$, namely $\{x_1, x_2, x_3, x_4\}$ is reduced to 1.58.

**Example 3.5    Diabetes diagnosis prediction and regression selection**

The Pima Indians dataset relates to a sample of $n = 532$ females who were tested for the presence or not of diabetes mellitus (i.e. a binary response y), with $n_1 = 177$ classed as positive ($y = 1$). Data on risk factors and clinical readings include PRG ($x_1$, number of times pregnant), PLAS ($x_2$, plasma glucose concentration in saliva), DBP ($x_3$, diastolic blood pressure), SKIN ($x_4$, forceps skin fold thickness), BMI ($x_5$), PEDIG ($x_6$, a diabetes pedigree function), and AGE in years ($x_7$). Logit regression is obtained by truncated logistic sampling

$$z_i \sim \text{logistic}(X_i\beta, 1) \qquad I(0, \infty) \qquad (y_i = 1)$$

$$z_i \sim \text{logistic}(X_i\beta, 1) \qquad I(-\infty, 0) \qquad (y_i = 0).$$

As well as main effects in the seven risk factors, the 21 first order interactions are also considered (in the sequence $x_1x_2, x_1x_3 \ldots x_1x_7, x_2x_3, x_2x_4, \ldots , x_6x_7$). All 28 predictors are standardised. An initial analysis without predictor selection adopts N(0,1000) priors on the $\beta$ parameters. A two chain run of 5000 iterations (with the final 4000 for inference) shows only 2 of the 28 predictors having significant effects, in the sense of 95% intervals entirely positive or negative. The sensitivity $n_{11}/n_1$ is estimated by monitoring the $n_{11}$ subjects with $y_i = 1$ and with $\pi_i > 0.5$, and obtained as 0.62, while specificity is estimated at 0.865. Probability thresholds other than 0.5 could of course be used in assessing sensitivity and specificity.

A second analysis involves predictor selection, with binary retention indicators $\gamma_j \sim \text{Bern}(0.5)$ combined with mildly informative prior variances on $\beta_j$

$$\beta_j \sim N(0, (s_\eta k)^2), \qquad j = 1, \ldots , 28.$$

where $s_\eta = sd(\eta_i) = sd(X_i\beta)$ is evaluated at each iteration. $k$ is an additional unknown following a uniform prior with upper and lower limits implied by Kuo and Mallick (1998),

$$k \sim U(0.5, 4).$$

The regression term has the form

$$\eta_i = \beta_0 + \gamma_1\beta_1 x_{1i} + \ldots + \gamma_7\beta_7 x_{7i} + \gamma_8\gamma_1\gamma_2\beta_8 x_{1i}x_{2i} + \gamma_9\gamma_1\gamma_3\beta_9 x_{1i}x_{3i} \ldots \cdot + \gamma_{28}\gamma_6\gamma_7\beta_{28} x_{6i}x_{7i},$$

where interaction predictors $(x_1x_2, x_1x_3, \ldots , x_6x_7)$ are only included when both the corresponding main effects are included, namely constrained selection (Farcomeni, 2010). So retention probabilities for the interaction effects are obtained by monitoring $\gamma_m\gamma_j\gamma_k$, for $m > 7$ and $\{j, k \in 1, \ldots , 7\}$. A two chain run of 10,000 iterations in OpenBUGS (with inferences based on the second half) shows $\Pr(\gamma_j = 1|y) = 1$ for four of the main effects (PRG, PLAS, PEDIG, and SKIN). The highest retention probability for an interaction is 0.56, for PLAS*SKIN. The sensitivity with threshold $\pi_i > 0.5$ is reduced to 0.57.

Even though none of the interactions are definitively retained under predictor selection, it is of interest to assess whether a different optimal model among the 'main effect' variables $x_1$

to $x_7$ is chosen when interactions are not considered as candidate predictors from the outset. A baseline model without selection produces four predictors (PRG, PLAS, BMI, PEDIG) having significant effects, in the sense of 95% intervals entirely positive or negative, with the 95% credible interval for AGE, namely (-0.02,0.60), also indicating potential relevance.

In a subsequent selection stage for this reduced candidate predictor scenario, two options on prior variances for $\beta_j$ are considered. The first is as above

$$\beta_j \sim N(0, (s_\eta k)^2), \qquad j = 1, \ldots, 7$$

$$k \sim U(0.5, 4),$$

while the second adopts a Cauchy prior, which in terms of a Student $t$ is

$$\beta_j \sim t(0, 6.25, 1), \qquad j = 1, \ldots, 7.$$

The BUGS code for this model (using the dlogis density option and constrained sampling) is

```
model { for (i in 1:N) {z[i] ~dlogis(eta[i],1) I(A[y[i]+1],B[y[i]+1])
eta[i] <- alpha+inprod(delta[], x.st[i, 1:P])
for (k in 1:P) {x.st[i,k] <- x[i,k]/sd(x[,k])}
pi[i] <- 1/(1+exp(-eta[i]))
# sensitivity threshold
thresh[i] <- step(pi[i]-0.5)
S.subj[i] <- thresh[i]*y[i]; C.subj[i] <- (1-thresh[i])*(1-y[i])}
# Priors
alpha ~dnorm(0,0.001);
for (j in 1:P) {gam[j] ~dbern(0.5); delta[j] <- gam[j]*beta[j]}
# Prior on beta coefficients 1
K ~dunif(0.5,4); sd.eta <- sd(eta[]); pr.beta <- 1/(K*K*sd.eta*sd.eta)
for (j in 1:P) {beta[j] ~dnorm(0,pr.beta)}
# Prior on beta coefficients 2
# for (j in 1:P) {beta[j] ~dt(0,0.16,1)}
# predictive classification
Sens <- sum(S.subj[])/177; Spec <- sum(C.subj[])/355}
```

Under the first prior option, the second half of a two chain run of 10 000 iterations shows marginal retention probabilities for PRG, PLAS, and PEDIG all exceeding 0.99, while that for BMI is 0.68. This retention pattern is similar to that in Holmes and Held (2006). The posterior mean for $k$ is 0.78, albeit with a spike at 0.5. For the second prior, convergence was not obtained in the selection indicators by 50 000 iterations of a two chain run, but the second half of the run shows $Pr(\gamma_j = 1|y) > 0.95$ for PRG, PLAS and PEDIG.

### Example 3.6    Asking price of chevrolet camaro

To predict the asking price ($y$, in standardised form) of a used Chevrolet Camaro (with $n = 31$ cases), independent variables on the car's age ($x_1$), mileage ($x_2$), condition ($x_3$-$x_4$), and seller status ($x_5$) are collected. The binary variable $x_3$ equals 1 if the car is in average condition, 0 if not, while $x_4$ equals 1 if the car is in poor condition, 0 if not. A binary variable $x_5$ equals 1 if the seller is a dealer and is equal to zero if the seller is a private individual. Models $M_1 = (1, x_1, x_2)$ and $M_2 = (1, x_1, x_2, x_3, x_4, x_5)$ are to be compared. Conjugate normal priors are assumed with residual precision, $\tau_k = \sigma_k^{-2} \sim Ga(1.5, 0.5)$, for models $k$, and regression coefficients $\beta_{jk}|\sigma^2 \sim N(0, 5\sigma^2)$.

The following Matlab code

```
load chev.txt; y = chev(:,1); x1 = chev(:,2); x2 = chev(:,3) ;
        x3 = chev(:,4) ; x4 = chev(:,5) ; x5 = chev(:,6); n=31;
% Model 1
X = [ones(n,1) x1 x2]; p = 3;
% prior hyperparameters;
a = 1.5; b = 0.5;
mu_beta = zeros(p,1); V_beta = 5*eye(p); invV_beta = inv(V_beta);
% Marg Lkd Calculation
D = (det(V_beta))*det( (invV_beta + X'*X));
E = eye(n) - X*inv(invV_beta + X'*X)*X';
C = (D^(-1/2))* (b^a)*gamma(n/2+a) /(gamma(a)*(2*pi)^(n/2));
ML1 = C*(b + 0.5*y'*E*y)^(-( n/2 + a));
disp('Log Marginal Likelihood'); log(ML1)
% Model 2
X = [ones(n,1) x1 x2 x3 x4 x5]; p = 6;
% prior hyperparameters;
a = 1.5; b = 0.5;
mu_beta = zeros(p,1); V_beta = 5*eye(p); invV_beta = inv(V_beta);
% Marg Lkd Calculation
D = (det(V_beta))*det( (invV_beta + X'*X));
E = eye(n) - X*inv(invV_beta + X'*X)*X';
C = (D^(-1/2))* (b^a)*gamma(n/2+a) /(gamma(a)*(2*pi)^(n/2));
ML2 = C*(b + 0.5*y'*E*y)^(-( n/2 + a));
disp('Log Marginal Likelihood'); log(ML2)
BF21 =ML2/ML1
```

provides analytic log marginal likelihoods of $-32.17$ and $-29.88$, with $BF_{21} = 9.90$.

The model weights procedure of Congdon (2007) is then applied using multivariate normal approximations (of dimension 3 and 6 respectively) to the posterior density of the regression coefficients under the two models. MVN ordinates from these approximations provide pseudo-priors $g_{k1}$ for the models $k = 1, 2$. The approximations are based on posterior means of, and covariances between, samples of the $\beta$ coefficients (intercepts included) obtained from separate initial runs of 50 000 iterations for each model. Estimated gamma posterior densities for $\tau_k$ from these initial runs provide pseudo priors $g_{k2}$ for the precision parameters. Subsequent analysis comparing the models in parallel[7] provides posterior mean weights (w[1] and w[2] in the code) of 0.0924 and 0.9076, with $BF_{21} = 9.83$.

Although model 2 has a higher posterior probability, its predictive performance has to be verified. A global posterior predictive check comparing $R^2$ for observed and replicate data is satisfactory for both models. However, observation specific predictive checks $d_i^{(t)} = I(y_{new,i}^{(t)} \leq y_i | \theta^{(t)})$ suggest underprediction at higher y-values. For model 1, the correlation between the resulting posterior predictive p-values (posterior means of $d_i^{(t)}$) and the responses is 0.42, and this is reduced only slightly to 0.34 under model 2. The most serious predictive discrepancy (a posterior predictive value over 0.99 in both models) is for observation 29, which has the highest observed price. The predictive deficiencies might indicate the need for nonlinear predictor effects or downweighting of extreme observations (see Chapter 4).

## 3.6   Multinomial, nested and ordinal regression

Many outcomes in social and health applications involve ordered or unordered polytomous variables. Examples include political or religious affiliation (unordered), severity of diagnosis (ordered), occupational rank (ordered), or commuting choice (unordered) (Amemiya, 1981). Underlying the observed category outcome may be envisaged a latent continuous variable $z$ (Albert and Chib, 1993), which may be conceptualised as an attitudinal, prestige or utility scale, with the maximum value over categories determining the observed category.

The multinomial logit and multinomial probit models generalise their binomial equivalents and are suitable – perhaps with modifications - to modelling multicategory outcomes which do not involve ordered categories. So just as the logit model for a binary outcome involves the log odds of a positive to a negative response, so a multinomial logit involves stipulating a baseline choice (say the first of $K$ possible outcomes) and comparing the probabilities of choices $2, 3, \dots, K$ against that of the baseline. Covariates may be defined for individuals $i$, for choices $j$, or particular features of choice $j$ unique to individual $i$. Thus in a travel mode choice example, the first type of variable might be individual income, the second might be the generic cost of alternative modes, and the third might be individual costs attached to different modes.

Consider a vector of covariates $X_i$ specific to individuals $i$ alone, and let $y_{ij} = 1$ if option $j$ is chosen with $y_{ik} = 0$, for all $k \neq j$. Then the fixed effects multiple logit model is obtained, with category 1 as reference, as

$$\Pr(y_{ij} = 1) = p_{ij} = \exp\left(\alpha_j + X_i\beta_j\right) / \left[1 + \sum_{k=2}^{K} \exp\left(\alpha_k + X_i\beta_k\right)\right], \qquad j > 1$$

$$\Pr(y_{i1} = 1) = p_{i1} = 1 / \left[1 + \sum_{k=2}^{K} \exp\left(\alpha_k + X_i\beta_k\right)\right],$$

or equivalently

$$\log\{p_{ij}/p_{i1}\} = \alpha_j + X_i\beta_j.$$

Also for $j$ and $k$ both exceeding 1 the odds ratio is

$$\log\{p_{ij}/p_{ik}\} = (\alpha_j - \alpha_k) + (\beta_j - \beta_k)X_i,$$

so that choice probabilities are governed by differences in coefficient values between alternatives, independent of all other alternatives in the choice set (the 'independence of irrelevant alternatives' assumption). The above formulation (involving only covariates constant across alternatives $j$) is known as a multinomial logit or MNL (e.g. Koop and Poirier, 1993; Fry and Harris, 1996), and focuses on attributes of subjects $i$ as the focus for inferences.

If instead choice is based on attributes $H_{ij}$ of the $j^{th}$ alternative specific for individual $i$, then a conditional logit model is obtained with

$$p_{ij} = \exp\left(H_{ij}\delta\right) / \sum_{k=1}^{K} \exp\left(H_{ik}\delta\right).$$

Dividing through by $\exp\left(\delta H_{ij}\right)$ gives

$$p_{ij} = 1 / \sum_{k=1}^{K} \exp\left([H_{ik} - H_{ij}]\delta\right).$$

In the conditional logit model, the coefficients $\delta$ are usually constant across alternatives and so choice probabilities are determined by differences in the values of characteristics between alternatives. A mixed model, combining features of both multinomial and conditional schemes, would include individual level attributes $X_i$ and alternative specific characteristics $H_{ij}$. Thus

$$\log (p_{ij}/p_{ik}) = (\alpha_j - \alpha_k) + X_i(\beta_j - \beta_k) + (H_{ij} - H_{ik})\delta.$$

Multiple logit models can be expressed in terms of a utility model for individual choice behaviour. Thus let $z_{ij}$ be the unobserved value or utility of choice $j$ to individual $i$, with

$$z_{ij} = U(X_i, G_j, H_{ij}, \varepsilon_{ij}),$$

where $G_j$ are known predictors for choice $j$ (e.g. climate in state $j$ for potential migrants), and $X_i$ and $H_{ij}$ are as above. The $\varepsilon_{ij}$ are random utility terms. Assuming additivity and separability of stochastic and deterministic components leads to

$$z_{ij} = \eta_{ij} + \varepsilon_{ij},$$

with a regression function such as

$$\eta_{ij} = \alpha_j + X_i\beta_j + H_{ij}\delta + G_j\zeta.$$

Then the choice of option $j$ means

$$z_{ij} > z_{ik}, \qquad k \neq j$$

and so

$$p_{ij} = \Pr(z_{ij} > z_{ik}).$$

Equivalently

$$y_{ij} = 1 \qquad if \qquad z_{ij} = \max(z_{i1}, z_{i2}, \dots, z_{iK}).$$

Assume the $\varepsilon_{ij}$ follow a type $I$ extreme value (double exponential) distribution with cdf

$$F(\varepsilon_{i.}) = \prod_{j=1}^{K} \exp[-\exp(-\varepsilon_{ij})],$$

and that the above additivity-separability assumptions hold, one obtains

$$\Pr(y_{ij} = 1 | X_i, H_{ij}, G_j) = \exp(\eta_{ij}) / \sum_{k=1}^{K} \exp(\eta_{ik})$$

with $\alpha_1 = \beta_1 = 0$ for unique identifiability.

### 3.6.1 Nested logit specification

A potentially limiting feature of the multinomial logit is the independence of irrelevant alternatives (IIA) assumption, since in practice, the presence of other alternatives (or the addition or deletion of alternatives) can affect substitution patterns – see Congdon (2000) for an application involving patient flows to hospitals. The nested logit model is one alternative adapted to account for departures from IIA and is useful when subsets of the alternatives are likely to share unobserved features (Poirier, 1996; Fischer and Aufhauser, 1988; Chib *et al.*, 2004).

Denote $M$ subsets of the alternatives as $\{S_1, \ldots, S_M\}$, respectively containing $K_1, \ldots, K_M$ sub-choices, with $\sum_{m=1}^{M} K_m = K$. The utility of choice $j$ in subset $m$ is then

$$z_{ijm} = \eta_{ijm} + \varepsilon_{ijm},$$

with the nested logit model derived by assuming a generalized extreme value (GEV) density for the $\varepsilon$ terms, with cumulative distribution:

$$P(\varepsilon_i) = \exp\left[-\sum_{m=1}^{M}\left(\sum_{j \in S_m} e^{-\varepsilon_{ijm}/\lambda_m}\right)^{\lambda_m}\right]$$

where the $\lambda_m$ are positive. The marginal density of each $\varepsilon_{ijm}$ is then a univariate extreme value, but the $\varepsilon_{ijm}$ are correlated within subsets, with higher values of $\lambda_m$ implying reduced correlation. If $\lambda_m = 1$ for all subsets then the nested logit reduces to the usual MNL. The choice probability for option $j$ in subset $g$ is given by

$$p_{ijg} = \frac{\exp(\eta_{ijg}/\lambda_g)\left[\sum_{k=1}^{K_g} \exp(\eta_{ikg}/\lambda_g)\right]^{\lambda_g - 1}}{\sum_{m=1}^{M}\left[\sum_{h=1}^{K_m} \exp(\eta_{ihm}/\lambda_m)\right]^{\lambda_m}}$$

The ratio of choice probabilities, when the choices are in differing subsets $a$ and $b$, is

$$p_{ija}/p_{ikb} = \frac{\exp(\eta_{ija}/\lambda_a)\left[\sum_{k=1}^{K_a} \exp(\eta_{ika}/\lambda_a)\right]^{\lambda_a - 1}}{\exp(\eta_{ikb}/\lambda_b)\left[\sum_{k=1}^{K_b} \exp(\eta_{ikb}/\lambda_b)\right]^{\lambda_b - 1}}.$$

Within subsets (that is, when $a = b$) this reduces to

$$p_{ija}/p_{ika} = \frac{\exp(\eta_{ija}/\lambda_a)}{\exp(\eta_{ika}/\lambda_a)}.$$

A common values for $\lambda$ may be assumed across subsets. However, if substitution patterns are likely to differ between subjects, one can introduce independent variables $W_i$ to predict subject specific $\lambda$ values, with log link, $\log(\lambda_i) = \omega W_i$.

Suppose $\eta_{ijm} = \eta_{im} + \delta_{ijm}$ can be subdivided into a between-nest component $\eta_{im}$ (based on predictors that vary between nests but not between alternatives within nests) and a between-alternatives component $\delta_{ijm}$ (based on predictors varying between alternatives within nests). Then one can express $p_{ijm}$ as the product of a marginal probability $\pi_{im}$ that subset $m$ is chosen, and a conditional probability $\pi_{ij|m}$ that option $j$ within subset $m$ is then selected:

$$p_{ijm} = \pi_{im}\pi_{ij|m}.$$

The two $\pi$ probabilities can be represented as multinomial logits, namely

$$\pi_{ij|m} = \exp(\delta_{ijm}/\lambda_m)/\left[\sum_{k=1}^{K_m}\exp(\delta_{ikm}/\lambda_m)\right],$$

$$\pi_{im} = \exp[\eta_{im} + \lambda_m I_{im}]/\sum_{h=1}^{M}\exp[\eta_{ih} + \lambda_h I_{ih}],$$

where $I_{im} = \log\left[\sum_{k=1}^{K_m}\exp(\delta_{ikm}/\lambda_m)\right]$ is called the inclusive value.

### 3.6.2    Ordinal outcomes

The multinomial and conditional logit models make no assumptions about the ordering of a categorical outcome. However, ordinal response or choice data are frequently encountered in the social and health sciences (Greene and Hensher, 2009). In opinion surveys, respondents are often asked to grade their views on a statement on scales from "strongly agree" to "strongly disagree". Health status or diagnostic gradings of disease are often measured on multicategory scales as "normal" to "definitely abnormal" (diagnosis) or as "good" to "poor" (health). Among statistical issues raised by such data are whether an underlying latent scale assumption need be invoked (Armstrong and Sloan, 1989); and whether or not the ordering of responses is relevant to stratifying regression relationships, for example with different slopes according to each ordinal response category (Anderson, 1984).

   The regression model for an observed ordinal response variable $y_i \in (1, \ldots, K)$ is usually taken to reflect an underlying continuous random variable $z$ such that

$$z_i = X_i\beta + \varepsilon_i,$$

where $\varepsilon_i$ commonly has a normal or logistic distribution function, $P(\varepsilon) = \Phi(\varepsilon)$, or $P(\varepsilon) = 1/(1 + \exp(-\varepsilon))$. The observed $y_i$ is defined according to the location on the $z$ scale: thus $y_i = j$ when $\theta_{j-1} \le z_i < \theta_j$. So with binary indicators $d_{ij} = 1$ if $y_{1i} = j$, $d_{ij} = 0$ otherwise, and $d_i = (d_{i1}, \ldots, d_{iK})$, one has

$$d_i \sim Mult(1, \mathbf{p}_i),$$

$$\mathbf{p}_i = (p_{i1}, \ldots, p_{iK}),$$

$$p_{ij} = \Pr(y_i = j) = \Pr(\theta_{j-1} \le z_i < \theta_j),$$

$$= \Pr(\theta_{j-1} \le X_i\beta + \varepsilon_i < \theta_j),$$

$$= P(\theta_j - X_i\beta) - P(\theta_{j-1} - X_i\beta)$$

$$= \gamma_{ij} - \gamma_{i,j-1},$$

where

$$\gamma_{ij} = \Pr(y_i \le j) = P(\theta_j - X_i\beta), \qquad j = 1, \ldots, K - 1$$

are cumulative probabilities over ranked categories, $\gamma_{ij} = p_{i1} + \ldots + p_{ij}$. Conversely the probabilities of an observation being in the $j^{th}$ ranked category are given by differencing the cumulative probabilities.

$$p_{i1} = \gamma_{i1}$$

$$p_{ij} = \gamma_{ij} - \gamma_{i,j-1}, \qquad j = 2, \ldots, K-1,$$

$$p_{iK} = 1 - \gamma_{i,K-1}.$$

If $P(\varepsilon)$ is a logistic, with

$$C_{ij} = \text{logit}(\gamma_{ij}) = \theta_j - X_i\beta,$$

and $\beta$ is uniform across response categories $j$, then the $\theta_j$ are the logits of belonging to categories $(1, \ldots, j)$ up to and including the $j^{th}$ (as against categories $j+1, \ldots, K$) for subjects with $X = 0$. The difference in cumulative logits for different values of $X$, for example $X_1$ and $X_2$, is independent of $j$. This is known as the 'proportional odds' property:

$$C_{1j} - C_{2j} = (X_2 - X_1)\beta.$$

Assuming $X_i$ excludes an intercept, the $K-1$ thresholds $\{\theta_1, \theta_2 \ldots, \theta_{K-1}\}$ are unknowns subject to the order constraint $\theta_1 \le \theta_2 \ldots \le \theta_{K-1}$. A convenient prior involves reparameterisation

$$\theta_j = \theta_{j-1} + \exp(\Delta_j) \qquad j = 2, \ldots, K-1,$$

$$\theta_1 = \Delta_1,$$

with the $\Delta$ parameters taken as unconstrained normal. A constrained prior is avoided in the procedure used in the program JAGS whereby a set of K-1 unranked parameters $\theta_0 = (\theta_{01}, \ldots, \theta_{0,K-1})$ are sampled (e.g. using a diffuse normal prior), and then the required ranked $\theta_j$ are obtained by sorting[8].

Residual assessments for ordinal regression are complicated by the presence of $K-1$ residuals for each observation, and by the fact that such residuals are correlated (Johnson and Albert, 1999). Predictive assessment involving replicates $y_{rep,i} \sim Mult(1, \mathbf{p}_i)$ may be combined with checks comparing $y_{rep,i}$ with $y_i$. Predictive and residual assessment may be simpler in the latent response scale using replicates $z_{rep,i}$ and residuals $\varepsilon_i = z_i - X_i\beta$ and $\varepsilon_{rep,i} = z_{rep,i} - X_i\beta$.

## Example 3.7    Infection after caesarian birth

As an example of multinomial logistic analysis, consider data from Fahrmeier and Tutz (1994) involving categorical predictors and an unordered categorical response relating to infection in 251 births involving Caesarian section; namely no infection, type I infection, and type II infection. The $P = 3$ risk factors are defined as NOPLAN = 1 (if the Caesarian is unplanned, 0 otherwise), ANTIB = 1 if antibiotics were given as prophylaxis, and FACTOR = 1 (if risk factors were present, 0 otherwise). Of the eight possible predictor combinations, seven were observed. The numbers of maternities under the possible combinations range from 2 (for NOPLAN = 0, ANTIB = 1, FACTOR = 0) to 98 (NOPLAN = 1, ANTIB = 1, FACTOR = 1). N(0,10) priors are assumed for the impacts $\beta_{kj}$ of the binary predictors on outcomes $j$, and a multinomial logistic model is adopted with

$$\log(p_{ij}/p_{i1}) = \alpha_j + \sum_{k=1}^{P} \beta_{kj}x_{ik}.$$

The estimates for this MNL model show antibiotic prophylaxis as decreasing the relative risk of infection type I more than type II. Hence $\beta_{22}$ is more highly negative than $\beta_{23}$. By contrast, the presence of risk factors in a birth increases the risk of type II infection more: the relative risk of type II infection with risk factors present is $\exp(a_3 + \beta_{33}) = \exp(-2.48 + 2.11) = 0.70$

times the infection risk when the risk factors are absent, whereas for type I infections the corresponding relative risk is about 0.42 (see Table 3.2).

In fact, MCMC sampling can provide a full density for such structural parameters, rather than obtaining 'point estimates' based on the posterior means of constituent parameters. This involves defining new quantities $s_1 = \exp(\alpha_3 + \beta_{33})$ and $s_2 = \exp(\alpha_2 + \beta_{32})$ and monitoring them in the same way as basic parameters. One can assess the probability that $s_1 > s_2$ by using the step() function in BUGS.

If there were not strong evidence that $s_1 > s_2$, the coefficients on NOPLAN, ANTIB and FACTOR might be equalised since Table 3.2 suggests homogeneity across the two infection types. This would involve three fewer parameters and may provide a better fit.

### Example 3.8    Central heating and cooling choice

As an example of nested logit modelling to take account of possible departures from the IIA axiom, consider data on choice between heating and central cooling systems for 250 new single-family houses in California, one of the datasets included in the mlogit package in R. The alternatives are

1. Central cooling and gas central heating (gcc)

2. Central cooling and electric central heating (ecc)

3. Central cooling and electric room heating (esc)

4. Heat pump (central cooling and heating) (hpc)

5. No central cooling, gas central heating (gc)

6. No central cooling, electric central heating (ec)

7. No central cooling, electric room heating (er).

A natural nesting is implied in the choice of options with and without central cooling, namely options 1 to 4 (nest 1) and options 5 to 7 (nest 2).

Choice is modelled as a function of attributes which differ by respondent ( $H$ variables). The $P = 7$ predictors are the installation cost for the heating portion of the system; the operating cost for heating; the installation cost for the cooling portion; the operating cost for cooling; interaction between annual income and electric room heating (options 3,7); interaction between annual income and cooling (options 1 to 4); and a binary variable coded 1

**Table 3.2**    Caesarian birth and infection.

| Parameter | Mean | St devn | 2.5% | 97.5% |
|---|---|---|---|---|
| $\alpha_2$ | −2.57 | 0.55 | −3.79 | −1.60 |
| $\alpha_3$ | −2.48 | 0.54 | −3.67 | −1.56 |
| $\beta_{12}$ | 1.06 | 0.51 | 0.07 | 2.05 |
| $\beta_{13}$ | 0.90 | 0.47 | 0.00 | 1.83 |
| $\beta_{22}$ | −3.40 | 0.65 | −4.73 | −2.19 |
| $\beta_{23}$ | −2.99 | 0.53 | −4.06 | −1.99 |
| $\beta_{32}$ | 1.71 | 0.57 | 0.68 | 2.91 |
| $\beta_{33}$ | 2.11 | 0.58 | 1.05 | 3.33 |

for options 1 to 4 and zero otherwise. Without nesting (i.e. the standard model), the choice between options is modelled for $K = 7$ according to a conditional logistic

$$y_{ik} \sim Mult(1, p_{ik}), \qquad p_{ik} = \exp(H_{ik}\delta) / \sum_{j=1}^{K} \exp(H_{ij}\delta),$$

where $\delta = (\delta_1, \dots, \delta_P)$ is constant across alternatives. Let $V_{ik} = H_{ik}\delta$. With nesting and assuming a common log-sum coefficient $\lambda$ across nests, one has

$$p_{ik} = \frac{\exp(V_{ik}/\lambda) \left[ \sum_{j=1}^{4} \exp(V_{ij}/\lambda) \right]^{\lambda-1}}{\left[ \sum_{j=1}^{4} \exp(V_{ij}/\lambda) \right]^{\lambda} + \left[ \sum_{j=5}^{7} \exp(V_{ij}/\lambda) \right]^{\lambda}} \qquad k = 1, \dots, 4$$

$$p_{ik} = \frac{\exp(V_{ik}/\lambda) \left[ \sum_{j=5}^{7} \exp(V_{ij}/\lambda) \right]^{\lambda-1}}{\left[ \sum_{j=1}^{4} \exp(V_{ij}/\lambda) \right]^{\lambda} + \left[ \sum_{j=5}^{7} \exp(V_{ij}/\lambda) \right]^{\lambda}} \qquad k = 5, \dots, 7.$$

A frequentist analysis of these data is useful for setting priors and initial values. These results (and the construction of the predictors used in the Bayesian analysis) are obtained by the following sequence of commands in R:

```
library(mlogit); data("HC")
HC <- mlogit.data(HC, varying = c(2:8, 10:16),
          choice = "depvar", shape = "wide")
cooling.modes <- attr(HC, "index")$alt %in% c("gcc", "ecc", "erc", "hpc")
room.modes <- attr(HC, "index")$alt %in% c("erc", "er")
HC$icca[!cooling.modes] <- 0; HC$occa[!cooling.modes] <- 0
HC$inc.cooling <- HC$inc.room <- 0
HC$inc.cooling[cooling.modes] <- HC$income[cooling.modes]
HC$inc.room[room.modes] <- HC$income[room.modes]
HC$int.cooling <- as.numeric(cooling.modes)
M1 <- mlogit(depvar~ich+och+icca+occa+inc.room+inc.cooling
                            +int.cooling|0,data=HC)
summary(M1)
M2 <- mlogit(depvar~ich+och+icca+occa+inc.room+inc.cooling+int.cooling|0, HC,
nests = list(cooling = c("gcc", "ecc","erc", "hpc"),
          other = c("gc", "ec", "er")), un.nest.el = TRUE, print.level = 0)
summary(M2)
```

The frequentist analysis shows a small gain in log-likelihood (from $-180.3$ to $-178.1$) with $\lambda$ estimated as 0.59, and 95% interval entirely under 1. While values of $\lambda$ exceeding 1 are sometimes obtained, here the prior on $\lambda$ in Bayesian analysis is confined to values under 1. MCMC estimation of the nested logit model involved two chains of 75 000 iterations with some delay in convergence in the less well identified predictor effects, namely $\delta_4$ and $\delta_7$. Table 3.3 summarises parameter estimates from the last 25 000 iterations, and also from the last 5000 iterations of a 25 000 iteration estimation of the standard conditional logistic model. The posterior mean for $\lambda$ slightly exceeds the frequentist estimate but the posterior density does not contain a spike at 1: such a spike would suggest the standard model was appropriate. However, the average mean deviance is reduced by only 3.2 at the cost of an extra parameter.

### Example 3.9    Political involvement: Ordinal regression

Consider data from Tarling (2009) on whether it is possible to influence political decisions in the subject's local area. There are $n = 9574$ subjects and $K = 4$ choices ($1 =$ definitely agree that it's possible to influence local decisions, $2 =$ tend to agree, $3 =$ tend to disagree that it's possible to influence local decisions, $4 =$ definitely disagree). So the latent scale is a measure of scepticism about ability to influence decisions. Predictors are number of years lived in neighbourhood, education level ($1 =$ higher education, $2 =$ secondary "A level", $3 =$ secondary "GCSE", $4 =$ no qualifications, with the last as reference), whether subject reads the local newspaper ($1 =$ does read, $0 =$ doesn't read), and civic participation ($1 =$ does participate, $0 =$ doesn't participate).

The core code for a proportional effects ordinal logistic regression, incorporating the constrained prior on $\theta$ discussed in Section 3.6.2, is as follows:

```
model { for (i in 1:n) { y[i]~dcat(pi[i,1:K]);
for (j in 1:K-1) { gam[i,j] <- 1/(1+exp(-theta[j] + eta[i,j]))
eta[i,j] <- beta[1]*tneig[i]+beta[2]*equals(educ[i],1)
+beta[3]*equals(educ[i],2)+beta[4]*equals(educ[i],3)
+beta[5]*equals(news[i],1)+beta[6]*equals(pciv[i],1)}
pi[i,1] <- gam[i,1]; pi[i,K] <- 1-gam[i,K-1]
for (j in 2:K-1) { pi[i,j] <- gam[i,j] - gam[i,j-1] }}
theta[1] ~dnorm(0,1) ; for (j in 2:K-1) {delta[j] ~dexp(1);
theta[j] <- theta[j-1]+delta[j]}
beta[1:p] ~dmnorm(b0[ ], B0[ , ])}.
```

Ordinal probit regression replaces the second line with

```
gam[i,j] <- phi(theta[j] - eta[i,j]).
```

**Table 3.3**    Heating choice, parameter estimates.

|  | Parameter | Mean | St dvn | MC error | 2.5% | 97.5% |
|---|---|---|---|---|---|---|
| Standard | $\delta_1$ | −0.87 | 0.06 | 0.006 | −0.98 | −0.76 |
| Conditional | $\delta_2$ | −1.36 | 0.12 | 0.013 | −1.60 | −1.16 |
| Logistic | $\delta_3$ | −0.35 | 0.07 | 0.008 | −0.48 | −0.22 |
|  | $\delta_4$ | −2.65 | 0.20 | 0.024 | −3.03 | −2.30 |
|  | $\delta_5$ | −0.60 | 0.05 | 0.005 | −0.69 | −0.50 |
|  | $\delta_6$ | 0.34 | 0.06 | 0.007 | 0.21 | 0.44 |
|  | $\delta_7$ | −6.15 | 0.69 | 0.087 | −7.09 | −5.27 |
|  | Log-likelihood | −183.3 | 1.4 | 0.1 | −186.8 | −181.4 |
| Nested | $\delta_1$ | −0.62 | 0.11 | 0.010 | −0.82 | −0.42 |
| Logistic | $\delta_2$ | −0.98 | 0.18 | 0.016 | −1.32 | −0.64 |
|  | $\delta_3$ | −0.28 | 0.10 | 0.009 | −0.45 | −0.08 |
|  | $\delta_4$ | −0.94 | 0.51 | 0.051 | −1.93 | −0.08 |
|  | $\delta_5$ | −0.42 | 0.08 | 0.007 | −0.57 | −0.28 |
|  | $\delta_6$ | 0.27 | 0.05 | 0.003 | 0.18 | 0.36 |
|  | $\delta_7$ | −6.48 | 1.58 | 0.159 | −9.03 | −4.35 |
|  | $\lambda$ | 0.68 | 0.13 | 0.011 | 0.44 | 0.94 |
|  | Log-likelihood | −181.7 | 1.8 | 0.1 | −185.9 | −179.0 |

Additional code, namely

```
yrep[i] ~dcat(pi[i,1:K]); pred.fit[i] <- equals(y[i],yrep[i])
```

with summation pred.fit.pct <- 100*sum(pred.fit[])/n is seeking to measure percentage success of predictive classification. For ordinal regression, the distance (gap in ranks) between the observation $y_i$ and the prediction $y_{rep,i}$ is also relevant, and one could compare different models in terms of probabilities of misclassification by one rank only, two ranks, etc.

Ordinal regression can also be implemented in WinBUGS14 using the option

```
y[i] ~dcat.aux(theta[1:KM],z[i],1),
```

where KM = K-1 is one less than the number of categories, and with z[i] ~dnorm(eta[i], 1) for probit, or z[i] ~dlogis(eta[i], 1) for logit, where $\eta_i = X_i\beta$ is the regression term. This leads on to simpler predictive assessments. MCMCpack in R can also be used for ordinal probit regression, and enables Bayes factor caculations using the Chib (1995) method.

Convergence is obtained early under either link and Table 3.4 summarises the parameter estimates. The more highly educated, those reading local news, and those participating in civil

**Table 3.4**  Perceptions of political influence (influences on negative perceptions).

| | Parameter Summary Ordinal Probit | | | | | |
|---|---|---|---|---|---|---|
| | Parameter | Mean | St devn | MC error | 2.5% | 97.5% |
| Length of residence in neighb'd | $\beta_1$ | 0.0041 | 0.0007 | 0.00003 | 0.0027 | 0.0054 |
| Higher education | $\beta_2$ | −0.54 | 0.03 | 0.001 | −0.61 | −0.48 |
| Secondary 'A level' | $\beta_3$ | −0.37 | 0.04 | 0.001 | −0.44 | −0.29 |
| Secondary GCSE | $\beta_4$ | −0.25 | 0.03 | 0.001 | −0.31 | −0.19 |
| Read local news | $\beta_5$ | −0.14 | 0.03 | 0.001 | −0.19 | −0.09 |
| Civic participation | $\beta_6$ | −0.26 | 0.02 | 0.001 | −0.31 | −0.22 |
| Percent correct classified | | 29.9 | 0.5 | 0.010 | 29.0 | 30.8 |
| Cut point 1 | $\theta_1$ | −1.73 | 0.04 | 0.002 | −1.80 | −1.66 |
| Cut point 2 | $\theta_2$ | −0.59 | 0.03 | 0.002 | −0.66 | −0.53 |
| Cut point 3 | $\theta_3$ | 0.30 | 0.03 | 0.002 | 0.23 | 0.37 |
| | Ordinal Logit | | | | | |
| | Parameter | Mean | St devn | MC error | 2.5% | 97.5% |
| Length of residence in neighb'd | $\beta_1$ | 0.0071 | 0.0012 | 0.00004 | 0.0047 | 0.0094 |
| Higher education | $\beta_2$ | −0.94 | 0.05 | 0.002 | −1.05 | −0.84 |
| Secondary 'A level' | $\beta_3$ | −0.63 | 0.06 | 0.002 | −0.75 | −0.50 |
| Secondary GCSE | $\beta_4$ | −0.43 | 0.05 | 0.002 | −0.53 | −0.33 |
| Read local news | $\beta_5$ | −0.25 | 0.04 | 0.002 | −0.33 | −0.16 |
| Civic participation | $\beta_6$ | −0.43 | 0.04 | 0.001 | −0.50 | −0.35 |
| Percent correct classified | | 30.0 | 0.5 | 0.025 | 29.1 | 30.9 |
| Cut point 1 | $\theta_1$ | −2.97 | 0.06 | 0.004 | −3.09 | −2.84 |
| Cut point 2 | $\theta_2$ | −1.00 | 0.06 | 0.003 | −1.11 | −0.88 |
| Cut point 3 | $\theta_3$ | 0.47 | 0.06 | 0.003 | 0.36 | 0.58 |

activities are less likely to have sceptical views regarding their influence over local decisions. The proportional odds assumption for ordinal logistic or ordinal probit regression is that the relationship between each pair of response categories is the same. So coefficients that describe the relationship between the lowest category versus all higher response categories are the same as those describing the relationship between the next lowest category and all higher categories. If proportional odds does not hold different regressions are needed to describe the relationship between each pair of response categories. There is in fact evidence of non-proportionality with the political influence data, and Exercise 3.9 involves assessing any improvement in fit through allowing category specific coefficients.

## Exercises

**3.1.** In Example 3.1, adapt the code to include a posterior predictive $p$-tests to assess skewness and kurtosis in the residuals. For example, the $p$-test for skewness would compare a skew measure for replicates

$$D_{rep}^{skew} = \frac{1}{n} \sum_{i=1}^{n} \varepsilon_{rep,i}^{3(t)} / \sigma^{3(t)} \quad \text{against} \quad D_{obs}^{skew} = \frac{1}{n} \sum_{i=1}^{n} \varepsilon_{i}^{3(t)} / \sigma^{3(t)}.$$

**3.2.** In Example 3.2 (student attainment), standardize the SATM-score predictor to provide values $x_i^s$. Then replace the conventional prior by a CMP prior at values $\tilde{x}_1 = (1, x_L^s) = (1, -1)$ and $\tilde{x}_2 = (1, x_U^s) = (1, 1)$. Assess sensitivity in the optimal threshold $\eta_d^T$ to this alternative prior.

**3.3.** Using data originally from Mullahy (1997) on smoking consumption (cigarettes smoked per day) by $n = 807$ subjects, assess the suitability of a Poisson regression using a posterior predictive $p$-test to assess overdispersion. Focal response points, especially at one pack per day ($y = 20$), may cast doubt on the Poisson assumption. The predictors are years of education, logarithm of income, logarithm of cigarette price, restaurant ban (binary), white or nonwhite (binary) and age. As discussed above, one option involves comparing $D = 0.5 \sum_{i=1}^{n} \{(y_i - \mu_i)^2 - y_i\}$ with its replicate data counterpart $D_{rep} = 0.5 \sum_{i=1}^{n} \{(y_{rep,i} - \mu_i)^2 - y_{rep,i}\}$.

**3.4.** Using data on Irish education transitions (http://lib.stat.cmu.edu/datasets/irish.ed) compare logit and probit regression using the augmented data method and the dbern.aux function. Take $P = 2$ predictors, namely sex and DVRT (Drumcondra Verbal Reasoning Test), with response $y = 0$ if leaving certificate not taken, and $y = 1$ if taken. Specifically obtain the number of subjects with probabilities $\Pr(z_{rep,i} < z_i | y)$ either under 0.1 or over 0.9. Also obtain the number of cases with elevated residuals $\varepsilon_i = z_i - X_i \beta$, namely with $\Pr(|\varepsilon_i| \| y) > 1.96$ under a probit regression, and $\Pr(|\varepsilon_i| / \kappa^{0.5}) > 2.02$ under a logistic regression. The basic code (not including these checks) for a probit regression using the dbern.aux function is

```
model { for (i in 1:n) {y[i] ~dbern.aux(z[i]);
z[i] ~dnorm(eta[i],1); eta[i] <-
b[1]+b[2]*x[i,1]+b[3]*x[i,2]
x[i,1] <- equals(sex[i],2); x[i,2] <- DVRT[i]}
for (j in 1:3) {b[j] ~dnorm(0,tau.beta)}}.
```

Also using the dbern.aux function assess the best subset of predictors (for probit regression only) via the jump.lin.pred function. In this case the basic code omits specifying the eta[] regression term, or a prior on the regression coefficients, though a prior precision on the coefficients (tau.beta) is included as a known input. Thus one has

```
model { for (i in 1:n) {y[i] ~dbern.aux(z[i])
z[i] ~dnorm(eta[i], 1)
x[i,1] <- equals(sex[i],2); x[i,2] <- DVRT[i]}
eta[1:n] <- jump.lin.pred(x[1:n, 1:2], p.ret, tau.beta)
# Under Inference/samples, enter model.id
# Then select jump/summarize and type model.id in "id node"
model.id <- jump.model.id(eta[1:n]); p.ret ~dbin(0.5,p)}
```

**3.5.** Apply the Kuo–Mallick model to predictor selection in the nodal involvement data (Example 3.4) using a uniform prior on $k$ between the extremes $k = 0.5$ and $k = 4$. So the variance in the normal prior for $\beta_j$ is

$$V_j = k(1.18)^2, \quad k \sim U(0.5, 4).$$

How does this affect the posterior model probabilities for $\{x_1, x_2, x_3\}$ and $\{x_1, x_2, x_3, x_4\}$.

**3.6.** In Example 3.6 use the WinBUGS jump RJMCMC interface to obtain the highest posterior probability model for the Chevrolet asking price data, assuming a prior $P_{ret} \sim$ Bin$(P, 0.5)$ on the number of retained predictors (with $P = 5$). This can be implemented by adapting the RJMCMC code from Example 3.3, including statements for the conjugate normal prior assumed in Example 3.6:

```
tau.beta.prec <- 5/tau
tau ~dgamma(1.5,0.5).
```

How far does this model remedy the predictive deficiencies identified in Example 3.6.

**3.7.** In Example 3.7, define quantities $s_1 = \exp(\alpha_3 + \beta_{33})$ and $s_2 = \exp(\alpha_2 + \beta_{32})$, and by monitoring them obtain the probability that $s_1 > s_2$.

**3.8.** In Example 3.8, compare models 1 and 2 (standard conditional logistic and nested logit) using LPML and DIC criteria, and also predictive classification success: how far predicted choice obtained by sampling replicate data matches actual choice (see Example 3.9 for an illustration). Also evaluate the nested logit with differing $\lambda$ coefficients between the two subsets.

**3.9.** In Example 3.9 (political involvement), compare predictive accuracy, DIC and LPML between the proportional odds logistic model and a model allowing the regression effect to differ by response category,

$$\gamma_{ij} = \Pr(y_i \le j) = P(\theta_j - X_i \beta_j), \quad j = 1, \ldots, K - 1.$$

**3.10.** Compare the suitability of ordinal logistic and ordinal probit regression for the political involvement data using the posterior predictive criteria

$$\Pr(z_{rep,i} > z_i | y).$$

The $z$ and $z_{rep}$ are readily obtained using the likelihood command

```
y[i] ~dcat.aux(theta[1:KM],z[i],1),
```

where KM = K-1 = 3 is one less than the number of categories, and with

```
z[i] ~dnorm(eta[i], 1),
```

for ordinal probit regression with regression term eta[i] excluding an intercept, and

```
z[i] ~dlogis(eta[i], 1),
```

for ordinal logit regression. Specifically one may obtain a tally of poorly fitted cases, where either $Pr(z_{rep,i} > z_i|y) < 0.1$ or $Pr(z_{rep,i} > z_i|y) > 0.9$.

# Notes

1. The code for the model in Example 3.1 is

```
model { for (i in 1:n) {y[i] ~dnorm(mu[i],tau)
mu[i] <- beta[1]+beta[2]*log(x[i])
eps[i] <- y[i]-mu[i]
# replicate responses and errors
yrep[i] ~dnorm(mu[i],tau); eps.rep[i] <- yrep[i]-mu[i]
# outlier measures
d.rep[i] <- step(yrep[i]-y[i]); d.eps[i] <-step(abs(eps.st[i])-1.96)
eps3[i] <- pow(eps[i],3); eps4[i] <- pow(eps[i],4)
eps.st[i] <- eps[i]/sig; eps.rep.st[i] <- (yrep[i]- mu[i])/sig}
# Check error normality
for (i in 1:n) {
# observed and expected cumulative proportions
C[i] <- rank( eps.st[], i )/n;
C.rep[i] <-rank(eps.rep.st[], i )/n
C.exp[i] <- phi( eps.st[i] ); C.rep.exp[i] <-phi(eps.rep.st[i] )
# posterior predictive check components and p-test
ch2[i] <- pow(C[i]-C.exp[i],2)/(C.exp[i]*(1-C.exp[i]))
ch2.rep[i] <-pow(C.rep[i]-C.rep.exp[i],2)/(C.rep.exp[i]*(1-C.rep.exp[i]))}
Ch2 <- sum(ch2[]); Ch2.rep <- sum(ch2.rep[])
p.nrm <- step(Ch2.rep - Ch2)
# Whites test for heteroscedasticity
for (i in 1:n) {eps2[i] <- eps[i]*eps[i];
      eps2.hat[i] <- inprod(b.eps[],X[,i])
e1[i] <- pow(eps2[i]-eps2.hat[i],2); e2[i] <-pow(eps2[i]-mean(eps2[]),2)
X[1,i] <- 1; X[2,i] <- log(x[i]); X[3,i] <- log(x[i])*log(x[i])}
for (j in 1:3) {for (k in 1:3) {XTX[j,k] <- inprod(X[j,], X[k,])}}
XTX.inv[1:3,1:3] <- inverse(XTX[,])
for (i in 1:3) {X.eps2[i] <- inprod(eps2[], X[i,]);
b.eps[i] <- inprod(X.eps2[], XTX.inv[i,])}
R2.eps <- (1-sum(e1[])/sum(e2[])); n.R2.eps <-n*R2.eps
# Whites test, replicate data
for (i in 1:n) {eps.rep2[i] <- eps.rep[i]*eps.rep[i]
eps.rep2.hat[i] <- inprod(b.eps.rep[],X[,i])
e1.rep[i] <- pow(eps.rep2[i]-eps.rep2.hat[i],2);
e2.rep[i] <- pow(eps.rep2[i]-mean(eps.rep2[]),2) }
for (i in 1:3) {X.eps.rep2[i] <- inprod(eps.rep2[], X[i,]);
b.eps.rep[i] <- inprod(X.eps.rep2[], XTX.inv[i,])}
R2.eps.rep <- 1-sum(e1.rep[])/sum(e2.rep[])
```

```
# compare to 5% cumulative chi-square for 2df
J.hx <- step(n.R2.eps-5.99)
# posterior predictive check
p.hx <- step(R2.eps.rep-R2.eps)
#priors
for (j in 1:2) {beta[j] ~dflat()}
log.sig ~dflat(); tau <- exp(-2*log.sig); sig <-exp(log.sig)
sig.rep <- sd(yrep[])
# testing error skew and kurtosis (against Normal values of 0 and 3)
D.sk <- sum(eps3[])/(n*pow(sig,3))
D.krt <- sum(eps4[])/(n*pow(sig,4))
J.sk <- step(D.sk); J.kt <- step(D.krt-3)}
```

2. The code for Example 3.2 is

```
model { for (i in 1:n) {y[i] ~dbern(p[i])
eta[i] <- beta[1]+beta[2]*(SATM[i]-mean(SATM[]))
logit(p[i]) <- beta[1]+beta[2]*(SATM[i]-mean(SATM[]))
# replicate data
yrep[i] ~dbern(p[i])
LL[i] <- y[i]*log(p[i])+(1-y[i])*log(1-p[i])
# inverse likelihood (to derive CPO)
G[i] <- 1/exp(LL[i])
y1[i] <- equals(y[i],1); y0[i] <- equals(y[i],0)
# posterior-predictive distribution
m[i] <- equals(y[i],yrep[i])
# threshold analysis
for (d in 1:D) {yattr[i,d] <- step(eta[i]-etaT[d])
n11[i,d] <- y1[i]*equals(yattr[i,d],y[i])
n00[i,d] <- y0[i]*equals(yattr[i,d],y[i])}}
# priors
for (j in 1:2) {beta[j] ~dnorm(0,0.01)}
# sensitivity, specificity at each threshold
for (d in 1:D) {sens[d] <- sum(n11[,d])/sum(y1[])
                spec[d] <-sum(n00[,d])/sum(y1[])
# combined measure (Bohning et al., 2008)
totacc[d] <- sens[d]+spec[d]}}
```

3. The code for the RJMCMC analysis is

```
model { for (i in 1:N) { y[i] ~dnorm(mu[i], tau)}
for (j in 1:P) {# Indicator for Inclusion of predictor j
Retain[j] <- 1-equals(beta.jump[j], beta.jump[P1])
Marker[j,j] <- 1; Marker[P1,j] <- 0
for (k in 1:j-1) { Marker[j,k] <- 0; Marker[k,j] <- 0}}
mu[1:N] <- jump.lin.pred(x.stand[1:N,1:P], P.ret, tau.beta.prec)
ModSel <- jump.model.id(mu[1:N])
beta.jump[1:P1] <- jump.lin.pred.pred(mu[1:N], Marker[1:P1,1:P])
# realised regression parameters
for (j in 1:P) { beta[j] <- beta.jump[j] - beta.jump[P1]}
alpha <- beta.jump[P1]
# prior on number of retained predictors
P.ret ~dbin(0.25, P);
# assumed precision of beta coefficients
tau.beta.prec <- 0.0001;
tau ~dgamma(1,0.001)}
```

To monitor model probabilities, set ModSel under Inference/Samples, then for ModSel use Jump/Summarize/Table.

4. Devising a BUGS code for the $g$-prior model is complicated by use of the prior

$$p(\tau) = 1/\tau$$

for $\tau$, which is not available in BUGS. One possible indirect approach is a discrete prior over a suitable range of prior values for $1/\tau$, possibly drawing on results of another form of estimation for $\tau$ to decide such a suitable range. The value of the Bayes factor relative to the null (constant only) model is estimated using the posterior mean for the coefficient of determination (R2 in the code below):

```
model {for (i in 1:n) {y[i] ~dnorm(mu[i],phi); mu[i] <-alpha+sum(reg[i,])
SSerr[i] <- pow(y[i]-mu[i],2); SStot[i] <-pow(y[i]-mean(y[]),2)
for (j in 1:P) {reg[i,j] <- beta[j]*x.st[i,j]
              x.st[i,j] <- (x[i,j]-x.m[j])/x.sd[j]}}
for (j in 1:P) {for (k in 1:P) { P.beta[j,k] <-
g.inv*tau*XX[j,k]
XX[j,k] <- inprod(x.st[,j],x.st[,k])}}
g.inv <- 1/g
R2 <- 1-sum(SSerr[])/sum(SStot[])
# grid prior on tau
for (i in 1:nbins){tau.prior[i] <- 1/(i*100)
tau.prior.sc[i] <- tau.prior[i]/sum(tau.prior[])}
k.bin ~dcat(tau.prior.sc[1:nbins]); tau <-tau.prior[k.bin];
alpha ~dflat(); beta[1:P] ~dmnorm(nought[1:P],P.beta[1:P,1:P])}
```

For the elementary school performance data in Example 3.3, maximum likelihood estimates of $\sigma^2$ are approximately 3200, and a discrete prior with 50 bins between $1/100$ and $1/5000$ could be adopted for $\tau$.

5. The code for the model selection focus in Example 3.3, including the discrete prior on $1/\sigma^2$, is

```
model {for (i in 1:n) { y[i] ~dnorm(mu[i],tau);
mu[i] <- alpha+sum(reg[i,])
for (j in 1:P) {reg[i,j] <- delta[j]*x.st[i,j]
x.st[i,j] <- (x[i,j]-x.m[j])/x.sd[j]}}
# realised coefficients
for (j in 1:P) {delta[j] <- gam[M,j]*beta[j]
for (k in 1:P) {P.beta[j,k] <- tau*XX[j,k]/g
XX[j,k] <- inprod(x.st[,j],x.st[,k])}}
# Priors
beta[1:P] ~dmnorm(nought[1:P],P.beta[1:P,1:P])
alpha ~dflat();
# prior on residual precision
for (i in 1:nbins){tau.prior[i] <- 1/(i*100)
tau.prior.sc[i] <- tau.prior[i]/sum(tau.prior[])}
k.tau ~dcat(tau.prior.sc[1:nbins]); tau <-tau.prior[k.tau]; sig2 <- 1/tau
# prior on g (set or unknown priors defined by g=gset or g=gfree
              respectively)
g <- gset
gset <- n
A ~dbeta(1,0.5); gfree <- A/(1-A)
# probability prior on models
```

```
# set or unknown model probability priors defined by pmodset[],
      pmodunk[] respectively
M ~dcat(pmod[1:K])
for (k in 1:K) {ModSel[k] <- equals(k,M)
pmod[k] <- pmodset[k]
pmodset[k] <- 1/K
pmodunk[k] <- Pmod[k]/sum(Pmod[]); Pmod[k] <-1/(1+pow(kap,dim[k]))}
kap ~dexp(1)
# Values for gam under alternative models (could also be read in as data)
gam[1,1] <- 1; gam[1,2] <- 1; gam[1,3] <- 0; gam[1,4] <- 1
gam[2,1] <- 1; gam[2,2] <- 1; gam[2,3] <- 1; gam[2,4] <- 1}
```

6. Code for the Congdon (2007) method in Example 3.4 (nodal involvement) is

```
model {for (i in 1:n) {y1[i] <- y[i]; y2[i] <-y[i]
# M1: p1=3 predictors
y1[i] ~dbern.aux(Z1[i]); Z1[i] ~dnorm(eta1[i], 1)
eta1[i] <- alpha1 + sum(reg1[i,])
# M2: p2=4 predictors
y2[i] ~dbern.aux(Z2[i]); Z2[i] ~dnorm(eta2[i], 1)
eta2[i] <- alpha2 + sum(reg2[i,])
# log likelihoods M1 and M2
for (j in 1:p1) {reg1[i,j] <- beta1[j]*x[i,j]}
for (j in 1:p2) {reg2[i,j] <- beta2[j]*x[i,j]}
pi1[i] <- phi(eta1[i]);
pi2[i] <- phi(eta2[i])
LLk1[i] <- y[i]*log(pi1[i]) + (1-y[i])*log(1-pi1[i])
LLk2[i] <- y[i]*log(pi2[i]) + (1-y[i])*log(1-pi2[i])}
# priors
alpha1 ~dnorm(0,0.01);
for (j in 1:p1) {beta1[j] ~dnorm(0.75,0.04)}
alpha2 ~dnorm(0,0.01);
for (j in 1:p2) {beta2[j] ~dnorm(0.75,0.04)}
# M1: prior ordinates of sampled parameters against prior
logPrior.alpha1 <- 0.5*(log(0.01/6.283)-0.01*pow(alpha1,2))
for (j in 1:p1) {logPrior.beta1[j] <-
0.5*(log(0.04/6.283)-0.04*pow(beta1[j]-0.75,2))}
# M2: prior ordinates of sampled parameters against prior
logPrior.alpha2 <- 0.5*(log(0.01/6.283)-0.01*pow(alpha2,2))
for (j in 1:p2) {logPrior.beta2[j] <-0.5*(log(0.04/6.283)-0.04*
                          pow(beta2[j]-0.75,2))}
# total log prior ordinates
pi[1] <- sum(logPrior.beta1[1:p1])+logPrior.alpha1
pi[2] <- sum(logPrior.beta2[1:p2])+logPrior.alpha2
# log likelihoods
LL[1] <- sum(LLk1[]); LL[2] <- sum(LLk2[])
# log ordinate estimated posterior den-
sity of th1[dimension d1=p1+1], th2[d2=1+p2]
th1[1] <- alpha1; for (j in 1:p1) {th1[j+1] <-beta1[j]}
th2[1] <- alpha2; for (j in 1:p2) {th2[j+1] <-beta2[j]}
g[1] <- -0.5*d1*log(2*3.1416) - 0.5*logdet(Sigma.th1[,]) -0.5*R1
g[2] <- -0.5*d2*log(2*3.1416) - 0.5*logdet(Sigma.th2[,]) -0.5*R2
Inv.Sigma.th1[1:d1,1:d1] <- inverse(Sigma.th1[,])
for (i in 1:d1) { resid1[i] <- th1[i] - mu.th1[i]
Q1[i] <- inprod(Inv.Sigma.th1[i,], resid1[])}
R1 <- inprod(resid1[],Q1[])
Inv.Sigma.th2[1:d2,1:d2] <- inverse(Sigma.th2[,])
```

```
for (i in 1:d2) { resid2[i] <- th2[i] - mu.th2[i]
Q2[i] <- inprod(Inv.Sigma.th2[i,], resid2[])}
R2 <- inprod(resid2[],Q2[])
# composite Log Lkd
for (j in 1:2) {eta[j] <- LL[j]+pi[j]-g[j];
eta.star[j] <- LL[j]+pi[j]-g[j]+log(0.5)}
max.eta.star <- ranked(eta.star[],2);
for (j in 1:2) {SL[j] <- eta.star[j]-max.eta.star; expSL[j] <- exp(SL[j])}
# model weights
w[1] <- expSL[1]/sum(expSL[]); w[2] <-expSL[2]/sum(expSL[])}
```

The first dataset (with estimated posterior density coefficients) is

```
list(n=53,p1=3,p2=4,d1=4,d2=5,mu.th1=c(-0.71,1.44,1.29,1.06),
Sigma.th1=structure(.Data=c(0.171,0.147,-0.055,-0.097,
0.147,0.439,0.020,0.033,
-0.055,0.020,0.221,0.008,
-0.097,0.033,0.008,0.186),.Dim=c(4,4)),
mu.th2=c( -0.796,1.65,1.255,0.979,0.546),
Sigma.th2=structure(.Data=c(0.170,0.127,-0.057,-0.088,-0.029,
0.127,0.500,-0.001,0.055,0.073,
-0.057,-0.001,0.225,0.008,-0.012,
-0.088,0.055,0.008,0.202,-0.036,
-0.029,0.073,-0.012,-0.036,0.203),.Dim=c(5,5)))
```

7. The model weights code for the Chevrolet data (with $p1 = 3$, $p2 = 6$) is

```
model {# log-lkd for model 1
for (i in 1:n) {y1[i] <- y[i]; y1[i] ~dnorm(mu1[i],tau1)
mu1[i] <- beta1[1]+beta1[2]*x1[i]+beta1[3]*x2[i]
LL1[i] <- 0.5*log(tau1/(2*3.14159))-0.5*tau1*pow(y1[i]-mu1[i],2)}
for (j in 1:3) {beta1[j] ~dnorm(0,tau.beta1[j])
v.beta1[j] <- 5*sig2.1; tau.beta1[j] <- 1/v.beta1[j];
                    pr.s.beta1[j] <- sqrt(v.beta1[j])}
tau1 ~dgamma(1.5,0.5); sig2.1 <- 1/tau1
# log prior ordinates for model 1
for (j in 1:p1) {prior.ord1[j] <- -0.5*log(2*3.1416)-log(pr.s.beta1[j])
-0.5*pow(beta1[j],2)/(pr.s.beta1[j]*pr.s.beta1[j]) }
prior.ord1[4] <- 1.5*log(0.5)-loggam(1.5)+0.5*log(tau1)-0.5*tau1
# log pseudo-prior ordinates for model 1
g.ord1[1] <- -0.5*p1*log(2*3.1416) - 0.5*logdet(Sigma.beta1[,])-0.5*R1
Inv.Sigma.beta1[1:p1,1:p1] <- inverse(Sigma.beta1[,])
for (i in 1:p1) { resid1[i] <- beta1[i] - mu.beta1[i]
Q1[i] <- inprod(Inv.Sigma.beta1[i,], resid1[])}
R1 <- inprod(resid1[],Q1[])
g.ord1[2] <-a.tau1*log(b.tau1)-loggam(a.tau1)
                    +(a.tau1-1)*log(tau1)-b.tau1*tau1
# log-lkd for model 2
for (i in 1:n) {y2[i] <- y[i]; y2[i] ~dnorm(mu2[i],tau2)
mu2[i] <- beta2[1]+beta2[2]*x1[i]+beta2[3]*x2[i]+beta2[4]*x3[i]
                    +beta2[5]*x4[i]+beta2[6]*x5[i]
LL2[i] <- 0.5*log(tau2/(2*3.14159))-0.5*tau2*pow(y2[i]-mu2[i],2)}
for (j in 1:p2) {beta2[j] ~dnorm(0,tau.beta2[j])
v.beta2[j] <- 5*sig2.2; tau.beta2[j] ~ -1/v.beta2[j];
                    pr.s.beta2[j] <- sqrt(v.beta2[j])}
tau2 ~dgamma(1.5,0.5); sig2.2 <- 1/tau2
```

```
# log prior ordinates for model 2
for (j in 1:6) {prior.ord2[j] <- -0.5*log(2*3.1416)-log(pr.s.beta2[j])
-0.5*pow(beta2[j],2)/(pr.s.beta2[j]*pr.s.beta2[j])}
prior.ord2[7] <- 1.5*log(0.5)-loggam(1.5)+0.5*log(tau2)-0.5*tau2
# log pseudo-prior ordinates for model 2
g.ord2[1] <- -0.5*p2*log(2*3.1416) - 0.5*logdet(Sigma.beta2[,]) -0.5*R2
Inv.Sigma.beta2[1:p2,1:p2] <- inverse(Sigma.beta2[,])
for (i in 1:p2) { resid2[i] <- beta2[i] - mu.beta2[i]
Q2[i] <- inprod(Inv.Sigma.beta2[i,], resid2[])}
R2 <- inprod(resid2[],Q2[])
g.ord2[2] <- a.tau2*log(b.tau2)-loggam(a.tau2)
                  +(a.tau2-1)*log(tau2)-b.tau2*tau2
# Combining loglikelihoods, log-priors, log importance
                  samples, and log prior model probs
TL[1] <- sum(LL1[])+sum(prior.ord1[1:4])-sum(g.ord1[1:2])+log(0.5)
TL[2] <- sum(LL2[])+sum(prior.ord2[1:7])-sum(g.ord2[1:2])+log(0.5)
# Composite log-likelihoods
eta[1] <- sum(LL1[])+sum(prior.ord1[1:4])-sum(g.ord1[1:2])
eta[2] <- sum(LL2[])+sum(prior.ord2[1:7])-sum(g.ord2[1:2])
# Scaling against the largest likelihood
maxL <- ranked(TL[],2);
for (j in 1:2) {SL[j] <- TL[j]-maxL; expSL[j] <- exp(SL[j])}
# model weights
w[1] <- expSL[1]/sum(expSL[]); w[2] <- expSL[2]/sum(expSL[])}
```

8. The use of JAGS for ordinal probit or ordinal logistic regression (and the implementation of constraints on the cut-points) is illustrated by the following simulated example involving ordinal logistic regression. It is assumed that the code, named ordlog.jag, would be in a working directory C://R files. The code for two predictors and $K = 4$ options is

```
model{ for(i in 1:100){ eta[i] <- beta[1]*x1[i] + beta[2]*x2[i]
logit(Q[i,1]) <- theta[1]-eta[i]
p[i,1] <- Q[i,1]
for(j in 2:3) {logit(Q[i,j]) <- theta[j]-eta[i]
            p[i,j] <- Q[i,j] - Q[i,j-1]}
            p[i,4] <- 1 - Q[i,3]
y[i] ~dcat(p[i,])}
# prior for cut-points
for(r in 1:3){ theta0[r] ~dnorm(0,1.0E-3)}
theta <- sort(theta0)
for(j in 1:2){beta[j] ~dnorm(0,1.0E-3)}}.
```

An illustrative calling sequence in R (using the rjags package) is then

```
require(rjags)
setwd("C://R files")
set.seed(123)
x1 <- rnorm(100); x2 <- rnorm(100)
z <- -1.0 + x1*1.0 - x2*1.5 + rnorm(100)
y <- z
y[z > 0] <- 1
y[z > 0 & z ~ 1] <- 2
y[z > 1 & z ~ 1.5] <- 3
y[z > 1.5] <- 4
jagsdf <- list(y=y,x1=x1,x2=x2)
params <- c("theta","beta")
```

```
inits <- list("theta0" = c(-0.5, 0, 0.5))
# Create model
mcmcmodel <- jags.model(file="ordlog.jag", data=jagsdf,
                             inits=inits, n.chains=3)
# Burn-in:
update( mcmcmodel, n.iter=1000 )
# parameter samples
samples = coda.samples(mcmcmodel,variable.names=params, n.iter=1000)
# diagnostic plots
plot(samples)
gelman.plot(samples)
```

# References

Albert, J. and Chib, S. (1993) Bayesian regression analysis of binary and polychotomous response data. *Journal of the American Statistical Association*, **88**, 657–667.

Amemiya, T. (1981) Qualitative response models – a survey. *Journal of Economic Literature*, **19**(4), 1483–1536.

Anderson, J. (1984) Regression and ordered categorical variables. *Journal of the Royal Statistical Society B*, **46**, 1–30.

Anderson, S. and Newell, R. (2003) Simplified marginal effects in discrete choice models. *Economics Letters*, **81**, 321–326.

Armstrong, B. and Sloan, M. (1989) Ordinal regression models for epidemiologic data. *American Journal of Epidemiology*, **129**, 191–204.

Bae, K. and Mallick, B.K. (2004) Gene selection using a two-level hierarchical Bayesian model. *Bioinformatics*, **20**(18), 3423–3430.

Baksh, M.F., Böhning, D. and Lerdsuwansri, R. (2011) An extension of an over-dispersion test for count data. *Computational Statistics and Data Analysis*, **55**(1), 466–474.

Barbieri, M.M. and Berger, J.O. (2004) Optimal predictive model selection. *Annals of Statistics*, **32**(3), 870–897.

Bedrick, E., Christensen, R. and Johnson, W. (1996) A new perspective on priors for generalized linear models. *Journal of the American Statistical Association*, **91**(436), 1450–1460.

Bedrick, E., Christensen, R. and Johnson, W. (1997) Bayesian binomial regression: Predicting survival at a trauma center. *The American Statistician*, **51**(3), 211–218.

Blum, M. and François, O. (2010) Non-linear regression models for approximate Bayesian computation. *Statistics and Computing*, **20**, 63–73.

Böhning, D., Böhning, W. and Holling, H. (2008) Revisiting Youden's index as a useful measure of the misclassification error in meta-analysis of diagnostic studies. *Statistical Methods in Medicine Research*, **17**, 543–554.

Bové, S. and Held, L. (2011) Hyper-g priors for generalized linear models. *Bayesian Analysis*, **6**(3), 387–410.

Chaloner, K. and Brant, R. (1988) A Bayesian approach to outlier detection and residual analysis. *Biometrika*, **75**(4), 651–659.

Chib, S. (1995) Marginal likelihood from the Gibbs output. *Journal of the American Statistical Association*, **90**, 1313–1321.

Chib, S., Seetharaman, P.B. and Strijnev, A. (2004) Model of brand choice with a no-purchase option calibrated to scanner-panel data. *Journal of Marketing Research*, **41**, 184–196.

Collett, D. (1991) *Modelling Binary Data*. Chapman and Hall, London, UK.

Congdon, P. (2000) A Bayesian approach to prediction using the gravity model, with an application to patient flow modelling. *Geographical Analysis*, **32**(3), 205–224.

Cui, W. and George, E. (2008) Empirical Bayes vs. fully Bayes variable selection. *Journal of Statistics and Planning Inference*, **138**, 888–900.

Dean, C.B. (1992) *Testing for overdispersion in Poisson and binomial regression models. *Journal of the American Statistical Association*, **87**(418), 451–457.

Dean, C. and Lawless, J. (1989) Tests for detecting overdispersion in poisson regression models. *Journal of the American Statistical Association*, **84**, 467–472.

Dellaportas, P. and Smith, A. (1993) Bayesian inference for generalized linear and proportional hazards models via Gibbs sampling. *Journal of the Royal Statistical Society C*, **42**, 443–459.

Dey, D., Chen, M.-H. and Chang, H. (1997) Bayesian approach for nonlinear random effects models. *Biometrics*, **53**, 1239–1252.

Efron, B., Hastie, T., Johnstone, I. and Tibshirani, R. (2004) Least angle regression. *Annals of Statistics*, **32**, 407–451.

Fahrmeir, L. and Lang, S. (2001), Bayesian inference for generalized additive mixed models based on Markov random field priors. *Applied Statistics*, **50**, 201–220.

Fahrmeir, L. and Tutz, G. (1994) *Multivariate Statistical Modelling Based on Generalized Linear Models*. Springer-Verlag, Berlin.

Fang, B. and Dawid, A. (2002) Nonconjugate Bayesian regression on many variables. *Journal of Statistical Planning and Inference*, **103**, 245–261.

Farcomeni, A. (2010) Bayesian constrained variable selection. *Statistica Sinica*, **20**, 1043–1062.

Fernandez, C. and Steel, M. (1999) Multivariate Student-t regression models: pitfalls and inference. *Biometrika*, **86**(1), 153–167.

Fernandez, C., Ley, E. and Steel, M.F. (2001) Benchmark priors for Bayesian model averaging. *Journal of Econometrics*, **100**(2), 381–427.

Fischer, M. and Aufhauser, E. (1988) Housing choice in a regulated market – a nested multinomial logit analysis. *Geographical Analysis*, **20**, 47–69.

Fry, T. and Harris, M. (1996) A Monte Carlo study of tests for the independence of irrelevant alternatives property. *Transportation Res B*, **30**, 19–30.

Gamerman, D. (1997) *Markov Chain Monte Carlo*. Chapman and Hall, London, UK.

Gelfand, A. (1996). Model determination using sampling based methods. In W. Gilks, S. Richardson and D. Spiegelhalter (eds), *Markov Chain Monte Carlo in Practice*, pp 145–157. Chapman and Hall/CRC, Boca Raton, FL.

Gelfand, A. and Dey, D. (1994) Bayesian model choice: Asymptotics and exact calculations. *Journal of the Royal Statistical Society B*, **56**(3), 501–514.

Gelfand, A. and Ghosh, S. (1998) Model choice: A minimum posterior predictive loss approach. *Biometrika*, **85**(1), 1–11.

Gelfand, A. and Ghosh, S. (2000) Generalized linear models: a Bayesian view. In D. Dey, S. Ghosh and B. Mallick (eds), *Generalized Linear Models: A Bayesian Perspective*, pp. 1–22. Marcel Dekker, New York, NY.

Gelman, A., Jakulin, A., Pittau, M. and Su, Y.-S. (2008) A weakly informative default prior distribution for logistic and other regression models. *Annals of Applied Statistics*, **2**, 1360–1383.

George, E. and McCullough, R. (1993) Variable selection via Gibbs sampling. *Journal of the American Statistical Association*, **88**, 881–889.

George, E. and McCulloch, R. (1997) Approaches for Bayesian variable selection. *Statistica Sinica*, **7**(2), 339–373.

Gerwinn, S., Macke, J. and Bethge, M. (2010) Bayesian inference for generalized linear models for spiking neurons. *Front. Comput. Neurosci*, **4**, 12.

Geweke, J. (1993) Bayesian treatment of the independent Student-t linear model, *Journal of Applied Econometrics*, **8S**, 19–40.

Greene, W. and Hensher, D. (2009) Modeling ordered choices. http://pages.stern.nyu.edu /~wgreene/DiscreteChoice/Readings/OrderedChoiceSurvey.pdf.

Griffiths, W., Hill, R. and Judge, G. (1993) *Learning and Practicing Econometrics*. Wiley, Chichester, UK.

Grün, B. and Leisch, F. (2009) Finite mixtures of generalized linear regression models. In Shalabh and Christian Heumann (eds), *Recent Advances in Linear Models and Related Areas*, pp 205–230. Springer, New York, NY.

Guy, C., Wrigley, N., O Brien, L. and Hiscocks, G. (1983) The Cardiff Consumer Panel, *UWIST papers in Planning Research*, 68.

Hastie, T. and Tibshirani, R. (1990) *Generalized Additive Models*. Chapman & Hall, London, UK.

Heumann, C. and Grenke, M. (2010) An efficient model averaging procedure for logistic regression models using a Bayesian estimator with Laplace prior. In T. Kneib and G. Tutz (eds), *Statistical Modelling and Regression Structures*, pp 79–90. Springer, New York, NY.

Holmes, C.C. and Held, L. (2006) Bayesian auxiliary variable models for binary and multinomial regression. *Bayesian Analysis*, **1**(1), 145–168.

Johnson, V.E. and Albert, J.H. (1999) *Ordinal Data Modeling*. Springer, New York, NY.

Jones, P.N. and McLachlan, G. (1992) Fitting finite mixture models in a regression context. *The Australian Journal of Statistics*, **34**, 233–240.

Kass, R. and Raftery, A. (1995) Bayes factors. *Journal of the American Statistical Association*, **90**, 773–795.

Kass, R.E. and Wasserman, L. (1995) A reference Bayesian test for nested hypotheses and its relationship to the Schwarz criterion. *Journal of the American Statistical Association*, **90**(431), 928–934.

Kass, R., Tierney, L. and Kadane, J.B. (1989). Approximate methods for assessing influence and sensitivity in Bayesian analysis. *Biometrika*, **76**(4), 663–674.

Koop, G. and Poirier, D. (1993) Bayesian analysis of logit models using natural conjugate priors. *Journal of Econometrics*, **56**, 323–340.

Kuo, L. and Mallick, B. (1998) Variable selection for regression models. *Sankhya*, **60B**, 65–81.

Lang, J. (1999) Bayesian ordinal and binary regression models with a parametric family of mixture links. *Computational Statistics and Data Analysis*, **31**(1), 59–87.

Lange, K., Little, R. and Taylor, J. (1989) Robust statistical modeling using the t-distribution, *Journal of the American Statistical Association*, **84**, 881–896.

Laud, P. and Ibrahim, J. (1995) Predictive model selection. *Journal of the Royal Statistical Society B*, **57**(1), 247–262.

Lawless, J. (1987) Negative binomial and mixed Poisson regression. *Canadian Journal of Statistics*, **15**(3), 209–225.

Leslie D., Kohn R. and Nott D. (2007) A general approach to heteroscedastic linear regression. *Statistics and Computing*, **17**(2), 131–146.

Liang, F., Paulo, R., Molina, G., Clyde, M. and Berger, J.O. (2008) Mixtures of g-priors for Bayesian variable selection. *Journal of the American Statistical Association*, **103**, 410–423.

Lindley, D.V. and Smith, A.F.M. (1972) Bayes estimates for the linear model (with discussion). *Journal of the Royal Statistical Society B*, **34**, 1–41.

Lunn, D.J., Whittaker, J.C. and Best, N. (2006) A Bayesian toolkit for genetic association studies. *Genetic Epidemiology*, **30**, 231–247.

Lunn, D.J., Best, N. and Whittaker, J. (2009) Generic reversible jump MCMC using graphical models. *Statistics and Computing*, **19**, 395–408.

Lykou, A. and Ntzoufras, I. (2013) On Bayesian lasso variable selection and the specification of the shrinkage parameter. *Statistics and Computing*, **23**, 361–390.

McCullagh, P. (1980) Regression models for ordinal data. *Journal of the Royal Statistical Society B*, **42**, 109–142.

McCullagh, P. and Nelder, J. (1989) *Generalized Linear Models*. Chapman & Hall/CRC, Boca Raton, FL.

Melnykov, V. and Maitra, R. (2010) Finite mixture models and model-based clustering. *Statistics Survey*, **4**, 80–116.

Meng, X.-L. (1994) Posterior predictive p-values. *Annals of Statistics*, **22**, 1142–1160.

Meyer, R.D. and Wilkinson, R.G. (1998). Bayesian variable assessment. *Communications in Statistics: Theory and Methods*, **27**, 2675–2705.

Mukhopadhyay, N., Ghosh, J. and Berger, J. (2005) Some Bayesian predictive approaches to model selection. *Statistics & Probability Letters*, **73**(4), 369–379.

Mullahy, J. (1997) Instrumental-variable estimation of count data models: applications to models of cigarette smoking behavior. *Review of Economics and Statistics*, **79**, 596–593.

Nelder, J. and Lee, Y. (1991) Generalized linear models in the analysis of Taguchi-type experiments. *Applied Stochastic Models in Data Analysis*, **7**, 103–120.

Ntzoufras, I. (2002). Gibbs variable selection using BUGS. *Journal of Statistical Software*, **7**, 1–19.

Ntzoufras, I., Dellaportas, P. and Forster, J. (1999) MCMC variable and link determination in generalised linear models. *Technical Report 65*, Department of Statistics, Athens University of Economics and Business.

O'Hara, R.B. and Sillanpää, M.J. (2009) A review of Bayesian variable selection methods: what, how and which. *Bayesian Analysis*, **4**(1), 85–117.

Park, T. and Casella, G. (2008). The Bayesian Lasso. *Journal of the American Statistical Association*, **103**(482), 681–686.

Peña, D., Zamar, R. and Yan, G. (2009) Bayesian likelihood robustness in linear models. *Journal of Statistical Planning and Inference*, **139**, 2196–2207.

Poirier, D. (1996) A Bayesian analysis of nested logit models. *Journal of Econometrics*, **75**(1), 163–181.

Raftery, A. (1995) Bayesian model selection in social research. In P. Marsden (ed.), *Sociological Methodology*. Blackwell, Oxford, UK.

Raftery, A. (1996) Approximating Bayes factors and accounting for model uncertainty. *Biometrika*, **83**, 251–266.

Rubin, D.B. (1984) Bayesianly justifiable and relevant frequency calculations for the applies statistician. *Annals of Statistics*, **12**, 1151–1172.

Schoenbach, V. and Rosamond, W. (2000) *Understanding the Fundamentals of Epidemiology*. Department of Epidemiology, UNC Chapel Hill, http://www.epidemiolog.net/evolving /TableOfContents.htm.

Sharples, L. (1990) Identification and accommodation of outliers in general hierarchical models. *Biometrika*, **77**, 445–453.

Smith, M. and Kohn, R. (1996) Nonparametric regression using Bayesian variable selection. *Journal of Econometrics*, **75**, 317–334.

Tibshirani, R. (2011) Regression shrinkage and selection via the lasso: a retrospective. *Journal of the Royal Statistical Society B*, **73**, 273–282.

Wang, X. and George, E. (2007) Adaptive Bayesian criteria in variable selection for generalized linear models. *Statistica Sinica*, **17**, 667–690.

Weiss, R. (1994) Pediatric pain, predictive inference, and sensitivity analysis. *Evaluation Review*, **18**, 651–677.

Winkelmann, R. and Zimmermann, K.F. (1995) Recent developments in count data modeling: theory and applications. *Journal of Economic Surveys*, **9**, 1–24.

Wood, S. and Kohn, R. (1998) A Bayesian approach to robust binary nonparametric regression. *Journal of the American Statistical Association*, **93**(441), 203–213.

Woodward, P. and Walley, R. (2009) Bayesian variable selection for fractional factorial experiments with multilevel categorical factors. *Journal of Quality Technology*, **41**, 228–240.

Wrigley, N. and Dunn, R. (1986) Diagnostics and resistant fits in logit choice models, in G. Norman (ed.), *Spatial Pricing and Differentiated Markets*. London Papers in Regional Science 16. Pion, London, UK.

Yang, A.-J. and Song, X.-Y. (2010) Bayesian variable selection for disease classification using gene expression data. *Bioinformatics*, **26**, 215–222.

Yi, N. and Xu, S. (2008) Bayesian LASSO for quantitative trait loci mapping. *Genetics*, **179**(2), 1045–1055.

Yuan, M. and Lin, Y. (2005) Efficient empirical Bayes model selection and estimation in linear models. *Journal of the American Statistical Association*, **100**, 1215–1225.

Zellner, A. and Moulton, B. (1985) Bayesian regression diagnostics with applications to international consumption and income data. *Journal of Econometrics*, **29**, 187–221.

Zhu, H., Ibrahim, J.G. and Tang, N. (2011) Bayesian influence analysis: a geometric approach. *Biometrika*, **98**(2), 307–323.

# 4

# More advanced regression techniques

## 4.1 Introduction

This chapter considers alternatives to standard regression schemes such as normal linear regression and generalised linear models when underlying assumptions are not met. For example, the assumptions of the normal linear model are often violated by the data, and inferences regarding regression impacts may be affected. These include the linearity assumption, namely that the conditional means of the response are a linear function of the predictors, and the assumptions that the error terms are normally distributed and have constant variance. More general formulations such as skew normal and heteroscedastic regression (Section 4.2), discrete mixture regression (Section 4.5), and non-linear regression (Section 4.6) are then needed.

Assumptions made under the generalised linear modelling approach are summarised by Breslow (1996) namely 'the statistical independence of the observations, the correct specification of the link and variance functions, the correct scale for measurement of the explanatory variables, and the lack of undue influence of individual observations on the fitted model'. Of these issues, those considered in detail in this chapter are specification of the variance function in terms of representing overdispersion (Section 4.3), and flexible link specification in Section 4.4.

Bayesian computing options for such generalised regression techniques include BUGS, R-INLA (e.g. Wang, 2013), and R packages such as bayesm, bayescount, tgp, mgcv, gamair, and MCMCglmm. BayesX, and the associated R package R2BayesX, are of particular utility in non-linear regression.

## 4.2 Departures from linear model assumptions and robust alternatives

The normal linear model for metric responses makes a number of simplifying assumptions, necessitating techniques to deal with non-constant variance or non-normality due to skewness, heavy tailed errors, or outliers. Errors may not be identically distributed because of the presence of outliers, which may unduly influence estimates of regression coefficients, leading to worse fit to the bulk of the observations (Rousseeuw, 1991). Outliers also tend to increase the estimated error variance. Heavy tailed errors without skewness may be accommodated by adopting a Student $t$ error assumption, with resistance to outliers obtained by varying the degrees of freedom parameter, while skew normal and skew $t$ regressions adapt to skewed errors.

Student $t$ regression can be implemented directly or via scale mixing (West, 1984; Geweke, 1993; Fonseca $et$ $al.$, 2008), using the result that a Student $t$ variable with mean $\mu$, variance $\sigma^2$, and degrees of freedom $v$ can be obtained by assuming $y \sim N(\mu, \sigma^2/\lambda)$ where $\lambda \sim$ $Ga(0.5v, 0.5v)$. For a regression analysis one may therefore assume the errors are distributed as $t(0, \sigma^2, v)$ achieved via gamma scale mixing

$$y_i \sim N(X_i\beta, \sigma^2/\lambda_i),$$

$$\lambda_i \sim Ga(0.5v, 0.5v),$$

with posterior proportional to

$$\pi(v)\pi(\beta)\pi(\sigma^2)\frac{(0.5v)^{0.5v}}{\Gamma(0.5v)} \prod_{i=1}^{n} \frac{\sqrt{\lambda_i}}{\sigma} \exp\left[-\frac{\lambda_i}{2\sigma^2}(y_i - X_i\beta)^2\right] \prod_{i=1}^{n} \lambda_i^{0.5v-1} \exp\left[-\frac{v\lambda_i}{2}\right].$$

Observations with low values of $\lambda_i$ have less influence on this density and hence on posterior estimates, and so the $\lambda_i$ may be used to indicate outlier status. As mentioned in Chapter 3, this approach may be used for probit regression with binary responses in combination with augmented data sampling. Thus

$$z_i \sim N(X_i\beta, 1/\lambda_i) \qquad I(0, \infty) \qquad y_i = 1$$

$$z_i \sim N(X_i\beta, 1/\lambda_i) \qquad I(-\infty, 0) \qquad y_i = 0$$

$$\lambda_i \sim Ga(0.5v, 0.5v).$$

Among other approaches applicable both to normal linear and generalised linear models are schemes to downweight influential cases or cases with high leverage (Heritier $et$ $al.$, 2009). For example, weighted linear normal regression proceeds with precisions $w_i/\sigma^2$, where the weights $w_i$ take account of realised standardised residuals or other case level measures of fit. Thus

$$y_i \sim N(\mu_i, \sigma^2/w_i),$$

$$w_i = \min(1, t_r/r_i),$$

$$r_i = \left|\frac{y_i - X_i\beta}{\sigma}\right|,$$

where $t_r$ specifies a threshold standardised residual where downweighting starts (e.g. $t_r = 2$ or 3). A weight function based on the Huber M-estimation technique (Fox, 2002) specifies $t_r = 1.345$, while bi-square weight functions specify $w_i = [1 - (r_i/t_r)^2]^2$ for $|r_i| \le t_r$, and $w_i = 0$ otherwise.

Modifications of the normal and Student $t$ densities have also been suggested to accommodate skew responses (and errors) and associated departures from the linear normal model (e.g. heteroscedasticity). Instead of transforming the data, such methods transform the error distributions to accommodate skewness (Sahu $et$ $al.$, 2003). For example, Chen $et$ $al.$ (1999)

derive a skewed random variable (positively or negatively skewed) as the sum of a symmetric random variable $\varepsilon$, and the product of skew coefficient $\delta$ and a positively skewed variable $z$,

$$y_i = \delta z_i + \varepsilon_i,$$

where $\varepsilon \sim F, z \sim G$, $F$ is the cumulative distribution function of a symmetric density, and $G$ is the cdf of a positive skew density. Let $\sigma_z^2$ and $\sigma^2$ be the variances of $z$ and $\varepsilon$. Then the standardised third moment of $y$ is

$$\mu_y^3 = E\left[\frac{y - E(y)}{\sigma_y}\right]^3 = \frac{\delta^3 \sigma_z^3 \mu_z^3}{\sigma_y^3}$$

where $\text{var}(y) = \sigma_y^2 = \delta^2 \sigma_z^2 + \sigma^2$, and $\mu_z^3$ is the standardised third moment of $z$ (Branco and Dey, 2002). If $z$ is positively skewed with $\mu_z^3 > 0$, then $y$ is positively or negatively skewed ($\mu_y^3 > 0$ or $\mu_y^3 < 0$) according as $\delta$ is positive or negative. A regression component is included with

$$y_i = X_i \beta + \delta z_i + \varepsilon_i.$$

If $F = \Phi$, and $G$ is the cdf of a half-standard normal distribution, with density

$$g(u) = \frac{2}{\sqrt{2\pi}} e^{-u^2/2}, \qquad u > 0,$$

then the conditional distribution of $y$ given $z$ is normal with mean $X_i \beta + \delta z_i$ and variance $\sigma^2$. The marginal distribution of $y$ is a skewed normal (Liseo and Loperfido, 2006) with

$$p(y_i | \delta, \sigma, \beta) = \frac{2}{\sqrt{\sigma^2 + \delta^2}} \phi\left(\frac{y_i - X_i\beta}{\sqrt{\sigma^2 + \delta^2}}\right) \Phi\left(\frac{\delta(y_i - X_i\beta)}{\sqrt{\sigma^2 + \delta^2}}\right). \qquad (4.1)$$

The skew-$t$ density and corresponding regression can be obtained by scale mixing (Branco and Dey, 2002; Liseo, 2004; Lachos et al., 2011). Thus

$$y_i = X_i \beta + u_i^{-0.5}(\delta z_i + \varepsilon_i),$$

where $u_i \sim \text{Ga}(0.5\nu, 0.5\nu)$. In BUGS this class of models can be fitted using the conditional likelihood or complete data likelihood (Chen et al., 1999), including the latent positive variables in the regression terms $X_i\beta + \delta z_i$, or by using the marginal likelihood in combination with special devices for non-standard likelihoods (e.g. the dloglik option in OpenBUGS). Sahu et al. (2003) set out a general class of multivariate skew elliptically symmetric distributions, containing standard forms such as the multivariate skew-normal and Student-$t$ distributions, while Azzalini and Regoli (2012) provide an overview of skew-symmetric distributions.

Fernandez and Steel (1998) address both skewness and heavy tails together via differential scaling of the error variance according to whether the realised residual in normal linear regression

$$e_i = y_i - X_i\beta = y_i - \mu_i$$

is negative or positive. For residuals exceeding zero, the error variance $\sigma^2$ is scaled by a positive factor $\gamma^2$, (i.e. the precision is scaled by $1/\gamma^2$), while for negative residual terms, it is necessary to scale the error variance by $1/\gamma^2$ (scale the precision by $\gamma^2$). The value $\gamma = 1$ corresponds to a symmetric (non-skewed) density, while values of $\gamma$ exceeding 1 correspond to positive skewness, and values of $\gamma$ under 1 correspond to negative skewness. For positively skewed residuals, values $\gamma > 1$ will produce a lower value of $\sigma^2$ as compared to an inflated variance estimate obtained without any allowance for skewness.

While error assumptions may be varied to accommodate such departures, transformations of the response or predictors may be used to induce a closer approximation to linear model assumptions. A well known method for skew responses or predictors involves skew-minimising transformation such as the Box–Cox transform, which applied to positive responses has the form:

$$y^{(\lambda)} = (y^{\lambda} - 1)/\lambda \qquad (\lambda \neq 0),$$

$$y^{(0)} = \log(y) \qquad (\lambda = 0)$$

$$y^{(\lambda)} = X_i \beta + \varepsilon_i,$$

$$\varepsilon_i \sim N(0, \sigma^2 I)$$

with Bayesian applications including Weiss (1994), Hoeting et al. (2002), and Gottardo and Raftery (2009). Setting $z = y^{(\lambda)}$, the log-likelihood is

$$\ln L(\beta, \sigma^2 | y) = -0.5\sigma^2(z - X\beta)(z - X\beta) - \frac{n}{2}\ln(2\pi\sigma^2) + (\lambda - 1)\sum_{i=1}^{n}\ln(y_i).$$

An extension for negative response data is

$$y^{(\lambda)} = (y + \lambda_2)^{\lambda_1} - 1)/\lambda_1 \qquad (\lambda_1 \neq 0),$$

$$y^{(0)} = \log(y + \lambda_2) \qquad (\lambda_1 = 0)$$

and simultaneous response and covariate Box–Cox transformation can also be investigated. Unless a simple transformation is indicated, analysing data in the transformed scale may, however, complicate interpretability. In BUGS, Box–Cox transformation models can be fitted using devices for non-standard likelihoods (e.g. the dloglik option in OpenBUGS). For positive variables (e.g. cost data), skewness may also be tackled by adopting flexible asymmetric densities, such as the Gamma density (Mitsakakis, 2012). For example, one may assume

$$y_i \sim Ga(\eta, \eta/\mu_i),$$

$$\log(\mu_i) = X_i \beta.$$

A further common violation of linear regression assumptions is caused by nonconstant error variances, for example when the variance of the residuals increases with the size of the fitted values. As for skewness, one solution to this involves transformation of the response. Direct modelling of changes in the variance $V(y_i | X_i)$ between observations is also possible (Boscardin and Gelman, 1996). Among possible schemes for such heteroscedasticity, consider

$$y_i = X_i \beta + \varepsilon_i = \mu_i + w_i \varepsilon_i,$$

where $\varepsilon_i \sim N(0, 1)$. Then if the variance of the residuals increases with the the fitted values one might set $w_i = \exp(\alpha \mu_i)$ or $w_i = |\mu_i|^{\xi}$, as under 'power of the mean' approaches (Carroll and Ruppert, 1988). Alternatively if heteroscedasticity is related to predictors (Aitkin, 1987; Cepeda and Gamerman, 2001) consider

$$y_i = \mu_i + \varepsilon_i$$

where

$$\varepsilon_i \sim N(0, \eta_i),$$

$$\log(\eta_i) = Z_i \gamma,$$

where $Z_i$ are additional predictors (or include some or all of the $X_i$), and heteroscedasticity would be shown by values of any $\gamma$ coefficient differing from zero. Dynamic linear priors for volatility in regression errors have been proposed, especially for time series but with relevance to other types of regression. For example, if the data are arranged in ascending order of $y$ then with $\kappa_i = \log(\sigma_i)$, a RW(1) prior would take $\kappa_i \sim N(\kappa_{i-1}, \sigma_\kappa^2)$.

### Example 4.1    Depression study

To illustrate sensitivity of inferences in metric data regression to alternative possible solutions to heteroscedasticity, consider an epidemiological study of depression and help-seeking behavior among adults (Afifi *et al.*, 2004). A CESD depression index was constructed by asking $n = 294$ subjects to respond to 20 items: 'I felt I could not shake off the blues ...,' 'My sleep was restless,' etc. Scores for each item were obtained in terms of weekly frequency: 'less than 1 time per week' (score 0), '1 to 2 days per week' (score 1), '3 to 4 days per week' (score 2), or '5 to 7 days' (score 3). Responses to the 20 items were summed to form a total score, with subjects obtaining scores over 15 classified as depressed.

A plot of the CESD score demonstrates clear positive skew. Nevertheless to demonstrate how inferences may be affected by applying standard techniques, a normal linear regression (y defined as the CESD score plus 1) is initially undertaken on seven predictors: gender (1 = female, 0 = male), age, education, income (log10 transformed), Catholic religion (binary), Jewish religion (binary) and no religion (binary). This shows the majority of the predictors to be significant influences on depression, with only Catholic religion and education having 95% credible intervals straddling zero. However, a plot (Figure 4.1) of the posterior mean residuals against the regression means $\mu_i$ casts doubt on the constant variance assumption, with error variance increasing with $\mu_i$.

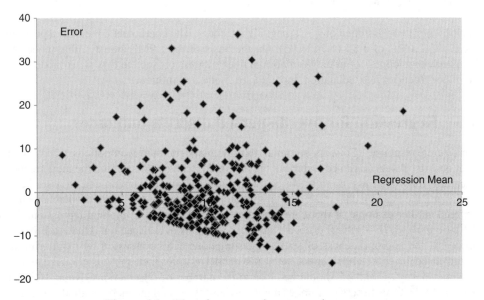

**Figure 4.1**    Plot of errors against regression means.

One way of correcting for the influence of skew (and associated heteroscedasticity) involves the Box–Cox transform regression. A defined likelihood coding in OpenBUGS is applied with a two chain run of 50 000 iterations, and with inferences from the last 45 000. This involves the OpenBUGS statements

```
model { for (i in 1:294) {d[i] <- 0; d[i] ~ dloglik(LL[i]);
LL[i] <- -log(sig)-0.918-0.5*e[i]*e[i]/sigma2+(lam-1)*log(y[i])
e[i] <- z[i]-mu[i]; z[i] <-(pow(y[i],lam)-1)/lam...}
```

A posterior mean (and 95% CRI) for $\lambda$ of 0.24 (0.13, 0.34) indicates the transformation $z = y^{0.25}$. The majority of predictors remain significant under the Box–Cox transform approach, while the DIC is much reduced (from 2095 to 1904). A skew-normal regression model can also be fitted in OpenBUGS with the statements

```
model { for (i in 1:294) {d[i] <- 0; d[i] ~ dloglik(LL[i])
LL[i] <- log(2/kap)-log(kap)-0.918 -0.5*pow((y[i] - mu[i])/kap, 2)
+ log(phi(delta*(y[i] -mu[i])/kap))...}
```

where $\kappa = \sqrt{\sigma^2 + \delta^2}$ in the marginal density $p(y_i|\delta, \sigma, \beta)$ in (4.1) above. A N(0,100) prior is assumed on $\delta$, and a Ga(1,0.001) prior on $\tau = 1/\sigma^2$. This model shows only two predictors (age and income) with significant effects. The posterior mean estimate of the skew parameter $\delta$ is 8.66 with 95% CRI from 8.13 to 9.23.

Finally, as one possible variance transformation model, it is assumed that $\sigma_i^2 = |\mu_i|^\xi$, although in OpenBUGS this transformation is applied using the precision, namely

```
model { for (i in 1:294) {y[i] ~ dnorm(mu[i],tau[i])
sigma2[i] <- 1/tau[i]; sig[i] <- sqrt(sigma2[i])
e[i] <- y[i]-mu[i]; LL[i] <- -log(sig[i])-1.838-0.5*e[i]*e[i]/
                    sigma2[i]; G[i] <- 1/exp(LL[i])
y[i] <- CESD[i]+1
tau[i] <- pow(abs(mu[i]),negxi)...}
```

A N(0,10) prior is assumed on $(-\xi)$ (negxi in the code). The coefficient $\xi$ then has a posterior mean (95% CRI) of 1.84 (1.74, 1.95). On the basis of the LPML statistic, this model has intermediate goodness of fit between normal linear regression and the Box–Cox model. Four predictors (gender, age, education, income) are significant influences.

## 4.3   Regression for overdispersed discrete outcomes

For discrete responses (Poisson, binomial, multinomial), standard assumptions regarding the relationship between variance and mean may not hold, with overdispersion the most frequent departure from such assumptions. In classical statistics, overdispersion can be assessed by comparing the Pearson chi-squared $\chi^2$ or the saturated deviance $D_{sat}$ to the available degrees of freedom. For example, if there are $p$ predictors in a Poisson or binomial regression, then the ratio $\phi = \chi^2/(n - p)$ will exceed 1 in the presence of overdispersion. Under a Bayesian approach, the appropriateness of standard assumptions can be assessed by monitoring the saturated deviance or chi-squared, since if a model is true, then the posterior mean saturated deviance $\overline{D}_{sat}$ and $\overline{\chi}^2$ should be close to the number of observations (Knorr-Held and Rainer, 2001). Overdispersion implies higher variability around the model's fitted values than is consistent with the assumed likelihood formulation. Thus is typically due to unobserved variation between cases that is not represented by the observed covariates: for a given setting of covariates, probabilities for binomial observations, or relative rates under a Poisson model, fluctuate randomly. Particular types of response pattern (e.g. an excess proportion of zero counts as compared to the expected Poisson or binomial frequency) may also cause overdispersion. Without correction for such extra-variability, regression parameter estimates may be biased, and their credible intervals will be too narrow, leading to incorrect inferences about significance.

For example, the Poisson regression model for count data

$$y_i \sim \text{Po}(\mu_i), \qquad \mu_i = \exp(\beta_0 + \beta_1 x_{1i} + \ldots + \beta_p x_{pi}),$$

assumes that the conditional mean and variance are equal, $E(y_i|x_i) = \text{Var}(y_i|x_i) = \mu_i$. However, overdispersion as compared to the Poisson assumption is often encountered. The conjugate hierarchical model for count data is the Poisson-gamma, and overdispersion can be handled by multiplicative unobserved gamma-distributed effects $\omega_i$ as in

$$y_i \sim \text{Po}(\mu_i \omega_i), \qquad \omega_i \sim \text{Ga}(\alpha, \alpha)$$

with the $\omega_i$ normalised to average 1 (so that the regression intercept is identified), and with variance $1/\alpha$. The conditional variance is now

$$\text{Var}(y_i|x_i) = \mu_i + \mu_i^2/\alpha,$$

with greater overdispersion for smaller $\alpha$. Equivalently one has a quadratic variance function

$$\text{Var}(y_i|x_i) = \mu_i + \phi\mu_i^2,$$

where $\phi = 1/\alpha = \text{Var}(\omega_i)$, with variance to mean ratio depending on $\mu_i$.

Integrating out the $\omega_i$ leads to a marginal negative binomial density (NB2 form) for the $y_i$, namely (Cameron and Trivedi, 1986; Greene, 2008)

$$p(y_i|x_i) = \frac{\Gamma(\alpha + y_i)}{\Gamma(\alpha)\Gamma(y_i + 1)} r_i^\alpha (1 - r_i)^{y_i}, \qquad r_i = \frac{\alpha}{\alpha + \mu_i}$$

In BUGS one would specify an NB2 regression as

```
y[i]~dnegbin(r[i], alpha)
r[i] <- alpha/(alpha+mu[i])
log(mu[i] ) <- inprod(beta[], x[i, 1:p]).
```

The overdispersion parameter $\alpha$ is assigned a prior limited to positive values, for example a gamma prior or uniform with range over positive values. The zic library in R is a possible alternative (e.g. see http://en.wikibooks.org/wiki/R_Programming/Count_Data_Models). The NB2 model can also be estimated in INLA (see Example 4.2).

An alternative gamma mixture scheme (denoted the NB1 model) leads to a linear variance function, and hence constant variance to mean ratio. Thus $y_i \sim \text{Po}(v_i)$ where

$$v_i \sim \text{Ga}(\mu_i \alpha, \alpha),$$

giving $\text{Var}(v_i) = \mu_i/\alpha$ and

$$V(y_i|x_i) = \mu_i + \mu_i/\alpha.$$

The NB1 likelihood is

$$p(y_i|x_i) = \frac{\Gamma(\alpha\mu_i + y_i)}{\Gamma(\alpha\mu_i)\Gamma(y_i + 1)} \left(\frac{\alpha}{\alpha + 1}\right)^{\alpha\mu_i} \left(\frac{1}{\alpha + 1}\right)^{y_i}.$$

Generalized negative binomial models have been suggested (Winkelmann and Zimmerman, 1995). Greene (2008) proposes an encompassing *NBR* model in which $\alpha$ in the NB2 likelihood is replaced by $\alpha\mu_i^{2-R}$, where the value $R = 1$ gives the NB1 likelihood, and $R = 2$ gives the NB2 likelihood. This model has conditional variance

$$V(y_i|x_i) = \mu_i + \mu_i^R/\alpha. \tag{4.2}$$

Binomial regression with overdispersion may occur when responses are arranged in clusters and responses from the same cluster are correlated: examples occur in teratological studies, when the observation unit is a litter of animals, and litters differ in terms of unknown genetic factors (Yamamoto and Yanagimoto, 1994). The conjugate approach to overdispersion involves a beta distributed success probability $p_i$, leading to a beta-binomial regression model. Thus with $y_i \sim \text{Bin}(n_i, p_i)$, one may assume (Albert, 1988)

$$p_i \sim \text{Beta}(\gamma\pi_i, (1 - \pi_i)\gamma),$$

with mean $\pi_i$ and variance

$$\pi_i(1 - \pi_i)/(\gamma + 1).$$

The regression on predictors involves a logit link for the $\pi_i$

$$\pi_i = \exp(X_i\beta)/(1 + \exp(X_i\beta)).$$

The variance of a beta-binomial response is of the form (Collett 2002, p. 201)

$$\text{Var}(y_i|x_i) = n_i p_i (1 - p_i) \left[1 + \frac{(n_i - 1)}{(\gamma + 1)}\right].$$

An alternative parameterisation is

$$p_i \sim \text{Beta}(\alpha_i, \beta_i),$$

$$\alpha_i = \exp(\kappa + X_i\beta); \beta_i = \exp(\kappa)$$

$$E(p_i) = \frac{\alpha_i}{\alpha_i + \beta_i} = \frac{\exp(X_i\beta)}{1 + \exp(X_i\beta)}.$$

In BUGS there is no option for the marginal beta-binomial density, and one must implement it from first principles, for example:

```
y[i] ~ dbin(p[i],n[i])
p[i] ~ dbeta(a1[i],a2[i])
a1[i] <- gamma*pi[i]; a2[i] <- gamma*(1-pi[i])
logit(pi[i]) <- beta0 + beta[1]*x[1,i] + ...beta[p]*x[p,i]
```

Nonconjugate hierarchical models are often adopted for overdispersed discrete data regression, such as with normal or Student *t* errors in the log link for count data or in the logit link regression for binomial data. As compared to the conjugate hierarchical model, this facilitates multiple or multilevel random effect models. Thus for count data one might specify

$$y_i \sim \text{Po}(v_i),$$

$$\log(v_i) = X_i \beta + \varepsilon_i,$$

with $\varepsilon_i \sim N(0, \sigma^2)$ under a normal errors assumption. One may then show

$$E(y_i|X_i) = v_i \exp(\sigma^2/2)$$

and

$$\text{Var}(y_i|X_i) = E_\varepsilon[\text{Var}(y_i|X_i, \varepsilon_i)] + \text{Var}_\varepsilon[E(y_i|X_i, \varepsilon_i)]$$

$$= v_i \exp(\sigma^2/2)\{1 + v_i \exp(\sigma^2/2)[\exp(\sigma^2) - 1]\}.$$

Taking $\phi = e^{\sigma^2} - 1$

$$\text{Var}(y_i|X_i) = E(y_i|X_i, \varepsilon_i)[1 + \phi E(y_i|X_i, \varepsilon_i)]$$

showing that the variance has a quadratic form.

For multinomial data, such as voting patterns $y_i = (y_{i1}, \dots, y_{iJ})$ for parties $j = 1, \dots, J$ by constituency $i$ (with $N_i = y_{i+}$), overdispersion may occur when choice or category probabilities vary between the $N_i$ individuals in each observation unit, but clusters of individuals within each unit have similar probabilities. The individual level factors associated with such clustering are not observed, so a random effect will proxy such unobserved factors. The conjugate approach for such heterogeneity is the multinomial-Dirichlet mixture. Thus with $\pi_i = (\pi_{i1}, \dots, \pi_{iJ})$,

$$y_i \sim \text{Mult}(N_i, [\pi_{i1}, \dots, \pi_{iJ}]),$$

$$\pi_i \sim \text{D}(\pi_{i1}, \dots, \pi_{iJ})$$

where the $\pi_{ij}$ may be related to predictors via multiple logit models (with one category as reference, see Chapter 3).

## 4.3.1 Excess zeroes

Overdispersion is often partly related to an excess proportion of zero observations as compared to that expected under standard count data models. To account for this data feature, a zero inflation process postulates that counts are either true zeroes or zeroes resulting from the density $p(y|\theta)$ governing the remaining observations. Let $t_i = 1$ for true zeroes, with $\Pr(t_i = 1) = \zeta$, as against density-generated zeroes with $t_i = 0$. Then

$$P(y_i = 0) = \Pr(t_i = 1) + p(y_i = 0|t_i = 0) \Pr(t_i = 0)$$

$$= \zeta + p(y_i = 0|t_i = 0, \theta)(1 - \zeta),$$

$$P(y_i = j) = p(y_i = j|t_i = 0, \theta) \Pr(t_i = 0) \qquad y_i > 0$$

Regressors may be relevant both to zero inflation, and to the parameters $\theta$ defining the count density $p(y_i = j|\theta)$, such as a Poisson, binomial, negative binomial, or multinomial regression (Hall, 2000; Gurmu and Dagne, 2012).

For count data, a frequently applied approach is zero inflated Poisson (or ZIP) regression, which with a logit regression for subject varying $\zeta_i$, regressors $X_i$, and Poisson mean $\mu_i = \exp(X_i\beta)$ implies

$$\zeta_i = \Pr(t_i = 1|X_i) = \frac{\exp(X_i\gamma)}{1 + \exp(X_i\gamma)},$$

$$p(y_i = 0|X_i) = \zeta_i + (1 - \zeta_i)e^{-\mu_i},$$

$$p(y_i = j|X_i, y_i > 0) = (1 - \zeta_i)e^{-\mu_i}\mu_i^{y_i}/y_i!,$$

with variance then

$$\operatorname{Var}(y_i|\zeta_i, \mu_i) = (1 - \zeta_i)[\mu_i + \zeta_i\mu_i^2] > \mu_i(1 - \zeta_i) = E(y|\zeta_i, \mu_i).$$

So the modelling of excess zeros implies overdispersion. A useful representation the zero inflated Poisson involves the mixed scheme (Ghosh *et al.*, 2006), whereby $t_i \sim dbern(\zeta_i)$ and $y_i|t_i \sim \mathrm{Po}(\mu_i(1 - t_i))$.

Hurdle regression (Mullahy, 1986) also involves a two-part model, with a binomial sub-model determining if the outcome is zero or not, and conditional on a non-zero outcome, a truncated-at-zero distribution for remaining observations (e.g. via a truncated Poisson or Negative binomial). Thus a logit-Poisson hurdle regression model has

$$p(y_i = 0|X_i) = \omega_i,    \tag{4.3}$$

$$\omega_i = \exp(X_i\gamma)/[1 + \exp(X_i\gamma)],$$

$$p(y_i|X_i, y_i > 0) = \frac{(1 - \omega_i)\exp(-\mu_i)\mu_i^{y_i}}{[1 - \exp(-\mu_i)]y_i!},$$

$$\mu_i = \exp(X_i\beta),$$

with log-likelihood terms

$$I(y_i = 0)\log(\omega_i) + I(y_i > 0)[\log(1 - \omega_i) - \mu_i + y_i\log(\mu_i) - \log(1 - \exp(-\mu_i)) - \log(y_i!)].$$

### Example 4.2    Doctor visits (SOEP)

Consider data on health care use from the German Socioeconomic Panel (SOEP), as analyzed by Riphahn *et al.* (2003). The data considered here are for $n = 14\,243$ male subjects, with the response variable doctor visits. The analysis includes coefficients $\{\delta_1, \ldots, \delta_7\}$ for time dummy variables at year points (1984, 1985, 1986, 1987, 1988, 1991, 1994), with a corner constraint $\delta_1 = 0$ on the first year. The other regression effects are for 15 predictors denoted (Age; Agesq; HSat; Handdum; Handper; Married; Educ; Hhninc; Hhkids; Self; Civil; Bluec; Working; Public; AddOn) in the BUGS code. Details on predictors are presented in Riphahn *et al.* (2003, p 393). The generalized Negative Binomial NBR with unknown power $R$ on $\mu_i$ in the conditional variance (Greene, 2008) is applied using the Poisson-Gamma equivalence to the Negative Binomial[1]. A Ga(1,0.001) prior is assumed on $\alpha$, and an exponential E(0.5) prior on $R$.

It may be noted that the NB2 may be estimated in INLA as well as BUGS, and so assist in setting initial values. The code (including predictor recodes) is

```
require(INLA)
D = read.table("docvis.txt",header=T)
```

```
D$ag <- D$age/10; D$ag2 <- D$agesq/100; D$ninc <- D$hhninc/10000;
                                   D$year.f <- factor(D$year)
f=docvis~year.f+ag+ag2+hsat+handdum+handper+ninc+hhkids+educ+
      married+working+bluec+self+civil+public+addon
m=inla(f,family="nbinomial",data=D,
control.family=list(hyper=list(theta=list(prior="gaussian",
                                   param=c(0,0.01)))))).
```

The parameter $\alpha$ used in specifying the NB2 in Section 4.3 above is denoted theta in INLA.

Predictive checks applied to the NB1 and NB2 models show that they underpredict large doctor visit counts. The NB2 has nearly 700 observations where the predictive probabilities based on the checks

$$C(y_{rep,i}^{(t)}, y_i) = I(y_i > y_{rep,i}^{(t)}) + 0.5I(y_i = y_{rep,i}^{(t)}),$$

exceed 0.95. This problem is eliminated in the NBR model, for which a 2500 iteration two chain run (1000 burn-in) provides an estimate for $R$ in (4.2) of 1.50 (1.44, 1.56) and for $\phi = 1/\alpha$ of 2.87 (2.68, 3.09). Other model extensions could include a two group discrete mixture over the intercept in the NB2 regression model, with one intercept representing a high visit rate among a minority of subjects (see Section 4.5).

### Example 4.3    Hospital stays (NMES)

Data analysed by Deb and Trivedi (1997) obtained from National Medical Expenditure Survey (NMES) have been shown to contain excess zeroes; see also Zeileis et al. (2008). The number of hospitalizations by $n = 4406$ subjects is the dependent variable (3541 subjects have response 0), and explanatory variables relate to health status namely excellent health (binary); poor health (binary), and number of chronic conditions; demographic status (age and male); and socio-economic status, namely years of education and private health insurance (binary). The zero inflated Poisson regression (without regression to predict subject varying $\zeta_i$) has core code

```
model { for (i in 1:4406) {y[i] ~ dpois(mu.c[i]);
mu.c[i] <- (1-t[i])*mu[i]; t[i] ~ dbern(zeta)
log(mu[i]) <- inprod(beta[], x[i,1:8])}
```

with a Be(1,1) prior assumed on $\zeta$. A two chain run of 10 000 iterations with convergence after around 3000 (and inference on the last 7000) provides a posterior mean for $\zeta$ of 0.53 (0.49, 0.57). Significant predictors are excellent health, poor health, total conditions, age, and health insurance. Using a predictive check appropriate to count data shows around 2% of subjects to be underpredicted (with posterior probability over 0.95 that $y_i$ exceeds $y_{rep,i}$), with no cases overpredicted. K-L calibration statistics (Section 2.2.6) show six cases (408, 827, 3534, 3698, 4143, 4154) to have $p_i^*$ exceeding 0.7.

A hurdle regression, including binary regression for the initial hurdle probability $\omega_i$, is implemented using the dloglik option in OpenBUGS with core code

```
model {for (i in 1:4406){d[i] <- 0;d[i] ~ dloglik(LL[i])
logit(omeg[i]) <- inprod(gam[],x[i,1:8]);
log(mu[i]) <- inprod(beta[],x[i,1:8])
LL[i] <- equals(y[i],0)*log(omeg[i])
+step(y[i]-1)*(log(1-omeg[i])-mu[i]+y[i]*log(mu[i])-log
                  (1-exp(-mu[i])))-logfact(y[i])}
```

Inferences are based on the last 4000 of a 5000 iteration two chain run. This model shows an improved LPML as compared to the ZIP model ($-2900$ as against $-2923$). Significant predictors of hospitalisation (after the initial hurdle) are again excellent health, poor health, total conditions, age, and health insurance, while significant predictors of the initial hurdle (with reversed effects) are excellent health, poor health, total conditions, and age.

## 4.4  Link selection

To provide a generalisation over default choices of link function in general linear models, several authors have considered the issue of choice of link functions in general linear models. Czado and Raftery (2006) and Czado (1997) discuss link families with shape parameters $(\psi_1, \psi_2)$ in addition to the linear predictor $\eta_i = X_i\beta$. These allow link modification according as $\eta_i$ is to the left or right of a reference value $\eta_0$.

Following Czado (1997), assuming that predictors are centred, one may set $\eta_0 = \beta_0$ where $\beta_0$ is the regression intercept. Denote the inverse link as $h(\eta)$, with argument $\eta$; for example, the inverse link for binary and binomial regression is the expit function,

$$h(\eta) = \frac{\exp(\eta)}{1 + \exp(\eta)}.$$

Omitting case identifiers, define $\eta_c = \eta - \eta_0 = \eta - \beta_0$, with $\eta = \beta_0 + \eta_c$. Then the fully extended argument (Czado and Raftery, 2006) to the inverse link is

$$G(\eta, \psi_1, \psi_2) = \begin{cases} \beta_0 - \frac{(-\eta_c+1)^{\psi_1}-1}{\psi_1} & \text{if } \eta_c > 0 \\ \beta_0 - \frac{(-\eta_c+1)^{\psi_2}-1}{\psi_2} & \text{if } \eta_c \leq 0 \end{cases}$$

involving both both left tail modification (via $\psi_2$) and right tail modification (via $\psi_1$) of the conventional term. Often modification of one tail only may be relevant, in which case the conventional link has default value $\psi = 1$. For example, a left tail modification would take

$$G(\eta, \psi) = \begin{cases} \eta = \beta_0 + \eta_c & \text{if } \eta_c > 0 \\ \beta_0 - \frac{(-\eta_c+1)^{\psi}-1}{\psi} & \text{if } \eta_c \leq 0 \end{cases}.$$

For binary/binomial and ordinal regressions, a number of differently shaped functions apart from the canonical logit are available and discrete mixtures over such links may alternatively be used, such as a mixture of the canonical symmetric logistic link and one or more asymmetric forms. An example (Lang, 1999) is a mixture over three forms for the inverse link, namely the left skewed extreme value (LSEV)

$$h_1(\eta) = 1 - \exp(-\exp(\eta)),$$

the logistic inverse link

$$h_2(\eta) = \frac{\exp(\eta)}{1 + \exp(\eta)},$$

and the right skewed extreme value (RSEV) inverse link

$$h_3(\eta) = \exp(-\exp(-\eta)).$$

The mixture has the form

$$h_\lambda(\eta) = w_1(\lambda)h_1(\eta) + w_2(\lambda)h_2(\eta) + w_3(\lambda)h_3(\eta),$$

where $\lambda$ is an extra parameter defining mixture proportions $w_1(\lambda) = \exp(-\exp(3.5\lambda + 2))$, $w_3(\lambda) = \exp(-\exp(-3.5\lambda + 2))$ and $w_2(\lambda) = 1 - w_1(\lambda) - w_3(\lambda)$. Negative values of $\lambda$ are obtained when the LSEV form is preferred, and positive values when the RSEV is preferred; $\lambda = 0$ corresponds to the logit link. Most variation in the mixture proportions $\{w_1(\lambda), w_2(\lambda), w_3(\lambda)\}$ occurs when $\lambda$ is in the interval $[-3, 3]$, and this leads to a normal prior on $\lambda$ with mean 0 and variance 9 as being essentially non-informative with regard to the appropriate link out of the three possible. For an ordinal response with cutpoint parameters $\theta_j$ and $\theta_{j-1}$, one has

$$p_{ij} = h_\lambda(\theta_j - X_i\beta) - h_\lambda(\theta_{j-1} - X_i\beta).$$

Since the interpretation and values of $\beta$ depend on the link, alternative definitions of covariate effects (with interpretation independent of link used) such as the LD50 may be monitored instead.

### Example 4.4    Byssinosis among cotton textile workers

Data on lung disease among textile workers involve three predictors: dust level of work place (values $1 = $ high, $= -0.5$ low); smoking status ($1 = $ smoker; $-1 = $ nonsmoker); length of employment ($1 = $ under 10 years; $-0.5 = $ greater lengths). An 18 cell breakdown based on these categorical predictors is used as the observation set (Czado, 1997). One expects positive effects of dust and smoking and a negative effect for shorter lengths of employment. A standard logit regression produces an average scaled deviance of 16.9.

To allow link selection, a uniform U($-2,2$) prior is assumed on $\psi$ under a left tail modification[2] (cf. Czado, 1997). A two chain run of 20 000 iterations with early convergence produces an improved average deviance of 10.0 as compared to the logit link (where $\psi = 1$). The posterior mean (95% interval) for $\psi$ of $-1.0$ ($-2$, 0.4) supports a modified link. Yates (1981) used a complementary log-log transform for these data. To demonstrate the discrete link mixture approach, a Dirichlet prior is adopted over two options only: $h_1(\eta) = 1 - \exp(-\exp(\eta))$, and $h_2(\eta) = \frac{\exp(\eta)}{1+\exp(\eta)}$. A two chain run of 50 000 iterations shows a slightly improved average deviance of 16.6 as against the standard logit, with the posterior weights on the two options inconclusive, namely $w_1 = 0.52$, $w_2 = 0.48$.

## 4.5    Discrete mixture regressions for regression and outlier status

Chapter 2 considered finite mixtures to describe heterogeneity in which latent classes $j = 1, \ldots, J$ differ only in their means $\mu_j$ and/or other population parameters (e.g. variances in a normal example). Finite mixture regressions introduce covariates to describe the relation between the mean $\mu_{ij}$ of subject $i$ on latent class $j$ and that subject's attributes profile. Predictors may also be used to model the prior probabilites $\lambda_{ij} = \Pr(L_i = j)$ of different categories of latent class indicators $L_i$, an option known as concomitant variable modelling. Finite mixtures may also be applied to different types of generalised linear regression (Grün and Leisch, 2008). Mixture regressions have been applied to modelling the behaviour or attitudes of individual human subjects with each individual's overall mean determined by their membership probabilities (Wedel *et al.*, 1993). Several applications of regression mixtures have been reported in consumer choice settings (Khalili and Chen, 2007) For example, Jones and

McLachlan (1992) consider consumer preference scales for different goods, which are related to product attributes (appearance, texture, etc.), and find sub-populations of consumers differing in the weight they attach to each attribute.

Consider first metric responses $y_i$ and normal linear regression framework with a $p$ dimensional vector of predictors $X_i$, and define latent membership indicators $L_i \in j = 1, \dots, J$. Were the indicators known,

$$y_i | X_i, L_i = j \sim N(X_i \beta_j, \tau_j),$$

where $\beta_j$ is a class specific regression vector of length $p$, and $\tau_j$ is the conditional variance. Denoting $\lambda_{ij} = \Pr(L_i = j | y)$, and $\mu_{ij} = X_i \beta_j$, the overall mean for subject $i$ is obtained as

$$\lambda_{i1} \mu_{i1} + \lambda_{i2} \mu_{i2} \dots + \lambda_{iJ} \mu_{iJ}.$$

The indicators $L_i$ may be sampled from a multinomial-Dirichlet prior without additional covariates, and with constant parameters over subjects $\lambda_{ij} = \lambda_j$. Alternatively, especially for $J$ preset, a multinomial regression can model the $L_i$ as functions of covariates $W_i$:

$$\lambda_{ij} = \exp(\phi_j W_i) / \left[ 1 + \sum_{k=2}^{J} \exp(\phi_k W_i) \right], \qquad j = 2, \dots, J.$$

$$\lambda_{i1} = 1 / \left[ 1 + \sum_{k=2}^{J} \exp(\phi_k W_i) \right].$$

For discrete responses (binomial, count, or multinomial) mixture regressions have utility in representing both departures from baseline assumptions (e.g. overdispersion) and regression effects between sub-populations (Cameron and Trivedi, 1986). For example, Wedel *et al.* (1993) argue for using a latent class Poisson regression mixture, both for modelling differential purchasing profiles among customers of a direct marketing company, and for representing over-dispersion. Similarly, Park and Lord (2009) consider discrete mixtures of negative binomial regression for vehicle accident data, with varying regression and dispersion parameters.

As mentioned in Chapter 2, there are the same problems in Bayesian analysis (as in frequentist analysis) concerning the appropriate number of components. Additionally, Bayesian sampling estimation may face the problems of empty classes at one or more iterations (e.g. no subjects are classified in the second of $J = 3$ groups), and the switching of labels unless the priors are constrained (Stephens, 2000). On the other hand, the introduction of predictors provides additional information that may improve identifiability. To counter label switching an ordering constraint may be applied to one or more of the intercepts, regression coefficients, variance parameters, or mixture proportions that ensures a consistent labelling. In some situations, it may be able to specify informative priors consistent with widely separated but internally homogeneous groups (Nobile and Green, 2000).

An alternative to constrained priors involves re-analysis of the posterior MCMC sample, for example, by random or constrained permutation sampling. Suppose unconstrained priors in the above discrete mixture linear regression were adopted, and parameter values $\theta_j^{(t)} = \{ \beta_j^{(t)}, \tau_j^{(t)} \}$ are sampled for the nominal group $j$ at iteration $t$. One may investigate first whether – after accounting for the label switching problem – there are patterns apparent on some of the parameter estimates which support the presence of sub-populations in the data. Random permutations of the nominal groups in the posterior sample from an unconstrained prior may be used to assess whether there are any parameter restrictions apparent in the output that may be associated with sub-populations (Fruhwirth-Schattner, 2001). Thus if there is only $p = 1$ predictor and the model is

$$y_i \sim \sum_{j=1}^{2} N(\mu_{ij}, \tau_j), \qquad \mu_{ij} = \beta_{0j} + \beta_{1j}x_i,$$

then a prior constraint which produces an identifiable mixture might be $\beta_{01} > \beta_{02}$, or $\beta_{11} > \beta_{12}$ or $\tau_1 > \tau_2$.

From the output of an unconstrained prior run with $J = 2$ groups, random permutation of the original sample labels means that the parameters nominally labelled as 1 at iteration $t$ are relabelled as 2 with probability 0.5, and if this particular relabelling occurs then the parameters at iteration $t$ originally labelled as 2 are relabelled as 1. Otherwise the original labelling holds. If $J = 3$ then the nominal group samples ordered $\{1, 2, 3\}$ keep the same label with probability $1/6$, change to $\{1, 3, 2\}$ with probability $1/6$, etc.

Let $\tilde{\theta}_{jk}$ then denote the relabelled group $j$ samples for parameters $k = 1, \ldots, K$. (A suffix for iteration $t$ is understood). The parameters relabelled as 1 (or any other single label among the $j = 1, \ldots, J$) provide a complete exploration of the unconstrained parameter space and one may consider scatter plots involving $\tilde{\theta}_{1k}$ against $\tilde{\theta}_{1m}$ for all parameter pairs $k$ and $m$. If some or all the plots involving $\tilde{\theta}_{1k}$ show separated clusters then an identifying constraint may be based on that parameter. To assess whether this is an effective constraint, the permutation method is applied based not on random reassignment but on the basis of reassignment to ensure the constraint is satisfied at all iterations.

## 4.5.1   Outlier accommodation

Discrete mixture models also provide a way of accommodating suspect data points, so that robust inferences are obtained for the majority of observations. A contaminated normal or Student $t$ model, in which outliers have shifted location and/or variances, provides one Bayesian approach to outliers (e.g. Verdinelli and Wasserman, 1991). For instance, assume a small probability such as $\pi_2 = 0.05$ that an outlier occurs, with $\pi_1 = 1 - \pi_2$, and

$$L_i \sim \text{Categoric}(\pi_1, \pi_2)$$

being a categoric indicator of outlier status. Then a contaminated normal density for $y_i$ with unknown intercept shift $D$ has the form

$$f(y_i|\mu, \sigma^2, \pi) = \pi_1 \phi(y_i|\mu, \sigma^2) + \pi_2 \phi(y_i|\mu + D, \sigma^2),$$

or conditioning on $L_i$

$$f(y_i|\mu, \sigma^2, \pi, L_i) = I(L_i = 1)\phi(y_i|\mu, \sigma^2) + I(L_i = 2)\phi(y_i|\mu + D, \sigma^2).$$

This approach can be extended to a small set of potential shift values, e.g. $D_i \in (D_1, D_2)$, to accommodate less extreme and more extreme outliers. One may adopt analogous methods for shifts in regression coefficients or shifts in error variance. Thus a variance shift model could take the form

$$f(y_i|\mu, \sigma^2, \pi, L_i) = I(L_i = 1)\phi(y_i|\mu, \sigma^2) + I(L_i = 2)\phi(y_i|\mu, D\sigma^2),$$

with $D > 1$ an unknown. Other options to protect against the effect of outliers or influential cases may be particular to the form of outcome. With a binary outcome, for instance, a prior may be set on a 'transposition', namely when $y_i = 1$ is the actual observation but the regression model provides much higher support for $\Pr(y_i = 0)$ than for $\Pr(y_i = 1)$. This scheme may be included in models allowing for misrecording or contaminated data (e.g. Copas, 1988).

**Example 4.5    Breastfeeding study with outlier accommodation**

An augmented data linear regression is applied to binary responses for $n = 139$ pregnant women, namely $y = 1$ if they planned to breastfeed their babies, and $y = 0$ for bottlefeeding. The data are of interest in containing several poorly fitted cases when standard regression techniques are applied, while application of robust regression methods may affect inferences on significant predictor effects (Heritier $et\ al.$, 2009). There are nine predictors, seven of which are binary: $x_1$, stage in pregnancy (beginning or end); $x_2$, how the subject was themselves fed as infants (1 = some/all breastfeeding, or 0 = only bottle); $x_3$, how the subject's friends fed their babies (some/all breastfeeding, or only bottle); $x_4$, if subject has a partner or not; $x_5$, age; $x_6$, age at which left full time education; $x_7$, ethnic group (1 = non-white, 0 = white); $x_8$, whether subject currently smokes (1 = yes, 0 = no), and $x_9$ whether subject has ever smoked. A conventional augmented normal linear regression (equivalent to probit regression) shows significant positive effects of $x_3$ and $x_7$, and a negative effect of $x_8$, though collinearity is indicated in an unexpected positive effect for $x_9$.

Augmented Student $t$ regression is then applied, namely

$$z_i \sim N(X_i\beta, 1/\lambda_i) \qquad I(A_i, B_i);$$

$$\lambda_i \sim Ga(0.5v, 0.5v),$$

with sampling limits determined by the observed $y$, and $1/v \sim U(0, 1)$. A two chain run of 5000 iterations (with inferences from the last 2500) gives an estimate for $v$ of 1.4. This model produces a major gain in fit (the DIC is reduced from 1702 to 1680). Inferences on predictor effects are unaffected though 11 cases have posterior mean $\lambda_i$ under 0.5, and subjects 12, 22, 78, 93 and 118 have $\lambda_i$ under 0.25.

A third analysis uses a weighted augmented data normal linear regression

$$z_i \sim N(X_i\beta, 1/w_i),$$

$$w_i = \min(1, t_r/r_i),$$

with $t_r = 1.345$, and $r_i = \left| \frac{z_i - X_i\beta}{\hat{\sigma}} \right|$. Although the theoretical variance is 1, the realised error variance may vary slightly from this, and so the standard deviation $\hat{\sigma}$ of the realised residuals $z_i - X_i\beta$ is included in the calculation of standardised residuals. Based on a two chain run of 10 000 iterations, it is apparent that this model produces no gain in fit over probit regression (with DIC of 1702), and no major changes in predictor inferences, though the effect of $x_6$ is enhanced, with the corresponding coefficient having an entirely positive 90% credible interval. The lowest posterior weights $w_i$ are for cases 12, 22, 78, 93 and 118.

A final analysis uses a two group discrete mixture on the error variance to accommodate potential outliers. The majority group have a known error variance $\sigma_1^2$ with prior probability $\pi_1 = 0.95$, while a minority outlier group have a higher (and unknown) error variance with prior probability $\pi_2 = 0.05$. Thus

$$z_i \sim N(X_i\beta, \sigma_{L_i}^2) \qquad I(A_i, B_i); \quad L_i \sim Categoric(\pi_{1:2})$$

The corresponding BUGS code fragment, including indicators OL[i] from which probabilities of being in the outlier group can be obtained, is

```
z[i] ~ dnorm(mu[i],tau[L[i]]) I(A[i],B[i])
L[i] ~ dcat(pi[1:2])
OL[i] <- equals(L[i],2).
```

A uniform prior $\sigma_2^2 \sim U(1, 10)$ is assumed on the unknown outlier group variance. From the final 4000 iterations of a two chain run of 5000 iterations, the posterior mean for $\sigma_2^2$ is obtained as 6.3, with five subjects having posterior probabilities exceeding 0.15 of being in the outlier group, namely subjects 12, 22, 78, 93 and 119. Predictor inferences are not changed.

### Example 4.6    Viral infections in potato plants

The data here are based on experiments on viral infections in potato plants according to total aphid exposure counts (Turner, 2000). The experiment was repeated 51 times. The data are in principle binomial, recording numbers of infected plants in a $9 \times 9$ grid with a single plant at each point. However, for reporting reasons, a normal approximation involving linear regression was taken. The outcome is then the totals of plants infected $y$, which vary from 0 to 24. A plot of the infected plant count against the number of aphids released ($x$) shows a clear bifurcation, with one set of $(y, x)$ pairs illustrating a positive impact of aphid count on infections, while another set of pairs shows no relation. Here two issues are paramount: first, possible relabelling in the MCMC sample, so that samples for the nominal group 1 (say) in fact are a mix of parameters from more than one underlying group, and second, assessment of the number of groups. A two group mixture, with means and conditional variances differing by group, is well identified by frequentist methods and bootstrap analysis shows marked improvement in fit over a single group model (Turner, 2000).

Here, MCMC estimation is initially applied to the model

$$y_i \sim \sum_{j=1}^{2} N(\mu_{ij}, \tau_j), \qquad \mu_{ij} = \beta_{0j} + \beta_{1j} x_i,$$

without parameter constraints. Thus there are $J = 2$ latent classess and $K = 3$ parameter sets (intercepts $\beta_{0j}$, slopes $\beta_{1j}$, and variances $\tau_j$). Using the output from 20 000 iterations with a single chain, 500 iteration burn-in, and null starting values, we apply (e.g. in a spreadsheet) the random permutation sampler. The relabelled group 1 parameters (every $10^{th}$ sample) are then plotted against each other (Figures 4.2 to 4.4). Both plots involving the slope, namely Figures 4.2 and 4.3, show well separated clusters of points, suggesting an identifiability constraint on the $\beta_{1j}$, such as $\beta_{11} > \beta_{12}$. Applying the constrained permutation sampler, again to the output from the unconstrained prior, shows that this is an effective way of identifying sub-populations, and one can either use the output from the constrained permutation to derive parameter estimates or formally apply the constraint in a new MCMC run. From the constrained permutation sampler the characteristics of the two groups are shown in Table 4.1.

In the present application, it was not possible to identify a stable solution with three groups that provided an improvement in log-likelihood over the two group solution. This was apparent both from MCMC sampling with a constraint on the slopes $\beta_{1j}$ and using the permutation sampler on the output from a three group model without a prior constraint. Plots of the intercepts on the slopes from the relabelled parameter iterates $\tilde{\theta}_{1k}$ ($k = 1, \ldots, 3$) showed no distinct groups for $J = 3$ but instead just 'scatter without form'.

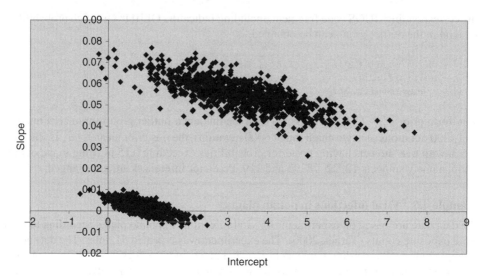

**Figure 4.2**    Plot of intercept vs slope for relabelled first group iterations.

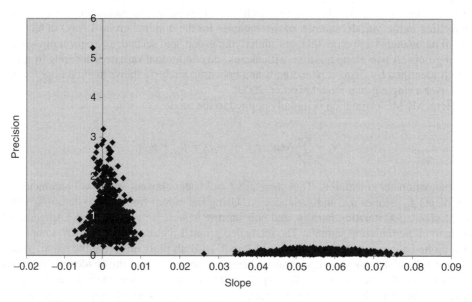

**Figure 4.3**    Slope against precision for relabelled first group iterations.

It remains to demonstrate the gain in fit for a two group as against a one group model. This might be possible to demonstrate using penalised fit measures such as the AIC statistics. Thus parallel sampling of a one and two group model (the latter with constrained prior) could be carried out, and the proportion of iterations obtained where the AIC is better for the two group than one group model, using the known number of parameters ($p_1 = 3, p_2 = 7$) in the models. Here a marginal likelihood approximation is obtained by monitoring the inverse likelihoods $g_i^{(t)} = 1/p(y_i|\theta^{(t)})$ (Dey *et al.*, 1997). This provides log pseudo marginal likelihood estimates

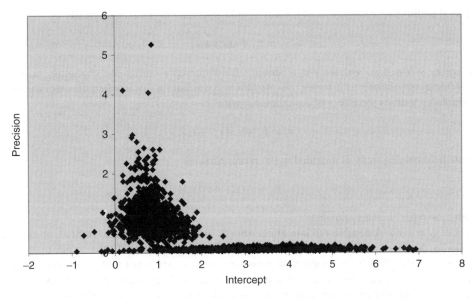

**Figure 4.4**    Intercept against precision for relabelled group 1 iterations.

**Table 4.1**    Potato virus data, parameter estimates.

|       | $\beta_{01}$ | $\beta_{02}$ | $\beta_{11}$ | $\beta_{12}$ | $1/\tau_1$ | $1/\tau_2$ |
|-------|--------|--------|--------|--------|--------|--------|
| Mean  | 0.868  | 3.324  | 0.002  | 0.055  | 0.900  | 0.096  |
| 2.5%  | 0.115  | 1.047  | −0.003 | 0.040  | 0.361  | 0.043  |
| 97.5% | 1.733  | 5.593  | 0.007  | 0.069  | 2.117  | 0.170  |

of −162.9 and −140.5 under the one and two group models, and so a clear preference for the two group model.

## 4.6    Modelling non-linear regression effects

The normal linear model and general linear models generally assume linear predictor effects in the regression mean, whereas non-linear predictor effects often occur. Polynomial regression may lack local flexibility in reproducing the relationship between the response and predictors. Greater flexibility is achieved under general additive regression models using smoothness prior approaches and basis function regression. These methods may also be applied to modelling other departures from standard assumptions, such as heteroscedastic variances (Wand, 2000; Yau and Kohn, 2003).

### 4.6.1    Smoothness priors for non-linear regression

Smoothness priors seek to estimate a correlated sequence of parameter values while penalizing abrupt jumps between successive values (Fahrmeir and Lang, 2001). These methods have utility in time series (see Chapter 6) and in regression applications. First consider a metric outcome $y_t$ with a single predictor $x_t$, and data are arranged so that $x_1 < x_2 < \dots < x_n$. A general additive regression model for this simple case may be stated generically as

$$y_t = \beta_0 + s(x_t) + \varepsilon_t,$$

where $\varepsilon_t \sim N(0, \sigma^2)$, and $s(x_t)$ is a smooth function representing the changing, possibly non-linear, impact of $x$ as it varies over its range. A variant on this is when the smooth in the variable $x$ modifies the effect of a predictor $z$, with

$$y_t = \beta_0 + z_t s(x_t) + \varepsilon_t,$$

while heteroscedasticity is included in the representation

$$y_t = \beta_0 + s_1(x_t) + s_2(x_t)e_t,$$

with $\varepsilon_t \sim N(0, 1)$, and $s_2$ positive.

Let $g_t = s(x_t)$. A flexible random effect prior is obtained using normal or Student random walks in the first, second or higher differences of the $g_t$. Commonly the $x_t$ are unequally spaced, and the prior for $g_t$ modifies successive precisions such that wider spaced points (in the covariate scale) are less tied to their predecessor than closer spaced points. Thus suppose spaces between predictor values are $\delta_1 = x_2 - x_1, \delta_2 = x_3 - x_2, \ldots, \delta_{n-1} = x_n - x_{n-1}$. A first order random walk smoothness prior, with normal errors, would be specified as

$$g_t \sim N(g_{t-1}, \delta_t \tau^2),$$

and a second order one would be

$$g_t \sim N(v_t, \delta_t \tau^2),$$

where $v_t = g_{t-1}(1 + \delta_t/\delta_{t-1}) - g_{t-2}(\delta_t/\delta_{t-1})$ (Fahrmeir and Lang, 2001). If there is equal spacing between successive covariate values, then the first and second order random walk priors reduce to $g_t \sim N(g_{t-1}, \tau^2)$ and $g_t \sim N(2g_{t-1} - g_{t-2}, \tau^2)$ respectively. A compromise between RW1 and RW2 dependence is achieved under

$$g_t \sim N(v_t, \delta_t \tau^2)$$

where

$$v_t = g_{t-1}[1 + (\delta_t/\delta_{t-1})\exp(-\alpha\delta_t)] - g_{t-2}[(\delta_t/\delta_{t-1})\exp(-\alpha\delta_t)].$$

Larger values of $\alpha > 0$, such that $\exp(-\alpha\delta_t)$ tends to zero, imply an approximate RW1 prior and less smoothness (Berzuini and Larizza, 1996). Under equal spacing this prior reduces to

$$g_t \sim N((1 + \eta)g_{t-1} - \eta g_{t-2}, \tau^2),$$

with $\eta \sim U(0, 1)$.

With smooths on several regressors $\{x_{1t}, x_{2t}, x_{3t}, \ldots, x_{pt}\}$, one approach to computation (e.g. in BUGS) is to supply an ordering index on each predictor at observation level $\{O_{1t}, O_{2t}, O_{3t}, \ldots\}$. So the regression becomes

$$y_t = \beta_0 + g_{O_{1t}} + g_{O_{2t}} + g_{O_{3t}} + \ldots + \varepsilon_t,$$

and separate smoothness priors are taken over the $\{M_1, M_2, M_3, \ldots, M_p\}$ unique values of each predictor. INLA may also be used for smooth regression functions in multiple predictors, without needing such indices (see Example 4.8). A frequent situation is the semi-parametric general additive form, with smooths on a subset of $K$ from all $p$ predictors, with the remainder modelled conventionally. Quite often there would be just a single regressor with a general additive form and the remainder included in a conventional linear combination. It may be noted that results may be sensitive both to the priors assumed for the initial smoothing values (e.g. $g_1$ in a first order random walk), and for the evolution variances $\tau_k^2$ (Fahrmeir and Lang, 2001, Section 2.2.3; Brezger and Lang, 2006, p 971). This is especially so for sparse data, such as for binary outcomes, and for smaller samples. Priors (e.g. uniform shrinkage priors) may also be set on the ratios (smoothing parameters) $\lambda_k = \tau_k^2/\sigma^2$ controlling the trade-off between goodness of fit to the data and the smoothness of $g_k = s(x_k)$.

Other smoothness priors have been proposed. For example, Carter and Kohn (1994) consider the signal plus noise model

$$y_t = s(x_t) + \varepsilon_t = g_t + \varepsilon_t,$$

where the $g_t$ are generated by a differential equation

$$\frac{d^2 g_t}{dt^2} = \tau \frac{dW_t}{dt},$$

with $W_t$ a Weiner process, and $\tau^2$ the variance of the smooth function. This leads to a bivariate state vector $f_t = \left(g_t, \frac{dg_t}{dt}\right)$ in which

$$f_t = F_t f_{t-1} + u_t,$$

where

$$F_t = \begin{pmatrix} 1 & \delta_t \\ 0 & 1 \end{pmatrix},$$

covariate differences $\delta_t$ are as above, and $\delta_1 = 0$. The $u_t$ are bivariate normal with mean zero and covariance $\tau^2 U_t$, where

$$U_t = \begin{pmatrix} \delta_t^3/3 & \delta_t^2/2 \\ \delta_t^2/2 & \delta_t \end{pmatrix}.$$

Wood and Kohn (1998) discuss the application of this prior with binary outcomes, which involves applying the latent variable model of Albert and Chib (1993).

## 4.6.2   Spline regression and other basis functions

There are a range of possible basis functions, and the linear predictor and polynomial regression are particular forms of basis function (Ruppert *et al.*, 2009). Basis functions typically rely on a relatively small set of representative points $\{c_k, k = 1, \ldots, K\}$ defining $K + 1$ intervals in the predictor space, with $K$ generally smaller than the number of observations $n$. Such points (called knots, centres or control points) are usually placed within the range $[x_{\min}, x_{\max}]$ of $x$. For example, for univariate $x_t$, Gaussian radial basis functions represent the unknown predictor effect as a weighted sum of terms

$$s(x_t) = \sum_{k=1}^{K} \beta_k \exp\left[-\frac{(x_t - c_k)^2}{2\kappa}\right],$$

while sigmoid functions have the form

$$s(x_t) = \sum_{k=1}^{K} \beta_k \frac{1}{1 + \exp(-a_{tk})},$$

where $a_{tk} = \frac{|x_t - c_k|}{\kappa}$, and thin plate splines (e.g. Clifford *et al.*, 2011) specify

$$s(x_t) = \sum_{k=1}^{K} \beta_k |x_t - c_k|^3.$$

The thin plate spline function generalises to higher dimensions with the argument in two dimensions being the distance $d_{ik} = \|x_i - c_k\|$ between $(x_{1i}, x_{2i})$ and centres $(c_{k1}, c_{k2})$, and the function taking the form $d[\log(d)]$ (Yau and Kohn, 2003).

Truncated polynomial spline regression on a single predictor $x_t$ has the form (Dennison *et al.*, 2002)

$$y_t = \beta_0 + \sum_{k=1}^{K} \beta_k (x_t - c_k)_+^q + \varepsilon_t,$$

where $\varepsilon_t \sim N(0, \sigma^2)$, and $q$ is a known positive integer. An alternative spline representation matches the degree $q$ of the truncated function $T(x_t) = \sum_{k=1}^{K} \beta_k (x_t - c_k)_+^q$ by a polynomial of order $q$ so that

$$y_t = \beta_0 + \alpha_1 x_t + \ldots + \alpha_q x_t^q + \sum_{k=1}^{K} \beta_k (x_t - c_k)_+^q + \varepsilon_t.$$

Low order powers such as $q = 1$ and $q = 3$ are typical, with $q = 1$ often being suitable for reproducing a true smooth function, even one subject to relatively sharp changes, given a large enough set of knots (Ruppert *et al.*, 2003, p. 68; Denison *et al.*, 2002, p. 52). Choice of knots is important: too few knots can produce oversmoothing, while too many result in overfitting. If knots are taken to have unknown locations within $[x_{min}, x_{max}]$, identification may rely on order constraints such as $c_k > c_{k-1}$, and analysis resembles time series with multiple change points.

Radial and truncated polynomial splines may be ill-conditioned (De Boor, 1972), and an alternative basis less prone to ill-conditioning involves B-splines. B-splines consist of sections of polynomial curves of the same degree $(q)$ connected at the knots $c_k$. A B-spline of degree $q$ consists of $q + 1$ polynomial curves and overlaps with $2q$ of its neighbours. For $K$ knots, there will be $K_q = K + q$ B-spline schedules with extra knots required outside the observed domain of $x$ to ensure $q$ overlapping B-splines in each interval. Let $B_k(x_t, q)$ be the value at $x_t$ of the $k^{th}$ B-spline of degree $q$. Successive B-spline values are defined recursively:

$$B_k(x_t, 0) = I(c_k \le x_t < c_{k+1}),$$

$$B_k(x_t, q) = \frac{x_t - c_k}{c_{k+q} - c_k} B_k(x_t, q - 1) + \frac{c_{k+q+1} - x_t}{c_{k+q+1} - c_{k+1}} B_{k+1}(x_t, q - 1),$$

with initial zero-degree polynomials being binary indicators defining intervals of $x_t$. For equally spaced knots a simplified recursion applies involving differences in truncated power splines. Then

$$y_t = \beta_0 + \sum_{k=1}^{K_q} \beta_k B_k(x_t, q) + \varepsilon_t.$$

### 4.6.3 Priors on basis coefficients

The $\beta_k$ coefficients in spline or B-spline regression may be modelled as fixed or random effects. If a fixed effects prior is assumed on the $\beta_k$, knot selection may be used to achieve parsimony (Wand, 2000). For example, in a model with only a truncated spline function, let $\gamma_k$ $(k = 1, \dots, K)$ be binary retention indices. Then one has

$$y_t = \beta_0 + \sum_{k=1}^{K} \gamma_k \beta_k (x_t - c_k)_+^q + \varepsilon_t,$$

with priors on $\gamma_k$ as discussed in Chapter 3, and realised coefficients obtained as the products $\delta_k = \gamma_k \beta_k$.

Random effect priors that use difference penalties on the coefficients lead to penalized spline or P-spline models (e.g. Eilers and Marx, 1996). Under the P-spline approach, the question of choosing the number or siting of the knots is alleviated, since providing enough knots are used, the penalty function should ensure that the resulting fits are very similar (Currie and Durban, 2002, p. 335). Options for random effects include an unstructured normal

$$\beta_k \sim N(0, \sigma_\beta^2),$$

which tends to shrink the $\beta_k$ as compared to a fixed effects prior, leading to a smooth fit (Wand, 2003). Alternatively, a random walk penalty may be applied to differences $\omega_k = \Delta^d \beta_k$, so that $\omega_k \sim N(0, \sigma_\omega^2)$. Taking $d = 1$ leads to

$$\beta_k \sim N(\beta_{k-1}, \sigma_\omega^2),$$

while a quadratic penalty in combination with a cubic B-spline leads to results similar to those under cubic smoothing splines (Currie and Durban, 2002).

### Example 4.7 Motorcycle data

The motorcycle dataset has been used by Silverman (1985) and others, and consists of observations of acceleration $y$ against time $x$. First consider the P-spline approach using a cubic B-spline, and model

$$y_t = \beta_0 + \sum_{k=1}^{K_q} \beta_k B_k(x_t, q) + \varepsilon_t.$$

with $\varepsilon_t \sim N(0, \sigma^2)$. $K = 19$ knots are taken at equally spaced points (3,6, ... ,57) in the range of x-values, and the resulting $K_q = 22$ B-spline coefficients obtained in R using the commands

```
library(splines); setwd("C://R files")
# columns headed y and x
D <- read.table("motorcycle.txt",header=T)
BS <- bs(D$x,knots=c(3,6,9,12,15,18,21,24,27,30,33,36,39,42,45,48,51,54,57),
Boundary.knots=c(0,60),degree=3,intercept=F).
```

A quadratic penalty function is assumed on the coefficients, namely

$$\beta_k \sim N(2\beta_{k-1} - \beta_{k-2}, \sigma_\omega^2),$$

with Ga(1,0.001) priors on $1/\sigma^2$ and $1/\sigma_\omega^2$, and a flat prior on the intercept. The random effects are modelled using the car.normal function, imposing a sum to zero constraint on the $\beta_k$ that improves identifiability in terms of separating the impacts of the intercept and the term

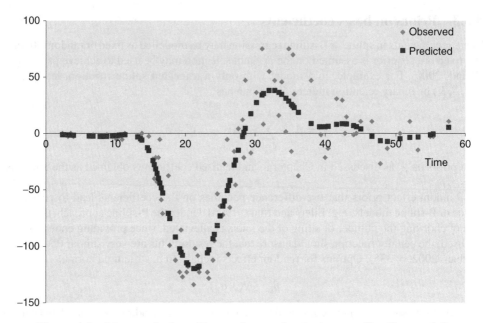

**Figure 4.5**   Motorcycle data. Observations and smooth curve (P-spline model).

$s(x_t) = \sum_{k=1}^{K_q} \beta_k B_k(x_t, q)$. Fitting in WinBUGS14, and with inferences from the 2nd half of a 10 000 run of two chains, gives a DIC of 872 and estimated smoothing parameter $\lambda = \sigma_\omega^2/\sigma^2$ of 0.35. The posterior means of the regression terms

$$\mu_t = \beta_0 + \sum_{k=1}^{K_q} \beta_k B_k(x_t, q)$$

are shown in Figure 4.5.

An alternative model involves a normal radial basis[3] function, namely

$$s(x_t) = \sum_{k=1}^{K} \beta_k \exp\left[-\frac{(x_t - c_k)^2}{2\kappa}\right],$$

with $\beta_k$ random. Again using the car.normal function and quadratic penalty on the $\beta_k$, and with a Ga(1,0.001) prior on $\kappa$, this leads to a lower DIC of 862.6. The posterior mean for $\kappa$ and $\lambda$ are respectively 8.6 and 0.83.

### Example 4.8   Kyphosis and age

Consider the kyphosis data from Hastie and Tibshirani (1990) on $n = 81$ patients receiving spinal surgery. The binary outcome $y_i$ relates to the post-surgical presence or otherwise of forward flexion of the spine from the vertical. Risk factors are $x_{1i} =$ the number of vertebrae level involved, $x_{2i} =$ the starting vertebrae level of the surgery, and $x_{3i} =$ age in months. Initially a probit model with linear effects in $X_i = (x_{1i}, x_{2i}, x_{3i})$ is applied using the latent dependent variable method discussed in Chapter 3. Thus

$$z_i \sim N(\mu_i, 1) \qquad I(0,) \qquad \text{if} \qquad y_i = 1,$$

$$z_i \sim N(\mu_i, 1) \qquad I(,0) \qquad \text{if} \qquad y_i = 0,$$

with mean $\mu_i = \beta_0 + \beta_1 x_{1i} + \beta_2 x_{2i} + \beta_3 x_{3i}$. Normal priors N(0,10) priors are set on $\beta_0, \beta_1$, and $\beta_2$ but an N(0,1) prior on $\beta_3$ as it is applied to ages which exceed 200 and too vague a prior may lead to numeric problems if a large coefficient is applied to a large age value. This analysis shows clear positive effects for $x_1$, and a negative effect for $x_2$, but a less clear positive (linear) effect of age.

A linear impact may in fact be doubted: if the ages are grouped into quintiles, the kyphosis rate (moment estimate) is lowest (0.05) for ages under 18 months, increases to 0.41 for ages 112–139, but declines to a low rate (0.07) for ages over 139 months To clarify possible non-linear effects, the impact of age is instead modelled via a general additive form, with mean now

$$\mu_i = \beta_1 x_{1i} + \beta_2 x_{2i} + s(x_{3i}).$$

In defining the smoothness prior for $s(x_3)$, the age variable is grouped according to the $n_g = 64$ distinct values. The group index appropriate to observation $i$ is represented by O3[i] in the BUGS data input. As a first option, the differential equation prior

$$\frac{d^2 g_t}{dt^2} = \tau \frac{dW_t}{dt} \qquad t = 1, \ldots, n_g$$

is used, with sampling $1/\tau^2$ from the gamma full conditional specified by Carter and Kohn (1994, p. 546), and prior values of 0.001 for index and shape parameters. The initial values in $f_t$ (for $t = 1$) are assigned N(0,10) priors. To ensure identifiability the intercept is omitted. A more computationally intensive alternative would be to centre the $g_{3t}$ at each iteration.

With inferences based on the second half of a two chain run of 500 000 iterations, a plot of posterior means (and 50% credible interval) of the smooth function $g_{3t}$ (Figure 4.6) shows a clear non-linear effect in age, with a negative effect most marked at youngest and oldest ages. The coefficients on $x_{1t}$ is somewhat enhanced. The average deviance improves from 65 to 60.6, but the extra parameters lead to an effective parameter estimate of 9.3, so the DIC deteriorates from 69 to 70.

An alternative smooth is obtained with a second order random walk prior on $g_{3t}$ namely

$$g_{3t} \sim N(v_t, \delta_t \tau^2), t = 1, \ldots, n_g,$$

$$v_t = g_{t-1}(1 + \delta_t/\delta_{t-1}) - g_{t-2}(\delta_t/\delta_{t-1}),$$

where the $\delta_t$ are differences between successive ordered distinct age values. Again to ensure identifiability the intercept is omitted. A two chain run of 500 000 iterations provides a DIC of 67.3, with a less pronounced fall off in the smooth at the highest ages as compared to the differential equation method (Figure 4.7).

The INLA package includes options for smooth regression effects based on RW1 or RW2 priors. For binary data, the family = "binomial" and data Ntrials = 1 options are used. The smooth $g_t$ may retain all distinct values of the predictor, or use the inla.group command to use grouped predictor values (with the option to specify the number of groups). Identification is aided by a constraint that the $g_t$ sum to zero. The code for an RW1 prior (with specified parameters in the prior for the precision) on the smooth age effect is as follows. The

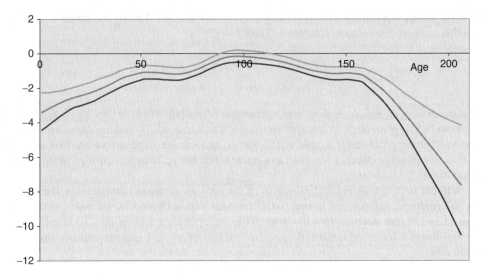

**Figure 4.6**   Differential equation prior, mean and 50% credible interval, age smooth.

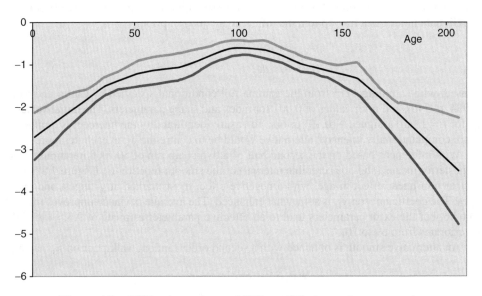

**Figure 4.7**   RW2 prior, mean and 50% credible interval, age smooth.

code includes options with and without grouping of the predictor values (formulas F1 and F2 respectively):

```
require(INLA); setwd("C://R files");
# columns headed y,x1,x2,x3 ( x1=num, x2=start, x3=age)
D = read.table("kyphosis.txt",header=T)
F1 = y ~ x1+x2+ f(x3,model="rw1",param=c(1,0.001))
M1 = inla(F1, family = "binomial", data = D,Ntrials=1)

# summarise random effect values
```

```
M1$summary.random$'x3'
plot(M1)
F2 = y ~ x1+x2+ f(inla.group(x3,n=20),model="rw1",param=c(1,0.001))
M2 = inla(F2, family = "binomial", data = D,Ntrials=1)
M2$summary.random$'inla.group(x3, n = 20)'
```

The posterior means of the $g_t$ series are obtained as the column headed mean in the $summary.random files, with the column headed id containing predictor values.

In R2BayesX numerical problems may occur in a model assuming smooth functions in all predictors due to the small number of observations with $y_t = 1$, and to the small number of distinct values for $x_1$ and $x_2$. This may be alleviated by reducing the number of knots, so limiting flexibility when there is only a small number of distinct covariate values. Thus one has

```
require(R2BayesX); setwd("C://R files");
D <- read.table("kyphosis.txt",header=T)
D=data.frame(D)
f <- y ~ sx(x1,nrknots=5) + sx(x2,nrknots=5) + sx(x3,nrknots=10)
b <- bayesx(f, family = "binomial", method = "MCMC", data=D, seed=123)
plot(b)
```

## 4.7   Quantile regression

Assumptions underlying normal linear regression and its focus on estimating the conditional mean of the response are often questionable in practical applications. Skewness or multimodality may cast doubt on the assumption of unimodal symmetric errors (Kottas and Gelfand, 2001), and inferences can be distorted by outlier observations. Thus Min and Kim (2004) consider different forms of non-Gaussian errors, with asymmetric and long-tailed distributions, and show that mean regression does not satisfactorily capture key properties of the data. Greater robustness may be obtained by regressing the median rather than the mean of the response on covariates (median regression), since the median is a more suitable centrality measure for data with a skewed response.

In classical statistics median regression involves minimisation of the absolute deviations, reducing the impact of outliers. Median regression for continuous reponses is a special case of quantile regression (Koenker, 2005), whereby

$$Q[y_i|X_i,p] = X_i\beta_p,$$

$$\Pr[y_i \leq X_i\beta_p] = p,$$

since other quantiles of the distribution of $y_i|X_i$ may be of interest. For example, one might consider different quantiles of the distribution of food spending according to income (Example 4.9).

Methods for Bayesian quantile regression include asymmetric Laplace likelihood (Yu and Moyeed, 2001), exponentially tilted empirical likelihood (Schennach, 2005), and Dirichlet process mixture median regression (Kottas and Gelfand, 2001). For $p \in (0, 1)$, quantile regression involves minimising $\sum_{i=1}^{n} \rho_p(y_i - X_i\beta_p)$, where $\rho_p(u) = u(p - I(u \leq 0))$. This minimisation can be implemented using an asymmetric Laplace distribution (ALD), with density function

$$\text{ALD}(y|\eta_p, \delta_p, p) = p(1 - p)\delta_p \exp[\delta_p(y - \eta_p)(p - I(y \leq \eta_p)],$$

**Table 4.2**   Quantile regression, food expenditure.

| Quantile | Parameter | Mean | St Devn | 2.5% | 97.5% |
|----------|-----------|------|---------|------|-------|
| 0.05 | $\beta_1$ | 4.62 | 0.05 | 4.52 | 4.72 |
|      | $\beta_2$ | 0.35 | 0.01 | 0.34 | 0.36 |
|      | $\delta$  | 10.79 | 0.71 | 9.46 | 12.20 |
| 0.25 | $\beta_1$ | 5.60 | 0.06 | 5.47 | 5.72 |
|      | $\beta_2$ | 0.47 | 0.01 | 0.44 | 0.50 |
|      | $\delta$  | 3.32 | 0.22 | 2.91 | 3.77 |
| 0.5 | $\beta_1$ | 6.32 | 0.06 | 6.21 | 6.44 |
|     | $\beta_2$ | 0.56 | 0.02 | 0.52 | 0.59 |
|     | $\delta$  | 2.67 | 0.17 | 2.34 | 3.02 |
| 0.75 | $\beta_1$ | 6.95 | 0.05 | 6.83 | 7.05 |
|      | $\beta_2$ | 0.64 | 0.02 | 0.61 | 0.67 |
|      | $\delta$  | 3.59 | 0.23 | 3.15 | 4.07 |
| 0.95 | $\beta_1$ | 7.61 | 0.04 | 7.54 | 7.70 |
|      | $\beta_2$ | 0.71 | 0.01 | 0.69 | 0.73 |
|      | $\delta$  | 12.37 | 0.80 | 10.85 | 13.96 |

which can be represented as a scale mixture of normals, so facilitating forms of Gibbs sampling. Thus for $y \sim \text{ALD}(\eta_p, \delta_p, p)$, one has

$$y = \eta + \xi W + \left[ \frac{2W}{\delta_p p(1-p)} \right]^{0.5} Z,$$

where $\xi = \frac{(1-2p)}{p(1-p)}$, $W \sim \exp(\delta_p)$, and $Z \sim N(0,1)$. Non-linear and spatial quantile regression are considered by Yue and Rue (2011).

### Example 4.9   Food expenditure

Consider data used in the R quantreg package on food expenditure y in relation to a single predictor household income (x) for a sample of n = 235 nineteenth century working class Belgian households. The values of y and x provided in quantreg are divided by 100, with $\eta_{pi} = \beta_{p1} + \beta_{p2}(x_i - \bar{x})$.

Using the Gibbs sampling approach based on the asymmetric Laplace likelihood, a quantile regression profile may be taken over P = 5 quantile values namely p = 0.05, 0.25, 0.5, 0.75 and 0.95. This involves creating P = 5 copies of the data in OpenBUGS[4]. The second half of a two chain run of 10 000 iterations provides non-overlapping values of the income effect $\beta_{p2}$ as in Table 4.2, ranging from 0.35 for p = 0.05 through to 0.71 for p = 0.95. Values of $\delta_p$ are highest for more extreme quantiles.

Quantile regressions may be carried out in INLA using the laplace likelihood, for example with p = 0.05 (and so setting alpha = 0.05 in control.family), one has

```
library(INLA); setwd("C://R files")
D <- read.table("expend.txt",header=T)
F = y ~ x
M = inla(F, family = "laplace",data=D,
control.family=list(alpha=0.05, param=c(1,0.001)),
```

```
        control.predictor =list(initial = 12),
control.inla = list(h=1e-3))
plot(M); summary(M).
```

# Exercises

**4.1.** In Example 4.1, estimate a variance transformation model with $\log(\sigma_i^2) = \xi_0 + \xi_1\mu_i$, and compare its fit with a constant variance normal linear regression using the log pseudo marginal likelihood (LPML). It may facilitate estimation (if using OpenBUGS or Win-BUGS) to use a transformation applied to the precision rather than variance. Identify the five cases with the highest influence statistics $K(y, y_{[i]})$, as described in Chapter 2, and assess the impact on the regression coefficients on excluding these subjects from the analysis.

**4.2.** Fit a Student $t$ linear regression to astronomy data from Rousseeuw (1991) relating to the star cluster CYG OBI, which contains 47 stars in the direction of Cygnus. The predictor (x[i] in the data in Exercise 4.2.odc) is the logarithmic temperature at the star surface and $y_i$ is the log of the star's light intensity. Compare the LPML of a model with $v$ set at 5 (Piche, 2011, page 2), and a model with $v$ as an unknown. For example, $v$ may be assigned an exponential prior (Geweke, 1993), or a uniform prior assumed on $1/v$, such as $1/v \sim U(1, 0.01)$ so that $v$ will be between 1 and 100.

**4.3.** In Example 4.3, apply the existing ZIP regression, but using a mixed predictive check based on replicate $t_{rep,i} \sim \text{Bern}(\zeta)$. This should show the proportion of underpredicted cases to be around 4%.

**4.4.** In Example 4.3 (hospital visits), apply a ZIP regression with the seven predictors now used to predict both $\zeta_i$ and $\mu_i$, and additionally with predictor selection (e.g. using Stochastic Search Variable Selection) in both sub-components. How far does this improve the LPML and the K-L measures on the six cases mentioned in the example.

**4.5.** In Example 4.5 (byssinosis incidence), modify the existing code (for the mixture family and logit options) to find the probabilities of 20% or more incidence in each of the 18 cells (analagous to an LD20 rate). The observed incidence for long employment, positive smoking status and dusty workplaces in fact exceeds 25% and a model better tuned to the data will predict this high incidence.

**4.6.** In Example 4.7 (motorcycle data) use a one-dimensional thin plate spline function with random coefficients to model the non-linear effect. Use the mixed replicate predictive scheme to assess the proportion of poorly fitted cases (e.g. posterior predictive probabilities that $y_i$ exceeds $y_{rep,i}$ that are under 0.05 or over 0.95). Assess improvements to predictive fit through (a) replacing the constant varance assumption for $\varepsilon_t$ by a scale mixture of normals (equivalent to Student $t$) with unknown degrees of freedom, and (b) modelling the error variances $h_t = \log(\sigma_t^2)$ in relation to $x_t$, also using a thin plate spline.

# Notes

1. The code for the power-unknown version of the Negative Binomial is as follows:

```
model { for (i in 1:14243) {DOCVIS[i] ~ dpois(nu[i]); nu[i]
                            <- omega[i]*mu[i];
```

```
omega[i] ~ dgamma(almu[i],almu[i])
almu[i] <- alph*pow(mu[i],Q)
r[i] <- almu[i]/(almu[i]+mu[i])
# Neg-Bin log-likelihood
LL[i] <- loggam(DOCVIS[i]+almu[i])-logfact(DOCVIS[i])
-loggam(almu[i])+almu[i]*log(r[i])+DOCVIS[i]*log(1-r[i])
G[i] <- 1/exp(LL[i])
# predictions
DOCVIS.rep[i] ~ dpois(nu[i]);
# predictive check
C[i] <- step(DOCVIS[i]-DOCVIS.rep[i])-0.5*equals(DOCVIS[i],
                                       DOCVIS.rep[i])
log(mu[i]) <- beta[1]+delta[Year[i]]+beta[2]*AGE[i]/10+beta[3]*
                                       AGESQ[i]/100
+beta[4]*HSAT[i]+beta[5]*HANDDUM[i]+beta[6]*HANDPER[i]+beta[7]*
                                       MARRIED[i]
+beta[8]*EDUC[i]+beta[9]*HHNINC[i]/10000+beta[10]*HHKIDS[i]+
                                       beta[11]*SELF[i]
+beta[12]*CIVIL[i]+beta[13]*BLUEC[i]+beta[14]*WORKING[i]
+beta[15]*PUBLIC[i]+beta[16]*ADDON[i]}
# Priors
for (j in 1:16) {beta[j] ~ dnorm(0,0.001)}
R ~ dexp(0.5); Q <- 2-R
delta[1] <- 0; for (j in 2:7) {delta[j] ~ dnorm(0,0.001)}
alph ~ dgamma(1,0.001); phi <- 1/alph; tLL <- sum(LL[])}
```

2. The BUGS program for left tail modification is

```
model { for (i in 1:18) {y[i] ~ dbin(p[i],n[i])
yhat[i] <- n[i]*p[i]; nyhat[i] <- n[i]*(1-p[i]);
ny[i] <- n[i]-y[i]
# deviance contribution
dv[i] <- y[i]*log((y[i]+0.001)/(yhat[i]+0.001))+ny[i]*
                log((ny[i]+0.001)/(nyhat[i]+0.001))
p[i] <- exp(eta.x[i])/(1+exp(eta.x[i]))
# extended argument to expit inverse link
eta.x[i] <- step(eta.c[i])*eta[i]+step(-eta.c[i])*
                (b0-(pow(eta.s[i],psi)-1)/psi)
eta[i] <- b0+b[1]*whigh[i]+b[2]*smk[i]+b[3]*elow[i]
eta.c[i] <- eta[i]-b0
# define eta.s=-eta.c+1 to ensure +ve number raised to
# power, since power calculation in eta.x is still done
# even if eta.c positive
eta.s[i] <- max(0.01,-eta.c[i]+1)}
# priors
for (j in 1:3) {b[j] ~ dnorm(0,0.01)}
b0 ~ dnorm(0,0.001); psi ~ dunif(-2,2)
D <- 2*sum(dv[])}
```

3. The BUGS code for the radial basis function applied to the motorcycle data is

```
model { for (i in 1:94) {y[i] ~ dnorm(mu[i],tau)
mu[i] <- beta0+inprod(beta[],B[i,])
# smooth in x
sx[i] <- inprod(beta[],B[i,])
for (k in 1:K) {B[i,k] <- exp(-(x[i]-c[k])*(x[i]-c[k])/(2*kap))}}
```

```
# quadratic penalty on coefficients
beta[1:K] ~ car.normal(adj[], w[], n[], tau.beta)
w[1] <- 1;    adj[1] <- 2;    n[1] <- 1
w[(K-2)*2 + 2] <- 1; adj[(K-2)*2 + 2] <- K-1; n[K] <- 1
for (i in 2:K-1) { w[2+(i-2)*2] <- 1; adj[2+(i-2)*2] <- i-1
w[3+(i-2)*2] <- 1;    adj[3+(i-2)*2] <- i+1; n[i] <- 2}
# other priors
beta0 ~ dflat(); kap ~ dgamma(1,0.001)
sig2[1] <- 1/tau.beta; sig2[2] <- 1/tau; lambda <- sig2[2]/sig2[1]
tau.beta ~ dgamma(1,0.001); tau ~ dgamma(1,0.001)}
```

4. The BUGS code for the profile quantile regression, with centred values of x, and multiple data copies in Y, is

```
model{ for (p in 1:P) {xi[p] <- (1-2*quant[p])/(quant[p]*(1-quant[p]))
for (i in 1:n) {Y[i,p] <- y[i]
eta[i,p] <- b[1,p] + b[2,p]*(x[i]-mean(x[]))
w[i,p] ~ dexp(delta[p])
mu[i,p] <- xi[p]*w[i,p] + eta[i,p]
tau[i,p] <- (quant[p]*(1-quant[p])*delta[p])/(2*w[i,p])
Y[i,p] ~ dnorm(mu[i,p],tau[i,p])}}
#priors for regression
for (p in 1:P) {b[1,p] ~ dflat()
b[2,p] ~ dnorm(0,0.00000001)
delta[p] ~ dgamma(1,0.001)}}
```

# References

Afifi, A., Clark, V. and May, S. (2004) *Computer Aided Multivariate Analysis*, 4th edn. Chapman and Hall, Boca Raton, FL.

Aitkin, M. (1987) Modelling variance heterogeneity in normal regression using GLIM. *Applied Statistics*, **36**(4), 332–339.

Albert, J. (1988) Computational methods using a Bayesian hierarchical generalized linear model. *Journal of the American Statistical Association*, **83**(404), 1037–1044.

Albert, J. and Chib, S. (1993) Bayesian analysis of binary and polychotomous response data. *Journal of the American Statistical Association*, **88**(422), 669–679.

Azzalini, A. and Regoli, G. (2012) Some properties of skew-symmetric distributions. *Annals of the Institute of Statistical Mathematics*, **64**(4), 857–879.

Berzuini, C. and Larizza, C. (1996) A unified approach for modeling longitudinal and failure time data, with application in medical monitoring. *IEEE Transactions on Pattern Analysis and Machine Intelligence*, **18**(2), 109–123.

Boscardin, W. and Gelman, A. (1996) Bayesian regression with parametric models for heteroscedasticity. *Advances in Econometrics*, **11**, A87–109.

Branco, M. and Dey, D. (2002) Regression model under skew elliptical error distribution. *Journal of Mathematical Science*, **1**, 151–169.

Breslow, N. (1996) Generalized linear models: checking assumptions and strengthening conclusions. *Statistica Applicata*, **8**, 23–41.

Brezger, A. and Lang, S. (2006) Generalized structured additive regression based on Bayesian P-splines. *Computational Statistics and Data Analysis*, **50**(4), 967–991.

Cameron, A. and Trivedi, P. (1986) Econometric models based on count data: comparisons and applications of some estimators and tests. *Journal of Applied Econometrics*, **1**, 29–54.

Carroll, R. and Ruppert, D. (1988) *Transformation and Weighting in Regression*. Chapman Hall, New York, NY.

Carter, C.K. and Kohn, R. (1994) On Gibbs sampling for state space models. *Biometrika*, **81**(3), 541–553.

Cepeda, E. and Gamerman, D. (2001) Bayesian modeling of variance heterogeneity in normal regression models. *Brazilian Journal of Probability and Statistics*, **14**(1), 207–221.

Chen, M.-H., Dey, D. and Shao, Q. (1999) A new skewed link model for dichotomous quantal response data. *Journal of the American Statistical Association*, **94**, 1172–1186.

Clifford, S., Low Choy, S., Hussein, T., Mengersen, K., Morawska, L. (2011) Using the Generalised Additive Model to model the particle number count of ultrafine particles. *Atmospheric Environment*, **45**(32), 5934–5945.

Collett, D. (2002) *Modelling Binary Data*, 2nd edn. Chapman & Hall/CRC, Boca Raton, FL.

Copas, J. (1988) Binary regression models for contaminated data. *J Royal Statistical Society Ser B*, **50**, 225–265.

Currie, I.D. and Durban, M. (2002) Flexible smoothing with P-splines: a unified approach. *Statistical Modelling*, **2**(4), 333–349.

Czado, C. (1997) On selecting parametric link transformation families in generalized linear models. *Journal of Statistical Planning and Inference*, **61**, 125–139.

Czado, C. and Raftery, A. (2006) Choosing the link function and accounting for link uncertainty in generalized linear models using Bayes factors. *Statistical Papers*, **47**, 419–442.

Deb, P. and Trivedi, P. (1997) Demand for medical care by the elderly: a finite mixture approach. *Journal of Applied Econometrics*, **12**(3), 313–336.

De Boor, C. (1972) On calculating with B-splines. *Journal of Approximation Theory*, **6**, 50–62.

Denison, D., Holmes C., Mallick B. and Smith, A. (2002) *Bayesian Methods for Non-linear Classification and Regression*. Wiley, Chichester, UK.

Dey, D., Chen, M. and Chang, H. (1997) Bayesian approach for non-linear random effects models. *Biometrics*, **53**, 1239–1252.

Eilers, P.H. and Marx, B.D. (1996). Flexible smoothing with B-splines and penalties. *Statistical Science*, **11**, 89–102.

Fahrmeir, L. and Lang, S. (2001) Bayesian inference for generalized additive mixed models based on Markov random field priors. *Journal of the Royal Statistical Society C*, **50**(2), 201–220.

Fernandez, C. and Steel, M. (1998) On Bayesian modelling of fat tails and skewness. *Journal of the American Statistical Association*, **93**, 359–371.

Fonseca, T., Ferreira, M. and Migon, H. (2008) Objective Bayesian analysis for the Student-t regression model. *Biometrika*, **95**, 325–333.

Fox, J. (2002) *An R and S Plus Companion to Applied Regression*. Sage Publications, London, UK.

Fruhwirth-Schnatter, S. (2001) Markov chain Monte Carlo estimation of classical and dynamic switching and mixture models. *Journal of the American Statistical Association*, **96**, 194–209.

Geweke, J. (1993) Bayesian treatment of the independent Student-t linear model. *Journal of Applied Econometrics*, **8**(S), 19–40.

Ghosh, S., Mukhopadhyay, P. and Lu, J. (2006) Bayesian analysis of zero-inflated regression models. *Journal of Statistical Planning and Inference*, **136**(4), 1360–1375.

Gottardo, R. and Raftery, A. (2009) Bayesian robust transformation and variable selection: a unified approach. *Canadian Journal of Statistics*, **37**(3), 361–380.

Greene, W. (2008) Functional forms for the negative binomial model for count data. *Economics Letters*, **99**, 585–590.

Grün, B. and Leisch, F. (2008) Finite mixtures of generalized linear regression models. In Shalabh and Christian Heumann (eds), *Recent Advances in Linear Models and Related Areas*. Physica Verlag, Heidelberg.

Gurmu, S. and Dagne, G. (2012) Bayesian approach to zero-inflated bivariate ordered probit regression model, with an application to tobacco use. *Journal of Probability and Statistics*, **2012**, Article ID 617678, doi:10.1155/2012/617678.

Hall, D. (2000) Zero-inflated Poisson and binomial regression with random effects: a case study. *Biometrics*, **56**(4), 1030–1039.

Hastie, T. and Tibshirani, R. (1990) *Generalized Additive Models*. CRC, Boca Raton, FL.

Heritier, S., Cantoni, E., Copt, S. and Victoria-Feser, M. (2009) *Robust Methods in Biostatistics*. Wiley, Chichester, UK.

Hoeting, J., Raftery, A. and Madigan, D. (2002) Bayesian variable and transformation selection in linear regression. *Journal of Computational and Graphical Statistics*, **11**(3), 485–507.

Jones, P. and McLachlan, G. (1992) Fitting finite mixture models in a regression context. *Australian Journal of Statistics*, **34**, 233–240.

Khalili, A. and Chen, J. (2007) Variable selection in finite mixture of regression models. *Journal of the American Statistical Association*, **102**(479), 1025–1038.

Knorr-Held, L. and Rainer, E. (2001) Projections of lung cancer mortality in West Germany: a case study in Bayesian prediction. *Biostatistics*, **2**, 109–29.

Koenker, R. (2005) *Quantile Regression*. Cambridge University Press, Cambridge, UK.

Kottas, A. and Gelfand, A. (2001) Bayesian semiparametric median regression modeling. *Journal of the American Statistical Association*, **96**, 1458–1468.

Lachos, V., Bandyopadhyay, D. and Garay, A. (2011) Heteroscedastic non-linear regression models based on scale mixtures of skew-normal distributions. *Statistics and Probability Letters*, **81**(8), 1208–1217.

Lang, J. (1999) Bayesian ordinal and binary regression models with a parametric family of mixture links. *Computational Statistics and Data Analysis*, **31**(1), 59–87.

Liseo, B. (2004) Skew-elliptical distributions in Bayesian inference. In M. Genton (ed.), *Skew-Elliptical Distributions and Their Applications: A Journey Beyond Normality*, pp. 153–171. Chapman & Hall/CRC, Boca Raton, FL.

Liseo, B. and Loperfido, N. (2006) A note on reference priors for the scalar skew-normal distribution. *Journal of Statistical Planning and Inference*, **136**(2), 373–389.

Liseo, B. and Loperfido, N. (2006) Default Bayesian analysis of the skew-normal distribution. *Journal of Statistical Planning and Inference*, **136**, 373–389.

Min, I. and Kim, I. (2004) A Monte Carlo comparison of parametric and non-parametric quantile regressions. *Applied Economic Letters*, **11**, 71–74.

Mitsakakis, N. (2012) Bayesian regression models for estimation of disease-specific net costs using aggregate data, presented at 40th Annual Meeting of the Statistical Society of Canada, http://ssc.ca/en/meetings/2012/abs/mrm.

Mullahy, J. (1986) Specification and testing of some modified count data models. *Journal of Econometrics*, **33**, 341–365.

Nobile, A. and Green, P.J. (2000) Bayesian analysis of factorial experiments by mixture modelling. *Biometrika*, **87**(1), 15–35.

Park, B. and Lord, D. (2009) Application of finite mixture models for vehicle crash data analysis. *Accident Analysis and Prevention*, **41**(4), 683–691.

Piche, R. (2011) *Robust Multivariate Linear Regression Using the Student-t Distribution*, Working Paper, Tampere University of Technology.

Riphahn, R., Wambach, A. and Million, A. (2003) Incentive effects in the demand for health care: a bivariate panel count data estimation. *Journal of Applied Econometrics*, **18**, 387–405.

Rousseeuw, P. (1991) Tutorial to robust statistics. *Journal of Chemometrics*, **5**, 1–20.

Ruppert, D., Wand, M.P. and Carroll, R.J. (2003) *Semiparametric Regression*. Cambridge University Press, Cambridge, UK.

Ruppert, D., Wand, M.P. and Carroll, R.J. (2009) Semiparametric regression during 2003–2007. *Electronic Journal of Statistics*, **3**, 1193–1256.

Sahu, K., Dey, D. and Branco, M.D. (2003) A new class of multivariate skew distributions with applications to bayesian regression models. *Canadian Journal of Statistics*, **31**, 129–150.

Schennach, S. (2005) Bayesian exponentially tilted empirical likelihood. *Biometrika*, **92**, 31–46.

Silverman, B.W. (1985) Some aspects of the spline smoothing approach to non-parametric regression curve fitting. *Journal of the Royal Statistical Society B*, **47**, 1–52.

Stephens, M. (2000) Dealing with label switching in mixture models. *Journal of the Royal Statistical Society B*, **62**(4), 795–809.

Turner, T.R. (2000) Estimating the propagation rate of a viral infection of potato plants via mixtures of regressions. *Journal of the Royal Statistical Society C*, **49**(3), 371–384.

Verdinelli, I. and Wasserman, L. (1991) *Bayesian Analysis* of outlier problems using the Gibbs sampler. *Statistics and Computing*, **1**(2), 105–117.

Wand, M. (2000) A comparison of regression spline smoothing procedures. *Computational Statistics*, **15**, 443–462.

Wand, M.P. (2003) Smoothing and mixed models. *Computational Statistics*, **18**, 223–249.

Wang, X.-F. (2013) Bayesian nonparametric regression and density estimation using integrated nested Laplace approximations. *Journal of Biometrics and Biostatistics*, **4**, e125.

Wedel, M., DeSarbo, W.S., Bult, J.R. and Ramaswamy, V. (1993) A latent class Poisson regression model for heterogeneous count data. *Journal of Applied Econometrics*, **8**(4), 397–411.

Weiss, R. (1994) Pediatric pain, predictive inference, and sensitivity analysis. *Evaluation Review*, **18**(6), 651–677.

West, M. (1984) Outlier models and prior distributions in Bayesian linear regression, *Journal of the Royal Statistical Society B*, **46**, 431–439.

Winkelmann, R. and Zimmermann, K. (1995) Recent developments in count data modeling: theory and application. *Journal of Economic Surveys*, **9**(1), 1–36.

Wood, S. and Kohn, R. (1998) A Bayesian approach to robust binary nonparametric regression. *Journal of the American Statistical Association*, **93**(441), 203–213.

Yamamoto, E. and Yanagimoto, T. (1994) Statistical methods for the beta-binomial model in teratology. *Environmental Health Perspectives*, **102**(Suppl 1), 25–31.

Yates, F. (1981) *Sampling Methods for Censuses and Surveys*, 4th edn. Hafner, New York, NY.

Yau, P. and Kohn, R. (2003) Estimation and variable selection in nonparametric heteroscedastic regression. *Statistics and Computing*, **13**(3), 191–208.

Yu, K. and Moyeed, R. (2001) Bayesian quantile regression. *Statistics and Probability Letters*, **54**, 437–447.

Yue, Y. and Rue, H. (2011) Bayesian inference for additive mixed quantile regression models. *Computational Statistics and Data Analysis*, **55**, 84–96.

Zeileis, A., Kleiber, C. and Jackman, S. (2008) Regression models for count data in R. *Journal of Statistical Software*, **27**(8), 1–25.

# 5

# Meta-analysis and multilevel models

## 5.1 Introduction

Meta-analysis and multilevel analysis are related techniques seeking to draw inferences regarding individual clusters or subjects from a broader collection of observations. The focus in Section 5.2 of this chapter is on hierarchical meta-analysis, namely methods for combining the results of independent studies to provide an estimate of a presumed common effect (e.g. treatment gain, environmental risk), while recognising heterogeneity between studies. Meta-analysis uses findings from sets of studies, assumed similar enough to be considered exchangeable, and typically uses summary statistics from originally multilevel or longitudinal studies, where subjects and clusters are analysed jointly.

The remaining sections in this chapter consider multilevel techniques (with longitudinal studies discussed in Chapter 7). Multilevel studies also adopt a hierarchical random effects approach (e.g. Goldstein, 2011), though fixed effects models for clustered data, particularly the varying intercept case, are sometimes used. The dual aims are then to allow for contextual or cluster effects on individual level outcomes, and conversely adjust contextual effects to fully reflect individual subject variation (e.g. Courgeau and Baccaini, 1997; Blakely and Woodward, 2000; Diez-Roux, 2004; Clarke, 2008). Multilevel analysis seeks to avoid ecological bias potentially present in studies simply of aggregate unit data, including meta-analysis (Berlin *et al.*, 2002).

Statistical inferences from multilevel analysis focus on the interplay between contextual and individual level impacts on the outcome, on partitioning variation in the outcome between levels in the hierarchy (Browne *et al.*, 2000), and on ranking of clusters after adjusting for the characteristics of individuals. Examples where this type of approach is relevant include pupil attainment within schools, psychopathological behaviour within families (Martinius, 1993), or illness or mortality rates among residents of different areas (Brodsky *et al.*, 1999). In these cases, pupils, offspring and community residents define the lower level, or level 1, of the data hierarchy, and the groups they are members of define the higher level or level 2. Further levels of aggregation are possible. The model will need to reflect typical features of multilevel data: similarities between pupils taught in the same school, or between residents

*Applied Bayesian Modelling*, Second Edition. Peter Congdon.
© 2014 John Wiley & Sons, Ltd. Published 2014 by John Wiley & Sons, Ltd.

of the same neighbourhood generate intra-cluster correlation, which if not represented will result in understatement of the variances of cluster effects.

As well as introducing known influences at both individual and cluster level, multilevel techniques use shared cluster random efects to represent unobserved contextual influences. Such random effects act both to borrow strength between clusters (i.e. provide more precise inferences), and to model intra-cluster correlation. A multilevel model will also provide estimates of cluster-specific regression estimates, namely varying intercepts and slopes, that use information from subjects within each cluster but also borrow strength from other clusters. Some contrasts in intercepts or slopes may be compositional – merely reflecting the aggregate effect of the composition of each cluster's subjects. However, there may additionally be genuine contextual effects: for instance, if the health experience or behaviour of an individual of a given type (e.g. as defined by age, class, etc.) varies across regions (Duncan *et al.*, 1999). A Bayesian estimation perspective may have advantages over classical methods which may understate random effect variances because not all sources of sampling uncertainty in parameters are allowed for. This may be particularly so for clusters varying in size, and for discrete outcomes. However, there may be the sensitivity to prior specifications, especially for small samples, or small numbers of clusters at higher levels. This may be the case, for instance, regarding covariation of the random effects at different levels (Daniels, 1999; Browne and Draper, 2000; Daniels and Kass, 1999).

This chapter starts by setting out a Bayesian perspective on meta-analysis (Section 5.2), and continues to consider application of multilevel concepts to both continuous and discrete data (Sections 5.3 and 5.4). Then Section 5.5 considers multilevel models including heteroscedasticity at one or more levels, with variances dependent on continuous or categoric regressors. Section 5.6 considers multivariate outcomes within a multilevel context. The programming focus is mainly on WinBUGS/OpenBUGS, but other options are available for Bayesian implementation, such as MLWIN, R-INLA, and R2BayesX, and for meta-analysis specifically, the R packages bspmma, bamdit and metamisc.

## 5.2    Meta-analysis: Bayesian evidence synthesis

Meta-analysis refers to methods for combining the results of independent studies to provide an estimate of a presumed common effect. While fixed effects models may be used to estimate the common effect without allowing for heterogeneity (classical meta-analysis), this may overstate the precision of the pooled effect. The hierarchical Bayesian approach (e.g. Welton *et al.*, 2012) recognises heterogeneity and estimates unknown treatment or study effects, defined by random deviations from an average pooled effect, also unknown. Bayesian methods may have advantages in handling issues which occur in meta-analysis, such as choice between alternative second stage priors, robust inference methods for assessing small studies or non Gaussian effects, and ease of predicting a treatment effect in a hypothetical new study (Smith *et al.*, 1995a; DuMouchel, 1996; Spiegelhalter *et al.*, 2004; Higgins *et al.*, 2009).

Input data to a meta-analysis typically consists of summary findings from a set of studies, with a range of possible outcomes including differences in proportions with health improved between treatment and control groups, logs of odds ratios for improvement, or logs of risk ratios. One might also apply meta-analysis to the slopes of dose–response curves (Berlin *et al.*, 1993; Smith *et al.*, 1995b), to correlation coefficients (Field, 2001), or effect measures (e.g. Cohen's *d*) derived from *t*-tests or *F*-tests. Extensions include meta-regression, where studies differ in design or confounders, analysis adjusting for average risk (possible, for example, when study observations include outcomes for both control and treatment groups), and meta-analysis which includes individual level as well as study characteristics (that is, a form of multilevel analysis) (Simmonds *et al.*, 2005; Askie *et al.*, 2011).

Further questions where a Bayesian method may be relevant include adjusting a meta-analysis for publication bias, and meta-analysis of multiple treatment studies. Thus, whereas most medical meta-analyses involve two treatment groups (or treatment vs control), Bayesian techniques can be used to compare either of the two main treatments with a common third treatment to improve estimation of the main treatment comparison (e.g. Hasselblad, 1998). Publication bias occurs if studies or trials for meta-analysis are based solely on a published literature review, so that there may be a bias towards studies that fit existing knowledge, or are statistically significant.

## 5.2.1    Common forms of meta-analysis

The canonical meta-analysis model for continuous data involves effect measures $y_i$, available for a set of $n$ studies, together with estimated standard error $s_i$ of the effect measure. Such effect measures may, however, often be derived from originally discrete data. For example, consider the log odds ratio for a certain outcome (e.g. smoking cessation) for an intervention as against placebo treatment. If numbers quitting smoking $a_i$ and $b_i$ are observed among sample numbers $r_i$ and $t_i$ under intervention and placebo respectively (e.g. Eisenberg *et al.*, 2008), then the moment estimator of the odds ratio is

$$OR_i = [a_i/(r_i - a_i)]/[b_i/(t_i - b_i)].$$

The log of this ratio, denoted $y_i = \log(OR_i)$, may (for moderate sample sizes) be taken as approximately normal with variance given by

$$s_i^2 = 1/a_i + 1/(r_i - a_i) + 1/b_i + 1/(t_i - b_i).$$

Similar approaches can be applied to logs of risk ratios $RR_i = (a_i/r_i)/(c_i/t_i)$, and to risk differences $RD_i = (a_i/r_i) - (c_i/t_i)$; see for example, Table 11.1 in Hox (2010).

Under a fixed effects model, data of this form may be modelled as

$$y_i \sim N(\mu, s_i^2),$$

where the estimate of $\mu$ is obtained as a weighted average of the $y_i$. The main two alternative fixed effect analyses are the inverse variance method with $1/s_i^2$ as weights, and the Mantel–Haenszel method which uses a weighting scheme originally derived for analysing stratified case–control studies and is more robust when data are sparse (e.g. White *et al.*, 1999). Under a random effects model by contrast, the results of different trials are often still taken as approximately normal, but the underlying effects may differ between trials, so that under an exchangeability assumption between studies one has

$$y_i \sim N(\theta_i, s_i^2),$$
$$\theta_i = \mu + \delta_i,$$

where the random deviations $\delta_i$ from the overall mean represent heterogeneity between studies, and have their own density. For example, with normal latent effects at stage 2, one has

$$\delta_i \sim N(0, \tau^2).$$

The rationale for random effects approaches is that at least some of the variability in effects between studies is due to random differences (e.g. different measurement of exposures, or differences in patient mix remaining after randomization). The normal hierarchical model leads to posterior means for $\theta_i$,

$$E[\theta_i|y] = w_i\mu + [1 - w_i]y_i$$

that average over the third stage mean $\mu$ and the study outcomes $y_i$ with weights $w_i = s_i^2/(s_i^2 + \tau^2)$ and $1 - w_i = \tau^2/(s_i^2 + \tau^2)$.

Prediction of the treatment effect in a hypothetical new study is obtained as

$$\theta_{new} \sim N(\mu, \tau^2),$$

while posterior predictive checks for model adequacy may consider the marginal model $y_i \sim N(\mu, s_i^2 + \tau^2)$ (Yuan and Little, 2009). Thus $S^2 = \sum_i \frac{(y_i-\mu)^2}{\tau^2+s_i^2}$ is compared to $X_n^2$, a chi-squared random variable with $n$ degrees of freedom.

A possible alternative at the second stage is a Student $t$ density with degrees of freedom $v$ either preset or an extra unknown. Thus

$$y_i \sim N(\theta_i, s_i^2),$$
$$\theta_i \sim t(\mu, \tau^2, v).$$

This may provide a useful sensitivity analysis when the number of studies is small or when outlier studies are suspected. This can be equivalently expressed using a scale mixture approach

$$y_i \sim N(\theta_i, s_i^2),$$
$$\theta_i \sim N(\mu, \tau^2/\lambda_i),$$
$$\lambda_i \sim Ga(0.5v, 0.5v),$$

where the gamma distributed scale factors $\lambda_i$ act as indicators of exchangeability (Verde, 2010).

Another strategy for discrete observations $y_i$ (e.g. infections among patient totals $N_i$) is a binomial likelihood (with probabilities $\pi_i$) or Poisson likelihood (with rates $N_i\lambda_i$) at the first stage. However, a second stage normal density is adopted for transformed probabilities or rates, namely $\theta_i = \text{logit}(\pi_i)$ or $\theta_i = \log(\lambda_i)$. So for binomial observations

Stage 1: $y_i \sim Bin(N_i, \pi_i); \text{logit}(\pi_i) = \theta_i;$

Stage 2: $\theta_i \sim N(\mu, \tau^2);$

Stage 3: Priors on $(\mu, \tau^2)$.

This approach extends to studies containing comparisons between treatment groups, or treatment vs placebo groups. Thus suppose numbers quitting smoking $y_{i1}$ and $y_{i2}$ are observed among sample numbers $N_{i1}$ and $N_{i2}$ under intervention and placebo groups. Then under binomial sampling one has $y_{ij} \sim \text{Bin}(N_{ij}, \pi_{ij})$, for $j=1,2$, and at the second stage

$$\text{logit}(\pi_{ij}) = \theta_{ij},$$
$$\theta_{i1} = \rho_i + \delta_i,$$
$$\theta_{i2} = \rho_i,$$
$$\rho_i \sim N(R, \sigma_\rho^2),$$
$$\delta_i \sim N(D, \sigma_\delta^2),$$

with priors on $(R, \sigma_\rho^2, D, \sigma_\delta^2)$ at the third stage. This prior assumes the treatment gain $\delta_i$ is unrelated to the underlying average risk $\rho_i$ in study $i$, but one might also take $(\rho_i, \delta_i)$ to be bivariate normal, or let $\theta_{i1}$ include a regression term $\beta(\rho_i - \bar{\rho})$ adjusting for the frailty effect. The treatment effect in a new study could be estimated by sampling new values of $R$ and $D$, with $\text{logit}(\pi_{new}) = R + D$ (cf. Welton et al., 2012, chapter 6).

More generally, one may let the underlying trial effects $\theta_i$ depend on known study characteristics or confounders $X_i$ (a technique called meta-regression), if these are plausible sources of some of the heterogeneity between study effects. This acts to improve the closeness to exchangeability, and estimate $\mu$ with greater precision. The $X_i$ exclude an intercept and are centred so that $\mu$ can be compared with an analysis not controlling for such characteristics. Thus for a hierarchical normal model

$$y_i \sim N(\theta_i, s_i^2),$$
$$\theta_i \sim N(\mu + X_i\psi, \tau^2),$$
$$\mu, \tau^2, \psi \sim \pi(\psi)\pi(\tau^2)\pi(\mu).$$

For instance, DuMouchel (1996) considers odds ratios $y_i$ from nine studies on the effects of indoor air pollution on child respiratory illness. These odds ratios were derived within each study from logistic regressions, either relating illness to domestic $NO_2$ concentrations, or to $NO_2$ surrogates (such as a gas stove). In deriving the odds ratio, only some studies adjusted for parental smoking, or for child's gender. Thus in the subsequent meta-analysis, indicators $X_i$ for each study describe which confounders were allowed for in deriving the odds ratio. There may also be grounds for regression to explain heteroscedasticity (e.g. Verde, 2010), for example according to aspects of study design, so that $\log(\tau_i^2)$ are related to study characteristics.

Validity of conclusions from meta-analysis is qualified by possible publication bias, generally assumed to take the form of more significant findings being more likely to be published (Copas and Shi, 2000; Rothstein et al., 2005; Carpenter et al. 2009). Proposals to allow for bias include the use of weight functions (Silliman, 1997; Larose and Dey, 1998), and specifying outcomes $y_i$ conditional on latent variables $Z_i$ underlying publication probabilities. Suppose $y$ has density $p(y|\theta)$ but due to selection bias it is not possible to sample from this distribution, and that the probability that an observation $y$ enters a sample is multiplied by a positive weight function $w(y|\gamma)$. Then the observations are actually a sample from the weighted distribution

$$p^w(y|\theta, \gamma) = w(y|\gamma)p(y|\theta) / \int w(y|\gamma)p(y|\theta)dy.$$

Examples of weight functions include weights proportional to effect size, $w_i = |y_i|^\gamma$ (Silliman, 1997). To illustrate a selection model where selection depends on the observed variance (Mavridis et al., 2012, 2013), one may represent the likelihood model as

$$y_i = \theta_i + s_i\varepsilon_i, \qquad \varepsilon_i \sim N(0, 1)$$

and suppose study $i$ is selected for publication if a latent threshold $Z_i > 0$, where

$$Z_i = \gamma_0 + \gamma_1/s_i + \delta_i, \qquad \delta_i \sim N(0,1),$$

$$\text{corr}(\delta_i \varepsilon_i) = \rho.$$

Equivalently the probability of selection is

$$\Pr(Z_i > 0) = \Phi(\gamma_0 + \gamma_1/s_i), \qquad (5.1)$$

where $\gamma_0$ reflects the overall chance of publication, and $\gamma_1$ reflects dependence of publication on observed sampling variances (which tend to be lower for larger studies). These parameters cannot be estimated, but, given prior guesses of publication probabilities at $\max(s_i)$ and $\min(s_i)$, they can be solved for. If $\rho = 0$, the model reduces to the usual random effects analysis, whereas for larger positive $\rho$, selected studies have $Z > 0$, so that $\delta$ and hence $\varepsilon$ are both likely to be higher, leading to a positive bias in $y$ values. Observed study effects conditional on $Z$ can be obtained as

$$y_i | Z_i \sim N(\theta_i + \rho s_i[Z_i - \gamma_0 - \gamma_1/s_i], s_i^2(1 - \rho^2)),$$

with second stage $\theta_i \sim N(\mu, \tau^2)$.

## 5.2.2 Priors for stage 2 variation in meta-analysis

Deriving an appropriate prior for the second stage variance $\tau^2$ in a hierarchical model for continuous outcomes may be problematic, since for small numbers of studies there may be increased sensitivity (Lambert *et al.*, 2005), and default use of flat priors may oversmooth – that is the true means $\theta_i$ are smoothed towards the global average to such an extent that the model approximates the fixed effects model. While not truly Bayesian, there are arguments to consider the actual variability in study effects as the basis for a sensible prior. Thus DuMouchel (1996) proposes a Pareto or log-logistic density

$$\pi(\tau) = s_0/(\tau + s_0)^2,$$

where $\frac{1}{s_0^2} = \frac{1}{n} \sum_{i=1}^{n} \frac{1}{s_i^2}$ is the harmonic mean of the known sampling variances. This is equivalent to a uniform prior on $s_0/(\tau + s_0)$. This prior is proper but highly dispersed with median $s_0$, but mean of infinity. The (1,25,75,99) percentiles of $\tau$ are $s_0/99, s_0/3, 3s_0, 99s_0$. In BUGS the Pareto for a variable $y$ is parameterised as

$$y \sim \alpha c^\alpha y^{-(\alpha+1)},$$

and to obtain the DuMouchel form involves setting $\alpha = 1, c = s_0$, and then $\tau = y - s_0$. Another possibility is a uniform prior on the average shrinkage

$$\omega = \frac{s_0^2}{s_0^2 + \tau^2} \sim U(0,1)$$

(e.g. Spiegelhalter *et al.* 2004, p. 172). The smaller is $\tau^2$ (and hence the larger is $\omega$) the closer the model approximates complete shrinkage to a common effect as in classical fixed effects pooling (obtained when $\tau^2 = 0$). Smaller values of $\omega$ (e.g. 0.1 or 0.2) might correspond to 'sceptical priors' in situations where exchangeability between studies, and hence the rationale for pooling under a meta-analysis, is in doubt.

One might also set a prior directly on $\tau^2$ directly without reference to the observed $s_i^2$. Introducing some degree of prior information may be relevant, and is natural under the inverse chi-squared density (sometimes called the scaled inverse chi-squared) with parameters $\{v, \lambda\}$. For $\tau^2$ a variance, this is equivalent to assuming

$$\tau^2 \sim IG\left(\frac{v}{2}, \frac{v\lambda}{2}\right),$$

where $\lambda$ (the scale parameter) is a prior guess at the mean variance, and $v$ (the degrees of freedom) can be viewed as a prior sample size parameter, with values $v = 1$, 2 or 3 being typical choices. By contrast, improper or highly diffuse priors may lead to identification or propriety problems (Paul et al., 2010). For example, the prior

$$p(\tau^2) \propto \frac{1}{\tau^2},$$

equivalent to taking $\tau^2 \sim IG(0, 0)$, can lead to improper posteriors in random-effects models. A just proper alternative, such as $\tau^2 \sim IG(c, c)$ with $c$ small (e.g. $c = 0.0001$) is often used, but in meta-analysis with few studies, a diffuse prior on the second stage variance tends to reduce the size of $\tau^2$ and produce results resembling classical pooling. Kelsall and Wakefield (1999) point out that this prior is not consistent with very small amounts of variability, and suggest a Gamma(0.5, 0.0005) prior on the second stage precision $1/\tau^2$.

Other options are a uniform prior on $\tau$, namely

$$\tau \sim U(0, T_\tau),$$

with $T_\tau$ generally specified as large (e.g. $T_\tau = 10$, $T_\tau = 100$, etc.), and half-normal or half-t priors on $\tau$, e.g.

$$\tau \sim N(0, s_\tau^2) \ I(0, ),$$

where $s_\tau^2$ is to be specified. In meta-analysis one typically assumes normal latent effects for a log transformed outcome (e.g log odds ratio, log relative risk), so that a priori, 90% of values of $\theta_i$ are postulated to be in the interval $\mu \pm 1.645\tau$. Suppose the $\theta_i$ are ranked with $\theta_{[5\%]}$ denoting the 5th percentile, and $\theta_{[95\%]}$ the 95th percentile. Then under a second stage normal prior, the prior gap between these percentiles is $\theta_{[95\%]} - \theta_{[5\%]} = 3.29\tau$, or equivalently

$$\frac{\exp(\theta_{[95\%]})}{\exp(\theta_{[5\%]})} = \exp(3.29\tau),$$

namely that the 95%-5% quantile ratio in the original scale (odds ratios, relative risks, etc.) is $\exp(3.29\tau)$. Taking $\tau = 0.5$ corresponds to a belief that the 95th percentile in the odds ratio (etc.) is 5.2 times the 5th percentile, and might accord with subject matter knowledge. A higher value such as $\tau = 1$ corresponds to a prior expectation of a 27-fold difference between the 95th and 5th percentiles, and may be implausible in some applications. So for the half-normal prior, and with $y_i$ representing log $(OR_i)$ or log $(RR_i)$ in study $i$, one might reasonably assume $s_\tau = 0.5$, namely

$$\tau \sim N(0, 0.25) \ I(0, ),$$

as values of $\tau > 1$ then have low probability.

**Example 5.1    Smoking cessation**

Consider data from Eisenberg *et al.* (2008, Figure 3) on smoking cessation treatments, and in particular on $n = 22$ studies comparing nicotine gum to placebo. The input data are log odds ratios $y_i$ and their associated sampling variances $s_i^2$. Three priors for the second stage variance/precision are compared, with a view to assessing sensitivity regarding the odds ratio derived from the pooled mean $\mu$, the second stage standard deviation $\tau$, and the odds ratio resulting from a treatment effect $\theta_{new}$ in a hypothetical new study. The first is a uniform prior on the average shrinkage:

$$\omega = \frac{s_0^2}{s_0^2 + \tau^2} \sim U(0, 1)$$

where $\frac{1}{s_0^2} = \frac{1}{n}\sum_{i=1}^{n} \frac{1}{s_i^2}$. The second option is a gamma prior $1/\tau^2 \sim Ga\left(\frac{\nu}{2}, \frac{\nu\lambda}{2}\right)$, with $\nu = 1$ degree of freedom and $\lambda = 0.1$. The third is a uniform prior on $\tau$ itself, namely $\tau \sim U(0, 10)$. Inferences are based on the last 90 000 iterations in two chain run of 100 000 iterations. All posterior predictive checks involving the summary statistic $S^2 = \sum_i \frac{(y_i - \mu)^2}{\tau^2 + s_i^2}$ are satisfactory, with $p = \int \Pr(X_n^2 < S^2)p(\theta|y)d\theta$ obtained as 0.43 under the first two priors and 0.39 under $\tau \sim U(0, 10)$. The first prior gives posterior mean (and 95% intervals) for OR $= \exp(\mu)$ of 1.70 (1.35, 2.15), for $\tau$ of 0.34 (0.15, 0.58) and for $OR_{new} = \exp(\theta_{new})$ of 1.82 (0.80, 3.68). Results with the second prior are virtually identical. The third prior provides a slightly higher level of heterogeneity with the posterior mean for $\tau$ being 0.37, but posterior means for OR and $OR_{new}$ similar at 1.71 and 1.85. This inferential stability may be related to the relatively large numbers of subjects included (over 5000) in the 22 studies.

Analysing these data with a binomial likelihood facilitates assessment of any frailty effect. Thus $y_{ij} \sim Bin(N_{ij}, \pi_{ij})$, for j=1 (treated) and j=2 (placebo), and logit$(\pi_{ij}) = \theta_{ij}$, where the intervention group latent effects $\theta_{i1}$, include a regression effect on frailty, representing a tendency to quit despite no intervention being made:

$$\theta_{i1} = \rho_i + \delta_i + \beta(\rho_i - \bar{\rho}),$$

$$\theta_{i2} = \rho_i,$$

$$\rho_i \sim N(R, \sigma_\rho^2),$$

$$\delta_i \sim N(D, \sigma_\delta^2).$$

Uniform priors are assumed for $\sigma_\rho$ and $\sigma_\delta$. In fact $\beta$ is not significant (with 95% interval from $-0.41$ to $0.30$), and the pooled treatment effect $e^D$ is similar to that obtained under a hierarchical normal model, namely 1.73 (1.34, 2.23).

**Example 5.2    Chronic infection rates in hip arthroplasties**

An illustration of exchangeability within groups is provided by data on re-infections after hip arthroplasties (Lange *et al.*, 2012, Figure 3). There is debate concerning the relative efficacy of two-stage revision or one-stage revision treatments in preventing re-infections. Lange *et al.* derive treatment effects $r_i$ from counts of re-infections $y_i$ among patients $N_i$ in $n = 38$ studies, ten of which are one-stage, the remainder being two-stage. Moment estimates of these rates are $r_i = y_i/N_i$, unless $y_i = 0$ in which case $r_i = (y_i + 0.5)/(N_i + 0.5)$ (four of the one-stage, and 8 of the two stage studies, have zero re-infections). Lange *et al.* assume different underlying means between one-stage and two-stage studies, and report a lower re-infection rate in the two-stage studies (10.4%) than in the one-stage studies (13.1%). These pooled rates reflect both the adjustment (in the moment estimators) for zero infections, and the weightings assigned to different studies: in the one-stage group, a single relatively large one-stage study (study 10 with y=29, N=183), is assigned 62% of the weights for this group, while accounting for 49% of the patients.

Here the data are analysed as Poisson outcomes $y_i \sim Po(N_i\lambda_i)$, with $\log(\lambda_i) = \theta_i$. Let $G_i$ denote the intervention type ($G_i = 1$ for one stage, $G_i = 2$ for two stage). Then an initial second stage normal model differentiates means and variances by treatment group, namely

$$\theta_i \sim N(\mu_{G_i}, \tau^2_{G_i}).$$

Uniform priors $\{\tau_j \sim U(0, 100), j = 1, 2\}$ are adopted on the second stage standard deviations, with flat priors on the $\mu_j$. The percent rate infection is obtained as $\rho_j = 100 \exp(\mu_j)$. Inferences are based on the last 9000 iterations of a two chain MCMC analysis involving 10000 iterations. This analysis in fact shows a lower infection rate in one stage studies (namely 7.65%) than in two stage studies (namely 8.65%), though there is an inconclusive 62% probability that $\rho_2 > \rho_1$. The variability in effects is greater in the one stage group (with posterior means $\tau_1 = 0.81$ and $\tau_2 = 0.22$).

A second analysis considers possible outlier status: Lange $et\ al.$ report a meta-regression showing a significant effect of study size on the infection rate in the one stage group. This may be due to the presence of study 10 with a rather higher re-infection rate than the other studies in this group. Hence the following scheme is adopted

$$\theta_i \sim N(\mu_{G_i}, \tau^2_{G_i}/\phi_i)$$

$$\phi_i \sim Ga(v_{G_i}/2, v_{G_i}/2)$$

namely a group specific model with scale mixing. For the group-specific degrees of freedom $\{v_j, j = 1, 2\}$, it is assumed that $1/v_j \sim U(0.5, 0.02)$, implying a range between $v_j = 2$ and $v_j = 50$. While posterior means of the $v_j$ are under 10, none of the $\phi_i$ conclusively indicates outlier status: the lowest being $\phi_{10}$ with posterior mean 0.93. The overall re-infection rates are estimated at 7.9% (one-stage) and 8.6% (two-stage).

A third analysis involves meta-regression on the year of the study, since Lange $et\ al.$ (2012, p. 68) mention a 'generally decreased risk of reinfection over time'. The independent variable $x_i$ is a centred transform of study year (namely the year minus 1989). Thus

$$\theta_i \sim N(\mu_{G_i} + x_i\psi, \tau^2_{G_i}),$$

with a N(0,10) prior on $\psi$. This shows a negative effect of year (albeit marginally non-significant), with a 95% interval for $\psi$ of (-0.068,0.001). Adjusting for this effect leads to mean reinfection rates of 6.7% (one-stage) and 8.7% (two-stage).

### Example 5.3     Aspirin use: Predictive cross-validation for meta-analysis

Predictive cross-validation of meta-analysis is important to assess model adequacy, and assess standard assumptions such as second stage normal random effects (Welton $et\ al.$, 2012, Chapter 6). Consider a meta-analysis of six studies of aspirin use after heart attack (DuMouchel, 1996), with the study effects $y_i$ being differences in percent mortality between aspirin and placebo groups. A Pareto prior for $\tau$ is based on the harmonic mean of the $s_i^2$ (the sampling variances of the differences). A standard hierarchical model is then

$$y_i \sim N(\theta_i, s_i^2)$$

$$\theta_i \sim N(\mu, \tau^2).$$

A two chain run to 100 000 iterations shows considerable posterior uncertainty in $\tau$, with 95% interval from 0.06 to 3.5. In five of the six studies the posterior standard deviation of $\theta_i$ is smaller than $s_i$, but for one study (study 6, the AMIS study), the posterior stanadrd deviation $sd(\theta_6) = 0.96$ exceeds $s_6 = 0.90$. Alternatively stated, there is greater uncertainty about the true effect for this study than if it had not been pooled with the other studies. Despite the potential influence of this study on inferences, the overall treatment effect $\mu$ has a posterior density concentrated on positive values, with the probability $\Pr(\mu > 0|y)$ being 0.95.

A full cross-validation approach to model assessment then involves study by study exclusion and considering criteria such as

$$U_k = \Pr(y_k^* < y_k|y_{[-k]}) = \int \Pr(y_k^* < y_k|\zeta, y_{[-k]})p(\zeta|y_{[-k]})d\zeta$$

where $\zeta$ represents all parameters, $y_{[-k]}$ is the data set omitting study $k$, and $y_k^*$ is the predicted outcome for the $k^{th}$ study when estimation is based on all studies but the $k^{th}$. If the model assumptions are adequate then the $U_k$ will be uniform over the interval $[0,1]$, and the quantities $Z_k = \Phi^{-1}(U_k)$ will be standard normal. A corresponding overall measure of adequacy is the Bonferroni statistic

$$Q = N \times \min_k(1 - |2U_k - 1|)$$

namely an upper limit to the probability that the most extreme $U_k$ could be as large as was actually observed. One may also sample the predicted latent study effect $\theta_k^*$ from the density $N(\mu_{[-k]}, \tau_{[-k]}^2)$, based on excluding the $k^{th}$ study. These have the same posterior mean (but greater variability) than the predictive pooled mean $\mu$ in the absence of the $k^{th}$ study.

Accordingly the $U_k$ statistic for study 6 is in the lowest 2% tail of its predictive distribution (with predictive probability 0.013). However, the Bonferroni statistic shows this may still be acceptable in terms of an extreme deviation among the studies, since $Q = 0.17$ (this is calculated from the posterior average $U_k$). While the AMIS study true mean is lower than the others, this procedure suggests it is not an outlier to such an extent as to invalidate the entire hierarchical model. The posterior mean $\theta_6^*$ for the AMIS study is about 2.45, with 95% interval 0.04 to 4.9, so that there is unambiguous mortality reduction were this study not included in the pooling; the remaining $\theta_k^*$, obtained when the AMIS study is included in the collection, all straddle zero (with inconclusive implications for treatment effectiveness).

### Example 5.4    Lung cancer and environmental smoking

Consider data from 37 studies reported by Hackshaw et al. (1997) on log relative risks $y_i$ (and standard errors $s_i$) of lung cancer among female non-smokers, according to partner smoking status. To allow for the operation of publication bias (e.g. Mavridis et al., 2013), paired values $(p_{max}, p_{min})$ are obtained, where $p_{max}$ is an elicited publication probability for the study with the highest standard error $s_{max} = \max(s_i)$, and $p_{min}$ is the elicited publication probability of the study with lowest standard error, $s_{min} = \min(s_i)$. These pairings assume $p_{max}$ uniform within a defined range, and to provide a sensitivity analysis, $p_{min}$ is taken as either 0.5 or 0.33 times $p_{max}$. The resulting $(\gamma_0, \gamma_1)$ values, as in (5.1), are obtained by solving the equations $p_{max} = \Phi(\gamma_0 + \gamma_1/s_{max})$, and $p_{min} = \Phi(\gamma_0 + \gamma_1/s_{min})$. A half normal prior is assumed for the second stage standard deviation $\tau$ with variance 0.25, while $\rho \sim U(-1, 1)$.

The case where $\rho = 0$ and there is no publication bias results in an estimate for pooled relative risk, RR = $\exp(\mu)$, of 1.24 with 95% interval (1.12,1.39) indicating a significantly elevated risk. Assuming $p_{max} \sim U(0.1, 0.9)$, and $p_{min} = 0.5p_{max}$, produces a lower pooled relative risk 1.15 (1.03, 1.31), while assuming $p_{max} \sim U(0.1, 0.9)$, and $p_{min} = 0.33 p_{max}$, produces a pooled estimate 1.12 (1.00, 1.28). Assuming a higher range of values for $p_{max}$ namely $p_{max} \sim U(0.5, 0.9)$, makes little difference. Thus assuming $p_{max} \sim U(0.5, 0.9)$, and $p_{min} = 0.5 p_{max}$, produces an estimate RR = 1.15 (1.02, 1.30), while assuming $p_{max} \sim U(0.5, 0.9)$, and $p_{min} = 0.33 p_{max}$, produces RR = 1.13 (0.99, 1.29). So the probability that the pooled relative risk RR exceeds 1 is reduced as the ratio of $p_{min}$ to $p_{max}$ is reduced.

### 5.2.3   Multivariate meta-analysis

Multivariate meta-analysis models are used to synthesize evidence on $J$ correlated endpoints such as disease-free survival and overall survival, or tests of diagnostic accuracy (e.g. sensitivity and specificity) (Mavridis and Salanti, 2013). In particular, focusing on the bivariate case, suppose $n$ studies contain $J = 2$ endpoints with summary outcome measures, $y_{ij}$, and associated variances, $s_{ij}^2$. Ideally a known study specific correlation $r_i$ is available between the measures (forming part of a bivariate first stage likelihood), but this may not always be provided (Riley *et al.*, 2008). Each summary statistic $y_{ij}$ is assumed to be an estimate of a true value $\theta_{ij}$ in each study, and the vector $(\theta_{i1}, \theta_{i2})$ is assumed to be drawn from a bivariate distribution with mean values $(\mu_1, \mu_2)$ and covariance $\Sigma$. The covariance is defined by between-study variances $\tau_j^2$ on each endpoint and a correlation $\rho$ between the two series of latent effects. So under a normal-normal prior, the first and second stage densities (when $r_i$ are observed) are

$$\begin{pmatrix} y_{i1} \mid \theta_{i1} \\ y_{i2} \mid \theta_{i2} \end{pmatrix} \sim BVN \left( \begin{bmatrix} \theta_{i1} \\ \theta_{i1} \end{bmatrix}, \begin{bmatrix} s_{i1}^2 & r_i s_{i1} s_{i2} \\ r_i s_{i1} s_{i2} & s_{i2}^2 \end{bmatrix} \right),$$

$$\begin{pmatrix} \theta_{i1} \\ \theta_{i2} \end{pmatrix} \sim BVN \left( \begin{bmatrix} \mu_1 \\ \mu_2 \end{bmatrix}, \begin{bmatrix} \tau_1^2 & \rho \tau_1 \tau_2 \\ \rho \tau_1 \tau_2 & \tau_2^2 \end{bmatrix} \right),$$

with the marginal model being

$$\begin{pmatrix} y_{i1} \mid \theta_{i1} \\ y_{i2} \mid \theta_{i2} \end{pmatrix} \sim BVN \left( \begin{bmatrix} \mu_1 \\ \mu_2 \end{bmatrix}, \begin{bmatrix} \tau_1^2 + s_{i1}^2 & r_i s_{i1} s_{i2} + \rho \tau_1 \tau_2 \\ r_i s_{i1} s_{i2} + \rho \tau_1 \tau_2 & s_{i2}^2 + \tau_2^2 \end{bmatrix} \right).$$

Student $t$ random effects with $v$ degrees of freedom can also assumed at the second stage, which if achieved via scale mixing involves unit specific scale factors $\lambda_i \sim Ga(\frac{v}{2}, \frac{v}{2})$ applied to $\Sigma$, so that the covariance matrix for unit $i$ is $\Sigma / \lambda_i$.

Another scenario (e.g. Verde, 2010) is that discrete observations $y_{ij}$ among totals $N_{ij}$ (e.g. counts of diagnosed true positives and true negatives among total diseased and non-diseased subjects respectively) be modelled as binomial (with probabilities $\pi_{ij}$) or Poisson (with rates $\mu_{ij} N_{ij}$). However, a second stage normal is adopted for transformed probabilities or rates, e.g. with $\theta_{ij} = \text{logit}(\pi_{ij})$ or $\theta_{ij} = \log(\mu_{ij})$. So for binomial observations

Stage 1: $y_{ij} \sim \text{Bin}(N_{ij}, \pi_{ij})$;

Stage 2: $\text{logit}(\pi_{ij}) = \theta_{ij}; \theta_i = (\theta_{i1}, \theta_{i2}) \sim N_2(\mu, \Sigma)$.

As for univariate meta-analysis, there may be sensitivity to the prior adopted for the covariance. The conjugate prior for the inverse of the covariance matrix is the Wishart distribution,

$$\Sigma^{-1} \sim \text{Wishart}(R, k),$$

where $R$ is a symmetric positive definite $J \times J$ matrix, and $k > J - 1$, with lower $k$ values representing higher uncertainty. The Wishart distribution has expectation $kR^{-1}$, so that $H = \frac{1}{k}R$ may be set as a prior guess at the unknown covariance matrix. In meta-analysis applications, the outcomes are often log transforms (e.g. of odds ratios, proportions etc.), and as discussed above, a prior expectation may be that values $\tau > 1$ on the second stage standard deviations are unlikely, while $\tau = 0.5$ or $\tau = 0.75$ might be more supported. Suppose one assumes prior variances for $\theta_{i1}$ and $\theta_{i2}$ of $0.75^2 = 0.56$. These could be taken as the diagonal terms $(h_{11}, h_{22})$ in $H$, so that the two diagonal terms in $R$ are 1.12. Off-diagonal terms in $R$ may be taken as 0 unless there is evidence regarding covariation.

Another possible strategy for bivariate meta-analysis is to reconstruct $\Sigma$ from separate priors on $\tau_1^2, \tau_2^2$, and $\rho$. For example, one might use gamma priors on $\phi_1 = 1/\tau_1^2$ and $\phi_2 = 1/\tau_2^2$, or half-normal priors on $\tau_1$ and $\tau_2$. For $\rho$, one option is a normal prior on the transformation $\rho^* = \text{logit}(\frac{\rho+1}{2})$, with $\rho^* \sim N(0, 2.5)$ corresponding approximately to a uniform prior on $U[-1, 1]$ for $\rho$ (Paul et al., 2010, p. 1328). To ensure $\Sigma$ is positive definite, one may apply tests such as all eigenvalues being positive (this can be done using the eigen.vals function in OpenBUGS). One may also adopt a version of the separation strategy of Barnard et al. (2000) which involves expressing $\Sigma$ as

$$\Sigma = V^{0.5}RV^{0.5},$$

where for $J$ outcomes, $V = diag(\tau_1, \ldots, \tau_J)$, and $R = L'L$ is obtained using a spherical transformation (Lu and Ades, 2009). If $L_j$ is the $j^{th}$ column of the $J \times J$ upper triangular matrix $L$, then the first $j$ elements of $L_j$ are

$$\ell_{j1} = \cos(\varphi_{j2}),$$
$$\ell_{j2} = \sin(\varphi_{j2})\cos(\varphi_{j3}),$$
$$\ldots,$$
$$\ell_{j,j-1} = \sin(\varphi_{j2})\sin(\varphi_{j3}) \ldots \cos(\varphi_{jj}),$$
$$\ell_{j,j} = \sin(\varphi_{j2})\sin(\varphi_{j3}) \ldots \sin(\varphi_{jj}),$$

where all unknown $\varphi$ coefficients are between 0 and $\pi$.

### Example 5.5   CT scan data

To demonstrate a bivariate meta-analysis application to diagnostic data, consider results from 51 clinical studies on the accuracy of computer tomography (CT) scans for the diagnosis of appendicitis. The sensitivity and specificity of the diagnostic procedure (denoted $s_i$ and $t_i$) refer respectively to unknown proportions of those with a disease who are correctly identified, and of those without a disease who are correctly identified. Let true positives be denoted $TP_i$ and total diseased as $D_i$, while true negatives are denoted $TN_i$ and those not diseased as $W_i$. The first analysis here adopts a binomial likelihood at the first stage and a bivariate normal at stage 2 in logit transformed diagnostic accuracy rates, so that

$$TP_i \sim \text{Bin}(D_i, s_i),$$
$$TN_i \sim \text{Bin}(W_i, t_i),$$
$$\text{logit}(s_i) = \theta_{i1}; \text{logit}(t_i) = \theta_{i2};$$
$$\begin{pmatrix} \theta_{i1} \\ \theta_{i2} \end{pmatrix} \sim BVN \left( \begin{bmatrix} \mu_1 \\ \mu_2 \end{bmatrix}, \begin{bmatrix} \tau_1^2 & \rho\tau_1\tau_2 \\ \rho\tau_1\tau_2 & \tau_2^2 \end{bmatrix} \right),$$

Parameters of interest from such a random effects model include the pooled sensitivity and specificity $\{s_g, t_g\}$, obtained by expit transforms of $\mu_1$ and $\mu_2$.

As mentioned above, prior variances for $\theta_{i1}$ and $\theta_{i2}$ of $0.75^2 = 0.56$ correspond to a prior expectation that values $\tau_j > 1$ of the second stage standard deviations are unlikely. A Wishart prior $\Sigma^{-1} \sim \text{Wishart}(R, k)$ with diagonal terms 1.12 for $R$ is accordingly adopted for the inverse of the covariance matrix

$$\Sigma = \begin{bmatrix} \tau_1^2 & \rho\tau_1\tau_2 \\ \rho\tau_1\tau_2 & \tau_2^2 \end{bmatrix}.$$

This analysis shows no evidence that trial-level sensitivity and specificity are correlated, with $\rho$ having 95% credible interval $(-0.49, 0.36)$. The global diagnostic accuracy rates $\{s_g, t_g\}$ both exceed 0.95.

A second analysis adopts a scale mixture of bivariate normals at level 2. As in Verde (2010), this analysis involves sums and differences in the transformed scale of the true positive rate (i.e. sensitivity) and the false positive rate $r_i$ (i.e. 1 minus the specificity). Thus

$$\eta_{i1} = \text{logit}(s_i) + \text{logit}(r_i),$$

$$\eta_{i2} = \text{logit}(s_i) - \text{logit}(r_i),$$

$$\begin{pmatrix} \eta_{i1} \\ \eta_{i2} \end{pmatrix} \sim \text{BVN}\left( \begin{bmatrix} \mu_1 \\ \mu_2 \end{bmatrix}, \frac{1}{\lambda_i} \begin{bmatrix} \tau_1^2 & \rho\tau_1\tau_2 \\ \rho\tau_1\tau_2 & \tau_2^2 \end{bmatrix} \right),$$

$$\lambda_i \sim \text{Ga}\left( \frac{\nu}{2}, \frac{\nu}{2} \right).$$

The global sensitivity, false positive rate, and specificity are obtained as

$$s_g = \exp(0.5\mu_1 + 0.5\mu_2)/[1 + \exp(0.5\mu_1 + 0.5\mu_2)],$$

$$r_g = \exp(0.5\mu_1 - 0.5\mu_2)/[1 + \exp(0.5\mu_1 - 0.5\mu_2)],$$

$$t_g = 1 - r_g.$$

For illustration, $\nu$ is set at 4, though obviously it could be taken as an unknown (e.g. by taking a uniform prior on $1/\nu$). While three studies have posterior mean $\lambda_i$ under 0.5, only study 47 has a 95% credible interval for $\lambda_i$ entirely under 1. The global diagnostic accuracy rates $\{s_g, t_g\}$ again both exceed 0.95.

## 5.3 Multilevel models: Univariate continuous outcomes

The nesting of multilevel observations allows considerable scope for differentiating or index-ing regression and error variance effects, guided both by subject matter indications and by statistical criteria such as model parsimony, sensitivity and identifiability. For example, sup-pose metric data on pupil attainment is arranged by school. This provides a two level data $y_{ij}$ in schools $j = 1, \ldots, J$ at level 2, and pupils $i = 1, \ldots, n_j$ within schools at level 1. Predic-tors may be defined at each level, say $x_{pij}$ ($p = 1, \ldots, P$) at level 1, and $z_{qj}$ ($q = 1, \ldots, Q$) at level 2; an example of the first type of predictor might be pupil ability or gender, and of the second, whether the school is mixed or denominational. In a multilevel regression model, pre-dictor effects may be specified at each level and are cumulative sources of explained residual variation in the outcome.

A variance components form of the normal linear multilevel model involves random inter-cept terms $u_{0j} \sim N(0, \tau^2)$ at level 2, and subject level residuals $e_{ij} \sim N(0, \sigma^2)$ at level 1, com-bined with fixed (homogeneous) impacts for level 1 predictors:

$$y_{ij} = \beta_0 + \sum_p \beta_p x_{pij} + u_{0j} + e_{ij}.$$

Combining the intercept and level 2 error gives

$$y_{ij} = \beta_{0j} + \sum_p \beta_p x_{pij} + e_{ij},$$

$$\beta_{0j} = \beta_0 + u_{0j}.$$

The conditional mean response for subjects within cluster $j$ with predictor profile $x_{pij}$ is

$$E\left(y_{ij} | X_{ij}, \beta_{0j}\right) = \mu_{ij} = \beta_{0j} + \sum_p \beta_p x_{pij}.$$

The interpretation of the regression coefficient $\beta_p$ for predictor $x_{pij}$ is then as a change in the expected response for a unit change in that predictor, with the cluster effect held constant. Generally the cluster effects (random intercepts) $\beta_{0j}$ are assumed exchangeable over level 2 units, that is unstructured 'white noise'. However, if the clusters were geographic areas one might envisage them being spatially correlated: that is, between-area dependencies should be considered as well as within-area correlations (Riva et al., 2007). One may well seek to explain varying intercepts in terms of the characteristics of clusters. Assume a single cluster level predictor $z_j$ is used to explain the varying $\beta_{0j}$. Then the second stage model becomes

$$\beta_{0j} \sim N(B_{0j}, \tau^2)$$

$$B_{0j} = \gamma_{00} + \gamma_{01} z_j.$$

The next stage in an exploratory modelling sequence might be to differentiate effects of level 1 predictors (i.e. allow for predictor heterogeneity) according to clusters $j = 1, \ldots, J$ namely:

$$y_{ij} = \beta_0 + u_{0j} + \sum_{p=1}^{P} (\beta_p + u_{pj}) x_{pij} + e_{ij},$$

where $u_{pj}$ are zero mean deviations at cluster level from the average impacts of level 1 vari-ables. Conflating the fixed and zero-centred random effects $u_{pj}$ leads to

$$y_{ij} = \beta_{0j} + \sum_{p=1}^{P} \beta_{pj} x_{pij} + e_{ij},$$

$$\beta_{pj} = \beta_p + u_{pj}, \qquad p = 0, \ldots, P$$

with conditional regression means $\mu_{ij} = \beta_{0j} + \sum_P^{p=1} \beta_{pj} x_{pij}$. If cluster predictors are relevant to explaining predictor heterogeneity, one has (for $Q = 1$)

$$\beta_{pj} = B_{pj} + u_{pj} = \gamma_{p0} + \gamma_{p1} z_j + u_{pj}, \qquad p = 0, \ldots, P,$$

with corresponding expanded model

$$y_{ij} = \gamma_{00} + \gamma_{01}z_j + u_{0j} + \gamma_{10}x_{1ij} + \gamma_{11}z_jx_{1ij} + u_{1j}x_{1ij} + \cdots + \gamma_{P0}x_{1ij} + \gamma_{P1}z_jx_{Pij} + u_{Pj}x_{Pij} + e_{ij},$$

showing that effects of level 1 variables $x_{ij}$ are modified (mediated) according to the value of the contextual variable $z_j$.

The Bayesian viewpoint means that there is no longer a need to partition a parameter effect into fixed and random components, and the focus is then on priors for varying intercepts and predictor slopes (Clayton, 1996). To enable borrowing of strength these effects are assumed to be random and follow a hierarchical prior, with the y-likelihood at stage 1 conditional on $\beta_{pj}$, the assumed hyperdensity for the effects $\beta_{pj}$ at stage 2, and assumptions regarding hyper-parameters at stage 3. The normal linear multilevel model stipulates multivariate normality at stage 2 in the cluster effects $\beta_j = \{\beta_{0j}, \beta_{1j}, \ldots, \beta_{Pj}\}$, combined with exchangeability between clusters. In applications where the clusters are geographic areas, the multivariate density for $\beta_j$ might allow for spatial correlation between units.

However, before considering a model allowing correlated predictor effects, a practical modelling strategy might initially consider simpler options. Thus the cluster effects might initially be taken as a sequence of independent normals

$$\beta_{pj} \sim N(B_p, \tau_p^2) \qquad p = 0, \ldots, P$$

where $B_p$ are uknown fixed effects, or if cluster predictors are relevant,

$$\beta_{pj} \sim N(B_{pj}, \tau_p^2)$$

$$B_{pj} = \gamma_{p0} + \gamma_{p1}z_{1j} + \cdots + \gamma_{pQ}z_{Qj}.$$

This stage would be preliminary to adopting a full $P + 1$ multivariate normal density for cluster effects $\beta_j = (\beta_{0j}, \beta_{1j}, \ldots, \beta_{Pj})$, namely

$$\beta_j \sim N_{P+1}(B, \Sigma),$$

in order to assess whether all slopes need to be random, and also to assess whether there are gains in fit through allowing correlated effects (MacNab et al., 2004). As considered further below, to ensure that heterogeneity genuinely derives from cluster level variation, any level 1 heteroscedasticity should be investigated (Snijders and Berkhof, 2008), as this may indicate the modelling of level 1 variances as a function of predictors.

The prior assumed for $\tau_p^2$ (univariate effects) or $\Sigma$ (correlated effects) may affect inferences on the amount of cluster variation and hence the extent of shrinkage to the mean population value over all clusters. Diffuse priors may be preferred, such as $1/\tau_p^2 \sim Ga(c, c)$, with $c$ small (e.g. $c = 0.001$), or a Wishart prior on the precision matrix, $\Sigma^{-1} \sim Wish(R, v)$, with $v = P + 1$, and a small constant (e.g. 0.01) along the diagonal of the scale matrix $R$ (O'Brien et al., 2007), or taking $R$ to be the identity matrix (O'Malley and Zaslavsky, 2008; Natarajan and Kass, 2000; Richardson, 1996, p. 411). Whereas in meta-analysis, the first stage sampling variances are often known, in multilevel models the stage 1 variance (e.g. in a normal multilevel model) is generally unknown. Hence the analysis may be sensitive to the prior assumptions made regarding both $\sigma^2$ and $\tau^2$. Interdependence between components of variation may be explicit in a dependent prior $\pi(\sigma^2)\pi(\tau^2|\sigma^2)$, when, for example, a conventional prior on $\sigma^2$ or $1/\sigma^2$ is combined with a uniform prior on the variance ratio $\frac{\tau^2}{\sigma^2+\tau^2}$ (Natarajan and Kass, 2000).

While the above notation implies a nested arrangement of the data, it is often convenient, especially with unequal $n_j$ in each cluster to arrange the data in terms of a single subject index $k = 1, \dots, N$ where $N = \sum_j n_j$. A vector of group membership indices $G_k$, with values between 1 and $J$ would also be of length $N$. A model with varying intercepts and slopes would be written

$$y_k = \beta_{0G_k} + \sum_{p=1}^{P} \beta_{pG_k} x_{pk} + e_k, \qquad k = 1, \dots, N,$$

$$\beta_{pj} = B_p + u_{pj}, \qquad p = 0, \dots, P.$$

This type of arrangement is also useful for crossed and multiple membership rather than simple nested data structures (e.g. Fielding and Goldstein, 2006; Rasbash and Browne, 2008). A crossed classification analysis is relevant when each subject may belong to more than one group variable, for example, pupil data classified both by school $G_{1k} \in (1, \dots, J_1)$ and by area of residence $G_{2k} \in (1, \dots, J_2)$. One may partition both intercept and slope variance (see Example 5.7) between the different classifications. For example, consider variation in both intercepts and predictor effects for two classifiers, as in

$$y_k = (\beta_0 + u_{01,G_{1k}} + u_{02,G_{2k}}) + (\beta_1 + u_{11,G_{1k}} + u_{12,G_{2k}})x_{1k} + (\beta_2 + u_{21,G_{1k}} + u_{22,G_{2k}})x_{2k} + \dots$$

$$+ (\beta_P + u_{P1,G_{1k}} + u_{P2,G_{2k}})x_{Pk} + e_k, \qquad k = 1, \dots, N,$$

where the random effects $\{u_{01j}, u_{11j}, u_{21j}, \dots, u_{P1j}; j = 1, \dots, J_1\}$ are defined over classifier 1, and $\{u_{02j}, u_{12j}, u_{22j}, \dots, u_{P2j}; j = 1, \dots, J_2\}$ are defined over classifier 2.

If subjects may belong to more than one level 2 cluster, for example, when education or residence histories can include shifts between schools or neighbourhoods, then a multiple membership model may be used. Suppose subject $k$ has $M_k$ level 2 affiliations with known weights $\{w_{k1}, w_{k2}, \dots, w_{kM_k}\}$, with $\sum_{m=1}^{M_k} w_{km} = 1$. The normal linear two level model with varying intercepts becomes

$$y_k = \beta_0 + \sum_{p=1}^{P} \beta_p x_{pk} + \sum_{m=1}^{M_k} w_{km} u_{0j} + e_k,$$

If there are also varying slopes then

$$y_k = \beta_0 + \sum_{p=1}^{P} \beta_p x_{pk} + \sum_{m=1}^{M_k} w_{km} \left( u_{0j} + \sum_{p=1}^{P} u_{pj} x_{pk} \right) + e_k$$

Multiple member schemes may also be used for moving average approaches to spatial effects (Langford et al., 1999).

Posterior predictive checks might focus on particular features of the data such as proportion of $y$ and $y_{rep}$ values exceeding a particular threshold $y_T$, or assumptions regarding errors such as multivariate normality or homoscedasticity. In particular, the mixed replicate checking procedure of Marshall and Spiegelhalter (2007) is relevant to ensure model predictions $y_{rep}$ are consistent with the actual data $y$. For the normal linear two level model, this procedure involves sampling new cluster effects $\beta_{rep,j}$ to provide replicate means

$$\mu_{rep,ij} = \beta_{rep,0j} + \sum_p \beta_{rep,pj} x_{pij},$$

and predictions $y_{rep,ij} \sim N(\mu_{rep,ij}, \sigma^2)$. Specifically the criteria

$$C_{ij}^{(t)} = I\left(y_{rep,ij}^{(t)} > y_{ij}\right)$$

are monitored at each MCMC iteration, and the proportions $p_{mix,ij} = \Pr(y_{rep,ij} > y_{ij}|y)$ obtained. Under normality assumptions regarding error terms at different levels, outlying data points, especially at level 2 and above, may unduly influence model parameter estimates and distort credible intervals, tending to make them too wide. Normality may be assessed by monitoring residuals at different levels and applying established tests or normal quantile plots to the posterior mean residuals. Possible level 1 heteroscedasticity can be assessed by checking whether posterior mean level 1 residuals (or squared residuals) show systematic trends according to the mean $\mu_{ij}$ or particular predictors.

Options for robust estimation when there are departures from standard normality assumptions, especially for cluster effects, include Student $t$ cluster effects with unknown degrees of freedom (Seltzer, 1993), and discrete mixtures (Rabe-Hesketh and Pickles, 1999) – see Example 5.8 for an illustration of such issues. Thus Carlin et al. (2001) consider panel binary data on smoking with subjects $j$ at level 2, and contrast normally distributed subject effects, with a two group mixture of subjects, with cluster membership $L_{ij}$ based on a subsidiary regression. For the normal linear two level model, a discrete mixture strategy, with $M$ groups at level 2, may involve a mixture hierarchical prior, so that there are $M$ distinct multivariate normal priors for cluster effects at level 2. Thus with predictors $W_{ij}$ of group membership, $m \in (1, \dots, M)$, and $\beta_{jm} = (\beta_{0jm}, \dots, \beta_{Pjm})$,

$$y_{ij} = \beta_{0jL_{ij}} + \sum_{p=1}^{P} \beta_{pjL_{ij}} x_{pij} + e_{ij},$$

$$L_{ij} \sim \text{Categorical}(\pi_{ij1}, \pi_{ij2}, \dots, \pi_{ijM}),$$

$$\beta_{mj} \sim N_{P+1}(B_m, \Sigma_m),$$

$$\pi_{ij1} = 1 \left/ \left[ 1 + \sum_{m=2}^{M} \exp(\phi_m W_{ij}) \right] \right.,$$

$$\pi_{ijm} = \exp(\phi_m W_{ij}) \left/ \left[ 1 + \sum_{m=2}^{M} \exp(\phi_m W_{ij}) \right] \right., \qquad m = 2, \dots, M.$$

A less heavily parameterised option is a discrete mixture in the level 1 regression, for example

$$y_{ij} = \beta_{0L_{ij}} + \sum_{p=1}^{P} \beta_{pL_{ij}} x_{pij} + e_{ijm},$$

$$L_{ij} \sim \text{Categorical}(\pi_1, \pi_2, \dots, \pi_M),$$

with residual variation varying by group.

### Example 5.6  Pupil popularity

Consider two level continuous data, namely pupil popularity scores $y_{ij}$ for $N = 2000$ pupils in $J = 100$ school classes (the clusters). At level 1 there are $P = 2$ predictors, gender $x_{1ij}$ (1=girl, 0=boy), and a measure of extraversion $x_{2ij}$. At level 2 there is a single predictor $z_j$, namely teacher experience. The data are arranged by subject in a 'single string' vector, with class memberships $\{G_k \in 1, 2, \ldots, 100\}$ in a vector of length $N$. An initial analysis (model 1) assumes univariate normal cluster effects of gender and extraversion, and also that these effects are related to teacher experience. So in nested data form the model (i.e. with $j = G_k$) is

$$y_{ij} = \beta_{0j} + \beta_{1j}x_{1ij} + \beta_{2j}x_{2ij} + e_{ij}, \qquad e_{ij} \sim N(0, \sigma^2);$$

$$\beta_{pj} \sim N(B_{pj}, \tau_p^2),$$

$$B_{pj} = \gamma_{p0} + \gamma_{p1}z_j, \qquad p = 0, \ldots, P$$

Equivalently

$$y_{ij} = \gamma_{00} + \gamma_{01}z_j + u_{0j} + \gamma_{10}x_{1ij} + \gamma_{11}z_jx_{1ij} + u_{1j}x_{1ij} + \gamma_{20}x_{2ij} + \gamma_{21}z_jx_{2ij} + u_{2j}x_{2ij} + e_{ij}.$$

Uniform priors are assumed on the standard deviations, $\sigma \sim U(0, 1000)$ and $\tau_p \sim U(0, 1000)$. Inferences are based on the final 9000 iterations in two chain runs of 10000 iterations. The posterior means (sd) of $\{\gamma_{00}, \gamma_{10}, \gamma_{20}\}$ (denoted gam0[] in the BUGS code) are $-1.16$ (0.23), 1.24 (0.08), 0.79 (0.03), while those for $\{\gamma_{01}, \gamma_{11}, \gamma_{21}\}$ are 0.22 (0.01), $-0.0004$ (0.005), and $-0.024$ (0.002). Hence moderator effects of teacher experience are apparent for the extraversion variable: specifically, higher teacher experience lessens the effect of pupil extraversion on pupil popularity. Posterior means (sd) of $\{\sigma, \tau_1, \tau_2, \tau_3\}$ are 0.75 (0.01), 0.53 (0.05), 0.053 (0.048), and 0.024 (0.017). The DIC is 4610 ($d_e = 100$), with LPML obtained by monitoring the inverse likelihoods estimated at $-468.3$.

A second analysis (model 2) assumes multivariate normal cluster effects $\beta_j = (\beta_{0j}, \beta_{1j}, \beta_{2j})$ (for intercept, gender and extraversion), and again that these effects are related to teacher experience, namely

$$y_{ij} = \beta_{0j} + \beta_{1j}x_{1ij} + \beta_{2j}x_{2ij} + e_{ij}, \qquad e_{ij} \sim N(0, \sigma^2);$$

$$\beta_j \sim N_3(B_j, \Sigma)$$

$$B_{pj} = \gamma_{p0} + \gamma_{p1}z_j, \qquad p = 0, \ldots, P$$

A Wishart prior with identity scale matrix ($R = I$) and 2 degrees of freedom is assumed on the precision matrix, namely $\Sigma^{-1} \sim W(R, 2)$. Estimates (again based on the last 9000 of a two chain run of 10000 iterations) are similar to the first analysis for the $\gamma$ coefficients. However, estimates of the cluster variance parameters do differ: posterior means (sd) of $\{\tau_1, \tau_2, \tau_3\}$ are 0.71 (0.12), 0.26 (0.03), and 0.16 (0.02). Fit criteria appear not to support a multivariate extension: the DIC rises to 4636 ($d_e = 170$) and the LPML falls to $-483.3$, illustrating the potential benefit of initial estimation of multilevel models under simplified covariance assumptions.

A third analysis (model 3) again adopts multivariate normal cluster effects but with $R = 0.01I$ in the Wishart prior for $\Sigma$. Under this prior, cluster variation is reduced: the posterior means (sd) of $\{\tau_1, \tau_2, \tau_3\}$ are 0.64 (0.12), 0.09 (0.03), and 0.06 (0.02). Fit criteria are improved to comparability with the univariate normal cluster effects model, with a DIC of 4610 ($d_e = 116$) and LPML $= -468.7$. Such results demonstrate possible sensitivity to specification of $R$ in the Wishart prior (Natarajan and Kass, 2000).

A final analysis uses a Cholesky decomposition of $\Sigma$, namely

$$\Sigma = RR^T = \begin{pmatrix} R_{11} & 0 & 0 \\ R_{21} & R_{22} & 0 \\ R_{31} & R_{32} & R_{33} \end{pmatrix} \begin{pmatrix} R_{11} & R_{21} & R_{31} \\ 0 & R_{22} & R_{32} \\ 0 & 0 & R_{33} \end{pmatrix}$$

$$= \begin{pmatrix} R_{11}^2 & \cdots & \cdots \\ R_{21}R_{11} & R_{21}^2 + R_{22}^2 & \cdots \\ R_{31}R_{11} & R_{31}R_{21} + R_{32}R_{22} & R_{31}^2 + R_{32}^2 + R_{33}^2 \end{pmatrix}.$$

$N(0, 1)$ priors constrained to positivity are adopted for $R_{jj}$ ($j = 1, 3$), with remaining $R_{jk}$ assigned unconstrained $N(0, 1)$ priors. Under this prior, the posterior means (sd) of $\{\tau_1, \tau_2, \tau_3\}$ are 0.69 (0.13), 0.11 (0.05), and 0.07 (0.02). Fit criteria are again broadly similar to the univariate normal cluster effects model, with a DIC of 4615 ($d_e = 116$) and LPML $= -469.2$. Such potential sensitivity for inferences regarding fit and cluster variation confirm the need for assessment of alternative priors as one component of multilevel models.

**Example 5.7    Cross-classification and pupil attainment**

This example considers cross classified data (Rabe-Hesketh and Skrondal, 2008) on secondary exam attainment for N=3435 pupils in Fife (Scotland), with $G_{1k}$ being secondary school attended ($J_1 = 19$), and $G_{2k}$ the primary school (up to age 12) attended before secondary school ($J_2 = 148$). An initial analysis uses mother's education (MED=1 if left school at 16 or later,=0 otherwise) and gender (1=F, 0=M) as predictors, and considers partitioned intercept variation only, as in

$$y_k = (\beta_0 + u_{01,G_{1k}} + u_{02,G_{2k}}) + \beta_1 MED_k + \beta_2 Gend_k + e_k.$$

Separate gamma Ga(1,0.001) priors are assumed on the variance components $\sigma^2$ (level 1), and $\tau_{01}^2$ and $\tau_{02}^2$ (level 2). Among relevant inferences are the respective intra-cluster correlation (ICC) coefficients due to each classification, for example, with regard to intercept variation, $ICC_{0j} = \tau_{0j}^2/(\tau_{01}^2 + \tau_{02}^2 + \sigma^2)$. The last 4000 of a two chain run of 5000 iterations show significant effects of gender and mother's education, and a greater relative share in intercept variation due to secondary school ($ICC_{01} = 0.11$, $\tau_{01} = 1.02$) than primary school ($ICC_2 = 0.034$, $\tau_{02} = 0.54$).

Schools may differ in how far they counteract the influence of socioeconomic background, and so a second model assumes varying impacts of mother's education over both primary and secondary school classifiers, namely

$$y_k = (\beta_0 + u_{01,G_{1k}} + u_{02,G_{2k}}) + (\beta_1 + u_{11,G_{1k}} + u_{12,G_{2k}})MED_k + \beta_2 Gend_k + e_k.$$

Separate gamma Ga(1,0.001) priors are assumed on the variance components $\sigma^2$ (level 1), and $\tau_{01}^2$, $\tau_{02}^2$, $\tau_{11}^2$ and $\tau_{12}^2$ (level 2). In fact, variation in the $\beta_1$ slope is relatively small, and DIC and LPML show only slight improvement in fit. The second half of a two chain sequence of 50000 iterations show the relative share in the slope variation due to secondary school, $\tau_{11}^2/(\tau_{11}^2 + \tau_{12}^2)$ averages 46%, with posterior means for $\tau_{11}$ and $\tau_{12}$ (tau1[] in the BUGS code) of respectively 0.071 and 0.098.

## 5.4    Multilevel discrete responses

Multilevel analysis of discrete outcomes is generally carried out in the appropriate linked regression, such as a log link for a Poisson dependent variable or logit link for a binomial variable (Goldstein, 2011, chapter 4). Stage 2 and 3 specification issues (regarding random cluster effects) are similar to those of the linear normal multilevel model, but random effects may also be needed to account for overdispersion.

For example, consider two level counts $y_{ij}$ for subjects $i = 1, \ldots n_j$ within $j = 1, \ldots J$ groups. Denoting the Poisson means as $\mu_{ij}$, and assuming a single level 1 predictor $x_{ij}$, a log-linear regression may be specified either as

$$\log(\mu_{ij}) = \beta_{0j} + \beta_j x_{ij},$$

or with a level 1 error term

$$\log(\mu_{ij}) = \beta_{0j} + \beta_j x_{ij} + e_{ij}, \qquad e_{ij} \sim N(0, V_{ij}).$$

The first form assumes extra-Poisson heterogeneity will be largely accounted for by group specific intercepts and slopes, while the second allows an unstructured error with constant variance $\sigma^2$ to account for residual heterogeneity beyond that associated with the Poisson regression. Such over-dispersion is apparent in posterior mean deviances exceeding the number of observations (Knorr-Held and Rainer, 2001). Level 1 variances $V_{ij}$ may depend on subject level or cluster level predictors.

For a binomial outcome, there is a similar choice. Thus suppose $y_{ij}$ is binomial, with $y_{ij} \sim B(R_{ij}, \pi_{ij})$, where $R_{ij}$ is the total number of subjects at risk, and $y_{ij}$ is the number of positive outcomes. Then a logit link model with cluster intercepts only is

$$\text{logit}(\pi_{ij}) = \beta_{0j} + \beta x_{ij},$$

while additional over-dispersion indicates the extended model

$$\text{logit}(\pi_{ij}) = \beta_{0j} + \beta x_{ij} + e_{ij}.$$

For suitably large counts, normal approximations to the binomial or Poisson may be used, with a variance function appropriate to the form of the data. Thus for a binomial outcome, $y_{ij} \sim N(\pi_{ij} R_{ij}, V_{ij})$,

$$V_{ij} = \phi^2 R_{ij} \pi_{ij}(1 - \pi_{ij}),$$

and with the regression for the $\pi_{ij}$ involving a logit or probit link. $\phi^2 \simeq 1$ would be expected if the level 1 variation were binomial, whereas heterogeneity beyond that expected under the binomial yields $\phi^2 > 1$.

As model checks, the mixed replicate checking scheme (Marshall and Spiegelhalter, 2007) is an approximation to full cross-validation checks and involves sampling new cluster effects $\beta_{rep,j}$ to provide replicate means $\mu_{rep,ij}$ or $\pi_{rep,ij}$. Replicate data, for example $y_{rep,ij} \sim \text{Po}(\mu_{rep,ij})$ for Poisson data, are then sampled. The criteria

$$C_{ij}^{(t)} = I\left(y_{rep,ij}^{(t)} > y_{ij}\right) + 0.5I(y_{rep,ij}^{(t)} = y_{ij}),$$

are then monitored, and proportions $p_{mix,ij} = \Pr(y_{rep,ij} > y_{ij}|y) + 0.5\Pr(y_{rep,ij} = y_{ij}|y)$ obtained. Posterior predictive $p$-tests might consider relevant test statistics (e.g. for overdispersion) evaluated both for actual and replicate data, or might compare predicted and actual proportions of counts under/over a particular threshold. Under such posterior predictive tests, the replicates are generally obtained by using the complete data replicates, $y_{rep,ij} \sim \text{Po}(\mu_{ij})$.

For multilevel binary responses, one has $y_{ij} \sim \text{Bin}(1, \pi_{ij})$, with a logit or probit link. There may be advantages in modelling the underlying latent data that generate the observed responses, denoted $z$, with $z_{ij} > 0$ equivalent to $y_{ij} = 1$, and $y_{ij} = 0$ equivalent to $z_{ij} \leq 0$. The data augmentation density depends on the assumed link. For example, a logit link implies truncated standard logistic sampling to generate the $z_{ij}$, namely

$$z_{ij} \sim Logistic(\eta_{ij}, 1) \quad I(A_{ij}, B_{ij}),$$

where $A_{ij} = -\infty$ or 0, and $B_{ij} = 0$ or $\infty$, according as $y_{ij} = 0$ or 1. Data augmentation leads to simpler residual checks, and may enable other assessments, such as choice of thresholds on the $z$ scale (apart from the default zero) that produce an optimal balance between sensitivity $T_1$ (proportion of unity responses correctly identified) and specificity $T_0$ (proportion of zero responses correctly identified). The latent response model with random cluster intercept may be expressed as

$$z_{ij} = \beta_{0j} + \beta X_{ij} + e_{ij},$$

where the variance of the $e_{ij}$ depends on the cumulative distribution function used to define the link. The logistic distribution for $e_{ij}$ implies a variance of $\pi^2/3$, so for a two-level logit random intercept model with $\text{var}(\beta_{0j}) = \tau^2$, the intraclass correlation coefficient for intercept variation is

$$\text{ICC}_0 = \tau^2/(\tau^2 + \pi^2/3).$$

The normal distribution for the level-1 residual $e_{ij}$ implies $\text{var}(e_{ij}) = 1$. So for a two-level random intercept probit model, the intraclass correlation coefficient becomes $\tau^2/(\tau^2 + 1)$.

**Example 5.8   School year repetition**

Consider binary data on 7516 Thai children repeating a primary grade, and drawn from a 1988 national survey of primary education considered by Raudenbush and Bhumirat (1992). Children are nested within 356 schools, with individual level variables being sex (= 1 for male), and pped (=1 if child had pre-primary education experience, 0 otherwise). The response is $y = 1$ if the child repeated a grade, 0 otherwise. There is a school-level variable msesc (mean socio-economic status score).

A two level augmented data probit model is initially applied with normally distributed school intercepts depending on msesc. Thus

$$z_{ij} = \beta_{0j} + \beta_1 sex_{ij} + \beta_2 pped_{ij} + e_{ij}, \qquad e_{ij} \sim N(0, 1),$$

$$\beta_{0j} \sim N(\gamma_{00} + \gamma_{01} msesc_j, \tau^2)$$

The analysis addresses the assumption of normality in the school effects $\beta_{0j}$, and considers predictive accuracy of unity and zero responses using replicate school effects (Marshall and Spiegelhalter, 2007). Replicate school effects are obtained simply as $\beta_{rep,0j} \sim N(\gamma_0 + \gamma_1 msesc_j, \tau^2)$, with corresponding predictions $z_{rep,ij} \sim N(\beta_{rep,0j} + \beta_1 sex_{ij} + \beta_2 pped_{ij}, 1)$. The default classification then defines the predicted binary outcome $y$ as $y_{rep,ij} = 1$ if $z_{rep,ij} > 0$, or $y_{rep,ij} = 0$ if $z_{rep,ij} \leq 0$. However, the latent scale approach allows one to consider alternative thresholds that may result in improved sensitivity $T_1$ (proportion of unity responses correctly predicted), or an improved total classification accuracy $T_{tot} = T_1 + T_0$.

A two chain run of 5000 iterations (with last 4000 for inferences) produces an average predictive sensitivity of 0.165 (based on the above mixed replicate predictive scheme), as compared to a specificity of 0.852. The total classification accuracy has mean $T_{tot} = 1.017$. The cluster variance $\tau^2$ posterior mean estimate is 0.50, with the ICC at 0.33. However, a normal quantile plot of the posterior mean $\beta_{0j}$ suggests some departure from normality, and a Shapiro–Wilk W test confirms this.

One strategy in such cases is to use a heavier tailed alternative for cluster effects, and a scale mixture of normals[1],

$$\beta_{0j} \sim N(\gamma_{00} + \gamma_{01} msesc_j, \tau^2 / \lambda_j)$$

$$\lambda_j \sim Ga(2.5, 2.5),$$

corresponding to a Student $t_5$ density is adopted instead. This raises the LPML from $-8682$ to $-8544$, and as an example of prior sensitivity, the posterior mean $\tau^2$ and the ICC are reduced to 0.36 and 0.26 respectively, while the posterior mean sensitivity is raised to 0.171 and total classification accuracy to 1.019.

A third model, and a flexible alternative to normal cluster effects, is provided by a DP mixture of normal effects, with a maximum $M = 30$ components. The $\beta_{0j}$ are written as

$$\beta_{0j} = \gamma_{00} + \gamma_{01} msesc_j + u_{0j},$$

and potential values of $u_{0j}$, $\{u_m^*, m = 1, \ldots, M\}$, are drawn from $G_0 = N(0, \tau_u^2)$, with $1/\tau_u^2 \sim Ga(1, 0.001)$. Allocation of potential values is based on a categorical variable $L_j \sim$ Categorical$(\pi_{1,2,\ldots,M})$, where the $\pi_m$ are obtained by a stick-breaking prior, $\pi \sim \text{GEM}(\alpha)$. For the mixed predictive replication procedure, one takes $u_{rep,m}^* \sim N(0, \tau_u^2)$. Although $\alpha$ can be taken unknown, here fixed values are considered. For $\alpha = 1$, the LPML is reduced as compared to the default cluster normality model, but the sensitivity is increased to 0.28, and total classification accuracy improves to 1.022 despite a reduction in specificity. Monitoring of the realised $u_{0j}$ suggests both positive skew and bimodality in the cluster effects. For $\alpha = 5$, the classification parameters are similar to those obtained using normal and Student-$t$ cluster effects.

A final analysis illustrates adoption of varying thresholds $K_s$ ($s = 1, \ldots, 8$) in addition to the default zero, to predict binary outcome $y$ as $y_{rep,ij} = 1$ if $z_{rep,ij} > K_s$, or $y_{rep,ij} = 0$ if $z_{rep,ij} \leq K_s$. The thresholds used, in combination with normal cluster effects, are $K = (-2, -1.5, -1, -0.8, -0.6, -0.4, -0.2, 0)$. The total classification accuracy (assessed from the nodes T.tot[s] in the code) is highest at $K_s = -1$, namely 1.028, obtained with $T_1 = 0.431$ and $T_0 = 0.597$.

## 5.5    Modelling heteroscedasticity

Regression models for continuous outcomes, whether single or multilevel, most frequently assume that the error variance is constant. In a multilevel analysis, for instance, this means that the level 1 variance is independent of explanatory variables at this level. It is quite possible however that the level 1 variance, denoted $V_{ij}$ where $e_{ij} \sim N(0, V_{ij})$, is related systematically to explanatory variables or other characteristics of the subjects. In discrete data models (e.g. Poisson or binomial) random effects at level 1 may be introduced if there is over-dispersion, and such errors may have a variance which depends on the explanatory variates. Heteroscedasticity may also be modelled at higher levels if one or more cluster variances are linked to cluster predictors $z_{qj}$. Proper specification of the random part of a multilevel model may be important in inferences regarding regression coefficients and cluster variances. For example, unrecognised level-one heteroscedasticity may lead to estimating a model with significant slope variance (Snijders and Bosker, 1999)

If the differences in variance are expected according to levels of a subject level categorical variable $C_{ij}$, then one might simply take variances specific to the levels $1, \ldots, K$ of $C_{ij}$. For instance, if $\phi_k = 1/\sigma_k^2$ denotes the inverse variance for the $k^{th}$ level of $C_{ij}$, then one might adopt a series of gamma priors

$$\phi_1 \sim Ga(a_1, b_1), \phi_2 \sim Ga(a_2, b_2), \ldots, \phi_K \sim Ga(a_K, b_K).$$

Equivalently $\log(V_{ij})$ can be regressed on a factor defined by the levels of $C_{ij}$. If the categories are mutually exclusive and $d_{ij} = I(C_{ij} = j)$, one has

$$\text{var}(e_{ij}) = d_{i1}\sigma_1^2 + d_{i2}\sigma_2^2 + \ldots + d_{iK}\sigma_K^2.$$

One might also model the heterosecdasticity as a general function of relevant predictors or the entire regression term. Consider a simple two level linear model, with

$$y_{ij} = \beta_0 + \beta_1 x_{1ij} + e_{ij}.$$

Then if earlier investigation shows dependence of estimated residuals on $\mu_{ij} = \beta_0 + \beta_1 x_{1ij}$ or $x_{ij}$, one may model the level 1 random effect as

$$e_{ij} = e_{0ij} + e_{1ij}x_{ij},$$

where $\text{var}(e_{0ij}) = \sigma_0^2$, $\text{var}(e_{1ij}) = \sigma_1^2$, $\text{cov}(e_{0ij}, e_{1ij}) = \sigma_{01}$. Hence

$$V_{ij} = \text{var}(e_{ij}) = \sigma_0^2 + 2\sigma_{01}x_{ij} + \sigma_1^2 x_{ij}^2.$$

The same principle applies at higher levels. Consider the simple random intercept model

$$y_{ij} = \beta_{0j} + \beta_1 x_{1ij} + e_{ij} = \beta_0 + u_{0j} + \beta_1 x_{1ij} + e_{ij},$$

$$u_{0j} = u_{00j} + u_{01j}z_j,$$

where $\text{var}(u_{00j}) = \tau_0^2$, $\text{var}(u_{01j}) = \tau_1^2$, $\text{cov}(u_{00j}, u_{01j}) = \tau_{01}$. Hence

$$\text{var}(u_{0j}) = \tau_0^2 + 2\tau_{01}z_j + \tau_1^2 z_j^2.$$

To assess the need for level 2 heteroscedasticity, one could monitor the realised level 2 residuals $u_{0j} = \beta_{0j} - \beta_0$ in the simpler model

$$y_{ij} = \beta_{0j} + \beta_1 x_{1ij} + e_{ij},$$

$$\beta_{0j} \sim N(\beta_0, \tau^2)$$

and plot them against the $z_j$.

### Example 5.9   Heteroscedasticity in the popularity data

This analysis continues Example 5.6, but considers whether the assumption of level 1 homoscedasticity is appropriate (see Exercise 5.4). Examination of the variance of the level 1 residuals in Example 5.6 according to category of $\mu_{ij}$ suggests variance dependence. With a Wishart prior on $\Sigma^{-1}$, and scale matrix $R = 0.01I$ (as in model 3 of Example 5.6), the variance of the standardised residuals $e_{ij}^s$ varies quadratically over categories (e.g. deciles) of $\mu_{ij}$. Among possible ways of representing this dependence, it is possible to make $V_{ij} = \text{var}(e_{ij})$ a direct function of $\mu_{ij}$ and $\mu_{ij}^2$, namely a form of 'power of the mean' model (Carroll and Ruppert, 1988).

However, initially a simpler strategy is adopted, namely dependence of $\text{var}(e_{ij})$ on the two predictors, gender and extraversion (linear and squared terms). The first model considered is then

$$y_{ij} = \beta_{0j} + \beta_{1j}x_{1ij} + \beta_{2j}x_{2ij} + e_{ij}, \qquad e_{ij} \sim N(0, V_{ij});$$

$$\beta_{pj} \sim N(\gamma_{p0} + \gamma_{p1}z_j, \Sigma),$$

$$\Sigma^{-1} \sim W(R, 3), \qquad R = 0.01I,$$

$$\log(V_{ij}) = \delta_1 + \delta_2 x_{1ij} + \delta_3 x_{2ij} + \delta_4 x_{2ij}^2.$$

However, this model shows a slight deterioration in fit, with the $\delta$ coefficients not significant. A second model relates the level 1 variance to the mean, namely

$$\log(V_{ij}) = \delta_1 + \delta_2 \mu_{ij} + \delta_3 \mu_{ij}^2.$$

This model is slow to converge using two chains, with initial values based on an earlier single chain run. BGR statistics on selected parameters (the $\gamma$ and $\delta$ coefficients, and the diagonal terms in $\Sigma$) show convergence taken from iteration 20 000. Inferences from iterations 30 000-50 000 show significant effects for $\delta_2$, with posterior mean (and 95% CrI), $-0.69$ $(-1.09, -0.50)$, and for $\delta_3$, namely 0.07 (0.05, 0.11). The DIC is reduced (as compared to the homoscedastic model) from 4610 to 4598 and the LPML is increased from $-469$ to $-463$. Of importance for inferences regarding partitioning of variance is that level 2 heterogeneity declines somewhat, with the posterior means of the standard deviations $\tau_1, \tau_2, \tau_3$ (the square roots of the diagonal terms of $\Sigma$) falling to 0.47, 0.08, and 0.05 as compared to (0.64, 0.09, 0.06) under the homoscedastic model. Estimates of the $\gamma_{p1}$ parameters are, however, similar to those under the homoscedastic model.

## 5.6   Multilevel data on multivariate indices

Frequently profiling or performance rankings of public sector agencies (schools, hospitals, etc.) will involve multiple indicators. Inferences about relevant summary parameters such as comparative ranks, or the probability that a particular institution exceeds the average, are readily obtained under the Bayes sampling perspective (Deely and Smith, 1998). Such inferences will often be improved by allowing for the interdependence between the indicators themselves, and also for features of the institutions and the individual subjects within institutions (e.g. the case-mix of patients in health settings, or intake ability of pupils in school comparisons) which influence performance on some or all of the indicators used. Similar gains in precision may occur in small area health profiling where multiple mortality or morbidity outcomes provide a firmer basis for defining health problem areas than a single outcome.

Suppose individual level data $y_{ijh}$ are available for variables $h = 1, \ldots, H$, clusters $j = 1, \ldots, J$ and subjects $i = 1, \ldots n_j$ within each cluster. Then the measurements on the different variables are considered as level 1 of the data hierarchy, subjects are at level 2, and the clusters (agencies, areas, etc.) at level 3. Consider observations on $H$ metric outcomes, and predictors $x_{pij}$ ($p = 1, \ldots, P$) at level 2, and $z_{qj}$ ($q = 1, \ldots, Q$) at cluster level. With vector observations and means, $y_{ij} = (y_{ij1}, y_{ij2}, \ldots, y_{ijH})$, and $\mu_{ij} = (\mu_{ij1}, \mu_{ij2}, \ldots, \mu_{ijH})$, a multivariate normal likelihood with outcome specific regression effects, but no cluster variation, is

$$y_{ij} \sim N_H(\mu_{ij}, \Sigma),$$

$$\mu_{ijh} = \beta_{h0} + \beta_{h1}x_{1ij} + \beta_{h2}x_{2ij} + \ldots \beta_{hP}x_{Pij},$$

with $\Sigma$ an $H \times H$ dispersion matrix representing dependencies in residuals $e_{ijh}$. Of particular interest is whether the level 1 covariation between different outcomes may be reduced or eliminated as cluster dependencies are allowed for, that is whether $\Sigma$ reduces to a diagonal matrix.

With random variability of regression coefficients over clusters, one has

$$\mu_{ijh} = \beta_{jh0} + \beta_{jh1}x_{1ij} + \cdots + \beta_{jhP}x_{Pij},$$

$$= \beta_{jh0} + \sum_{p=1}^{P} \beta_{jhp}x_{pij},$$

or separating out fixed and random effects,

$$\mu_{ijh} = (B_{h0} + u_{jh0}) + \sum_{p=1}^{P}(B_{hp} + u_{jhp})x_{pij}.$$

If cluster predictors are relevant, one has (for $Q = 1$)

$$\beta_{jhp} = B_{hp} + u_{jhp} = \gamma_{hp0} + \gamma_{hp1}z_j + u_{jhp}, \qquad p = 0, \ldots, P.$$

The random effects $\beta_{jhp}$ (or deviations $u_{jhp}$) may be modelled as correlated between predictors $p$ within outcomes $h$, between outcomes $h$ within predictors $p$, or between both predictors and outcomes within clusters.

Thus with $\beta_j = (\beta_{j10}, \beta_{j20}, \ldots \beta_{jH0}; \beta_{j11}, \beta_{j21}, \ldots, \beta_{jH1}; \ldots; \beta_{j11}, \beta_{j21}, \ldots, \beta_{jHP})$ and $B_j = (B_{10}, B_{20}, \ldots, B_{H0}; B_{11}, B_{21}, \ldots, B_{H1}; \ldots; B_{11}, B_{21}, \ldots, B_{HP})$, the most general form of covariation is $\beta_j \sim N_{HP}(B_j, T)$, where the diagonal submatrices of $T$ are of dimension $P + 1$ and represent between predictor covariance in effects $\beta_{jhp}$ within clusters and within outcomes $1, 2, \ldots, H$. The off-diagonal submatrices elements represent covariance in effects both between predictors and between outcomes.

## Example 5.10   Emergency COPD admissions and lung cancer incidence

Consider count data for emergency hospitalisations $y_{k1}$ for COPD (with $k$ a single string index concatenating indices $i$ and $j$), and for lung cancer incidence $y_{k2}$, over $N = 6757$ small areas in England (Middle Level Super Output areas, or MSOAs). These small areas (indexed $i = 1, \ldots, n_j$) are nested within $J = 326$ administrative Local Authority (LA) divisions, with $\{C_k \in 1, \ldots J; k = 1, \ldots N\}$ denoting the LA division that the $k^{th}$ MSOA belongs to.

The analysis involves transformed rates $\{z_{kh} = \log (y_{kh}/E_{kh}), k = 1, \ldots, 6757; h = 1, 2\}$ where $E_{kh}$ are expected event totals, and the distributional assumption (Breslow, 1984)

$$z_{kh} \sim N\left(\mu_{kh}, \sigma_h^2 + 1/y_{kh}\right)$$

is adopted. A measure of small area deprivation $x_{1k}$ is available, and between cluster (LA) variation in both intercepts and deprivation effects is envisaged. Three models are compared.

The first model allows correlation between outcomes $h$ within parameter types $p$ (intercepts vs slopes), namely

$$\mu_{kh} = \beta_{C_k,h0} + \beta_{C_k,h1}x_{1k}.$$

Separate bivariate normal priors are adopted for LA specific intercepts $\beta_{0j} = (\beta_{j10}, \beta_{j20})$, and LA-specific deprivation slopes $\beta_{1j} = (\beta_{j11}, \beta_{j21})$, so that

$$\beta_{0j} \sim N(B_0, T_0), \qquad \beta_{1j} \sim N(B_1, T_1),$$

where $T_0$ and $T_1$ are $2 \times 2$ covariance matrices. The fixed effects parameters are $B_0 = (B_{10}, B_{20})$ and $B_1 = (B_{11}, B_{21})$, with $B_{11}$ and $B_{21}$ being the average deprivation effects on risks for COPD emergencies and for lung cancer respectively. The precision matrices $T_0^{-1}$ and $T_1^{-1}$ are assigned Wishart priors with identity scale matrices and two degrees of freedom.

A second model specifies correlation between parameter types within outcomes, where separate bivariate normal priors are adopted for the COPD parameters denoted $\beta_{j1} = (\beta_{j10}, \beta_{j11})$, and the lung cancer parameters denoted $\beta_{j2} = (\beta_{j20}, \beta_{j21})$. A third model allows correlation both between outcomes and between parameter types, and is represented again as

$$\mu_{kh} = \beta_{C_k,h0} + \beta_{C_k,h1} x_{1k},$$

but now with the assumption that LA effects $\beta_j = (\beta_{j10}, \beta_{j11}, \beta_{j20}, \beta_{j21})$ follow a multivariate normal density of order 4.

Inferences are based on the last 1500 iterations of a two chain sequence of 2500 iterations. For model 1, the LPML is 9900 and the DIC is 4995. The average deprivation effects for COPD emergencies and for lung cancer $(B_{11}, B_{21})$ have means (95% CrI) respectively 0.045 (0.038,0.051) and 0.024 (0.017,0.030). The correlation between outcomes within intercepts (obtained from $T_0$) averages 0.73, but the correlation between outcomes within slopes is not significant. The second model has worse LPML and DIC (respectively 9823 and 5125), but shows substantively interesting contextual effects, with negative correlations between slopes and intercepts within outcomes: small area deprivation effects are higher in LAs with lower emergency and incidence rates (these are r.beta1 and r.beta2 in the BUGS code). The most general[2] model (model 3) has a slightly improved DIC as compared to model 1, but a lower LPML at 9895, so highlighting the primacy of correlations between outcomes within parameters, though negative correlations between slopes and intercepts within outcomes (C.beta[1,2] and C.beta[3,4] in the BUGS code) remain apparent.

## Exercises

**5.1.** For the aspirin use data in Example 5.3, consider a scale mixing model as an alternative to normal random effects at the second stage. This is equivalent to a Student $t$ second stage. Thus

$$y_i \sim N(\theta_i, s_i^2),$$

$$\theta_i \sim N(\mu, \tau^2/\varphi_i),$$

$$\varphi_i \sim Ga(0.5\nu, 0.5\nu),$$

where $\nu$ is an extra unknown degrees of freedom parameter. One possible prior is a uniform on $1/\nu$. How does this affect the posterior density for $\mu$ as compared to a normal hierarchical model without scale mixing, and are any studies apparent as significant outliers (with 95% intervals for $\varphi_i$ entirely under 1).

**5.2.** In the bivariate meta-analysis of data on true positives and negative CT scan diagnoses (Example 5.5), adopt a scale mixture of normals for the $\theta_i = (\theta_{i1}, \theta_{i2})$ in the first model, namely

$$TP_i \sim Bin(D_i, s_i),$$

$$TN_i \sim \text{Bin}(W_i, t_i),$$

$$\text{logit}(s_i) = \theta_{i1}; \text{logit}(t_i) = \theta_{i2};$$

$$\begin{pmatrix} \theta_{i1} \\ \theta_{i2} \end{pmatrix} \sim BVN \left( \begin{bmatrix} \mu_1 \\ \mu_2 \end{bmatrix}, \frac{1}{\lambda_i} \begin{bmatrix} \tau_1^2 & \rho\tau_1\tau_2 \\ \rho\tau_1\tau_2 & \tau_2^2 \end{bmatrix} \right),$$

$$\lambda_i \sim \text{Ga}\left(\frac{v}{2}, \frac{v}{2}\right),$$

with $v$ an unknown, with prior $\frac{1}{v} \sim U(0.5, 0.02)$. Are any studies classed as outliers using this strategy?

5.3.   In Example 5.6 (popularity data), and retaining separate univariate normal priors (i.e. model 1) on the cluster effects at stage 2, assess the impact on inferences, such as on posterior means of $\{\sigma, \tau_1, \tau_2, \tau_3\}$, of adopting uniform priors on the shrinkage parameters $\sigma^2/(\sigma^2 + \tau_p^2)$.

5.4.   In Example 5.6, and assuming a Wishart prior ($R = 0.01I$ as the scale matrix) on the second stage precision matrix (model 3), assess the assumption of level 1 homoscedasticity. It is suggested to monitor $\mu_{ij}$, and realised values of the standardised level 1 residuals $e_{ij}^s = (y_{ij} - \mu_{ij})/\sigma$, and obtain their posterior means. One can then see whether the residual variance changes according to grouped categories of $\mu_{ij}$ (e.g. deciles). Alternatively one may plot posterior mean $e_{ij}^s$, or their squares, against the posterior means of $\mu_{ij}$ or against the predictors. One may also estimate linear regressions of the squares of the posterior mean $e_{ij}^s$ against polynomial functions of the posterior mean $\mu_{ij}$. Analysis of this kind suggests a quadratic relation between the error variance and the posterior mean $\mu_{ij}$.

5.5.   In the bivariate multilevel analysis of the health outcome data (Example 5.10), assess normality of the cluster effects in the third model. For example, one may obtain posterior means of $\{\beta_{j10}, \beta_{j11}, \beta_{j20}, \beta_{j21}\}$, and assess normality using normal quantile plots and omnibus tests, such as the univariate and multivariate Shapiro–Wilk tests available in R. Your analysis should reject univariate normality in the slope effects, and also multivariate normality.

5.6.   Consider data involving replicate measures i=1,2 of anchovy larvae counts $y_{ij}$ over $j = 1, \ldots, 49$ larvae pairs (Booth et al., 2003). Predictors in a negative binomial regression are logarithms of water volumes (in cubic meters) and day when the readings were made. A code allowing for a random pair effect and homogeneous predictor effects is

```
model { for (j in 1:49) { u0[j] ~dnorm(0,tau0)
for (i in 1:2) { y[j,i] ~dnegbin(p[j,i],alpha)
p[j,i] <- alpha/(alpha+mu[j,i])
log(mu[j,i]) <- beta.0+beta[1]*log(vol[j,i])+beta[2]*day[j] + u0[j]}}
    alpha ~dgamma(1, 1);          beta.0 ~dnorm(0,0.000001)
tau0~dgamma(1,0.01)
for (k in 1:2){ beta[k] ~dnorm(0,0.001)}}
```

Assess the assumption of normality in the pair (cluster) random effects $u_{0j}$. Consider gains in fit, and other inferences (e.g. regarding the density of the cluster effects) under a two group discrete mixture allowing distinct $\beta_{km}$ coefficients (k=0,1,2) by group $m$, distinct negative binomial parameters $\alpha_m$, and distinct pair variances $\tau_{0m}^2$.

# Notes

1. The BUGS code for the Student $t$ cluster effects model for the primary school repetition data is

```
model { for (i in 1:7516) {ystar[i] ~dnorm(eta[i],1) I(A[i],B[i])
e[i] <- ystar[i]-eta[i]; LL[i] <- -0.5*pow(e[i],2);
 invLk[i] <- 1/exp(LL[i])
A[i] <- -10*equals(y[i],0); B[i] <- 10*equals(y[i],1)
eta[i] <- beta0[school[i]]+beta[1]*sex[i]+beta[2]*pped[i]
# predictive check (zero threshold)
D1[i] <- equals(y[i],1); D0[i] <- equals(y[i],0)
y.rep[i] <- step(z.rep[i])
C0[i] <- equals(y[i],y.rep[i])*equals(y[i],0);
 C1[i] <- equals(y[i],y.rep[i])*equals(y[i],1)
z.rep[i] ~dnorm(eta.rep[i],1)
eta.rep[i] <- beta0.rep[school[i]]+beta[1]*sex[i]+beta[2]*pped[i]}
# Cluster model
for (j in 1:356) {beta0[j] ~dnorm(mu[j],inv.tau[j])
inv.tau[j] <- inv.tau2*lam[j]
lam[j] ~dgamma(2.5,2.5)
beta0.rep[j] ~dnorm(mu[j],inv.tau[j])
mu[j] <- gam00+gam01*msesc[j]}
# Stage 3 Priors
inv.tau2 ~dgamma(1,0.001); tau2 <- 1/inv.tau2
Beta0 ~dnorm(0,0.001); gam00 ~dnorm(0,0.001); gam01 ~dnorm(0,0.001)
for (j in 1:2) {beta[j] ~dnorm(0,0.001)}
# intra-school correlation
ICC <- tau2/(1+tau2)
# classification rates according to whether y=0 or y=1
T0 <- sum(C0[])/sum(D0[]); T1 <- sum(C1[])/sum(D1[]); T.tot <- T0+T1}
```

Under the Dirichlet process mixture, the code for the cluster model and associated parameters becomes

```
# Cluster model
for (j in 1:356) {L[j] ~dcat(p[1:M]);
# realised cluster effect
    uL[j] <- u[L[j]]
# intercept in school j
    beta0[j] <- u[L[j]]+gam00+gam01*msesc[j]
# mixed replicate
    beta0.rep[j] <- u.rep[L[j]]+gam00+gam01*msesc[j]
    for (k in 1:M) {Clus[j,k] <- equals(L[j],k)}}
# truncated Dirichlet process
    alpha <- 1; V[M] <- 1
    for (k in 1:M-1){ V[k] ~dbeta(1,alpha)}
    p[1] <- V[1]
    for (j in 2:M) { p[j] <- V[j]*(1-V[j-1])*p[j-1]/V[j-1]}
# total realised clusters
    Mstar <- sum(NonEmp[])
for (j in 1:M) {NonEmp[j] <- step(sum(Clus[,j])-1)
                u[j] ~dnorm(0,inv.tau.u);    u.rep[j] ~dnorm(0,inv.tau.u)}
# identified level 2 intercept
    gam0.s <- mean(uL[])+gam0
# Stage 3 Priors
```

```
inv.tau.u ~dgamma(1,0.001); Beta0 ~dnorm(0,0.001);
                            gam00 ~dnorm(0,0.001);
    gam01 ~dnorm(0,0.001); for (j in 1:2) {beta[j] ~dnorm(0,0.001)}
# (effective) cluster variance and intra-school correlation
    tau2 <- sd(uL[])*sd(uL[]); rho <- tau2/(1+tau2).
```

2. The code for the most general bivariate multilevel model involves a stacked notation $\{beta[j,1] \ldots ,beta[j,4]\}$ for the cluster effects $\{\beta_{j10}, \beta_{j11}, \beta_{j20}, \beta_{j21}\}$. The level 1 correlation is assessed indirectly via the calculation r.L1, obtained from the residuals $e_{kh} = z_{kh} - \mu_{kh}$. The code is as follows

```
model {for (k in 1:N) {e12[k] <- e[k,1]*e[k,2]; e1.2[k] <- e[k,1]*e[k,1];
                                          e2.2[k] <- e[k,2]*e[k,2]
mu[k,1] <- beta[C[k],1]+beta[C[k],2]*x[k]
mu[k,2] <- beta[C[k],3]+beta[C[k],4]*x[k]
for (h in 1:2) {z[k,h] ~dnorm(mu[k,h],inv.sig2[k,h])
z[k,h] <- log(y[k,h]/E[k,h]); e[k,h] <- z[k,h]-mu[k,h]
inv.sig2[k,h] <- 1/(sig2[h]+1/y[k,h])
LL[k,h] <- 0.5*(log(inv.sig2[k,h])-inv.sig2[k,h]*pow(e[k,h],2))
# for obtaining LPML
G[h,k] <- 1/exp(LL[k,h])}}
# level 1 residual correlation between outcomes
r.L1 <- sum(e12[])/sqrt(sum(e1.2[])*sum(e2.2[]))
# correlation between parameters within outcomes, and
# correlation between outcomes within parameters
for (j in 1:J) {beta[j,1:4] ~dmnorm(B[j,1:4],Inv.T.beta[,])
B[j,1] <- Beta0[1]; B[j,2] <- Beta1[1]
B[j,3] <- Beta0[2]; B[j,4] <- Beta1[2]}
for (h in 1:2) {Beta0[h] ~dnorm(0,0.001); Beta1[h] ~dnorm(0,0.001)}
for (h in 1:4) {tau.beta[h] <- sqrt(T.beta[h,h])}
for (m in 1:4) {R[h,m] <- equals(h,m);
# correlations between cluster effects
C.beta[h,m] <- T.beta[h,m]/(tau.beta[h]*tau.beta[m])}}
Inv.T.beta[1:4,1:4] ~dwish(R[,],4);
 T.beta[1:4,1:4] <- inverse(Inv.T.beta[,]);
for (j in 1:2) {sig[j] ~dunif(0,10); sig2[j] <- sig[j]*sig[j]}}
```

# References

Askie, L., Ballard, R., Cutter, G., Dani, C., Elbourne, D., Field, D., Hascoet, J., Hibbs, A., Kinsella, J., Mercier, J. *et al.* (2011) Inhaled nitric oxide in preterm infants: an individual-patient data meta-analysis of randomized trials. *Pediatrics*, **128**(4), 729–739.

Barnard, J., McCulloch, R. and Meng, X.L. (2000) Modeling covariance matrices in terms of standard deviations and correlations, with application to shrinkage. *Statistica Sinica*, **10**, 1281–1311.

Berlin, J., Longnecker, M. and Greenland, S. (1993) Meta-analysis of epidemiologic dose-response data. *Epidemiology*, **4**(3), 218–228.

Berlin, J., Santanna, J., Schmid, C., Szczech, L. and Feldman, H. (2002) Individual patient- versus group-level data meta-regressions for the investigation of treatment effect modifiers: ecological bias rears its ugly head. *Statistics in Medicine*, **21**(3), 371–387.

Blakely, T. and Woodward, A. (2000) Ecological effects in multilevel studies. *Journal of Epidemiology and Community Health*, **54**(5), 367–374.

Booth, J.G., Casella, G., Friedl, H. and Hobert, J. (2003) Negative binomial loglinear mixed models. *Statistical Modelling*, **3**, 179–191.

Breslow, N.E. (1984) Extra-Poisson variation in log-linear models. *Applied Statistics*, **33**, 38–44.

Brodsky, A., O'Campo, P. and Aronson, R. (1999) PSOC in community context: multilevel correlates of a measure of psychological sense of community in low-income, urban neighborhoods. *Journal of Community Psychology*, **27**(6), 659–679.

Brooks, S. and Gelman, A. (1998) General methods for monitoring convergence of iterative simulations. *Journal of Computational and Graphical Statistics*, **7**, 434–455.

Browne, W. and Draper, D. (2000) Implementation and performance issues in the Bayesian and likelihood fitting of multilevel models. *Computational Statistics*, **15**, 391–420.

Browne, W., Draper, D., Goldstein, H. and Rasbash, J. (2000) Bayesian and likelihood methods for fitting multilevel models with complex level-1 variation. *Computational Statistics and Data Analysis*, **39**(2), 203–225.

Carlin, J.B., Wolfe, R., Brown, C.H. and Gelman, A. (2001) A case study on the choice, interpretation and checking of multilevel models for longitudinal binary outcomes. *Biostatistics*, **2**(4), 397–416.

Carpenter, J., Schwarzer, G., Rücker, G. and Künstler, R. (2009) Empirical evaluation showed that the Copas selection model provided a useful summary in 80 of meta-analyses. *Journal of Clinical Epidemiology*, **62**(6), 624–631.

Carroll, R. and Ruppert, D. (1988) *Transformation and Weighting in Regression*. Chapman and Hall, New York, NY.

Clarke, P. (2008) When can group level clustering be ignored? Multilevel models versus single-level models with sparse data. *Journal of Epidemiology and Community Health*, **62**(8), 752–758.

Clayton, D. (1996) Generalized linear mixed models. In W. Gilks, S. Richardson and D. Spiegelhalter (eds), *Markov Chain Monte Carlo in Practice*. Chapman & Hall, London, UK.

Copas, J. and Shi, J. (2000) Meta-analysis, funnel plots and sensitivity analysis. *Biostatistics*, **1**, 247–262.

Courgeau, D. and Baccaini, B. (1997) multilevel analysis in the social sciences. *Population*, **52**(4), 831–863.

Daniels, M. (1999) A prior for the variance in hierarchical models. *Canadian Journal of Statistics*, **27**, 567–578.

Daniels, M. and Kass, R. (1999) Nonconjugate Bayesian estimation of covariance matrices and its use in hierarchical models. *Journal of the American Statistical Association*, **94**, 1254–1263.

Deely, J. and Smith, A. (1998) Quantitative refinements for comparisons of institutional performance. *Journal of the Royal Statistical Society A*, **161**, 5–12.

Diez Roux, A. (2004) The study of group-level factors in epidemiology: rethinking variables, study designs, and analytical approaches. *Epidemiology Review*, **26**, 104–111.

DuMouchel, W. (1996) Predictive cross-validation of Bayesian meta-analyses. *Bayesian Statistics*, **5**, 107–127.

Duncan, C., Jones, K. and Moon, G. (1999) Smoking and deprivation: are there neighbourhood effects? *Social Science and Medicine*, **48**(4), 497–505.

Eisenberg, M., Filion, K., Yavin, D., Bélisle, P., Mottillo, S., Joseph, L. and Pilote, L. (2008) Pharmacotherapies for smoking cessation: a meta-analysis of randomized controlled trials. *Canadian Medical Association Journal*, **179**(2), 135–144.

Everson, P. and Morris, C. (2000) Inference for multivariate normal hierarchical models. *Journal of the Royal Statistical Society B*, **62**(2), 399–412.

Field, A. (2001) Meta-analysis of correlation coefficients: a Monte Carlo comparison of fixed- and random-effects methods. *Psychology Methods*, **6**(2), 161–180.

Fielding, A. and Goldstein, H. (2006) *Cross-classified and Multiple Membership Structures in Multilevel Models: An Introduction and Review*, DFES Research Report RR791.

Goldstein, H. (2011) *Multilevel Statistical Models*, 4th edn. Wiley, Chichester, UK.

Hackshaw, A., Law, M. and Wald, N. (1997) The accumulated evidence on lung cancer and environmental tobacco smoke. *British Medical Journal*, **315**(7114), 980–988.

Hasselblad, V. (1998) Meta-analysis of multitreatment studies. *Medical Decision Making*, **18**, 37–43.

Higgins, J., Thompson, S. and Spiegelhalter, D. (2009) A re-evaluation of random-effects meta-analysis. *Journal of the Royal Statistical Society A*, **172**(1), 137–159.

Hox, J. (2010) *Multilevel Analysis – Techniques And Approaches*, 2nd edn. Routledge, London, UK.

Kelsall, J. and Wakefield, J. (1999) Discussion of 'Bayesian models for spatially correlated disease and exposure data' by N. Best, l. Waller, A. Thomas, E. Conlon and R. Arnold. In J.M. Bernardo, J.O. Berger, A. Dawid and A. Smith (eds), *Bayesian Statistics 6*. Oxford University Press, London, UK.

Knorr-Held, L. and Rainer, E. (2001) Projections of lung cancer mortality in West Germany: a case study in Bayesian prediction. *Biostatistics*, **2**(1), 109–129.

Lambert, P., Sutton, A., Burton, P., Abrams, K. and Jones, D. (2005) How vague is vague? A simulation study of the impact of the use of vague prior distributions in MCMC using WinBUGS. *Statistics in Medicine*, **24**(15), 2401–2428.

Lange, J., Troelsen, A., Thomsen, R. and Søballe, K. (2012) Chronic infections in hip arthroplasties: comparing risk of reinfection following one-stage and two-stage revision: a systematic review and meta-analysis. *Clinical Epidemiology*, **4**, 57–73.

Langford, I., Leyland, A., Rasbash, J. and Goldstein, H. (1999) Multilevel modelling of the geographical distribution of diseases. *Journal of the Royal Statistical Society C*, **48**(2), 253–268.

Larose, D. and Dey, D. (1998). Modeling publication bias using weighted distributions in a Bayesian framework. *Computational Statistics and Data Analysis*, **26**(3), 279–302.

Lu, G. and Ades, A. (2009) Modeling between-trial variance structure in mixed treatment comparisons. *Biostatistics*, **10**(4), 792–805.

MacNab, Y., Qiu, Z., Gustafson, P., Dean, C., Ohlsson, A. and Lee, S. (2004) Hierarchical Bayes analysis of multilevel health services data: a Canadian neonatal mortality study. *Health Services and Outcomes Research Methodology*, **5**(1), 5–26.

Marshall, E. and Spiegelhalter, D. (2007) Identifying outliers in Bayesian hierarchical models: a simulation-based approach. *Bayesian Analysis*, **2**(2), 409–444.

Martinius, J. (1993) The developmental approach to psychopathology in childhood and adolescence. *Early Human Development*, **34**(1–2), 163–168.

Mavridis, D. and Salanti, G. (2013) A practical introduction to multivariate meta-analysis. *Statistical Methods in Medical Research*, **22**(2), 133–158.

Mavridis, D., Sutton, A., Cipriani, A. and Salanti, G. (2012) A Bayesian selection model for publication bias with informative priors. http://www.mtm.uoi.gr/.

Mavridis, D., Sutton, A., Cipriani, A. and Salanti, G. (2013) A fully Bayesian application of the Copas selection model for publication bias extended to network meta-analysis. *Statistics in Medicine*, **32**(1), 51–66.

Natarajan, R. and Kass, R. (2000) Reference Bayesian methods for generalized linear mixed models. *Journal of the American Statistical Association*, **95**(449), 227–237.

O'Brien, S., Shahian, D., DeLong, E., Normand, S., Edwards, F., Ferraris, V., Haan, C., Rich, J., Shewan, C., Dokholyan, R. *et al.* (2007) Quality measurement in adult cardiac surgery: part 2–Statistical considerations in composite measure scoring and provider rating. *Annals of Thoracis Surgery*, **83**(4 Suppl), S13–26.

O'Malley, J. and Zaslavsky, A. (2008) Domain-level covariance analysis for multilevel survey data with structured nonresponse. *Journal of the American Statistical Association*, **103**(484), 1405–1418.

Paul, M., Riebler, A., Bachmann, L., Rue, H. and Held, L. (2010) Bayesian bivariate meta-analysis of diagnostic test studies using integrated nested Laplace approximations. *Statistics in Medicine*, **29**(12), 1325–1339.

Rabe-Hesketh, S. and Pickles, A. (1999) Generalised, linear, latent and mixed models. In H. Friedl, A. Bughold and G. Kauermann (eds), *Proceedings of the 14th International Workshop on Statistical Modelling*, pp. 332–339. Statistical Modelling Society, Graz.

Rabe-Hesketh, S. and Skrondal, A. (2008) *Multilevel and Longitudinal Modeling Using Stata*, 2nd edn. Stata Press, College Station, TX.

Rasbash, J. and Browne, W. (2008) Non-hierarchical multilevel models. In J. De Leeuw and E. Meijer (eds), *Handbook of Quantitative Multilevel Analysis*, pp. 301–334. Springer, New York.

Raudenbush, S. and Bhumirat, C. (1992) The distribution of resources for primary education and its consequences for educational achievement in Thailand. *International Journal of Educational Research*, **17**, 143–164.

Richardson, S. (1996) Measurement error. In W. Gilks, S. Richardson and D. Spiegelhalter (eds.), *Markov Chain Monte Carlo in Practice*, pp. 401–417. Chapman & Hall, London, UK.

Riley, R., Thompson, J. and Abrams, K. (2008) An alternative model for bivariate random-effects meta-analysis when the within-study correlations are unknown. *Biostatistics*, **9**, 172–186.

Riva, M., Gauvin, L. and Barnett, T. (2007) Toward the next generation of research into small area effects on health: a synthesis of multilevel investigations published since July 1998. *Journal of Epidemiology and Community Health*, **61**(10), 853–861.

Rothstein, H., Sutton, A. and Borenstein, M. (2005) *Publication Bias in Meta-Analysis: Prevention, Assessment and Adjustments*. Wiley, New York, NY.

Seltzer, M. (1993) Sensitivity analysis for fixed effects in the hierarchical model – a Gibbs sampling approach. *Journal of Educational Statistics*, **18**(3), 207–235.

Silliman, N.P. (1997) Hierarchical selection models with applications in meta-analysis. *Journal of the American Statistical Association*, **92**(439), 926–936.

Simmonds, M., Higgins, J., Stewart, L., Tierney, J., Clarke, M. and Thompson, S. (2005) Meta-analysis of individual patient data from randomized trials: a review of methods used in practice. *Clinical Trials*, **2**(3), 209–217.

Smith, T., Spiegelhalter, D. and Thomas, A. (1995a) Bayesian approaches to random-effects meta-analysis: a comparative study. *Statistics in Medicine*, **14**, 2685–2699.

Smith, S., Caudill, S., Steinberg, K. and Thacker, S. (1995b) On combining dose-response data from epidemiological studies by meta-analysis. *Statistics in Medicine*, **14**, 531–544.

Snijders, T. and Berkhof, J. (2008) Diagnostic checks for multilevel models. In J. de Leeuw and E. Meijer (eds), *Handbook of Multilevel Analysis*, pp. 141–175. Springer, New York, NY.

Snijders, T. and Bosker, R. (1999) *Multilevel Analysis. An Introduction to Basic and Advanced Multilevel Modeling*. Sage, London, UK.

Spiegelhalter, D.J., Abrams, K. and Myles, J.P. (2004) *Bayesian Approaches to Clinical Trials and Health-care Evaluation*. Wiley, Chichester, UK.

Verde, P. (2010) Meta-analysis of diagnostic test data: a bivariate Bayesian modeling approach. *Statistics in Medicine*, **29**(30), 3088–3102.

Welton, N., Sutton, A., Cooper, N., Abrams, K. and Ades, A. (2012) *Evidence Synthesis for Decision Making in Healthcare*. Wiley, Chichester, UK.

White, A.R., Resch, K. and Ernst, E. (1999) A meta-analysis of acupuncture techniques for smoking cessation. *Tobacco Control*, **8**(4), 393–397.

Yuan, Y. and Little, R. (2009) Meta-analysis of studies with missing data. *Biometrics*. **65**(2), 487–496.

# 6

# Models for time series

## 6.1 Introduction

Many scientific disciplines raise issues in representing and forecasting series of observations generated in time. Often the series, although varying continuously in time, is observed at discrete time points, $t = 1, \ldots, T$. Bayesian perspectives are relevant since increasingly time series models are framed hierarchically in terms of hyperparameters and latent state variables, both conditional on the observations. Recent overviews of time series modelling with a Bayesian perspective include Prado and West (2010), Steel (2008), Migon *et al.* (2005), Johannes and Polson (2009), De Pooter *et al.* (2006) and Geweke and Whiteman (2006). General time series computing options in R are discussed by McLeod *et al.* (2012), while specifically Bayesian packages for time series models (possibly only for certain model classes) include BUGS, R-INLA, tsbugs, stochvol, BayesGARCH, MSBVAR and dlm.

The goals of time series models include smoothing an irregular series, forecasting series into the medium or long-term future, and causal modelling of variables moving in parallel through time. Time series analysis exploits the temporal dependencies both in the deterministic (regression) and stochastic (error) components of the model. In fact dynamic regression models are defined when model components are indexed by time, and a lag appears on one or more of them in the model specification (Bauwens *et al.*, 2000). For instance, a dynamic structure on the exogenous variables leads to a distributed lag model, and a dynamic structure may also be specified for endogenous variables, error terms, or variances of the errors.

While simple curve fitting (e.g. in terms of polynomials in time) may produce a good fit it does not facilitate prediction outside the sample and may be relatively heavily parameterised. By contrast, models accounting for the dependence of a response (or error) on its previous values may be both parsimonious and effective in prediction. The ARMA models developed by Box and Jenkins (1976) and Zellner (1971), from classical and Bayesian perspectives respectively, are often effective for forecasting purposes (see Section 6.2), but dynamic linear and varying coefficient models (Section 6.4) have perhaps greater flexibility in modelling non-stationary series, and are interpretable in terms of latent processes (e.g. trend, seasonal effects) driving the series (Petris *et al.*, 2009).

*Applied Bayesian Modelling*, Second Edition. Peter Congdon.
© 2014 John Wiley & Sons, Ltd. Published 2014 by John Wiley & Sons, Ltd.

Bayesian methods have been widely applied in time series contexts and have played a significant role in recent developments in discrete data time series (Section 6.3), dynamic linear models (Section 6.4), and stochastic volatility models (Section 6.5). They may have advantages in situations where non-standard distributions or latent variable representations are more realistic. In ARMA models a Bayesian perspective may facilitate approaches not limited to stationarity, so that stationarity and non-stationarity are assessed as alternative models for the data series. A Bayes approach may also assist in analysis of shifts in time series where likelihood methods may either be complex or inapplicable (Section 6.6).

Time series model assessment often involves cross-validatory principles as well as standard measures of fit to all observations. Within sample fit measures include DIC or BIC criteria (Berg et al., 2004), marginal likelihood approximations (Gelfand and Dey, 1994; Geweke and Whiteman, 2006), and predictive loss criteria (e.g. Gelfand and Ghosh, 1998). Cross-validation (Nandram and Petrucelli, 1997; Chu and Xin, 2007; Petris et al., 2009; Geweke and Amisano, 2010) may use $k$-step-ahead predictive densities, $p(y_{t+k}|y_1, y_2, \ldots, y_t)$, (with $t + k \leq T$), namely predictions within the span of the observed series, or use a training sample covering only part of the observed series.

## 6.2    Autoregressive and moving average models

A starting point in dynamic regression models is provided by considering dynamic structures in the outcomes. Autoregressive process models describe data driven dependence in an outcome over successive time points. For continuous data $y_t$, observed at times $t = 1, \ldots, T$ the simplest autoregressive dependence in the outcomes is of order 1 or AR(1), as in

$$y_t = \mu + \rho y_{t-1} + u_t, \qquad t = 2, \ldots, T$$

where $\mu$ represents the level of the series, and $\rho$ represents autocorrelation between successive observations. Additional dependence on lagged observations $y_{t-2}, y_{t-3}, \ldots, y_{t-p}$ leads to AR(2), AR(3), $\ldots$, AR(p) processes. After accounting for observation driven serial dependence, the errors may be taken as exchangeable normal, $u_t \sim N(0, \sigma^2)$ with constant variance, and $cov(u_s, u_t) = 0$. If $|\rho| < 1$, the process is stationary (see below) with variance $\sigma^2/(1 - \rho^2)$ and long run mean $\mu_e = \mu/(1 - \rho)$, and the series will tend to revert to its mean level after undergoing a shock.

A conditional likelihood for this model (conditioning on the first observation) with a normal prior for $\rho$ and inverse gamma prior for $\sigma^2$ provides conjugate posteriors with simple updating. However, no conjugate prior is available for a full likelihood, namely the likelihood including the first observation under stationarity (with $-1 < \rho < 1$), or the likelihood referring to a pre-series latent observation $y_0$ when no stationarity constraint is imposed (Prado and West, 2010; Marriott et al., 1996). For an AR($p$) model not conditioning on $\{y_1, \ldots, y_p\}$ there are $p$ implicit latent values, $y_0, y_{-1}, \ldots, y_{1-p}$ in the full likelihood model without stationarity.

Classical estimation and forecasting with the AR($p$) model rest on stationarity, namely that the process generating the series is the same whenever observation starts: so the vectors $(y_1, \ldots, y_k)$ and $(y_t, \ldots, y_{t+k})$ have the same distribution for all $t$ and $k$. Specifically, under weak stationarity, expectations $E(y_t)$ and covariances $C(y_t, y_{t+k})$ are independent of $t$. For the stationary AR($p$) model to be applicable, an observed series may require initial transformation and differencing to eliminate trend. This may be combined with a variance stabilising transformation, e.g. $y_t^* = \log(y_t)$.

With the $B$ operator denoting a backward shift by one time unit, a first difference ($d = 1$) in $y_t$ is defined as

$$z_t = y_t - y_{t-1} = y_t - By_t = (1 - B)y_t,$$

with an AR(1) model in $z_t$ then represented as

$$z_t - \rho z_{t-1} = z_t(1 - \rho B) = u_t.$$

An AR($p$) process in $z_t$ involves a $p$th order polynomial in $B$, so that

$$z_t(1 - \rho_1 B - \rho_2 B^2 - \ldots \rho_p B^p) = u_t,$$

with alternative notation $\rho(B)z_t = u_t$. The process is stationary if the roots of $\rho(B)$ lie outside the unit circle.

In a similar way, vector autoregressive models are used to represent multivariate autoregressive dependence through time, with each series depending both on its own past and the past values of the other series (Canova, 2007, Ch. 10). A vector autoregressive order $p$, or VAR($p$), model for $K$ centred metrical variables $Y_t = (y_{1t}, y_{2t}, \ldots, y_{Kt})'$ follows

$$Y_t = \Phi_{p1} Y_{t-1} + \ldots + \Phi_{pp} Y_{t-p} + U_t,$$
$$U_t \sim N_K(0, V),$$

where the matrices $\Phi_{p1}, \ldots, \Phi_{pp}$ are each $K \times K$. For $K = 2$, $\Phi_{p1}$ would consist of own-lag coefficients relating $y_{1t}$ and $y_{2t}$ to the lagged values $y_{1,t-1}$ and $y_{2,t-1}$ respectively, and cross-lag coefficients relating $y_{1t}$ to $y_{2,t-1}$ and $y_{2t}$ to $y_{1,t-1}$.

In the AR($p$) model, an outcome depends on its past values and a random error or innovation term $u_t$. If the impact of $u_t$ is not fully absorbed in period $t$, there may be moving average dependence in the error term. Thus for centred data, the model

$$z_t - \rho_1 z_{t-1} = u_t - \theta_1 u_{t-1},$$

defines a first order moving average MA(1) process in $u_t$ combined with AR(1) dependence in the data themselves.

More generally an ARIMA($p, d, q$) model is defined by dependence up to lag $p$ in the observations, by $q$ lags in the error moving average, and by differencing the original observation ($y_t$) $d$ times. An ARMA($p, q$) model in $y_t$ therefore retains the original data without differencing. In the ARMA($p, q$) representation

$$\rho(B)y_t = \theta(B)u_t,$$

the process is stationary if the roots of $\rho(B)$ lie outside the unit circle, and invertible if the roots of $\theta(B)$ lie outside the unit circle. If $p = q = 1$, the series is stationary and invertible if $|\rho_1| < 1$ and $|\theta_1| < 1$. Note that an extended AR($p$) with $p$ large will often approximate an ARMA($p, q$) representation, and permit easier estimation (Prado and West, 2010).

In a distributed lag regression, lagged values in predictors are introduced into the regression. A distributed lag model for centred data (Bauwens *et al.*, 2000; Baltagi, 2011) and a single predictor has the form

$$y_t = \sum_{m=0}^{M} \beta_m x_{t-m} + u_t,$$

while lags in both $y$ and $x$ leads to an autoregressive distributed lag (ADL or ARDL) model:

$$\rho(B)y_t = \beta(B)x_t + u_t.$$

The latter form leads into recent model developments in terms of error correction models.

## 6.2.1  Dependent errors

In the specifications above, the errors $u_t$ are assumed temporally uncorrelated with diagonal covariance matrix. However, if correlation exists between successive errors then the covariance matrix is no longer diagonal. Let $\epsilon_t$ be correlated errors with

$$y_t = X_t\beta + \epsilon_t,$$

and suppose that an AR($p$) transformation of the $\epsilon_t$ is required

$$\gamma(B)\epsilon_t = u_t,$$

in order that $u_t$ is unstructured with constant variance, where $\gamma(B) = 1 - \gamma_1 B - \gamma_2 B^2 - \dots \gamma_p B^p$. More general schemes involve ARMA($p, q$) errors with

$$\epsilon_t - \gamma_1\epsilon_{t-1} - \gamma_2\epsilon_{t-2} \cdots - \gamma_p\epsilon_{t-p} = u_t - \theta_1 u_{t-1} - \theta_2 u_{t-2} \cdots - \theta_q u_{t-q}.$$

Often appropriate is a model with AR(1) errors $\epsilon_t$ (see Example 6.4), namely

$$y_t = X_t\beta + \epsilon_t,$$

$$\epsilon_t = \gamma\epsilon_{t-1} + u_t.$$

To facilitate estimation, the AR(1) error model may be re-expressed in non-linear autoregressive form (De Pooter *et al.*, 2006), for observations $t > 1$ subsequent to the first, and with iid errors $u_t$,

$$y_t = \gamma y_{t-1} + \alpha - \alpha\gamma + X_t\beta - X_{t-1}\gamma\beta + u_t = \gamma(y_{t-1} - X_{t-1}\beta) + \alpha(1 - \gamma) + X_t\beta + u_t.$$

The intercept in the original model is obtained by dividing the intercept in the transformed data model by $1 - \gamma$.

## 6.2.2  Bayesian priors in ARMA models

Among the questions involved in specifying priors for ARMA model parameters are whether stationarity and invertibility constraints are formally included, whether a full or conditional likelihood approach is used, and assumptions made about the innovation errors. Consider the AR($p$) model,

$$\rho(B)y_t = u_t.$$

Unlike classical approaches, a Bayesian analysis of this model is not confined to stationary processes (Steel, 2008; Koop *et al.*, 1995), and may be applied to observations $y_t$ without pre-differencing, with the probability of stationarity assessed by the proportion of iterations $s = 1, \dots, S$ where stationarity in the coefficients $\rho^{(s)}$ at iteration s actually holds. A significant probability of non-stationarity might then imply the need for differencing, different error assumptions, or model elaboration, for example, to a higher order AR model (Naylor and Marriott, 1996).

For $p > 1$, Schur's theorem may be used to check non-stationarity in an AR($p$) regression (in conjunction with priors not constrained to stationarity) within an MCMC sampling sequence. Thus a high probability of non-stationarity occurs if any of the $p$ coefficients has posterior mean NonStat[k] exceeding 0.9 in the following BUGS code:

```
model {         a[1,1] <- -1
                  for (k in 1:p)  { a[k+1,1] <- rho[k]
       for (j in 1:p+1-k) {  b[j,k] <- a[1,k]*a[j,k]-a[p+2-k,k]*a[p+3-k-j,k]
                  a[j,k+1] <- b[j,k]}
                  NonStat[k] <- step(-b[1,k])}}
```

In the absence of a stationarity constraint, one option is a diffuse or reference prior on $\{\rho, \tau\}$, where $\tau = 1/\sigma^2$ and $\rho = (\rho_1, \ldots, \rho_p)$. An example is Jeffrey's prior (Zellner, 1971), with

$$\pi(\rho, \tau) \propto \tau^{-1}.$$

As with any non-informative prior, potential problems of identifiability may be increased, whereas identifiability generally improves as just proper or informative proper priors are adopted. Kleibergen and Hoek (2000) discuss local identifiability issues for the ARMA(1, 1) model when diffuse priors are used.

Normal–gamma conjugate priors (Broemeling and Cook, 1993) for the AR($p$) model involve a gamma prior for $\tau$, with $\rho|\tau$ then multivariate normal $N(r_0, \tau\Sigma_0)$, where $r_0$ is the prior mean on $\rho$, and $\Sigma_0$ a $p \times p$ positive definite matrix. A straightforward analysis is defined by conditioning on the first $p$ observations $Y_1 = \{y_1, y_2, \ldots, y_p\}$, so avoiding the specification of a prior on the latent pre-series value. The likelihood then only relates to observations $Y_2 = \{y_{p+1}, y_{p+2}, \ldots, y_n\}$, namely

$$p(Y_2|Y_1, \rho, \tau) \propto \tau^{0.5(n-p)} \exp\left(-0.5\tau \sum_{t=p+1}^{n} [\rho(B) y_t]^2\right).$$

Naylor and Marriott (1996) and Marriott *et al.* (1996) discuss full likelihood analysis of the ARMA model using proper but diffuse priors on latent pre-series values $Y_0 = (y_0, y_{-1}, \ldots, y_{1-p})$ and $E_0 = (u_0, u_{-1}, \ldots, u_{1-q})$. For instance, if the observed series is assumed normal with mean $\mu$ and conditional variance $\sigma^2$, the pre-series values $Y_0$ may be taken as Student $t$ with low degrees of freedom, having mean $\mu$ but residual variance larger by a factor $\kappa \geq 1$, namely $\kappa\sigma^2$. If there are several pre-series values, a multivariate $t$ would be used.

### 6.2.2.1   Enforcing stationarity and alternatives

One option to enforce stationarity involves rejection sampling, for example, with samples of AR($p$) coefficients $\rho$ accepted (as a block) if they lie in the acceptable region (Chib and Greenberg, 1994). Another option consistent with stationarity is reparameterisation in terms of the partial correlations $r_j$ of the AR($p$) process (Marriott and Smith, 1992; Marriott et al., 1996), with stationarity equivalent to restrictions that $|r_k| < 1$ for $k = 1, 2, \ldots, p$.

In the AR($p$) model let $\rho^{(p)} = (\rho_1^{(p)}, \rho_2^{(p)}, \ldots, \rho_p^{(p)})$ with $\rho_j^{(p)}$ the $j$th AR coefficient. The transformations linking the partial autocorrelation coefficients $r_j$ to the AR coefficients $\rho_j$ are then for $k = 2, \ldots, p$, and $i = 1, \ldots, k-1$,

$$\rho_k^{(k)} = r_k,$$

$$\rho_i^{(k)} = \rho_i^{(k-1)} - r_k \rho_{k-i}^{(k-1)}.$$

So for $p = 3$ the transformations would be

$$\rho_3^{(3)} = r_3,$$

$$\rho_1^{(3)} = \rho_1^{(2)} - r_3\rho_2^{(2)} = \rho_1^{(2)} - r_3r_2 \text{ (for } k = 3, \ i = 1),$$

$$\rho_2^{(3)} = \rho_2^{(2)} - r_3\rho_1^{(2)} = r_2 - r_3 \ \rho_1^{(2)} \text{ (for } k = 3, \ i = 2),$$

$$\rho_1^{(2)} = \rho_1^{(1)} - r_2\rho_1^{(1)} = r_1 - r_2r_1 \text{ (for } k = 2, \ i = 1).$$

Hence the stationary AR(3) coefficients are

$$\rho_1^{(3)} = \rho_1^{(2)} - r_3 \ r_2 = r_1 - r_2r_1 - r_3r_2,$$

$$\rho_2^{(3)} = r_2 - \ r_3 \ \rho_1^{(2)} = r_2 - \ r_3(r_1 - r_2r_1),$$

$$\rho_3^{(3)} = r_3.$$

To implement this transformation, $U(-1, 1)$ priors may be assumed directly on the $r_j$. Alternatively letting $r_k^*$ be normal or uniform on the real line, the $r_k$ are obtained as $r_k^* = \log([1 + r_k]/[1 - r_k])$. The $r_k^*$ may also be assigned beta priors, with transformation then $r_k = 2r_k^* - 1$. Priors on the $\theta$ coefficients consistent with invertibility may be obtained by a parallel reparameterisation, for $k = 1, \ldots, q$ and $i = 1, \ldots, k - 1$ (Monahan, 1984; Marriott *et al.*, 1996).

If a stationarity constraint is not imposed, then a flat prior may be chosen for $\{\rho_1, \ldots, \rho_p\}$, but proper priors, such as

$$\rho_j \sim N(0, 1), \quad j = 1, \ldots, p$$

provide a relatively vague but proper alternative (Prado and West, 2010). Adopting priors on the $\rho_j$ that had larger variances would neglect the typical pattern of autoregressive coefficients on the endogenous variable, with values exceeding 1 being uncommon except in short term explosive series. More specialised priors on coefficients at successive lags may be applied, as in vector autoregressions. Thus Litterman (1986) specifies a prior mean of unity for the first own-lag coefficient (relating $y_{kt}$ to $y_{k,t-1}$), but zero prior means for subsequent own-lag coefficients, and for all cross variable lags.

Akaike (1986) discusses smoothness priors on differences in successive distributed lag coefficients $\beta_m$, and the same idea can be applied to autoregressive coefficients. Distinctive approaches to coefficient shrinkage take account of expected decay in coefficients at higher lags (Prado and West, 2010). Thus smoothness priors of the form

$$\rho_1 \sim N(0, \omega)$$

$$\rho_k \sim N(\rho_{k-1}, \omega/\delta_k), \quad \delta_k > 1, \quad k > 1,$$

may be used to induce smoothness at higher lags. A known form such as $\delta_k = k^2$ may be applied, or a prior adopted such as

$$\delta_k = 1 + \eta_k,$$

$$\eta_k \sim Ga(1, b)$$

with $b$ small.

For AR coefficients modelled as fixed effects, model selection may be applied using procedures such as those of George and McCulloch (1993) or Kuo and Mallick (1998), and discussed in Chapter 3. For instance, if $J_k = 1$ or $J_k = 0$ according as the $k$th autoregressive coefficient $\rho_k$ is included or excluded in an AR($p$) regression, then one might set prior probabilities that $J_k = 1$ which decline as the lag $k$ increases (Barnett *et al.*, 1996).

### 6.2.2.2   Representing outliers or shifts

Standard assumption of homoscedastic errors in the AR($p$) model may need to be adapted to accommodate outliers, since, especially in economic time series, they may reflect aspects of economic behaviour which should be included in the specification (Thomas, 1997). Greater robustness may be obtained by adopting a normal mixture distribution or Student $t$ errors to replace the usual normal error assumption. Thus let $\Delta$ be the small probability of an outlier (e.g. $\Delta = 0.05$), and let the binary indicator

$$J_t \sim \text{Bernoulli}(\Delta)$$

govern whether the observation $t$ is an outlier. Then one alternative (West, 1996) to the standard AR(1) model is an innovation outlier model

$$y_t = \mu + \rho y_{t-1} + u_t,$$

where $u_t \sim N(0, K_t \sigma^2)$, and where random or fixed effect parameters $K_t > 1$ inflate the variance when $J_t = 1$, leading to the error scheme

$$u_t \sim (1 - \Delta)N(0, \sigma^2) + \Delta N(0, K_t \sigma^2).$$

If a Student $t$ density is used as a prior for the innovations, then an appropriate option is the scale mixture form, with weights $w_t$ averaging 1 to scale the overall precision parameter, and with low weights (e.g. under 0.5) indicating possible outliers.

One may also define additive outliers (Barnett *et al.*, 1996) corresponding to occasional shifts in the observational series, as well as innovation outliers, as in

$$y_t = X_t \beta + o_t + \epsilon_t,$$

$$\epsilon_t - \gamma_1 \epsilon_{t-1} = u_t,$$

$$o_t \sim N(0, K_{1t} \sigma^2),$$

$$u_t \sim N(0, K_{2t} \sigma^2),$$

where $K_{1t}$ is either 0 or positive (corresponding to times when an additive outlier does or does not occur), and $K_{2t}$ is either 1 or greater than 1. One might then model the two outliers jointly, for instance via a set of possible paired values for $K_t = \{K_{1t}, K_{2t}\}$. So $K_t$ might consist of (0, 1), (3, 1), (10, 1), (0, 3), (0, 10) with selection among pairs based on a multinomial indicator $J_t$, with prior probabilities biased towards the null option (0, 1). For example, prior probabilities on the preceding options might be (0.9, 0.025, 0.025, 0.025, 0.025).

Another approach to additive outliers is illustrated by a random level-shift autoregressive (RLAR) model

$$y_t = \mu_t + \epsilon_t,$$

$$\mu_t = \mu_{t-1} + \delta_t \eta_t,$$

$$\epsilon_t = \gamma_1 \epsilon_{t-1} + \gamma_2 \epsilon_{t-2} + \ldots + u_t,$$

where $\delta_t$ is Bernoulli with probability $\Delta$, governing the chance of a level shift at time $t$, the terms $\eta_t \sim N(0, \xi^2)$ describe the shifts, and $u_t$ are iid $N(0, \sigma^2)$ (McCulloch and Tsay, 1994). The shift variance $\xi^2$ is taken as a large multiple (e.g. 10) of the noise variance $\sigma^2$. The probability of a shift $\Delta$ is beta with parameters favouring low probabilities, for instance $\Delta \sim$ Beta(5, 95).

### 6.2.3  Further types of time dependence

Other forms of time dependence in the observations may be combined with autoregression on previous values, while still assuming the errors are uncorrelated, especially if stationarity is not assumed a priori.

Thus one feature of many time series is periodic fluctuations (see Examples 6.6 and 6.7). Suppose a series of length $T$ contains $K$ cycles (timed from peak to peak or trough to trough), so that the frequency of cycles per unit of time, is $f = K/T$. An appropriate model for a series with a single cycle is then

$$y_t = A\cos(2\pi ft + P) + u_t,$$

where $A$ is the amplitude and $P$ the phase of the cycle, and period $1/f$, namely the number of time units from peak to peak. To allow for several ($r$) frequencies operating simultaneously the preceding may be generalised to

$$y_t = \sum_{j=1}^{r} A_j \cos(2\pi f_j t + P_j) + u_t$$

For stationarity to apply, the $A_j$ may be taken as uncorrelated with mean 0 and the $P_j$ as uniform on $(0, 2\pi)$. Because of the relationship $\cos(2\pi f_j t + P_j) = \cos(2\pi f_j t)\cos(P_j) - \sin(2\pi f_j t)\sin(P_j)$, the model is equivalently written

$$y_t = \sum_{j=1}^{r} \{\alpha_j \cos(2\pi f_j t) + \beta_j \sin(2\pi f_j t)\} + u_t$$

where $\alpha_j = A_j \cos P_j, \beta_j = -A_j \sin P_j$.

The AR(1) model may also be extended to capture trend by adding dependence in time $t$ (e.g. linear growth) and/or lags in $\Delta y_t$ (Schotman, 1994, Marriott et al., 2004); see Example 6.2. These modifications are intended to improve specification and ensure the assumption of iid errors $u_t$. An example is the extended model

$$y_t = \mu + \rho y_{t-1} + \lambda t + \phi_1 \Delta y_{t-1} + \phi_2 \Delta y_{t-2} + u_t,$$

where $\lambda t$ models a linear trend (Hoek et al., 1995). Bauwens et al. (2000, p. 166) consider an alternative non-linear form of the AR model, also involving a trend in time $\lambda t$. For an AR(1) model this is expressed as

$$(1 - \rho B)(y_t - \mu - \lambda t) = u_t,$$

which can be rewritten as

$$y_t = \rho y_{t-1} + \rho \lambda + (1 - \rho)(\mu + \lambda t) + u_t.$$

When lags in $\Delta y_t$ are introduced, an extended version of the non-linear form is illustrated by the model

$$y_t = \rho y_{t-1} + \rho \lambda + (1 - \rho)(\mu + \lambda t) + \phi_1 \Delta y_{t-1} + \phi_2 \Delta y_{t-2} + u_t.$$

Bayesian tests of non-stationarity in the basic AR(1) model or in these extended versions may follow the classical procedure in testing explicitly for the simple hypothesis that $\rho = 1$ versus the composite alternative that $|\rho| < 1$. Thus a prior for $\rho$ may exclude explosive values, but put prior mass on the unit root $\rho = 1$ (Hoek et al., 1995). Other values of $\rho$ are uniformly distributed with mass $1/(2A)$ between $[1 - A, 1)$ where $1 \geq A > 0$. For instance, taking $A = 1$

gives the prior

$$\pi(\rho) = 0.5 \qquad \rho = 1$$

$$\pi(\rho) = 0.5 \qquad \rho \epsilon [0, 1).$$

Tests of non-stationarity may also (Lubrano, 1995a) compare the composite alternatives

$$H_0: \rho \geq 1 \text{ as against } H_1: \rho < 1.$$

The posterior probability that $\rho \geq 1$ is then a test for non-stationarity.

### Example 6.1   Oxygen inhalation

This example considers an AR(1) model

$$y_t \sim N(c + \rho y_{t-1}, \sigma^2),$$

without a prior stationary constraint, applied to undifferenced data[1] on a burns patient (Broemeling and Cook, 1993). The series consists of 30 readings $y_t$ of the volume of oxygen inhaled at 2 minute intervals (Figure 6.1). Two forms of just proper gamma prior for $\tau = 1/\sigma^2$ are considered, the first being $\tau \sim Ga(0.001, 0.001)$, which approximates $\pi(t) \propto \tau^{-1}$, and the second being $\tau \sim Ga(1, 0.001)$, which approximates a uniform prior. The prior for $\mu$ is taken as $N(300, 10^8)$, weakly reflecting the average level of the observations, and the prior for $\rho$ is $\rho \sim N(0, 1)$. Since the outcome is positive, normal sampling for $y$ is truncated below at zero.

A conditional likelihood (conditioning on $y_1$) is initially adopted, with three chains with dispersed starting values (null start values, and values provided by 2.5th and 97.5th percentiles of a trial run). Inferences are from the 2nd half of a two chain sequence of 20 000 iterations. With the prior $\tau \sim Ga(0.001, 0.001)$, posterior means (sd) for $\mu$ and $\rho$ are 407 (16) and 0.287 (0.18) respectively. The probability of non-stationarity is negligible at 7E-4. With the

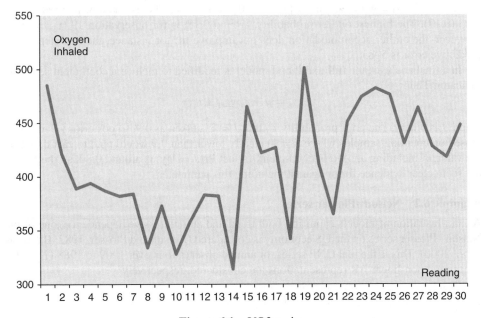

**Figure 6.1**   VO2 series.

alternative prior $\tau \sim Ga(1, 0.001)$, $\rho$ is estimated as 0.295 (0.18) and posterior summaries are very similar for other parameters.

The Durbin–Watson statistic (DW in Program 6.1) is used to assess auto-correlation in the residuals of the fitted model and is approximately 2 when there is no correlation. The 95% intervals for the Durbin–Watson statistic include the value of 2 under both priors for $\tau$. Another option to assess the validity of the iid assumption on $u_t$ is an analysis of the posterior residual estimates, $\hat{u}_t = y_t - \hat{y}_t$. For example, an effectively zero correlation parameter between $\hat{u}_t$ and $\hat{u}_{t-1}$ would be consistent with white noise.

In terms of fit criteria, the conditional likelihood approach yields, with $\tau \sim Ga(0.001, 0.001)$, a LPML of $-155.7$ (Gelfand and Dey, 1994), a predictive loss criterion (PLC) of 139 230 (Gelfand and Ghosh, 1998), and a mean square prediction error (one-step-ahead) of 4868. The individual CPOs, scaled to a maximum of 1, show the lowest scaled CPO for $y_{19}$, namely 0.036.

For a full likelihood approach, again without a stationary constraint (model 2), it is necessary to specify priors on the latent pre-series data or errors. Define $m_t = c + \rho y_{t-1}$. Then $y_0$ is assumed to be Student $t$, $y_0 \sim t(\overline{m}, \kappa_0 \sigma^2, 2)$, where $\overline{m}$ is an estimator of the overall mean of the data, namely the realised average of the $m_t$ over readings $t = 2, \dots, T$ (i.e. those readings where $m_t$ is based on observed $y_t$). The multiplier $\kappa_0$ is assigned a $Ga(0.01, 0.01)$ prior, constrained to values over 1. From OpenBUGS, the posterior mean (sd) of $\rho$ is obtained as 0.324 (0.17), and the posterior mean for $\kappa_0$ as 1.04. The one step predictive error as compared to the conditional likelihood model (and defined over the same time points) now stands at 4752, the predictive loss criterion is now 136 190, and the LPML (over observations 2, $\dots$, 30) is increased to $-150.6$. The pre-series value is estimated at 456 (sd = 95) compared to $y_1 = 485$.

A further elaboration (model 3) retains the latent $y_0$, and also allows for outliers, with a normal mixture replacing the standard error assumption. Thus

$$u_t \sim (1 - \Delta)N(0, \sigma^2) + \Delta N(0, \kappa \sigma^2),$$

where $\kappa > 1$ inflates the variance for outliers. The prior outlier probability $\Delta$ is set to 0.05, and a discrete prior with ten values $\{1, \dots, 10\}$ is set on $\kappa$. The estimate of $\rho$ is reduced to 0.23 (0.17). The highest outlier probability, around 0.18, is for observation 19 ($y_{19} = 501$). However, the outlier accommodation does not improve fit; for instance, the one-step-ahead predictive error is 5083.

In a final analysis, the full likelihood model is modified to include a coefficient selection indicator. Thus

$$y_t = \mu + J\rho y_{t-1} + u_t,$$

where $J$ is binary. The prior probability $\pi_J$ that $J = 1$ is taken as 0.9, so posterior values of $\pi_J$ (based on counting samples where $J = 1$) clearly lower than 0.9 would tend to cast doubt on the value of including an AR(1) coefficient. In fact $\Pr(J = 1|y)$ is around 0.80, so there does not seem clear evidence for or against including this parameter.

### Example 6.2    Nelson–Plosser series

As an illustration of models capturing both trend and autoregressive dependence, one of the Nelson–Plosser series on the US economy is considered (Nelson and Plosser, 1982; Bauwens *et al.*, 2000). This is the real GNP series, of annual observations over 1909 to 1988 ($T = 80$). Linear and non-linear AR representations are considered, respectively

$$y_t = \mu + \rho y_{t-1} + \lambda t + \phi_1 \Delta y_{t-1} + \phi_2 \Delta y_{t-2} + u_t,$$

$$y_t = \rho y_{t-1} + \rho \lambda + (1 - \rho)(\mu + \lambda t) + \phi_1 \Delta y_{t-1} + \phi_2 \Delta y_{t-2} + u_t,$$

with full likelihoods referring to latent pre-series values assumed. Student $t$ innovation errors via a normal scale mixture with variance $\sigma^2$, $v$ degrees of freedom, and weights $w_t \sim \text{Ga}(0.5v, 0.5v)$ are adopted. A gamma prior for the degrees of freedom is taken with sampling constrained to [1,100], so encompassing both the Cauchy density ($v = 1$) and an effectively normal density ($v = 100$). A factor $\kappa > 1$ scales up the main residual variance in a way appropriate for the Student $t$ distributed pre-series latent data, while the mean for pre-series values follows the principle set out in Example 6.1.

Summaries are based on the second half of two chains with 10 000 iterations. The GNP series is stationary according to the linear model, with zero probability that $\rho > 1$ (Table 6.1). However, GNP is marginally non-stationary (with probability of 0.15 that $\rho > 1$) under the non-linear model. Fit measures are inconclusive: the linear model has better one-step-ahead prediction error but lower LPML (114.0 vs 115.6) than the non-linear model. Figure 6.2 plots $y_t$ against the weights of the scale mixture from the non-linear model: lower weights apply to observations at odds with the remainder of the data. The lowest weight (posterior mean of 0.56) is for the depression year 1932.

As a third modelling approach, an additive outlier version[2] of the linear model is applied, namely

$$y_t = \mu + \rho(y_{t-1} - o_{t-1}) + \lambda t + \phi_1 \Delta z_{t-1} + \phi_2 \Delta z_{t-2} + u_t,$$

$$o_t = \delta_t \eta_t,$$

$$\delta_t \sim \text{Bern}(\Delta),$$

$$\Delta \sim \text{Be}(1, 19).$$

**Table 6.1**  Alternative models for GNP series.

| Linear AR | Mean | Standard deviations | 2.5% | 50% | 97.5% |
|---|---|---|---|---|---|
| $Prob(\rho > 0)$ | 0 | | | | |
| $\lambda$ | 0.005 | 0.002 | 0.002 | 0.005 | 0.008 |
| $v$ | 21.4 | 24.7 | 2.5 | 8.9 | 89.4 |
| $\phi_1$ | 0.32 | 0.09 | 0.15 | 0.32 | 0.51 |
| $\phi_2$ | −0.10 | 0.07 | −0.23 | −0.10 | 0.07 |
| $\rho$ | 0.84 | 0.05 | 0.75 | 0.84 | 0.94 |
| *Non-linear* | | | | | |
| $Prob(\rho > 0)$ | 0.154 | | | | |
| $\lambda$ | 0.031 | 0.029 | −0.051 | 0.032 | 0.090 |
| $v$ | 28.4 | 28.4 | 2.6 | 15.0 | 94.2 |
| $\phi_1$ | 0.33 | 0.09 | 0.15 | 0.33 | 0.51 |
| $\phi_2$ | −0.11 | 0.07 | −0.25 | −0.11 | 0.03 |
| $\rho$ | 0.95 | 0.06 | 0.80 | 0.98 | 1.01 |
| *Linear AR, Additive Outlier Model* | | | | | |
| $Prob(\rho > 0)$ | 0.001 | | | | |
| $\lambda$ | 0.005 | 0.001 | 0.002 | 0.005 | 0.008 |
| $v$ | 21.2 | 24.3 | 2.2 | 9.4 | 88.5 |
| $\phi_1$ | 0.35 | 0.10 | 0.17 | 0.35 | 0.54 |
| $\phi_2$ | −0.11 | 0.07 | −0.24 | −0.11 | 0.04 |
| $\rho$ | 0.85 | 0.05 | 0.76 | 0.85 | 0.94 |

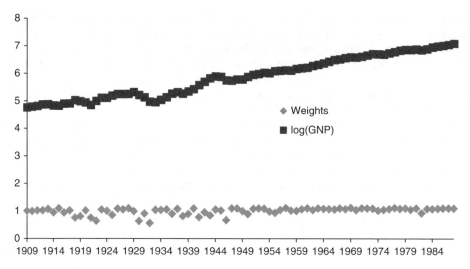

**Figure 6.2**   GNP series and scale mixture weights, 1909–1988.

with $o_t$ for years preceding the series taken as zero, and $\text{var}(\eta_t) = 10\ \text{var}(u_t)$. The posterior mean for $\Delta$ is 0.036, and taking evidence of outlier status as $\Pr(\delta_t = 1|y) > 5\Delta$, the years 1918, 1921, and 1938 have probabilities $\Pr(\delta_t = 1|y)$ exceeding this threshold. This model has the lowest one step predictive error and also the highest LPML (namely 116.1).

### Example 6.3   Wholesale prices

Consider data from Enders (2004, p. 87) on the US Wholesale Price Index (WPI), specifically quarterly data from 1960Q1, with $y_1 = 30.7$, through to 1990Q4, with $y_{224} = 116.2$. Consider an ARMA(1, 1) model in the first differences $z_t = y_{t+1} - y_t$. The model may be written

$$z_t = c + \phi z_{t-1} + \theta u_{t-1} + u_t = \mu_t + u_t, \qquad u_t \sim N(0, \sigma^2).$$

The errors at may be represented recursively as

$$u_t = z_t - c - \phi z_{t-1} - \theta u_{t-1},$$

starting with

$$u_1 = z_1 - c - \phi z_0 - \theta u_0.$$

To set the recursion going, one may generate latent pre-series values $\{z_0, u_0\}$ using a diffuse prior on $z_0$ and assuming $u_0 \sim N(0, \sigma^2)$. The log likelihood values for quarter $t$ are then

$$\log(L_t) = -0.5 \log(2\pi\sigma^2) - 0.5u_t^2/(2\sigma^2),$$

and one may use the dloglik option in OpenBUGS. Alternatively (method 2) one may represent the lagged residual as $u_{t-1} = z_{t-1} - \mu_{t-1}$ (e.g. Marriott *et al.*, 1996) so that

$$z_t \sim N(c + \phi z_{t-1} + \theta(z_{t-1} - \mu_{t-1}), \sigma^2).$$

Flat priors are adopted on $c$ and $z_0$, a uniform prior on $\sigma$, and priors on $\phi$ and $\theta$ are constrained to stationarity and invertibility. The second halves of two chain runs of 10 000 iterations provide estimates (95% CRI) for $c, \phi, \theta$ and $\sigma$ as in Table 6.2.

**Table 6.2** ARMA estimates, WPI data.

| Method 1 | Mean | 2.5% | 97.5% |
|---|---|---|---|
| $c$ | 0.13 | 0.02 | 0.29 |
| $\phi$ | 0.84 | 0.64 | 0.96 |
| $\theta$ | -0.33 | -0.56 | -0.04 |
| $\sigma$ | 0.75 | 0.66 | 0.85 |
| Method 2 | Mean | 2.5% | 97.5% |
| $c$ | 0.14 | 0.01 | 0.30 |
| $\phi$ | 0.83 | 0.64 | 0.96 |
| $\theta$ | -0.32 | -0.56 | -0.02 |
| $\sigma$ | 0.74 | 0.65 | 0.85 |

Method 2 has the advantage of allowing forecasts outside the sample period, and forecasts of $z_t$ for the four quarters of 1991 are all positive, consistent with a continuing rise in prices. The posterior means for the four quarters of 1991 are 2.25, 2.01, 1.82 and 1.65. It may be noted that if priors on $\phi$ and $\theta$ are adopted that are not constrained to stationarity, such as $N(0, 1)$ priors on $\phi$ and $\theta$, or on $\delta = \phi - \theta$ (Kleibergen and Hoek, 2000), then posterior means on $\theta$ are closer to zero.

### Example 6.4  Investment levels by firms

Maddala (1979) compares several procedures for predicting investment levels by a set of US firms, where autocorrelated errors are apparent. Maddala's investigation derives from an earlier study by Grunfeld and Griliches (1960) on the validity of using aggregate data to draw inferences about micro-level economic functions (e.g. consumption and investment functions). The autoregressive errors model is applied to gross investment in years $t$ ($t = 1$ in 1935 through to $t = 20$ in 1954) by a particular firm, General Motors, which is related to lagged levels of value $V_{t-1}$ and capital stock $C_{t-1}$.

The first model (model 1) assumes AR(1) dependence in the errors combined with a stationary assumption, namely a uniform prior on the autoregressive parameter, $\rho \sim U(-1, 1)$. Then for years $t > 1$,

$$y_t = \beta_0 + \beta_1 V_{t-1} + \beta_2 C_{t-1} + \epsilon_t,$$

$$\epsilon_t - \rho \epsilon_{t-1} = u_t,$$

with $u_t \sim N(0, \tau^{-1})$ being iid. This model can be re-expressed (for $t > 1$) as

$$y_t = \rho y_{t-1} + \beta_0 (1 - \rho) + \beta_1 (V_{t-1} - \gamma V_{t-2}) + \beta_2 (C_{t-1} - \gamma C_{t-2}) + u_t.$$

The model for year 1 can then be written

$$y_1 = \beta_0 + \beta_1 V_{t-1} + \beta_2 C_{t-1} + \epsilon_1$$

$$\epsilon_1 \sim N(0, 1/\tau_1), \quad \tau_1 = (1 - \rho^2)\tau.$$

One-step-ahead forecasts include known $X_{t+1}$.

The posterior estimates of the parameters $\beta_1$ and $\beta_2$ (from the second half of a two chain run of 10 000 iterations) are close to the maximum likelihood estimates cited by Maddala (1979). They show both coefficients to be above zero, with $\beta_1$ and $\beta_2$ having means and 95% credible

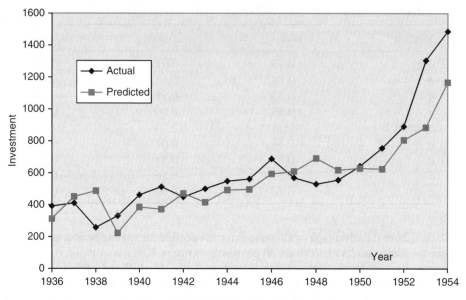

**Figure 6.3**    Within series forecasts.

interval 0.079 (0.041, 0.117) and 0.41 (0.25, 0.56) respectively. The autoregressive coefficient $\rho$ is estimated to have mean 0.77 with 95% credible interval (0.38, 0.99). The density plot for $\rho$ shows a spike at 1 indicating that the stationarity assumption may be in doubt.

A second model (model 2) accordingly avoids assuming stationarity in the errors, with prior $\rho \sim N(0, 1)$ on the autoregressive parameter. The model for the first year observation now involves a separate prior

$$y_1 \sim N(\mu_1, V_1),$$

with $V_1$ set large, and mean

$$\mu_1 = \beta_0(1 - \rho) + \beta_1 V_{t-1} + \beta_2 C_{t-1} + M_0,$$

where $M_0$ is a composite parameter for the missing term $\rho(y_0 - \beta x_0)$. There is a 23% chance of $\rho$ exceeding 1, and its mean is 0.88 with 97.5% point 1.21. The within series one-step-ahead forecasts under this model are shown in Figure 6.3, and trace the observed series reasonably well. The LPML is very similar between the two models ($-113.1$ under model 2, as against $-112.8$ under model 1).

**Example 6.5    Swedish unemployment and production: Bivariate forecasts**

To illustrate vector autoregressive models consider a bivariate series on unemployment and industrial production (Kadiyala and Karlsson, 1997). This consists of quarterly data for Sweden between 1978.1 and 1992.4, with the production index in logged form. A lag four model in each component of the bivariate outcome is adopted. It may be anticipated that some of the cross-variable lags are negative, with increased unemployment expected as production falls. There are also seasonal influences on both variables, and therefore seasonal dummies are included as exogenous predictors. The analysis uses $T = 52$ quarters of the series from 1979.1 to 1991.4, conditioning on 1978.1 to 1978.4, and then predictions made for 1992.

Experimentation here with an undifferenced series analysis showed poor identifiability and convergence on the cross-variable lag effects. Identifiability was improved by modelling aspects of trend, such as by polynomials in time to take account of the upward shift in unemployment in the early 1990s, following a decline during the 1980s. However, convergence remained problematic. By contrast, analysis of differenced series was less subject to convergence problems. For this analysis $N(0, 1)$ priors are taken on all lags, including first own lags, without stationarity constraints. Priors on seasonal effects take the first quarter as reference (with coefficient zero).

In model 1, all cross variable lags (unemployment on production and vice versa) are included and the residuals $u_{kt}$ are taken as bivariate normal. With a two chain run of 10 000 iterations (and burn-in of 1000), the clearest defined lags are for unemployment on its own third and fourth lags, with posterior means (standard deviations) of 0.34 (0.15) and 0.28 (0.16). All cross lags of unemployment on production are negative, with the mean of the second cross lag being around $-1.20$ (s.d. $= 0.78$). Lags of production on unemployment are insignificant (close to zero). The cross-variable correlation in the residuals is estimated at around $-0.15$ (with 95% credible interval ($-0.44, 0.18$). Short term forecasts of unemployment for the four quarters of 1992 match the observed rise.

In a reduced model (model 2) the lags of production on unemployment are set to zero. This produces an improvement in marginal likelihood (and in other fit criteria) and no great change in the pattern of remaining coefficients or the forecasts for 1992.

## 6.3    Discrete outcomes

Time series models for discrete, especially count, data have attracted considerable recent interest – see Jung *et al.* (2006), Fokianos (2011), Jung and Tremayne (2011), McKenzie (2003), Fokianos (2012), and the references listed at https://ites.google.com/site/integervaluedtimeseries/bibliography/latest-papers. Bayesian perspectives are illustrated by Brandt and Sandler (2012), Cargnoni *et al.* (1997), Farrell *et al.* (2007), McCabe and Martin (2005), Yelland (2009) and Silva *et al.* (2005).

For such outcomes dependence on past observations and predictors may be handled by adapting metric variable methods within the appropriate regression link. Thus for Poisson counts, $y_t \sim \text{Po}(\mu_t)$, autoregressive dependence on previous values in the series could be specified directly in the observations, with for first order lag being

$$\log(\mu_t) = \rho y_{t-1} + X_t \beta,$$

and with non-stationarity implied by $\rho > 0$ (Fahrmeir and Tutz, 2001, p. 244; Fokianos, 2012). So in an MCMC framework stationarity can be assessed by the proportion of iterations with $\rho$ positive.

For Poisson (or binomial) time series regression, it might be sensible that the lagged outcome is in the same form as the link transform. Thus a symmetry with the log link for Poisson counts $y_t \sim \text{Po}(\mu_t)$, involves a lag in the log of a transformed $y_{t-1}$, denoted $y'_{t-1}$,

$$\log(\mu_t) = X_t \beta + \rho \log(y'_{t-1})$$

with transformations being $y'_{t-1} = y_{t-1} + c$ with $c > 0$, or $y'_{t-1} = \max(c, y_{t-1})$ with $0 < c < 1$ (Cameron and Trivedi, 1999). Either $c$ is unknown or assigned a default value such as $c = 1$. The autoregression may be extended to include lags in $v_t = \log(\mu_t)$ (Fokianos, 2012).

Analogues to autoregressive errors may be considered, such as a lag 1 model for count response (Zeger and Qaqish, 1988)

$$\log(\mu_t) = \beta x_t + \gamma[\log(y'_{t-1}) - X_{t-1}\beta].$$

Autoregression on both past observations and means is included in autoregressive conditional Poisson (ACP) models (e.g. Fokianos et al., 2009), with Poisson means in an ACP($p$, $q$) model specified as

$$\mu_t = \omega + \sum_{j=1}^{p} \alpha_j y_{t-j} + \sum_{k=1}^{q} \beta_j \mu_{t-k}$$

with all parameters positive. Under an ACP(1, 1) model

$$\mu_t = \omega + \alpha y_{t-1} + \beta \mu_{t-1},$$

stationarity is obtained when $\alpha + \beta < 1$, and defining $\Delta = 1 - (a + \beta)^2$, the unconditional mean and variance are given by

$$E(y_t) = \mu = \omega/(1 - (\alpha + \beta))$$

$$\text{Var}(y_t) = \mu(\Delta + \alpha^2)/\Delta.$$

Autoregression may also be applied in the errors for overdispersed count data (Jung et al., 2006; Jung and Tremayne, 2011), leading to a specification such as

$$\log(\mu_t) = X_t\beta + \epsilon_t,$$

$$\epsilon_t = \gamma \epsilon_{t-1} + u_t \ (t > 1)$$

with $u_t$ being iid, and with stationarity requiring $|\gamma| < 1$.

For binary data, one may specify direct lags in the response. So for $y_t \sim \text{Bern}(\pi_t)$, a logit regression with a single lag in $y_t$ is

$$\text{logit}(\pi_t) = X_t\beta + \rho y_{t-1}.$$

Examples of binary logit or probit time series regression occur in predictions of economic recession (e.g. Kauppi and Saikkonen, 2008; Startz, 2008; Sephton 2009). For such models lags in the predicted probability (or the discrepancy between actual and predicted probabilities) as well as the lagged response have been found to improve fit, as, for example, in

$$\text{logit}(\pi_t) = X_t\beta + \rho_1 y_{t-1} + \rho_2 \pi_{t-1},$$

or

$$\text{logit}(\pi_t) = X_t\beta + \rho y_{t-1} + \theta(y_{t-1} - \pi_{t-1}).$$

Exercise 6.6 exemplifies an application of this kind.

For multi-category data with $K$ categories there are $K - 1$ free category probabilities and these might be related to lagged values on up to $K - 1$ dummy variables (or predicted probabilities). This leads to models similar to VAR($p$) models for multivariate metric outcomes, in that there are own category lags and cross category lags (Pruscha, 1993; Kedem and Fokianos, 2002, Ch 3). For example, consider the scheme

$$y_t \sim \text{Categorical}(\pi_t),$$

where both $y_t$ and $\pi_t$ are of dimension $K$. The probability $\pi_{tk}$ that the $t$ th value of the series is in category $k$ is a function of lagged binary indicators $D_{t-1,k}$ ($= 1$ if $y_{t-1} = k$) via a multiple logit regression. Thus with first category as reference,

$$\pi_{tk} = \exp(\eta_{tk}) / \left[ 1 + \sum_{k=2}^{K} \exp\left(\eta_{tk}\right) \right]$$

where

$$\eta_{tk} = X_t \beta_k + \rho_{k2} D_{t-1,2} + \cdots \rho_{kK} D_{t-1,K},$$

and the lag coefficients represent impacts of transition between states.

Another approach, especially for binary, ordinal and categorical time series, invokes an underlying metric variable. This approach has been suggested for Poisson count data (van Ophem, 1999) but a more frequent application is for binary data (Keenan, 1982; Kedem and Fokianos, 2002, p. 50). Thus positive values of the latent series $z_t$ correspond to $y_t = 1$ and negative values to $y_t = 0$. One might then posit an underlying smooth signal $\theta_t$, such that

$$z_t = \theta_t + u_t,$$
$$\theta_t = \rho \theta_{t-1} + u_t,$$

with $|\rho| < 1$ corresponding to stationarity. An AR1 model in the latent data would be

$$z_t = \mu + \rho z_{t-1} + u_t,$$

with $u_t \sim N(0, 1)$, but may lead to endogeneity, and autoregressive dependence in errors considered as an alternative (Pang, 2010).

### 6.3.1 INAR models for counts

Integer valued autoregressive (INAR) schemes are oriented to discrete outcomes and have an affinity with ARMA models for metric outcomes (McKenzie, 1986; McCabe and Martin, 2005). In particular their specification often includes devices to ensure stationarity (e.g. of the underlying mean count through time), though a Bayesian approach may make this constraint less necessary. INAR schemes introduce dependence of the current count $y_t$ on previous counts $y_{t-1}$, $y_{t-2}$, ..., etc., and also allow an integer valued innovation series $w_t$. The autoregressive component of the model can be seen as a survival model to time $t$ for each particle in the previous total counts $y_{t-1}, y_{t-2}, \ldots$, etc.

Thus for an INAR(1) model, one considers the chance $\rho$ that each of the $y_{t-1}$ particles survives through to the next period. The autoregressive component of the INAR(1) model for $y_t$ is represented as

$$S_t = \sum_{k=1}^{y_{t-1}} \mathrm{Bern}(\rho).$$

Equivalently $S_t$ is binomial with denominator $y_{t-1}$, and $\rho$ the probability of survival. This process is known as binomial thinning and denoted as $\rho^\circ y_{t-1}$. If $y_{t-1} = 0$ then there is no first order lag autoregressive component in the model for $y_t$. In BUGS, the binomial can still be used when $y_t = 0$ so one can code an INAR(1) model using the binomial, rather than program the full thinning operation.

An INAR(2) process would refer to preceding counts $y_{t-1}$ and $y_{t-2}$, and involve two survival probabilities, $\rho_1$ and $\rho_2$. For an INAR($p$) process, stationarity is defined by

$$\sum_{k=1}^{p} \rho_k < 1$$

(Cardinal *et al.*, 1999). For overdispersed data, the survival probabilities, such as $\rho_1$ in an INAR(1) model, may be taken as time varying (McKenzie, 1986).

As well as survival of existing particles there is a birth process, analogous to the innovations of an ARMA model for a metric outcome. Thus new cases $w_t$ are added to surviving cases from previous periods, with $y_t$ then the summation of separate survival and birth processes

$$y_t = \rho^{\circ} y_{t-1} + w_t$$

where to ensure stationarity $w_t$ is Poisson with mean $\theta(1 - \rho)$. Both $\rho$ and $\theta$ may depend on covariates (Enciso-Mora *et al.*, 2009), using logit and log links respectively.

One might also consider unconstrained Poisson densities for $w_t$ (not tied to the survival parameter $\rho$), especially if there is over-dispersion. An example is a mixed Poisson for $w_t$, with two possible means $\lambda_1$ and $\lambda_2$, in an analysis of epileptic seizure counts (Franke and Seligmann, 1993). Switching in the innovation process at time $t$ is determined by latent binary variables $Q_t$. One might also envisage a pseudo INAR approach, namely a conditional Poisson with mean $\mu_t$ composed of a survival term $\rho^{\circ} y_{t-1}$, and an additional series $\omega_t$ following a positive density (e.g. a gamma). For $y_t \sim \text{Po}(\mu_t | \rho, \omega_t)$ this leads to

$$\mu_t = \rho^{\circ} y_{t-1} + \omega_t; \quad \omega_t \sim \text{Ga}(b(1 - \rho), b).$$

### 6.3.2   Evolution in conjugate process parameters

Different sources of randomness may be modelled using conjugate hierarchical schemes (e.g. Poisson–gamma, binomial–beta) with evolution at both levels of the hierarchy. One may thereby capture both changes in the underlying level and changes in the distribution of observations around that level. Thus with Poisson sampling (e.g. Ord *et al.*, 1993), $y_t \sim \text{Po}(\mu_t)$, the means $\mu_t$ are gamma distributed

$$\mu_t \sim \text{Ga}(a_t, b_t),$$

with parameters $(a_t, b_t)$ related to previous parameters $(a_{t-1}, b_{t-1})$ via a continuation parameter $\phi$. This takes values between 0 and 1 and applies to both scale and index:

$$a_t = \phi a_{t-1}; \quad b_t = \phi b_{t-1,}$$

with initial values $a_0, b_0$ assigned a positive valued prior (e.g. log-normal or gamma). To avoid improper priors for later time periods, one may modify this scheme to include a small positive constant – indicating the minimum prior scale and index in the prior for the $\mu_t$. This might be taken as an extra parameter or preset. Thus

$$a_t = \phi a_{t-1} + c; \quad b_t = \phi b_{t-1} + c.$$

It is possible to drop the constraint $\phi < 1$ and assess the probability that $\phi$ is in fact consistent with information loss (i.e. with accumulated discounting of past observations as $\phi < 1$ implies).

This approach may be extended to multivariate count series $\{y_{kt}, k = 1, \ldots, K, t = 1, \ldots, T\}$ by modelling the total count at time t

$$Y_t = \sum_{k=1}^{K} y_{kt}$$

in the same way as a univariate count with parameter $\phi_1$. The disaggregation to the individual series is modelled via a multinomial-Dirichlet model, with the evolution of the Dirichlet parameters governed by a second parameter $\phi_2$.

Gamma distributed evolution parameters are also included in models for multiple count series $y_{1t}, y_{2t}, \ldots, y_{Kt}$ that specify latent factors $F_{mt} (m = 1, \ldots, M; M < K)$ to account for interdependence in the original series (Jorgensen et al., 1999). For example, with $M = 1$, components of the multivariate series are assumed conditionally independent given $F_t$, so that

$$y_{kt} \sim \text{Po}(v_{kt} F_t),$$

where $v_{kt} = \exp(X_{kt} \beta_k)$ includes regressors $X_k$ relevant to each outcome. The latent process $F_t$ evolves as a gamma Markov process

$$F_t \sim \text{Ga}(c_t, d_t).$$

Evolution of $\{c_t, d_t\}$ may follow a stationary scheme as above, or a non-stationary scheme, for example
$$c_t = F_{t-1}/\omega; \quad d_t = 1/[b_t \omega]$$

where $\omega$ is a variance parameter, and the mean of $F_t$ given preceding values $F_0, \ldots, F_{t-1}$ is $b_t F_{t-1}$. So one may modify the impact of the preceding latent value by regressing $b_t$ on covariates $z_t$, as in $\log(b_t) = \alpha z_t$. The $z_t$ are defined as differences $z_t = Z_t - Z_{t-1}$, where $z_t$ are viewed as long term influences, and $X_{kt}$ as short term.

### Example 6.6    Benefit claims

Consider monthly data on injury claimants for cuts and lacerations analysed by Freeland and McCabe (2004), Enciso-Mora et al. (2009) and others. There are $T = 120$ observations between January 1985 and December 1994. Here an INAR lag-1 model with and without seasonal covariates is estimated first. Without covariates one has a constant mean $\lambda$, and data generating process
$$y_t = \rho^{\circ} y_{t-1} + w_t, \quad w_t \sim \text{Po}(\lambda),$$

here with priors $\lambda \sim \text{Ga}(1, 0.001)$, and $\rho \sim \text{Be}(1, 1)$. Denoting $S_t = \rho^{\circ} y_{t-1}$ the likelihood is

$$\prod_{t=2}^{T} (e^{-\lambda} \lambda^{w_t})/w_t! \binom{y_{t-1}}{S_t} \rho^{S_t} (1 - \rho)^{y_{t-1} - S_t}$$

The analysis conditions on the first observation, though it would be possible to set an initialising Poisson prior on $y_0$.

In BUGS, the code[3] for this model constrains the binomial survival component at time $t$ to be less than the value of $y_t$. The remaining part of the likelihood, the Poisson part for the innovations $w_t = y_t - \rho^{\circ} y_{t-1} = y_t - S_t$, is specified using the new sampling distribution option, which in WinBUGS results in

```
for (t in 2:T) {h[t] <- 0
h[t] ~ dpois(negLL[t])
negLL[t] <- -LL[t]
LL[t] <- -lambda + (y[t]-S[t])*log(lambda)-logfact(y[t]-S[t])
S[t] ~ dbin(rho,y[t-1]) I(,y[t])}
```

The second half of a two chain run of 10 000 iterations provides estimates (with 95% CRI) for $\rho$ of 0.43 (0.32, 0.53) and for $\lambda$ of 3.53 (2.88, 4.24). Out-of-sample forecasts for the 12 months in the subsequent year, $t = T + 1, ..., T + 12$, have means 6.24 ($t = 121$), 5.65 ($t = 122$), 5.91 ($t = 123$), stabilising thereafter at between 6.0 and 6.2. An EPD (expected predictive deviance) measure is used to assess fit and has posterior mean 407.

There may also be seasonal variation, which may be modelled by a mix of sine and cosine terms, $\sin(2\pi jt/12)$ and $\cos(2\pi jt/12)$ ($j = 1, ... , J$) in a regression for time-varying $\lambda_t$. With $J = 1$ a two chain ruin of 10,000 iterations provides coefficient estimates on the sine and cosine terms of $-0.26$ ($-0.39, -0.13$) and $-0.31$ ($-0.47, -0.15$) respectively, similar to those in Table II of Freeland and McCabe (2004). The EPD is reduced to 399, with forecasts in the subsequent year now showing quite considerable seasonal variation.

Also consider an APC(1,1) model

$$\mu_t = \omega + \alpha y_{t-1} + \beta \mu_{t-1},$$

with all coefficients positive, and stationarity ensured by $\alpha + \beta < 1$. A prior ensuring the latter constraint involves a three category Dirichlet (in turn represented by a series of gamma densities). From the second half of a two chain run of 10 000 iterations, the posterior mean for $\beta$ is obtained as 0.01, while that for $\alpha$ is 0.56. Out-of-sample forecasts for the subsequent year have means 5.5 ($t = 121$), 5.8 ($t = 122$), stabilising thereafter at between 5.9 and 6.2. The EPD measure is 386. One may include predictors (excluding an intercept since the level of the series is already modelled by $\omega$) via

$$y_t \sim Po(v_t),$$
$$v_t = \exp(X_t\gamma)\mu_t,$$
$$\mu_t = \omega + \alpha y_{t-1} + \beta \mu_{t-1},$$

and including the seasonal sine and cosine terms in $X_t$ gives an EPD of 390, with the coefficient on the sine term having 95% credible interval ($-0.31, -0.08$), while there are doubts about the necessity for the cosine term, with 95% interval ($-0.22, 0.00$).

### Example 6.7    Polio infections in the USA

Considered a time series $y_t$ of new polio infections per month in the USA between January 1970 and December 1987 ($T = 216$). A question raised with these data is the existence or otherwise of a linear trend in time, after accounting for seasonal variations. The latter are represented by sine and cosine terms, namely

(a) cosine of annual periodicity, beginning with 1 in January 1970 (i.e. frequency 1/12)

(b) sine of annual periodicity, beginning with 0 in January 1970

(c) cosine of semi-annual periodicity, beginning with 1 January 1970 (frequency 1/6)

(d) sine of semi-annual periodicity, beginning with 0 in January 1970

We consider a model including a linear trend, periodic impacts and lags up to order 5 in the counts themselves (i.e. in $y_{t-1}$ through to $y_{t-5}$). This model therefore has 11 parameters (intercept, trend, seasonals and lags).

To illustrate cross-validation forecasts with this approach, estimation is to the end of 1983 ($T0 = 168$) and forecasts made beyond then. Estimation conditions on the first five observations, so that no model is required for the latent data values $(y_0, y_{-1}, \ldots, y_{-4})$. From a two chain run of 5000 iterations, with

$$y_t \sim \text{Po}(\mu_t) \qquad t = 6, \ldots, T0$$

estimated coefficients (posterior means and standard deviations) are

$$\log(\mu_t) = -0.14 \quad -0.0032t \quad -0.22\cos(2\pi t/12) \quad -0.47\sin(2\pi t/12)$$
$$\qquad\quad (0.19) \quad\ (0.0015) \quad\ (0.11) \qquad\qquad\qquad (0.11)$$
$$+\ 0.13\cos(2\pi t/6) - 0.37\sin(2\pi t/6) + 0.084y_{t-1} + 0.038y_{t-2} - 0.044y_{t-3}$$
$$\quad (0.11) \qquad\qquad (0.11) \qquad\qquad (0.03) \qquad\ (0.04) \qquad (0.04)$$
$$+\ 0.028y_{t-4} + 0.079y_{t-5}$$
$$\quad (0.04) \qquad (0.03)$$

The 95% credible interval for the linear trend is confined to negative values, namely $(-0.0062, -0.0003)$, and so supports a downward trend.

Some of the coefficients are not significant and a second option in model 1 allows for coefficient selection or exclusion using the Kuo–Mallick (1998) method. Inclusion rates are found to be lowest for lags 3 and 4, and in fact are below 5% for all coefficients except the time term, the sine effects and the first lag in $y$. The consequent forecasting model (model 2) sets lag 3 and 4 coefficients in $y_t$ to zero. The actual counts of cases beyond 1983 are generally 0 or 1, exceeding 1 in only three of the 48 months, and the posterior mean forecast counts (y.pred[] in the BUGS code) are all between 0.3 and 1.5, in line with the trend to lower incidence.

As an illustration of the latent factor approach of Section 6.3.2 to these data, consider again the series from 1970 to 1983 (there is only one series so $K = 1$). Taking $Z_t = t$ to model the linear trend, the long term model reduces to using a constant $\Delta Z_t = Z_{t+1} - Z_t = 1$, the coefficient on which represents the trend coefficient. The short term effects are modelled as the coefficients of $\cos(2\pi t/12), \sin(2\pi t/12), \cos(2\pi t/6)$ and $\sin(2\pi t/6)$. The initial values of $F_t$ and $b_t$ are set to 1 for identifiability, while the standard deviation $\omega^{0.5}$ is assigned a uniform $(0, 10)$ prior[4]. Additionally the intercept is omitted from the regression $v_t = \exp(X_t\beta)$ because, under the assumption $y_t \sim \text{Po}(v_t F_t)$, the level of the overall Poisson mean is affected by the average of the $F_t$ series.

From the second half of a two chain run of 50 000 iterations, the posterior mean of $\omega$ is obtained as around 0.065. The trend coefficient $\alpha$ is not significant under this latent factor model, possibly because of the non-stationary nature of the latent process $F_t$. For example, the average of the posterior mean $F_t$ for times $t = 1, \ldots, 84$ is 1.26, while for $t = 85, \ldots, 168$ their average is 1.08. The seasonal sine terms are found to have significant negative effects, as in other analyses of these data (Chan and Ledolter, 1995). However, a plot of the $F_t$ reveals significant stochastic variation even after taking account of seasonal effects.

## 6.4   Dynamic linear and general linear models

In contrast to ARMA methods, dynamic linear models and their extensions (e.g. Durbin and Koopman, 2000; Harvey et al., 2004; Petris et al., 2009; Fahrmeir and Kneib, 2011; West,

2013) based on state space priors aim to directly capture features of time series, such as trend and seasonality. Such models are also useful in representing volatility and time varying regression relationships. Dynamic linear models represent observed time series $\{y_t, t = 1, \dots, T\}$ as conditional on unobserved continuous states $\theta_t$ (at the second stage of a hierarchical prior) which in turn condition on hyperparameters such as variance of the states, regression parameters, etc. The observation $y_t$ is assumed independent of the past observations given the knowledge of $\theta_t$, so that temporal dynamics are represented by the state parameters. Both observations and states may be multivariate, and a univariate outcome may be represented in terms of multivariate states.

Such models consist of two sets of equations, a measurement or observation equation linking the observations to unobserved state variables, and one or more transition (or state) equations describing the evolution of the state variables. In normal linear DLMs (Reis *et al.*, 2006) the observation equation specifies how latent states and predictors specified by a matrix $F_t$ (of dimension $p$) together impact on the outcomes. The state equation includes a transition matrix $G_t$ of dimension $p \times p$ describing how successive latent state values are related. Thus

$$y_t = F_t \theta_t + \epsilon_t, \qquad \epsilon_t \sim N(0, V_t),$$

$$\theta_t = G_t \theta_{t-1} + \omega_t, \qquad \omega_t \sim N(0, W_t),$$

where the evolution in $\theta_t$ is generally described by a low order Markov sequence, and a constant variance assumption, $\{V_t = V, W_t = W\}$ will often be made. Denoting $\theta = (\theta_1, \dots, \theta_T)$ the full set of unknowns in this scheme is then $(\theta, V, W)$. Separate prior assumptions are needed for the initial state values that set the system in motion. Thus for a first order Markov scheme for $\theta_t$ starting at $t = 1$, the initial conditions generally consist of a diffuse prior, $\theta_1 \sim N(a, R)$.

Letting $D_t$ denote the history of the process at $t$, then conditional on known parameters $\theta_{t-1}$, values of $\{\theta_t | D_t\}$ may in fact be updated recursively via the Kalman filter (West, 2013). Updating at time $t$ involves prior, predictive and posterior distributions, respectively

$$p(\theta_t | D_{t-1}) = \int p(\theta_t | \theta_{t-1}) p(\theta_{t-1} | D_{t-1}) d\theta_{t-1},$$

$$p(y_t | D_{t-1}) = \int p(y_t | \theta_t) p(\theta_t | D_{t-1}) d\theta_t,$$

$$p(\theta_t | D_t) \propto p(\theta_t | D_{t-1}) p(y_t | D_{t-1}).$$

Widely used dynamic linear models include the local level model, where the observed data is represented as an underlying signal plus noise,

$$y_t = \theta_t + \epsilon_t, \qquad \epsilon_t \sim N(0, V),$$

$$\theta_t = \theta_{t-1} + \omega_t, \qquad \omega_t \sim N(0, W),$$

with the first order random walk scheme on the signal penalising large differences $\theta_t - \theta_{t-1}$ (Fahrmeir and Lang, 2001). The state equation under a second order random walk has the form

$$\theta_t = 2\theta_{t-1} - \theta_{t-2} + \omega_t,$$

penalising large deviations from the linear trend $2\theta_{t-1} - \theta_{t-2}$. The variance $W$ of the latent series is often expected to be smaller than that of the observations, with signal to noise ratio $\lambda = W/V$ then under 1, and with lower values of $\lambda$ correspond to greater smoothing. So a prior (e.g. gamma) on $\lambda$ might be taken that favours small positive values, or the prior on $W$ expressed as a ratio of the observation variance $W = \lambda V$ (Migon *et al.*, 2005). A further option is a uniform $U(0, 1)$ prior on the ratio $W/(W + V)$. Autoregressive priors in the latent series, such as

$$\theta_t \sim N(\gamma \theta_{t-1}, W)$$

are also used.

The local level model can be extended to include a stochastic trend under the local linear trend model, with

$$y_t = \theta_t + \epsilon_t, \qquad\qquad \epsilon_t \sim N(0, V),$$

$$\theta_t = \theta_{t-1} + \delta_{t-1} + \omega_{1t}, \qquad \omega_{1t} \sim N(0, W_1),$$

$$\delta_t = \delta_{t-1} + \omega_{2t}, \qquad\qquad \omega_{2t} \sim N(0, W_2).$$

when $W_2 = 0$, the slope $\delta$ is fixed and the trend becomes a random walk with drift.

Taking account also of seasonal fluctuations leads to the basic structural model with changing level, trend and seasonal components $\gamma_t$ (e.g. Lau *et al.*, 2012), so that

$$y_t = \theta_t + \gamma_t + \epsilon_t,$$

$$\theta_t = \theta_{t-1} + \delta_{t-1} + \omega_{1t},$$

$$\delta_t = \delta_{t-1} + \omega_{2t},$$

$$\gamma_t = -\sum_{j=1}^{S-1} \gamma_{t-j} + \omega_{3t},$$

where $S$ is the number of seasons (e.g. $S = 12$ for months, $S = 4$ for quarters) and the errors $\epsilon_t, \omega_{1t}, \omega_{2t}$ and $\omega_{3t}$ are uncorrelated over time and independent of each other. The seasonal effects $\gamma_t$ are stochastic but sum to zero. The seasonal component may also be modelled in trigonometric form.

Other forms of dynamic linear model include time varying regression coefficients. As mentioned by Migon *et al.* (2005), a stochastic relationship may be appropriate given potential omission of other predictor variables, non-linearity in the functional relationship, or structural shifts in the relationship. A particular example of stochastic regression is time varying autoregression (Prado and West, 2010), as in a TVAR(1) model with a RW1 prior on the coefficients

$$y_t = \rho_t y_{t-1} + \epsilon_t,$$

$$\rho_t = \rho_{t-1} + \omega_t.$$

A time series regression including stochastic autoregressive and regression effects is then illustrated by the scheme

$$y_t = \alpha_t + X_t \beta_t + \rho_t y_{t-1} + \epsilon_t,$$

$$\alpha_t = \alpha_{t-1} + \omega_{1t},$$

$$\beta_t = \beta_{t-1} + \omega_{2t},$$

$$\rho_t = \rho_{t-1} + \omega_{3t},$$

with the state equations in $\alpha_t, \beta_t$ and $\rho_t$ defined for $t = 2, \dots, T$, and $\omega_{1t} \sim N(0, W_1)$, $\omega_{2t} \sim N(0, W_2)$, $\omega_{3t} \sim N(0, W_3)$. It is also possible to assume that the $\omega_{kt}$ follow a multivariate form, with a constant dispersion matrix. The system may be set in motion by vague priors on the intercept, and $\beta$ coefficients, but with a less diffuse prior on the initial autoregressive parameter, such as $\rho_1 \sim N(0, 1)$. For example, one might assume the first period regression parameter has a prior

$$\beta_1 \sim N(b_1, C_1),$$

where $b_1$ and $C_1$ are both known (typical values might be $b_1 = 0, C_1 = 1000$).

To accommodate changing volatility a discounting procedure can be applied to represent increasing uncertainty (Ameen and Harrison, 1985; Ferreira and Gamerman, 2000; Migon et al., 2005). For example, a prior could be set for observation and/or state precisions at time 1, and then discounting applied to subsequent precisions. So for the initial observation error variance and state variance one might take

$$V_1^{-1} \sim \text{Ga}(s_1, t_1), W_1^{-1} \sim \text{Ga}(s_2, t_2)$$

Subsequent precisions are downweighted by factors $\delta_V$ and $\delta_W (0 < \delta_V < 1, 0 < \delta_W < 1,)$, so that for $t > 1$

$$V_t^{-1} = \delta_V V_{t-1}^{-1},$$
$$W_t^{-1} = \delta_W W_{t-1}^{-1}.$$

A few known values (0.9, 0.95, 0.99) for each discount factor may be tried (Pole et al., 1994) and fit compared, since the likelihood is often flat over unknown values.

## 6.4.1    Further forms of dynamic models

Another class, non-linear Gaussian models (West and Harrison, 1997, Ch. 13; Tanizaki and Mariano, 1998; Prado and West, 2010) involves systems where either or both of the means of the observation and state equations depend on a non-linear function of the state variables. However, the errors for both equations remain Gaussian. An example is provided by the log-transformed Ricker model for population change (e.g. in ecological applications) with observed populations $y_t = \exp(z_t)$, and latent populations $N_t = \exp(\theta_t)$ modelled as

$$z_t = \theta_t + \epsilon_t,$$
$$\theta_t = \theta_{t-1} + \beta_0 - \beta_1 \exp(\theta_{t-1}) + \omega_t,$$

where $\omega$ and $\epsilon$ are normal, and $\beta_0$ is the maximum per capita growth rate (Gao et al., 2012).

Dynamic models for discrete responses are encompassed under the dynamic generalised linear model framework (Ferreira and Gamerman, 2000; Migon et al., 2005). For $y_t$ following the exponential family

$$f(y_t|\theta_t, \phi_t) = \exp[\{y_t\theta_t - b(\theta_t)\}/a(\phi_t) + c(y_t, \phi_t)]$$

with $\mu_t = E[y_t|\theta_t, \phi_t]$, the observation model with link $g$ is

$$g(\mu_t) = F_t\theta_t + \epsilon_t$$

where $\epsilon_t$ are random effects (e.g. for representing excess dispersion in relation to Poisson or multinomial assumptions), and the state equations are

$$\theta_t = G_t\theta_{t-1} + \omega_t \quad t = 2, \dots, T.$$

For example, a DGLM approach to count time series (e.g. in environmental epidemiology) could specify (e.g. Chiogna and Gaetan, 2002)

$$y_t \sim Po(E_t \exp(\theta_t + X_{2t}\beta)),$$

$$\theta_t = \lambda_t + X_{1t}\gamma_t,$$

$$\lambda_t = 2\lambda_{t-1} - \lambda_{t-2} + \omega_{1t},$$

$$\gamma_t = \gamma_{t-1} + \omega_{2t},$$

where $E_t$ are expected events, the covariates $X_{2t}$ represent perturbing factors, while $X_{1t}$ represent measured direct environmental risks, and $\lambda_t$ represent unmeasured risk factors.

Suitable MCMC sampling schemes vary according to the form of model. Full conditionals for the normal linear DLM with $V_t = V$ are set out by Migon *et al.* (2005), while those for the basic structural model with Gaussian outcome have been set out by Fruhwirth-Schnatter (1994), and Carter and Kohn (1994), including block updating for the state vector. Other computational considerations are relevant to identifiability of models involving state space priors. For example, random walk priors do not usually specify a mean for the series $\theta_t$, so if the level of the data is represented by another parameter, centring the $\theta_t$ at each MCMC iteration assists in stable convergence. Alternatively the mean of the $\theta_t$ may be taken as a measure of location.

Assuming normal errors in either observation or state equations may not be robust to sudden shifts in the series. Alternatives include a Student $t$ density achieved using scale mixing or a discrete mixtures of normals. For example, a scale mixture prior on the first order random walk prior for $\theta_t$ in the local level model is

$$\theta_t = \theta_{t-1} + \omega_t$$

where

$$\omega_t \sim N(0, W/\lambda_t),$$

$$\lambda_t \sim Ga(0.5v, 0.5v)$$

and $v$ is the Student degrees of freedom parameter. Alternatively a discrete mixture may be applied with known probabilities on components, as for two groups

$$\omega_t \sim (1 - \pi)N(0, W) + \pi N(0, \varphi W)$$

where $\pi = 0.05$ or $0.01$, and $\varphi$ is large (say between 10 and 100) to accommodate outliers.

### Example 6.8    Non-linear latent series

This example illustrates the detection of a signal in noisy data when the form of the signal is known. Thus consider a simulated series generated to follow the truncated and asymmetric form (for $t = 1, \ldots, 200$)

$$y_t = \theta_t + \epsilon_t \qquad \epsilon_t \sim N(0, 1)$$

with true signal

$$\theta_t = (12/\pi) \exp(-[t - 130]^2/2000).$$

The maximum value of the true series is just under 4 at $t = 130$, with the true series being effectively zero for $t < 50$.

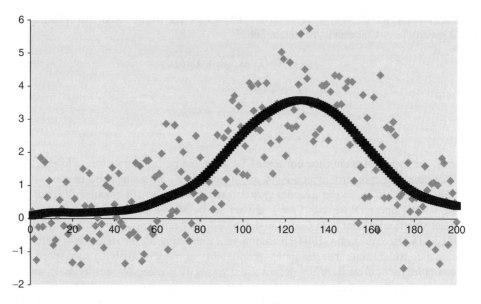

**Figure 6.4**    Reconstructing latent signal from simulated data.

The simulated data are now taken as observations with an unknown data generation process. For detecting the signal, an RW(2) prior is adopted with

$$\theta_t = 2\theta_{t-1} - \theta_{t-2} + \omega_t$$

with $\omega_t \sim N(0, W)$. Two alternative priors are set on $W$ conditional on $V = \text{var}(\epsilon)$, one a uniform prior $U(0, 1)$ on $B = V/[V + W]$, the other an exponential prior on the signal to noise variance ratio, $\lambda \sim E(\lambda_0)$, where $W = \lambda V$, and $\lambda_0 \sim E(1)$. The latter prior is neutral regarding the balance in anticipated variability in the signal as compared to that around the observations. A Ga(1, 0.001) prior on $1/V$ is assumed in both cases, while $N(0, 100)$ priors are adopted for the two initial points in the signal. A corner constraint on the $\theta_t$ is included for improved identification, with a separate intercept then needed, so that the observation equation is expressed as

$$y_t = \alpha + (\theta_t - \theta_1) + \epsilon_t.$$

The median value of W obtained under the first prior, from the second half of a two chain run to 100 000 iterations, stands at 1.31E-4; by contrast, Kitagawa and Gersch (1996) report a value of 0.79E-4 for data generated by this process. The median observational variance is estimated at 1.07. The true underlying series, estimated with means $\alpha + (\theta_t - \theta_1)$, is reproduced satisfactorily (Figure 6.4). Convergence is assessed using selected $\theta_t$ and the variance terms.

The alternative prior setup leads to similar results. The second half of a two chain run of 100 000 iterations provides posterior medians for W and V of 1.35E-4 and 1.08 respectively. $\lambda_0$ has an estimated posterior mean of 1.99.

This model may be fitted in R-INLA, where estimation is constrained so that the $\theta_t$ sum to zero, and so a fixed effect intercept is part of the output. The following sequence of commands produces a posterior mean for W of 3.8E-4. Choosing RW1 instead of RW2 to represent the latent level produces a notably more ragged signal (on plotting the fitted values in F1) with W estimated at 0.037.

```
library(INLA); setwd("C://R files");
# column labels time (1,...,200), and y
D <- read.table("simdata.txt",header = T)
f1 <- y ~ f(time,model = "rw2",param = c(1,0.001))
m1 = inla(f1,family = "gaussian",control.predictor = list( compute = T),
data = D,control.family = list(param = c(1,0.001)))
prec.marg = m1$marginals.hyperpar$"Precision for time"
pm1 <- inla.expectation(function(x) 1/x, prec.marg)
pm2 <- inla.expectation(function(x) (1/x)^2, prec.marg)
psd = sqrt(pm2 - pm1^2)
# posterior mean and sd of signal variance
print(c(mean = pm1, sd = psd))
F1 <- m1$summary.fitted.values
```

### Example 6.9    Arctic Sea Ice

As an illustration of a smoothing model for a time series with seasonal effects as well as a secular trend (decline) over time, consider $n = 396$ monthly readings of the extent of Arctic sea ice cover $y_t$ (millions of sq km, source: http://neptune.gsfc.nasa.gov/csb) from January 1979 through to December 2011 ($T = 396$ observations). Possible alternatives allow for seasonal effects and random changes in level over time, or seasonal effects combined with a local linear trend.

The data show negative skew, and a skew normal model with seasonal effects and a local linear trend is fitted in R-INLA. Thus

$$y_t \sim SN(m_t, \sigma^2, \lambda),$$

$$m_t = \mu_t + s_t,$$

$$\mu_t \sim N(\mu_{t-1} + \delta_{t-1}, \ \sigma_\mu^2),$$

$$\delta_{t-1} \sim N(\delta_{t-2}, \ \sigma_\delta^2),$$

$$s_t \sim N(-s_{t-1} - s_{t-2} - \ldots - s_{t-11}, \ \sigma_s^2),$$

with default Ga(1, 0.001) priors on all precision (inverse variance) parameters. The code[4] uses the data augmentation strategy described in Ruiz-Cárdenas *et al.* (2012). The skew normal model is compared to a normal model and has a lower DIC (namely −5249 compared to −4799), and higher estimated marginal likelihood (−5752 vs −6334). Figure 6.5 shows the estimated trend $\mu_t$ and the forecast trend for a further 12 months after 2011. An acceleration in ice cover depletion seems to occur just after mid-way in the series, namely from 1998 onwards.

Similar results are obtained in BUGS using a representation with corner constrained $\mu_t$ and $\delta_t$ series, in order to improve identifiability (as discussed in Example 6.8). It is also advisable in BUGS to centre the data, and provide initial values for all parameters (e.g. zero throughout for the $\mu_t$ and $\delta_t$). The following code assumes a normal response but to alleviate skewness one might transform the data before centring. A Fourier representation (with $S = 2$) is used for monthly seasonal effects.

```
model {for (t in 1:T) {# yhat: predictions with original location
                    yhat[t] <- m.y[t] + 11.63
                # centred data
                    y[t] ~ dnorm(m.y[t],tau[1]);
                    mu.c[t] <- beta0 + mu[t] - mu[1]
                    m.y[t] <- beta0 + mu[t] - mu[1]
```

```
                                       +sum(seas[t,1:S])
for (s in 1:S) {seas[t,s] <- beta[2*s-1]*cos(6.28*s*t/12)
                           +beta[2*s]*sin(6.28*s*t/12)}}
# level and trend
for (t in 2:T){mu[t] ~ dnorm(m.mu[t],tau[2])
                   m.mu[t] <- mu[t-1]+delta[t-1]-delta[1]
                   delta.c[t-1] <- delta[t-1]-delta[1]
                   delta[t] ~ dnorm(delta[t-1],tau[3])}
# initial conditions
     mu[1] <- mu1;    mu1 ~ dnorm(0,0.001);   delta[1] <- delta1;
                                   delta1 ~ dnorm(0,0.001)
     for (j in 1:3) {tau[j] ~ dgamma(1,0.01)}
     beta0 ~ dnorm(0,0.001)
for (s in 1:S) {beta[2*s-1] ~dnorm(0,0.001);   beta[2*s] ~ dnorm(0,0.001)}}.
```

An alternative for representing random walks for the trend component is to use the carnormal prior in BUGS, which produces centred effects. Hence the overall trend is represented as

$$\delta_t = \delta_0 + \delta_t^c,$$

where $\delta_0$ is a fixed effect representing the average trend, and $\delta_t^c$ is modeled using the carnormal. The relevant part of the code (with $TM = 395$) becomes

```
# trend
for (t in 2:T){mu[t] ~ dnorm(m.mu[t],tau[2])
                   delta[t-1] <- delta0+delta.c[t-1]
                   m.mu[t] <- mu[t-1]+delta0+delta.c[t-1]}
w[1] <- 1;              adjt[1] <- 2;                        num[1] <- 1
w[(TM-2)*2+2] <- 1;       adjt[(TM-2)*2+2] <- TM-1;    num[TM] <- 1
for (t in 2:TM-1) {w[2+(t-2)*2] <- 1;       adjt[2+(t-2)*2] <- t-1
                  w[3+(t-2)*2] <- 1;          adjt[3+(t-2)*2]
                                 <- t+1;         num[t] <- 2}
delta.c[1:TM] ~ car.normal(adjt[],w[],num[],tau[3])
```

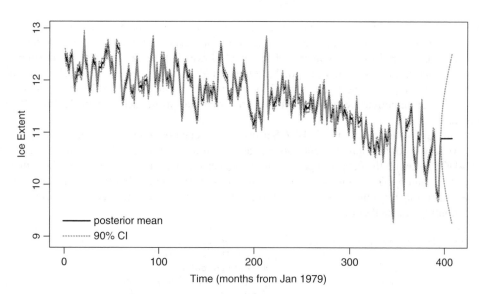

**Figure 6.5**  Arctic ice cover 1979–2011 change in level.

## Example 6.10    Impact of TV advertising

This example illustrates evolving regression impacts, and also involves a binomial outcome. Thus a study on the impact TV advertising involved asking a set weekly total of 66 individuals a 'yes or no' question regarding awareness of an advert for a chocolate-bar, and was continued over 171 weeks (Migon and Harrison, 1985). The number of positive answers $r_t$ in week $t$ is modelled as binomial with logit link to a single covariate, $x_t$ = weekly expenditure on advertisements. Thus

$$r_t \sim \text{Bin}(66, \pi_t),$$

$$\text{logit}(\pi_t) = \alpha_t + \beta_t x_t,$$

with the parameters evolving according to

$$\alpha_t = \alpha_{t-1} + \omega_{\alpha t},$$

$$\beta_t = \beta_{t-1} + \omega_{\beta t}.$$

In model 1, univariate RW(1) priors for $\alpha_t$ and $\beta_t$ (with variances $\sigma^2_\alpha$ and $\sigma^2_\beta$) are adopted, with diffuse $N(0, 1000)$ priors for the initial conditions $\alpha_1$ and $\beta_1$. Separate gamma priors are assumed on the precisions $1/\sigma^2_\alpha$ and $1/\sigma^2_\beta$. From the second halves of two chain runs of 15 000 iterations, posterior mean estimates of $\sigma^2_\alpha$ and $\sigma^2_\beta$ are respectively 0.0023 and 0.00045. The plot of the parameters themselves (Figure 6.6) shows a decrease in the mean (i.e. awareness level) though the positive impact of advertising is more or less stable.

An alternative model (model 2) adopts a bivariate normal prior for the evolution of $\alpha_t$ and $\beta_t$. A Wishart prior is adopted for the inverse covariance matrix, with diagonal elements 0.001 assumed. This model yields very similar estimates for the two sets of evolving parameters themselves. The second half of a two chain run of 15 000 iterations shows a significant effect of expenditure on awareness: the posterior mean of the average of the $\beta_t$ over all periods is

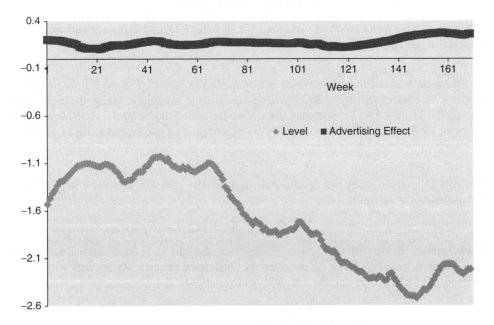

**Figure 6.6**    Awareness level and advertising effect.

0.23 with 95% interval (0.17, 0.28). There is, however, no significant correlation between the evolving intercepts and slopes.

## 6.5   Stochastic variances and stochastic volatility

There are many instances, including the dynamic coefficient models just discussed, where it may be necessary to model observed time series $y_t$ or make forecasts, when the variance, rather than the level, of the series, is stochastic over time (Shephard, 2008). Such situations are exemplified by financial time series $S_t$ (e.g. stock prices, exchange rates) where large forecast errors tend to occur in clusters, when the series are unsettled or rapidly changing. This is known as volatility clustering and suggests dependence between successive values of the squared errors. In many applications of such models the series has effectively a zero mean; for example, in many financial time series the ratio of successive values $S_t/S_{t-1}$ averages 1, and series defined by the log of these ratios $r_t = \log(S_t/S_{t-1})$ or by $r_t = S_t/S_{t-1} - 1$ will then average zero. In share price applications, the $r_t$ are known as returns, and there may be correlations between volatility and share prices: the term 'leverage effects' refers to a tendency for negative returns to be associated with increased volatility (Bollerslev et al., 2006).

### 6.5.1   ARCH and GARCH models

Following Engle (1982) consider a time series regression

$$y_t = X_t \beta + \epsilon_t, \qquad t = 1, \dots, T,$$

in which the error variances for subsets of the full period are different. One way to accommodate heteroscedasticity through time is through an autoregressive conditional heteroscedastic or ARCH model (e.g. Ruppert, 2011) such that

$$y_t = X_t \beta + \epsilon_t = X_t \beta + \sqrt{h_t} u_t,$$

where the $u_t$ have mean zero and variance 1. In an ARCH(1) model, the scale parameters $h_t$ depends on lag-1 squared errors

$$h_t = \omega + \alpha \epsilon_{t-1}^2.$$

To ensure $h_t$ is positive, $\omega$ and $\alpha$ are constrained to be positive and the further restriction $0 \leq \alpha \leq 1$ ensures that the ARCH series is covariance stationary. While the most usual assumption is $u_t \sim N(0, 1)$, one may also consider $u_t \sim$ Student $(0, 1, v)$ (Bauwens and Lubrano, 1998). An ARCH(2) model involves dependence on two lagged terms $\epsilon_{t-k}^2$. Thus

$$h_t = \omega + \alpha_1 \epsilon_{t-1}^2 + \alpha_2 \epsilon_{t-2}^2.$$

The ARCH(1) error series has conditional mean $E(\epsilon_t | \epsilon_{t-1}) = E(u_t | \epsilon_{t-1})(\omega + \alpha \epsilon_{t-1}^2)^{0.5} = 0$, and a conditional variance

$$V_t = \mathrm{var}(\epsilon_t | \epsilon_{t-1}) = E(u_t^2 | \epsilon_{t-1})[\omega + \alpha \epsilon_{t-1}^2] = \omega + \alpha \epsilon_{t-1}^2,$$

that is heteroscedastic with respect to $\epsilon_{t-1}$. However, the ARCH(1) unconditional variance is time invariant, namely $\omega/(1 - \alpha)$, provided the conditional variance is stationary with $\alpha < 1$, while for an ARCH($p$) model, the unconditional variance is

$$\omega/(1 - \alpha_1 - \dots - \alpha_p),$$

provided $\sum_{k=1}^{p} \alpha_k < 1$.

In the GARCH model (Vrontos *et al.*, 2000; Asai, 2006; Ardia and Hoogerheide, 2010), the conditional variance depends on lagged values of both $h_t$ and $\epsilon_t^2$, with a GARCH($p$, $q$) error model being

$$h_t = \omega + \alpha_1 \epsilon_{t-1}^2 + \ldots + \alpha_p \, \epsilon_{t-p}^2 + \gamma_1 h_{t-1} + \ldots + \gamma_q h_{t-q},$$

A GARCH(1, 1) model would be

$$y_t = X_t \beta + u_t \sqrt{h_t},$$

$$h_t = \omega + \alpha \epsilon_{t-1}^2 + \gamma h_{t-1},$$

where $u_t \sim N(0, 1)$, $\{\omega, \alpha, \gamma\}$ are all positive, and for stationarity $\alpha + \gamma < 1$. The latter constraint may be imposed with a Dirichlet prior on $(\gamma, \alpha, 1 - \gamma - \alpha)$.

For the GARCH($p$, $q$) stationarity requires $(\Sigma_{j=1}^{q} \gamma_j + \Sigma_{k=1}^{p} \alpha_k) < 1$, while the constraint $(\Sigma_{j=1}^{q} \gamma_j + \Sigma_{k=1}^{p} \alpha_k) = 1$ leads to an integrated GARCH or IGARCH error scheme, under which volatility is not mean-reverting and the conditional variance is non-stationary. Note that other definitions of the GARCH($p$, $q$) are possible, such as the definition used in BayesGARCH (Ardia and Hoogerheide, 2012).

$$h_t = \omega + \alpha_1 y_{t-1}^2 + \ldots + \alpha_p y_{t-p}^2 + \gamma_1 h_{t-1} + \ldots + \gamma_q h_{t-q}.$$

## 6.5.2   State space stochastic volatility models

Another option for modeling changing variances is known as stochastic volatility, generally within the state-space framework (Kitagawa and Gersch, 1996; Lopes and Polson, 2010). Thus in

$$y_t = X_t \beta + \epsilon_t,$$

with $\epsilon_t \sim N(0, V_t)$, one may assume the evolution of $\Delta^k \log V_t$ follows a non-stationary random walk process. For example taking $k = 1$, and setting $g_t = \log V_t$ gives a first order random walk which may follow a normal or Student form:

$$g_t \sim N(g_{t-1}, \sigma_g^2).$$

A generalisation, usually for series in return form (involving ratios of successive values), allows autoregressive dependence in latent variables $\kappa_t$ which represent the latent volatility process. Thus for a series with zero mean and no regressors, one might specify first order autoregressive dependence in the latent log variances (Harvey *et al.*, 1994; Meyer and Yu, 2000; Prado and West, 2010, section 7.5), with

$$y_t = \exp(\kappa_t / 2) u_t,$$

$$\kappa_t = \mu + \phi(\kappa_{t-1} - \mu) + \eta_t,$$

where $u_t \sim N(0, 1)$, and $\eta_t \sim N(0, \sigma_\eta^2)$. If $|\phi| < 1$, then the $\kappa_t$ are stationary with $\kappa_1 \sim N(\mu, \sigma_\eta^2 / (1 - \phi^2))$ (Kim *et al.*, 1998).

Another option is denoted the unobserved ARCH model (Shephard, 1996; Giakoumatos *et al.*, 2005), in which an ARCH model still holds but is observed with error. This is generally classified as a stochastic volatility approach. For a zero mean observation series and no covariates, a measurement error model combined with an ARCH model leads to

$$y_t \sim N(\lambda_t, \sigma^2),$$

$$\lambda_t \sim N(0, h_t),$$

$$h_t = \omega + \alpha \lambda_{t-1}^2,$$

with the restriction $0 < \alpha \le 1$ ensuring covariance stationarity.

One may also consider transformations of centred returns $y_t$ (Kim *et al.*, 1998; Prado and West, 2010, p. 219; Lopes and Polson, 2010) to reproduce a linear model,

$$\log(y_t^2) = h_t + \log(\epsilon_t^2),$$

where $\log(\epsilon_t^2)$ has a log $\chi_1^2$ density, with mean $-1.27$ and variance $4.94$. To approximate this density a mixture of normals with known means, variances and mixing proportions is necessary (see Example 6.11). Additional latent categorical variables are introduced for each time point to sample from this mixture.

For multivariate series (e.g. of several exchange rates) subject to volatility clustering, factor models may be used to capture correlated volatility (Harvey *et al.*, 1994; Pitt and Shephard, 1999). For instance for two mean zero series $y_{tk}$, $k = 1, 2$, one factor $f_t$, and no covariates, one might have

$$y_{t1} = \beta_1 f_t + \omega_{t1},$$

$$y_{t2} = \beta_2 f_t + \omega_{t2},$$

with $f_t$ and the $\omega_{tk}$ evolving in line with stochastic volatility. Thus $f_t \sim N(0, \exp(\kappa_t^f))$, $\omega_{t1} \sim N(0, \exp(\kappa_t^{\omega_1}))$, $\omega_{t2} \sim N(0, \exp(\kappa_t^{\omega_2}))$, with first order autoregressive dependence in the latent log variances leading to

$$\kappa_t^f = \rho^f \kappa_{t-1}^f + \eta_{t-1}^f,$$

$$\kappa_t^{\omega_1} = \rho^{\omega_1} \kappa_{t-1}^{\omega_1} + \eta_{t-1}^{\omega_1},$$

$$\kappa_t^{\omega_2} = \rho^{\omega_2} \kappa_{t-1}^{\omega_2} + \eta_{t-1}^{\omega_2}.$$

### Example 6.11    Intel share price

Let returns be defined as $r_t = S_t / S_{t-1} - 1$ for monthly Intel stock prices $S_t$ from January 1973 through to December 2003 ($T = 372$). Then consider log returns $Y_t = \log(r_t + 1)$, centred to have average zero, $y_t = Y_t - \bar{Y}$. The first three modeling options involve simple SV, ARCH and GARCH models. Thus as a first option, consider the autoregressive stochastic volatility (ARSV) model with

$$y_t = \exp(\kappa_t/2)u_t,$$

$$\kappa_t = \mu + \phi(\kappa_{t-1} - \mu) + \eta_t,$$

where $u_t \sim N(0, 1)$, $\eta_t \sim N(0, \sigma_\eta^2)$ and $|\phi| < 1$. The prior on $\phi$ follows Meyer and Yu (2000), with $\phi^* \sim Be (20, 1.5)$, and then $\phi = 2\phi^* - 1$. The second half of a two chain run of 10,000 iterations provides posterior mean (and 95% CRI) for $\phi$ of 0.94 (0.89, 0.98), with LPML of 245 and DIC of $-494$.

The second option is an ARCH(1) model, with

$$y_t = \sqrt{h_t}\, u_t = \epsilon_t$$

$$h_t = \omega + \alpha \epsilon_{t-1}^2 = \omega + \alpha y_{t-1}^2,$$

with volatility terms including $h_1$ depending on a latent pre-series value $y_0$. The second half of a two chain run of 10 000 iterations provides an LPML of 214 and DIC of −429, with $\alpha$ having estimate 0.38 (0.16, 0.68).

For the third option, OpenBUGS is used to estimate a GARCH(1, 1) model with

$$h_t = \omega + \alpha\epsilon_{t-1}^2 + \gamma h_{t-1},$$

with starting values based on classical estimates obtained from the fGarch package in R. The commands are

```
library(fGarch)
D = read.table("intel.txt",header = T)     # column headers date and rtn
intc = log(D$rtn + 1)
z = intc-mean(intc)
m = garchFit(~garch(1,1),data = z)
```

with estimates 0.08 and 0.86 for $\alpha$ and $\gamma$ respectively. The estimates from OpenBUGS for $\alpha$ and $\gamma$ are respectively 0.08 (0.03, 0.18) and 0.83 (0.70, 0.90). This model has a lower DIC (−452) and higher LPML (of 226) than the ARCH(1) model but does not fit as well as the SV model.

Now consider an extension of the ARSV model to include leverage effects (Meyer and Yu, 2000; Yu, 2005), namely correlation between observation errors and volatility. Thus

$$y_t = \exp\left(\frac{\kappa_t}{2}\right) u_t$$

$$\kappa_t = \mu + \varphi(\kappa_{t-1} - \mu) + \sqrt{\sigma}_\eta w_t,$$

$$\begin{pmatrix} u_t \\ w_t \end{pmatrix} \sim N\left(\begin{pmatrix} 0 \\ 0 \end{pmatrix}, \begin{pmatrix} 1 & \rho \\ \rho & 1 \end{pmatrix}\right),$$

resulting in a bivariate normal distribution for $y_t$ and $\kappa_t$ with

$$y_t \sim N\left(\frac{\rho}{\sigma_\eta} \exp\left(0.5\kappa_t\right) [\kappa_t - m_t], \exp(\kappa_t)[1 - \rho^2]\right).$$

This model yields a lower DIC, namely −514, than the ARSV model, but the LPML is reduced to 243. The leverage coefficient has a negative posterior mean of −0.27, but an inconclusive 95% CRI of (−0.55, 0.06).

Finally consider the mixture approximation (Kim et al., 1998, p 371; Prado and West, 2010, p. 218) to the linear model with redifined response $y_t = \log(r_t^2)$, namely

$$y_t = \kappa_t + v_t,$$

$$v_t = \log(\omega_t)/2,$$

$$\omega_t \sim \chi_1^2,$$

$$\kappa_t = \mu + \phi(\kappa_{t-1} - \mu) + \eta_t,$$

with $p(v_t)$ approximated by a discrete normal mixture with known mixture proportions, means and variances. Discrete categorical variables $J_t \in \{1, \ldots, 7\}$ sample the appropriate value from this mixture for each time point. The analysis is in OpenBUGS, and Figure 6.7 shows both the absolute value of the percent returns, and the estimated volatility process (the posterior mean $\kappa_t$).

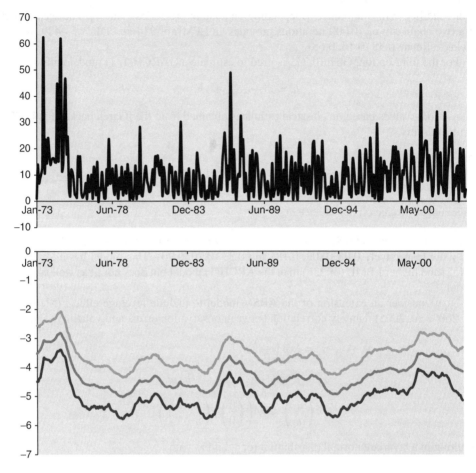

**Figure 6.7**    (a) Absolute percent returns. (b) Volatility process, mean and 95% interval.

## 6.6    Modelling structural shifts

State space models are designed to accommodate gradual or smooth shifts in time series parameters, though robustification to irregularity can be achieved (e.g. via scale-mixing). Similarly, for ARMA techniques, innovation and additive outliers can be introduced (Tsay, 1988). Often, however, there are temporary or permanent shifts in time series parameters that occur abruptly, and a more appropriate model allows for changes in parameter regimes (Piger, 2011). Essentially such models partition the series of observations $\{y_1, \ldots, y_T\}$ by a sequence of changepoints $\tau_1 < \tau_2 < \cdots < \tau_m$, with observations homogeneous within segments and heterogeneous across segments (Caron *et al.*, 2012). Relevant references from a Bayesian perspective include Albert and Chib (1993), Pesaran *et al.* (2006), Barry and Hartigan (1993), Ruggeri and Sivaganesan (2005), Koop and Potter (2009), Ray and Tsay (2002), Chib (1998), Lai and Xing (2011), Held *et al.* (2006), Giordani *et al.* (2007), Wang and Zivot (2000), Mira and Petrone (1996), Geweke and Jiang (2011), Caron *et al.* (2012), Maheu and Song (2012), Fearnhead (2006), and Bauwens *et al.* (2011), with applications including Spirling (2007), Thomson *et al.* (2010) and Achcar *et al.* (2008).

Shifts or change-points may be defined for the level or other components (e.g. trend) in a series, for variances, or for regression parameters (Section 6.6.3). Here we consider examples of models for shifts in mean, trend and variance, with switching between regimes defined in various ways (e.g. according to a uniform prior on the change-points, or by Bernoulli shift indicators).

## 6.6.1 Level, trend and variance shifts

To illustrate how shifts in means and trend might be modelled, consider observations $y_t$ for times $t = 1, \ldots, T$ where an intervention (or interventions) might have caused one or more changes in level or trend (e.g. Elliott and Shope, 2003; Held $et\ al.$, 2006; Thomson $et\ al.$, 2010; Eckley $et\ al.$, 2011). Thus for continuous responses one might postulate $y_t \sim N(\mu_t, \sigma^2)$, with piecewise constant levels and trends

$$\mu_t = \alpha_0 + \sum_{j=1}^{m} \alpha_j I(t \geq \Delta_j) + \beta_0 t + \sum_{j=1}^{k} \beta_j (t - \theta_j)_+,$$

where $(t - \theta_j)_+ = (t - \theta_j)$ if $t \geq \theta_j$ and 0 otherwise. This model allows for

(a) shifts in level at $m$ change points $\Delta_j$;

(b) shifts in linear trend between $k$ trend change-points $\theta_j$,

where both change-points $\{\Delta_j, \ldots, \Delta_m, \theta_1, \ldots, \theta_k\}$, and numbers of change-points $(m, k)$ may be unknown. Posterior probabilities that a particular time point is a level or trend shift can be obtained by monitoring whether shifts in level or slope occurred at each MCMC iteration.

Simpler models might assume a fixed number of change-points and only one possible type of shift. Thus for counts $y_t$ assumed Poisson or negative binomial one might assume a single change point $\Delta$ for a change in level, such that

$$y_t \sim Po(\mu_1) \qquad t = 1, \ldots, \Delta;$$

$$y_t \sim Po(\mu_2) \qquad t = \Delta + 1, \ldots, T;$$

$$\Delta \sim U(1, T).$$

with gamma priors, $\mu_1 \sim Ga(a, \ b)$ and $\mu_2 \sim Ga(c, \ d)$, the full conditionals for these parameters are also gamma, namely

$$\mu_1 \sim Ga(a + \sum_{t=1}^{\Delta} y_t, b + \Delta); \quad \mu_2 \sim Ga(c + \sum_{t=\Delta+1}^{T} y_t, d + T - \Delta).$$

This approach may be extended to multiple change points, raising issues about whether the uniform priors on each change point should have restricted ranges or have support over an equal number of time-points (Koop and Potter, 2009).

For a continuous outcome assumed normal, the regimes $\{(\mu_j, \ \sigma_j^2), j = 1, \ldots, m+1\}$ may include both means and variances with coincident change-points in both parameters. Shifts in the level of variability may also occur according to a separate switching scheme, and non-stationarity that might otherwise have been attributed to changes in level may be seen as due to heteroscedasticity (Tsay, 1988; McCulloch and Tsay, 1994). As well as uniform priors for change points, one may allow breaks to occur randomly according to a Bernoulli process $d_t$ (Geweke and Jiang, 2011). For example with $y_t = \mu_t + \epsilon_t$, changes in level could be accommodated by letting

$$\mu_t = \mu_{t-1} + v_t d_{1t},$$

where binary indicators $d_{1t} = 1$ if a level shift occurs, and $v_t$ models the shift that occurs, conditional on $d_{1t} = 1$. The $v_t$ may be taken as normal with mean zero and low precision $\tau_v$. Suppose the errors $\epsilon_t$ follow an ARMA scheme such as an AR(1) sequence $\epsilon_t = \gamma \epsilon_{t-1} + u_t$, where $u_t \sim N(0, V_t)$. Shifts in $V_t$ can occur independently according to

$$V_t = V_{t-1}(1 + \omega_t d_{2t}),$$

where positive variables $\omega_t$, capturing proportional variance changes, might be taken as gamma distributed. The probabilities $\Pr(d_{1t} = 1) = \eta_1$, and $\Pr(d_{2t} = \eta_2)$ may be assigned beta priors favouring low values, with the relative importance of the mean and variance shifts then reflected in the posterior sizes of $\eta_1$ and $\eta_2$.

### 6.6.2   Latent state models including historic dependence

Lags in the observed series or latent states may be relevant to the likelihood of subsequent structural shifts. A widely applied approach to regime change involves a categorical latent state determined by a Markov sequence (e.g. Leroux and Puterman, 1992; Chib, 1996, 1998; Frühwirth-Schnatter, 2006). Suppose for each time point the process is in one of $m$ states $\{s_t\}(t > 1)$, as determined by an $m \times m$ stationary Markov chain $P = \{p_{ij}\}$ where for $t > 1$

$$p_{ij} = \Pr[s_t = j | s_{t-1} = i],$$

with parameters in each row of $P$ assigned a Dirichlet prior. The first state (namely $s_1$) is determined by drawing from a separate multinomial with $m$ categories. Given $s_t = j$, the observation follows the $j$th of the $m$ possible density components, where such components might differ in means, variances or other summary shape parameters. This scheme might be applied within different forms of time series model (e.g. autoregressive in observations or errors, state space models, etc.).

Switching based on autoregression in the observations is illustrated by a Poisson branching process for disease count data $y_t$ (e.g. infectious disease). Thus the observations are produced by latent endemic and epidemic components,

$$y_t = x_t + z_t,$$

with $x_t \sim Po(v_t)$, and with

$$z_t | y_{t-1} \sim Po(\lambda_t y_{t-1})$$

defined by autoregression on the previous observed count. Outbreaks are identified by non-stationarity in the autoregression, namely times where $\lambda_t > 1$. The $v_t$ capture stable aspects of disease occurrence (e.g. seasonal effects), as in, for example,

$$\log(v_t) = \gamma_0 + \gamma_1 \cos(2\pi t/S) + \gamma_2 \sin(2\pi t/S),$$

where $S$ is the number of time points (e.g. $S = 12$ or $S = 52$) defining seasonal effects. The $\lambda_t$ are necessarily positive, and may be taken as piecewise constant between change-points $\Delta_j$ (Held *et al.*, 2006).

### 6.6.3  Switching regressions and autoregressions

Switching regression and autoregression models originate in classical statistics with studies such as Quandt (1958), Tong(1983), and Tsay (1989), with Bayesian applications including Geweke and Terui (1993), Chen and Lee (1995), Lubrano (1995b), and Koop and Potter (2009). Choice between regimes may be determined by a threshold function $K_t$ (e.g. Koop and Potter, 1999) that drives discrete switching, or a smooth transition function, for example using a cumulative distribution function between 0 and 1, such as the logit (Bauwens *et al.*, 2000).

Thus define a binary step function $d_t$ if a time trend function exceeds a threshold $\tau$ and zero otherwise. If the trend were measured by the linear term t, then

$$K_t = t - \tau < 0 \Rightarrow d_t = 1$$

$$K_t = t - \tau > 0 \Rightarrow d_t = 2$$

with corresponding regimes incorporating shifts in regression parameters and the residual variance

$$y_t = X_t \beta_{d_t} + u_t, \quad u_t \sim N(0, \varphi_{d_t}).$$

A threshold function might also be defined by lags on the outcome, such as in the step function

$$K_t = y_{t-1} - \delta < 0 \Rightarrow d_t = 1$$

$$K_t = y_{t-1} - \delta > 0 \Rightarrow d_t = 2$$

with $\delta$ unknown. The regression might include lags in both the response and the threshold, as in a self-exciting threshold autoregressive model (Prado and West, 2010), with a SETAR(1) model illustrated by

$$y_t = \beta_{0,d_t} + \rho_{d_t} y_{t-1} + u_t,$$

$$u_t \sim N(0, \varphi_{d_t}),$$

with $d_t$ obtained as above. The appropriate lag $r$ in $y_t$, such that $d_t = 2$ if $y_{t-r} > \delta$, may be an additional unknown (Geweke and Terui, 1993). Similarly, a multiple regime SETAR could specify

$$y_t = \beta_0^{(1)} + \rho^{(1)} y_{t-1} + u_t \text{ if } y_{t-1} \leq \delta_1$$

$$= \beta_0^{(2)} + \rho^{(2)} y_{t-1} + u_t \text{ if } \delta_1 < y_{t-1} \leq \delta_2$$

$$\ldots$$

$$= \beta_0^{(J)} + \rho^{(J)} y_{t-1} + u_t \text{ if } \delta_{J-1} < y_{t-1} \leq \delta_J,$$

where $-\infty < \delta_1 < \ldots < \delta_J = \infty$.

A smooth transition function in the binary regime case might take the form

$$\Delta_t = \exp(\phi\{y_{t-1} - \delta\})/[1 + \exp(\phi\{y_{t-1} - \delta\})]$$

or

$$\Delta_t = \exp(\phi\{t - \tau\})/[1 + \exp(\phi\{t - \tau\})]$$

where $\varphi > 0$ governs the smoothness of the transition.

## Example 6.12    Global temperatures, 1850–2010

Consider data on global annual temperature anomalies. These are for 1850–2010 ($T = 161$) and from http://cdiac.esd.ornl.gov/trends/temp/jonescru/data.html. The data $y_t$ are annual averages of monthly observations relative to the 1961–1990 mean. Observational variances $V_t$ are obtained as the within year variances between months divided by 12. Two hierarchical models with normal scale mixing at the second stage are applied, since outlier years may be present as well as genuine shifts in level or trend. In the first model, $m = 3$ uniformly distributed change points in level are assumed. Taking $m = 4$ led to a less well identified analysis, and provided no further gain in fit. It is assumed that

$$y_t \sim N(\theta_t, V_t),$$

$$\theta_t \sim N(\mu_t, \sigma^2/\lambda_t),$$

$$\lambda_t \sim \text{Ga}(0.5v, 0.5v),$$

$$1/v \sim U(0.01, 0.5),$$

$$\mu_t = \alpha_0 + \sum_{j=1}^{m} \alpha_j I(t \geq \Delta_j),$$

$$\Delta_1 \sim U(0, \Delta_2), \Delta_2 \sim U(\Delta_1, \Delta_3), \Delta_3 \sim U(\Delta_2, T).$$

Inferences are based on the second half of a two chain run of 50 000 iterations. Low $\lambda_t$ values are estimated for years 1878 (warm outlier), and 1956 (cold outlier). Change points in level $\Delta_j$ are identified relatively precisely. Thus posterior mean (and standard deviations) for $\Delta_1$, $\Delta_2$ and $\Delta_3$ are obtained as 1929 (1.84), 1978 (1.16), and 1996 (0.66). The corresponding four regimes have respective levels (posterior mean and standard deviations) −0.36 (0.012), −0.12 (0.016), 0.11 (0.025) and 0.41 (0.026). Despite its simplicity this model performs relatively well with a satisfactory predictive check probability of 0.56, based on chi-square statistics comparing observations $y_t$, and predictions $y_{new,t}$, against latent states $\theta_t$. The DIC is obtained as −552, and an EPD measure (based on squared deviations between $y_t$ and $y_{new,t}$) estimated at 0.319.

Inferences about change points were not affected by adopting alternative uniform priors namely

$$\Delta_j \sim U(\Delta_{j-1}, \Delta_{j-1} + T), \quad j > 1,$$

or

$$\Delta_j \sim U(\Delta_{j-1}, \Delta_{j-1} + cT), \quad j > 1,$$

where $c$ is unknown, which allocate prior weight to change-points outside the range of the observations. For example[6], taking the prior $c \sim U(1/T, 2)$ (Koop and Potter, 2009), still leads to change years 1929, 1978 and 1996.

A second model, including trend shifts, adapts the approach of Thomson et al. (2010), and requires the RJMCMC option (jump add-on) in WinBUGS (Lunn et al., 2009). With $y_t \sim N(\theta_t, V_t)$, one has

$$\theta_t \sim N(\mu_t, \sigma^2/\lambda_t),$$

$$\lambda_t \sim \text{Ga}(0.5v, 0.5v),$$

$$1/v \sim U(0.01, 0.5),$$

$$\mu_t = \alpha_0 + \sum_{j=1}^{m} \alpha_j I(t \geq \Delta_j) + \beta_1 t + \sum_{j=1}^{k} \beta_{j+1}(t - \theta_j)_+,$$

with change-points $\{\Delta_j, \ldots, \Delta_m\}$ and $\{\theta_1, \ldots, \theta_k\}$ in level and trend respectively. The numbers of change-points $(m, k)$ are unknown, though the maximum number of change points in both level and trend is set at 5. Priors on the precision of the level and trend changes are preset, and follow the procedure of Thomson *et al.* (2010, p. 1435).

The second half of a two chain run of 100 000 iterations produces a lower DIC of −558, and an EPD of 0.318. The lowest $\lambda_t$ are for 1956 (cold outlier) and 1998 (warm outlier). The posterior mean numbers of shifts in level and trend are 3.5 and 3.1 respectively, and the years with highest probabilities of level shift are 1964 (with a shift probability of 0.96), with lesser level shift probabilities for the years 1946, 1896 and 1877. The highest probability for a trend shift is 0.30 in 1945.

### Example 6.13   Hepatitis A in Germany

Consider weekly counts $y_t$ (2001 through to 2004, with $T = 209$ observations) of Hepatitis A incidence in Germany from the SurvStat@RKI database (see Figure 6.8). The goal is to identify major shifts (especially epidemic onsets) as against stable occurrence levels. A modified version of the survival-birth process described in section 6.6.2 is applied. Thus $y_t = x_t + z_t$, with

$$x_t \sim \text{Po}(v_t),$$

$$\log(v_t) = \beta_0 + \beta_1 \cos(2\pi t/52) + \beta_2 \sin(2\pi t/52) + u_t,$$

$$u_t \sim N(0, 1/\tau_u)$$

$$z_t | y_{t-1} \sim \text{Po}(\lambda_t y_{t-1})$$

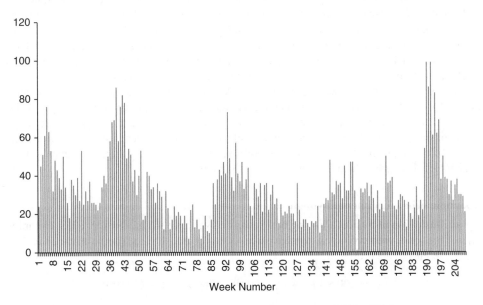

**Figure 6.8**   Hepatitis A incidence in Germany 2001–2004.

with autoregressive parameters $\lambda_t$ taken as positive, and following a gamma hyperdensity with level $\alpha_k$ chosen from a discrete set of $m = 5$ possible values. A prior is not set on change-points per se, but these can be identified by weeks with $\lambda_t \geq 1$. Thus with $K_t$ denoting a categorical variable with $m = 5$ possible values

$$\lambda_t \sim \mathrm{Ga}(a_\lambda \Lambda_t, a_\lambda)$$
$$a_\lambda \sim E(1)$$
$$\Lambda_t = \alpha[K_t]$$
$$K_t \sim \mathrm{Categoric}(\pi_{1:5})$$
$$\pi_{1:5} \sim D(0.2, 0.2, 0.2, 0.2, 0.2).$$

Exponential priors with mean 1 on potential levels $\alpha_k$ are adopted (Held et al., 2006, p. 427).

The BUGS defined likelihood device is applied to reproduce the survival-birth process, as in the code[7] fragment

```
{for (t in 2:T) {h[t] <- 0; h[t] ~ dpois(negLL[t])
# birth process, z[t] =y[t]-x[t]
            LL[t] <- -mu.w[t] + (y[t]-x[t])*log(mu.w[t])
               -logfact(y[t]-x[t]); negLL[t] <- -LL[t]
            mu.w[t] <- lam[t]*y[t-1]
# survivor process
            x[t] ~ dpois(nu[t]) I(,y[t]) ...}
```

The last 5000 iterations of a two chain run of 15 000 in OpenBUGS show the highest shift probabilities $\Pr(\lambda_t \geq 1)$ for times $t = 22, 170, 189$ and $190$, respectively 0.92, 0.80, 0.86, and 0.99. The prior $\lambda_t \sim \mathrm{Ga}(a_\lambda \Lambda_t, a_\lambda)$ on $\lambda_t$ has equal mass either side of 1, so values of $\Pr(\lambda_t \geq 1)$ exceeding 0.9 are strong evidence in favour of shifts.

A second model providing a different perspective assumes a three class latent state variable, with shifts between Poisson means determined by a Markov chain with a $3 \times 3$ transition matrix $P$. Dirichlet priors for each row of $P$ are adopted, such that $p_{i,1:3} \sim D(1, 1, 1)$. The same prior is used for the first period latent state. For the Poisson means $\nu_j$ associated with latent state $s_t = j$, $\mathrm{Ga}(1, 1)$ priors are stipulated, with an identifiability constraint ranking the means such that the first is largest (i.e. state 1 represents high incidence levels).

The second half of a two chain run of 10 000 iterations produces successive posterior means for the Poisson parameters (mean weekly incidence totals) of 64.6, 34.2, and 17.7. The low incidence state (which accounts for 31% of all time points) has a low probability $P_{31} = 0.026$ of shifting to the high incidence state, while the medium incidence state has a probability $P_{21} = 0.051$ of shifting to high incidence. Shift probabilities into the high incidence state are highest for times $t = 189$ at 0.98, and for $t = 93$ at 0.97.

### Example 6.14    US unemployment

As an illustration of models allowing both mean and variance shifts, consider a US unemployment time series with six month average percent rates $U_t$ transformed (both variance stabilised and differenced) according to

$$y_t = 100 \times \ln(1 + U_{t+1}/100) - 100 \times \ln(1 + U_t/100).$$

Monthly data from 1954 to 1992 inclusive provides $T = 77$ transformed observations for a stationary autocorrelated error model

$$y_t = \mu_t + \epsilon_t,$$

$$\epsilon_t = \gamma \epsilon_t + u_t,$$

$$u_t \sim N(0, V_t).$$

Assume binary indicators $d_{1t}$ and $d_{2t}$ for shifts in means $\mu_t$ and residual variances $V_t$ respectively. These are Bernoulli with unknown probabilities $\eta_1$ and $\eta_2$. Shifts in means occur according to

$$\mu_t = \mu_{t-1} + v_t d_{1t},$$

where $v_t$ are zero-mean effects, while shifts in the variance of $u_t$ occur according to

$$V_t = V_{t-1}(1 + \omega_t d_{2t}).$$

where $\omega_t$ are positive variables. The relative size of estimated $\eta_1$ and $\eta_2$ may be affected by the variances assumed for the $v_t$ and $\omega_t$. With $\mu_1 = \beta_0$, the autoregressive errors assumption (taken together with the potential for level shifts) means that the observation model can be restated as

$$y_1 = \beta_0 + \epsilon_1,$$

$$\epsilon_1 \sim N(0, V_1/(1 - \gamma^2)),$$

$$y_t = \gamma y_{t-1} + \beta_0(1 - \gamma) + v_t d_{1t} + u_t \quad t > 2.$$

A relatively low prior chance of shifts in either mean or variance is assumed, with priors $\eta_1 \sim Be(1, 19)$ and $\eta_2 \sim Be(1, 19)$, and with updates from the conditional densities. Proportional shifts in the variance have a gamma prior favouring a concentration around an average of 1, namely

$$\omega_t \sim Ga(5, 5).$$

As to the variance of the $v_t$ this can be preset, for example as a multiple (e.g. 10 times) the residual variance from classical ARMA models (Rosenberg and Young, 1999). Fitting ARMA(1, 1), ARMA(2, 2) and similar models showed a residual variance around 0.25 and so taking $v_t \sim N(0, 2.5)$ is one option. One might also take the variance of the $v_t$ as unknown.

The second half of a two chain run to 2500 iterations with $v_t \sim N(0, 2.5)$ provides posterior probability estimates of $\eta_1 = 0.074$ for level shifts, and $\eta_2 = 0.068$ for variance shifts. However, the estimated period by period probabilities that $d_{1t} = 1$ and $d_{2t} = 1$ suggest significant level shifts are more likely that significant variance shifts. The probabilities $Pr(d_{1t}|y)$ have highest values (respectively 0.84 and 0.87, as compared to the prior probabilities of 0.05) at extreme values in the transformed series, namely at times $t = 8$ and $t = 42$. The lag 1 parameter $\gamma$ autocorrelation parameter is estimated as 0.39. The LPML is $-64$ with the smallest CPO, around 1% of the maximum CPO, being for observation 42.

For comparison a state space regime switching model in the (untransformed) percentage unemployment rates (see Figure 6.9) is also fitted. This series is slightly different from the transformed series in the location of extreme values. This model involves a first order random walk model in the underlying signal, and switching can occur in the level of the signal and

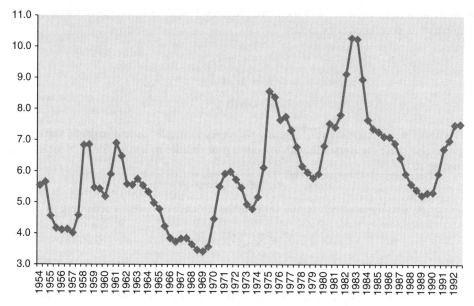

**Figure 6.9**   US unemployment (% rate).

also in the observation residual variance. Thus for $t = 2, \dots, T$ with $T = 78$

$$y_t \sim N(f_t, V_{s_{2t}}),$$
$$f_t \sim N(\mu_{s_{1t}} + f_{t-1}, 1/\tau_f).$$

Changes between latent states $s_{1t} \in \{1, 2\}$ and $s_{2t} \in \{1, 2\}$ occur according to $2 \times 2$ Markov chains $P_1$ and $P_2$. Dirichlet priors for each row of $P_1$ and $P_2$ are adopted, such that $p_{1i,1:2} \sim D(1, 1)$ and $p_{2i,1:2} \sim D(1, 1)$ (Prado and West, 2010). The same priors are adopted for the latent states in the first period. For the means $\mu_j$ associated with latent state $s_{1t}$, normal priors include an identifiability constraint ranking the means such that the first is largest. For the variances $V_j$ associated with latent state $s_{2t}$, log-normal priors include an identifiability constraint ranking the variances such that the first is largest. To improve identifiability a corner constraint is included in the state equation, so that

$$f_t \sim N(\mu_{s_{1t}} + f_{t-1} - f_1, 1/\tau_f)$$

and the regime means are obtained as $\mu^*_j = \mu_j + f_1$.

The second half of a two chain run of 10 000 iterations shows high level shift probabilities coincident with short-run upturns in unemployment, such as in 1958 (times $t = 9, 10$) and 1961 ($t = 15, 16$), and also with subsequent declines in unemployment rates (e.g. in 1959 and 1962). Volatility in variance is also detected around 1958 (at times $t = 9$ and $t = 11$), and there is a considerable contrast between the regime standard deviations (sqrtV in the code), namely 1.1 versus 0.06. Of the 78 periods, 21 are spent in the higher level regime (with $\mu^*_1 = 5.7$ as against $\mu^*_2 = 4.7$ in the low level regime).

### Example 6.15   Consumption function for France

To illustrate regression switching, consider a stochastic consumption function for France using log transformed values of consumption and income for 116 quarterly points from 1963Q1 to

1991Q4. The outcome is the first difference $\Delta \log C_t$ in logged consumption and is related to the comparable income variable $\Delta \log Y_t$, and to the lag 4 difference in log consumption. Conditioning on the first five observations gives for $t = 6, \dots, 116$

$$\Delta \log C_t = \delta + \eta D.1969Q2 + \gamma \Delta \log C_{t-4} + \beta \Delta \log Y_t + (\rho - 1)[\log C_{t-1} - v \log Y_{t-1}] + u_t,$$

where $u_t \sim N(0, \sigma^2)$. The term D.1969Q2 (when $t = 26$) reflects a short run distortion due to the wage increases following the Matignon negotiations of 1969. As to the error correction term this reflects a long term equilibrium between consumption and income

$$\log C^* = \log K + v \log Y^*.$$

Providing $v$, the long run elasticity, is 1, then $C^* = KY^*$ and the propensity to consume $K$ is constant. In a dynamic (first differences) framework, the long term model is expressed as

$$\Delta \log C^* = v \Delta \log Y^*$$

This implies that a shift in the propensity to consume (i.e. a change in $K$) only changes the constant in the above function, with all other regression effects unaffected by the switching.

The observed propensity to consume $K$ in this period in France is distorted not just by the 1969 negotiations but by a longer run upward movement from around 0.77 in 1978 to 0.89 10 years later. The question is whether, despite this apparent trend, an underlying constant propensity can be obtained by suitable parameterisation of the consumption function, and by switching of the regression constant in this function. Bauwens et al. (2000) fit the above consumption model without any switching mechanism and find that no equilibrium is defined, so then allow a single permanent shift in the propensity to consume via a step function. This leads to $v$ under 1, a situation incompatible with a constant $K$. One may also investigate whether a single, but non-permanent, shift in the mean propensity to consume restores a stable consumption function as defined by $v = 1$ (or by a 95% credible interval for $v$ including 1). This implies a return to a previous equilibrium after the temporary transition from equilibrium.

To this end consider a double parameter transition function such that $I(\tau_1, \tau_2) = 1$ for quarters t between $\tau_1$ and $\tau_2$ (both set within the period spanned by the observations) and $I(\tau_1, \tau_2) = 0$ otherwise, so that

$$\Delta \log C_t = \delta + \kappa I(\tau_1, \tau_2) + \eta D69.2 + \gamma \Delta \log C_{t-4} + \beta \Delta \log Y_t$$
$$+ (\rho - 1)[\log C_{t-1} - v \log Y_{t-1}] + u_t.$$

A revised consumption function consistent with $v = 1$ is estimated with this transition function and Bauwens et al. (2000) obtain a final equation (where $v$ is set to 1) as follows (posterior means and SDs):

$$\Delta \log C_t = -0.0088 \quad - 0.0071\ I(\tau_1, \tau_2) + 0.019\ D69.2 - 0.26\Delta \log C_{t-4}$$
$$\qquad\ \ (0.0041) \quad (0.0017) \qquad\qquad (0.0067) \qquad\quad (0.078)$$
$$+ 0.23\Delta \log Y_t - 0.11[\log(C_{t-1}/Y_{t-1})]$$
$$\ \ (0.069) \qquad\quad (0.022)$$

with the mean of $\tau_1$ estimated as 1973.3, and of $\tau_2$ as 1984.1. The prior ranges (within $t = 6, \dots, 116$) for these two threshold parameters $\tau_1$ and $\tau_2$ are $U(29, 61)$ and $U(62, 98)$. These ranges are separated for identifiability, and chosen by trial and error (Bauwens et al., 2000, p. 250).

Estimates with this model may be sensitive to prior specifications on the single or double break points, $\tau_1$ and $\tau_2$. Thus uniform priors over the full range of times (6 to 116) may give different estimates to priors restricted to an interior subinterval (e.g. 20 to 100). Similarly a gamma prior on the $\tau_k$, such as $\tau_k \sim Ga(0.6, 0.01)$ with average 60, approximately half way through the period, and with sampling constrained to the range (6, 116) might be used, combined with a constraint $\tau_2 > \tau_1$.

There are also possible identification and convergence problems entailed in the non-linear effects of $\rho$ and $v$ in the revised consumption function, when $v$ is a free parameter. Here $\rho$ is allowed to be outside the interval $[-1, 1]$. One way to deal with the identifiability problem is to introduce a conditional prior for $v$ given $\rho$ or vice versa (Bauwens et al., 2000, p. 142), and so $\rho$ is taken to be a linear function of $v$, namely $a_1 + a_2 v$. Then $N(0, 1)$ priors are adopted on $a_1$ and $a_2$, and on all regression coefficients, with the exception of $v$ which is assigned an $N(1, 1)$ prior, constrained to positive values.

We first estimate the revised consumption function with a single break point (i.e. a permanent shift in the propensity to consume), with $v$ taken as a free parameter, and a Ga(0.6, 0.01) prior on the breakpoint $\tau_1$. Convergence in all parameters in a three chain run occurs after 15 000 iterations and from iterations 15 000–20 000, a 95% credible interval for $\rho$ of $\{1.02, 1.19\}$ is obtained and LPML of 390. The density for $v$ is concentrated below unity, with 95% interval $\{0.86, 0.95\}$. The density for $\tau_1$ is negatively skewed and has some minor modes; however, there is a major mode at around $t = 85$ to $t = 90$, with the posterior median at 87 (i.e. 1984.3).

To fit the revised consumption function with $v$ still a free parameter, and two breakpoints (i.e. a temporary shift in the propensity to consume) constrained within the intervals (29, 61) and (62, 98), are used in conjunction with gamma priors (model 3 in the BUGS code). Taking wider intervals within which sampled values may lie, such as (7, 61) for $\tau_1$, causes convergence problems. The Gelman–Rubin scale reduction factor on $v$ remains at around 1.2 after iteration 8500 in a two chain run, and the 95% interval from 5000 iterations thereafter is $\{0.88, 1.02\}$, including the equilibrium value of 1.

## Exercises

**6.1.** Consider data on 3 month market yield on US Treasury securities from January 1983 to December 2012 ($T = 360$) (from wikiposit). Classical estimation (e.g. using the arima routine in R) shows an AR(3) model to have significant lags at $t = 1$ and $t = 3$, but not at $t = 2$, while an AR(2) model has significant lags at $t = 1$ and $t = 2$. Consider the smoothness prior

$$\rho_1 \sim N(0, \omega),$$

$$\rho_k \sim N(\rho_{k-1}, \omega/\delta_k), \quad \delta_k = (1 + \Delta^{k-1}), \quad k > 1,$$

with $\Delta > 0$. The following code applies this scheme, using a half Cauchy prior on $\Delta$, while checking for non-stationarity (since stationarity is not imposed in the prior). The code also provides empirical estimates of the residual autocorrelation function, which is one aspect of model adequacy.

Thus with $P = 3$, the BUGS code is

```
model { for (t in P+1:T) {y[t] ~ dnorm(mu[t],tau)
e[t] <- y[t]-mu[t]; e2[t] <- e[t]*e[t]
mu[t] <- beta0+sum(AR.terms[t,1:P])
for (j in 1:P) {AR.terms[t,j] <- beta[j]*y[t-j]}}
```

```
# smoothness prior
beta[1] ~ dnorm(0,w[1])
for (j in 2:P) {beta[j] ~ dnorm(beta[j-1],w[j])}
for (j in 1:P) {w[j] <- inv.omega*pow(1+Delta,j-1)}
B ~ dunif(0,1000); invB2 <- 1/(B*B);
z ~ dnorm(0,invB2); gam ~ dgamma(0.5,0.5);
Delta <- abs(z)/gam
# ACF estimates
for (t in twoP+1:T) {for (j in 1:P) {e.lag[t,j] <- e[t]*e[t-j]}}
for (j in 1:P) {auto.r[j] <- sum(e.lag[twoP+1:T,j])/
                                    sum(e2[twoP+1:T])}
beta0 ~ dnorm(0,0.001); tau ~ dgamma(1,0.001);
inv.omega ~ dgamma(1,0.001)
# stationarity checks
   a[1,1] <- -1;   for (kk in 1:P) { a[kk+1,1] <- beta[kk]
      for (j in 1:P+1-kk) { b[j,kk] <- a[1,kk]
         *a[j,kk]-a[P+2-kk,kk]*a[P+3-kk-j,kk]
                              a[j,kk+1] <- b[j,kk]}
                  NonStat[kk] <- step(-b[1,kk])}}}
```

Compare estimates from this model to:

(a) a model with stationarity imposed (e.g. using priors on the partial correlations of the AR($p$) process);

(b) a model where the coefficients $\delta_k > 1$ are allowed to vary between lags $k > 1$, subject to $\delta_k > \delta_{k-1}$.

**6.2.** Apply the same model sequence used in Example 6.2 to the extended velocity series (1869–1988) (e.g. Koop and Steel, 1994). This series is identified as non-stationary (with significant probability that $\rho > 1$) by Bauwens *et al.* (2000). The data for this series are included in Exercise 6.2.odc.

**6.3.** In Example 6.3, estimate an MA(1) model in the differences $z_t$ of the wholesale prices index using both methods applied in that example.

**6.4.** In Example 6.6 (benefit claimants), re-estimate the INAR models to include the a latent preseries data point, by setting a prior $y_0 \sim Po(\omega_0)$, where $\omega_0$ may be gamma, log-normal, etc.

**6.5.** In Example 6.6 (benefits claimants), estimate the following autoregressive model (Fokianos, 2012)

$$y_t \sim Po(\mu_t),$$

$$v_t = \log(\mu_t),$$

$$v_t = X_t\beta + \rho_1 \log(y_{t-1} + 1) + \rho_2 v_{t-1},$$

where $X_t$ includes the seasonal covariates used in Example 6.6, and where the $\rho$ parameters may take both positive and negative values.

**6.6.** Consider the prediction of months classed as economic recession ($y_t = 1$) as against growth ($y_t = 0$) for the US (the data are contained in the file Exercise 6.6.odc) for the 657 months between April 1953 and December 2007 (Nyberg, 2008). Relevant predictors are S&P500 stock market returns $r_t$ (percent differences between successive months) and term spread, $sp_t$, namely long-term minus short-term interest rates.

Consider a probit regression with

$$y_t \sim \text{Bern}(\pi_t),$$

$$\pi_t = \Phi(\eta_t).$$

Compare the fit and predictions (within sample and out-of-sample) of two models:

(a) the first uses predictors $r_{t-6}$ and $\text{sp}_{t-6}$ (i.e. six month lags in returns and term spread), a one month lag in the response (i.e. $y_{t-1}$), and a stationary AR1 normal error

$$\eta_t = \beta_0 + \beta_1 r_{t-6} + \beta_2 \text{sp}_{t-6} + \beta_3 y_{t-1} + e_t, \quad t = 7, \ldots, 645,$$

$$e_t \sim N(\rho e_{t-1}, 1/\tau_e), \quad -1 < \rho < 1.$$

This model conditions on the first six months data, and requires generation of an initial value for $e_6$;

(b) a model with predictors $r_{t-6}$ and $\text{sp}_{t-6}$, and one month lags in both the response (i.e. $y_{t-1}$) and the predicted probability, but no error, namely

$$\eta_t = \beta_0 + \beta_1 r_{t-6} + \beta_2 \text{sp}_{t-6} + \beta_3 y_{t-1} + \beta_4 \pi_{t-1}.$$

This model involves setting a prior for $\pi_6$. In sample predictions can be made by sampling predictive replicates $y_{rep,t} \sim \text{Bern}(\pi_t)$, checking whether $y_t$ and $y_{rep,t}$ are the same and then measuring the overall match (combining sensitivity and specificity). Out-of-sample predictions can be compared with the actual observed 12 points for the year 2007 ($t = 646, \ldots, 657$). For the first model a possible code (note that $y$ values are only taken as known up to $t = 645$, namely the end of 2006) is

```
model { e[6] ~ dnorm(0,tau1); tau1 <- tau*pow(1-rho,2)
# out-of-sample (next year ahead)
for (t in 646:657) {y[t] ~ dbern(p[t])
pr.out[t-645] <- equals(y[t],y2007[t-645])
p[t] <- phi(b[1]+b[2]*sp[t-6]+b[3]*r[t-6]+b[4]*y[t-1]+e[t])
e[t] ~ dnorm(e.m[t],tau); e.m[t] <- rho*e[t-1]}
# in-sample
for (t in 7:645) {y[t] ~ dbern(p[t]);     y.rep[t] ~ dbern(p[t])
                         pr.in[t] <- equals(y[t],y.rep[t])
p[t] <- phi(b[1]+b[2]*sp[t-6]+b[3]*r[t-6]+b[4]*y[t-1]+e[t])
e[t] ~ dnorm(e.m[t],tau); e.m[t] <- rho*e[t-1]}
# predictive match
Pr.match[1] <- sum(pr.in[7:645])/639;
Pr.match[2] <- sum(pr.out[1:12])/12
# priors
tau ~ dgamma(1,0.001); rho ~ dunif(-1,1)
for (j in 1:4) {b[j] ~ dnorm(0,0.001)}}
```

Possible initial values (e.g. Nyberg, 2008) are 3 for the coefficient on $y_{t-1}$, and $-2$ for the intercept.

**6.7.** In Example 6.8, fit a true series evolving according to an RW(3) model such that

$$\theta_t = 3\theta_{t-1} - 3\theta_{t-2} + \theta_{t-3} + \omega_t,$$

and assess fit against the RW(2) model.

**6.8.** The observed Intel monthly returns for 2004 (the first comparing January 2004 with December 2003) are −0.047, −0.042, −0.069, −0.054, 0.112, −0.033, −0.117, −0.125, −0.058, 0.109, 0.007, and 0.045. Compare the models in Example 6.11 in terms of their out-of-sample predictive performance for these future 12 observations.

**6.9.** In Example 6.13 (Hepatitis A) estimate a latent category Markov chain model with $m = 4$ states, and identify the time points with highest probabilities of shifts into the highest incidence state.

**6.10.** In Example 6.14 (US unemployment), fit model A with an unknown variance for the level shifts $v_t$. How does this affect the ratio of maximum to minimum CPO?

**6.11.** Consider the lynx data available in the R package tsDyn. These data consist of annual totals of Canadian lynx trapped in the Mackenzie River district of NW Canada during 1821–1934. Define $y_t$ as a log10 transformation of the original series. The regression model to be considered is

$$y_t = \beta_0 + \beta_1 y_{t-1} + \beta_2 y_{t-1} + u_t, \qquad u_t \sim N(0, \sigma^2)$$

where the regression coefficients and the residual variance are subject to different regimes, and SETAR threshold functions are defined as

$$K_t = y_{t-2} - \delta_j, \quad j = 1, \dots, J - 1.$$

Compare models with regimes involving $J = 2$ and $J = 3$ states, for example using the LPML, and obtain the proportion of periods spent in each regime. A code for the two regime case (conditioning on $y_1$ and $y_2$, and without a stationarity constraint on the AR parameters) is

```
model { for (t in 3:114) {y[t] ~ dnorm(mu[t],tau[S[t]])
LL[t] <- 0.5*log(tau[S[t]])-0.5*tau[S[t]]*pow(y[t]-mu[t],2)
InvLk[t] <- 1/exp(LL[t])
mu[t] <- b[S[t],1]+b[S[t],2]*y[t-1]+b[S[t],3]*y[t-2]
# define latent states according to
# whether threshold function applies
S[t] <- 1+step(y[t-2]-delta)}
# prior on threshold (uses minimum and maximum of transformed data)
delta ~ dunif(1.591,3.845)
log.tau[1] ~ dnorm(0,0.01); log.tau[2]   ~ dnorm(0,0.01)
for (k in 1:2) {tau[k] <- exp(log.tau[k])
for (j in 1:3) {b[k,j] ~ dnorm(0,0.001)}}}
```

with an example initial values file being

```
list(b=structure(.Data=c(0,0,0,0,0,0),.Dim=c(2,3)),log.tau=c(0,0),
                                                       delta=3).
```

# Notes

1. These data kindly provided by Peyton Cook.

2. The code for this model assumes that $\{y[1], y[2], y[3]\}$ in the program are the three latent data values $\{y_{-2}, y_{-1}, y_0\}$, with $y$ [4] to $y$ [83] corresponding to the actual observations for 1909–1988. The code is

```
model  { # outlier model
for (t in 4:T+3) { delta[t] ~ dbern(Delta)
                     eta[t] ~ dnorm(0,tau.eta)
                     o[t] <- eta[t]*delta[t]}
                     Delta ~ dbeta(1,19);            tau.eta <- tau.G/10
# main model
for (t in 4:T+3) { y[t] ~ dnorm(m[t],tau[t])
                     y.new[t]  ~ dnorm(m[t],tau[t])
                     e[t] <- pow(y[t]-y.new[t],2)
# weights for scale mixture
                     w[t] ~ dgamma(nu.2,nu.2)
                     tau[t] <- w[t]*tau.G;
m[t] <-  mu+o[t]+lambda*t+rho*(y[t-1]-o[t-1])
+phi[1]*(y[t-1]-y[t-2])+phi[2]*(y[t-2]-y[t-3])
# log likelihood and inverse likelihood
                     LL[t] <- 0.5*log(tau[t]/6.28)-0.5*tau[t]
                                     *pow(y[t] - m[t],2)
# CPO estimated by inverse of posterior average of InvLk[]
                     InvLk[t] <- 1/exp(LL[t])}
# one step ahead predictions
for (t in 5:T+3) { m.p[t-1] <- m[t-1]+lambda
                     y.one[t] ~ dnorm(m.p[t-1],tau[t])
                     e2.one[t] <-pow(y[t]-y.one[t],2)}
# assess stationarity
                NSTAT <- step(rho-1)
# Predictive error
                     E[1] <- sum(e2.one[5:T+3])/(T-1)
                     E[2] <- sum(e[4:T+3])
# Priors
rho ~ dnorm(0,1);       mu ~ dnorm(0,0.001)
for (j in 1:2) {phi[j] ~ dnorm(0,1)}
nu ~ dgamma(1,0.001) I(1,100);   nu.2 <- nu/2
tau.G ~ dgamma(1,0.001)
# trend coefficient
lambda ~ dnorm(0,1)
# pre-series values
for (t in 1:3) {y[t] ~ dt(mu.r,tau1,2); o[t] <- 0}
                     mu.r <- mean(m[7:T+3])
tau1 <- tau.G/kappa
km ~ dgamma(0.01,0.01);        kappa <- km+1}
```

3. The BUGS code for the INAR model without covariates (Example 6.6) is

```
model {for (t in 2:T) {h[t] <- 0; h[t] ~ dpois(negLL[t]);
                                     negLL[t] <- -LL[t]
# h[t] ~ dloglik(LL[t])   # OpenBUGS own likelihood
LL[t] <- -lambda+(y[t]-x[t])*log(lambda)-logfact(y[t]-x[t])
x[t] ~ dbin(rho,y[t-1]) I(,y[t])}
# out of sample predictions
xnew[T+1] ~ dbin(rho,y[T]); wnew[T+1] ~ dpois(lambda);
ynew[T+1] <- xnew[T+1]+wnew[T+1]
for (t in T+2:T+12) {wnew[t] ~ dpois(lambda)
xnew[t] ~ dbin(rho,ynew[t-1]);   ynew[t] <- xnew[t]+wnew[t]}
# in-sample predictions
ynew[1] <- y[1]
for (t in 2:T) {xnew[t] ~ dbin(rho,ynew[t-1]);
```

```
                  wnew[t]  ~ dpois(lambda)
                  ynew[t]  <- xnew[t] + wnew[t]
                  ch[t]    <- step(ynew[t] - y[t]) - 0.5*equals(ynew[t],y[t])
                  yts[t]   <- y[t] + eps; ytsnew[t]  <- ynew[t] + eps;
                  epd[t]   <- 2*(y[t]*log(yts[t]/ytsnew[t])
                                          - (y[t] - ynew[t]))}
              EPD <- sum(epd[2:T])
# priors
lambda ~ dgamma(1,0.001); rho ~ dbeta(1,1)}
```

4. The code for the latent factor model in Example 6.7 is

```
model {for (t in 1:168) {y[t] ~ dpois(m[t])
# Log-Likelihood
          L[t] <- -m[t] + y[t]*log(m[t]);
          m[t] <- nu[t]*F[t]
          log(nu[t]) <- beta[1]*x[t,1] + beta[2]*x[t,2]
                      + beta[3]*x[t,3] + beta[4]*x[t,4]
          F[t] ~ dgamma(c[t],d[t]);          d[t] <- 1/(omega*b[t])}
for (t in 2:168) {c[t] <- F[t-1]/omega
# trend model (z[t] = t)
          log(b[t]) <- alpha *(z[t] - z[t-1])}
F.0 <- 1; c[1] <- F.0/omega
b.1 <- 1; b[1] <- b.1
omega <- omegr*omegr; omegr ~ dunif(0,10)
alpha ~ dnorm(0,0.01)
for ( j in 1:4) {beta[j] ~ dnorm(0,0.01)}}
```

5. The relevant R-INLA code (Example 6.9) is

```
require(INLA);
# columns headed year, month, y
D <- read.table("arctic_sea_ice.txt",header = T)
y <- D$y
# extend data to include forecast horizon
f <- 12;  D <- c(y,rep(NA,f))
n = length(D); m = n-1
# data for augmented structure
Y = matrix(NA, n+m, 2); Y[1:n,1] = D; Y[1:m + n,2] = 0
# model indices
i = c(1:n, 2:n)                          # T_t
j = c(rep(NA,n), 1:m)               # T_{t-1}
wt1 = c(rep(NA,n), rep(-1,m))      # weights for j
l = c(rep(NA,n), 1:m)               # \delta_{t-1}
wt2 = c(rep(NA,n), rep(-1,m))      # weights for l
w1 = c(rep(NA,n), 2:n)             # w_{1,t}
q = c(1:n, rep(NA,m))             # S_t
# model
formula = Y ~ f(q, model = "seasonal", season.length = 12,initial = 4) +
              f(l, wt2, model = "rw1",initial = 4, constr = F) +
    f(i, model = "iid", initial = -10, fixed = TRUE)
    + f(j, wt1, copy = "i") + f(w1, model = "iid") -1
# skew-normal and Gaussian assumptions
rsn = inla(formula, data = data.frame(i,j,wt1,l,wt2,q,w1),family = rep("sn",2),
              quantiles = c(0.025, 0.05, 0.5, 0.95, 0.975), control.compute
                        = list(graph = T, dic = T),
```

```
                control.family=list(list(),list(initial=10, fixed=T)),
                                control.predictor=list(compute=T))
    rgn=inla(formula, data=data.frame(i,j,wt1,l,wt2,q,w1),
                                family=rep("gaussian",2),
                quantiles=c(0.025, 0.05, 0.5, 0.95, 0.975),
                    control.compute=list(graph=T, dic=T),
                control.family=list(list(),list(initial=10, fixed=T)),
                    control.predictor=list(compute=T))
# Plot of Underlying Level with 90%CI
R <- range(rsn$summary.random$i[1:n, 4:8])
plot(rsn$summary.random$i[1:n,2], type="l",
 ylim=R, col="red", xlim=c(1,n),ylab="Ice Extent",
                xlab="Time (months from Jan 1979)")
lines(rsn$summary.random$i[1:n,5], col="blue", lty=3)
lines(rsn$summary.random$i[1:n,7], col="blue", lty=3)
legend("bottomleft",legend=c("posterior mean","90% CI"),
                col=c( "red","blue"),lty=c(1,1,2),bty="n")
title("Figure 6.5 Arctic ice cover 1979-2011 change in level")
# Plot of seasonal effect
rang <- range(rsn$summary.random$q[1:n, 4:8])
plot(rsn$summary.random$q[1:n,2], type="l",
 ylim=rang, col="red", xlim=c(1,n),ylab="z",xlab="time")
lines(rsn$summary.random$q[1:n,5], col="blue", lty=3)
lines(rsn$summary.random$q[1:n,7], col="blue", lty=3)
legend("topleft",legend=c("observed","posterior mean","90% CI"),
            col=c("black", "red","blue"),lty=c(1,1,2),bty="n")
#title("Seasonal term")
```

6. The BUGS code for this option to analyse temperature shifts (Example 6.12) is

```
model{ for(t in 1:T) {y[t]   ~ dnorm(theta[t],obs.prec[t])
                                # observation model
                        theta[t] ~ dnorm(mu[t],tau[t])
                                # underlying signal
                        tau[t] <- tau.sig*lam[t]
                                # scale mixing
                        lam[t] ~ dgamma(nu.2,nu.2)
        mu[t]   <- alph0+alph[1]*step(t-Del[1])+alph[2]*step(t-Del[2])
                                +alph[3]*step(t-Del[3])}
# prior on change-points
        cT <- c*T;
        c ~ dunif(0.0062,2);
        Del[1] ~ dunif(1,cT);
        for (j in 2:m) {T.Del[j] <- Del[j-1]+cT; Del[j]
                        ~ dunif(Del[j-1],T.Del[j])}
        for (j in 1:m) {Year.shift[j] <- 1849+Del[j]}
# regime levels
        for (j in 1:m) {alph[j]~dnorm(0,0.0001)}
        alph0 ~ dnorm(0,0.0001);
        Lev[1] <- alph0;
        for (j in 2:m+1) {Lev[j] <- Lev[j-1]+alph[j-1]}
# df for scale mixing
        nu.2 <- nu/2; nu <- 1/kap; kap ~ dunif(0.01,0.5)
# prior on signal precision
        tau.sig ~ dgamma(1,0.001)
# predictive check
for (t in 1:T) {ynew[t] ~ dnorm(theta[t],obs.prec[t])
```

```
                   epd[t]  <- pow(y[t]-ynew[t],2)
                   chi.2[1,t]<-(y[t]-theta[t])*(y[t]-theta[t])/theta[t]
                   chi.2[2,t]<-(ynew[t]-theta[t])*(ynew[t]-theta[t])
                                                        /theta[t] }
     chi2.sum[1]<-sum(chi.2[1,]);
     chi2.sum[2]<-sum(chi.2[2,])
     check <-step(chi2.sum[1]-chi2.sum[2]);
     EPD <- sum(epd[]) }
```

7. The full code for the survival-birth process for the Hepatitis incidence data (Example 6.13) is

```
model {for (t in 2:T) {h[t]  <- 0; h[t]  ~ dloglik(LL[t])
LL[t]  <- -mu.w[t]+(y[t]-x[t])*log(mu.w[t])-logfact(y[t]-x[t]);
                                        negLL[t]  <- -LL[t]
                   mu.w[t]  <- lam[t]*y[t-1]
# survivor process
                   x[t]  ~ dpois(nu[t]) I(,y[t])
# stable patterns
                   log(nu[t])  <- beta0+sum(seas[t,1:S])+u[t]
                   u[t]  ~ dnorm(0,tau)
for (s in 1:S) {seas[t,s]  <- beta[2*s-1]*cos(6.28*s*t/52)+beta[2*s]
                                        *sin(6.28*s*t/52)}
# autoregressive component
                   lam[t]  ~ dgamma(m.lam[t],a.lam)
                   m.lam[t]  <- a.lam*mu.lam[t]
                   mu.lam[t]<-  alph[K[t]]
                   shift[t]  <- step(lam[t]-1)
                   K[t]  ~ dcat(pi[1:m])}
# priors
        tau ~ dgamma(1,0.01); beta0 ~ dnorm(0,0.01)
        for (k in 1:m) {alph[k] ~ dexp(xi); omeg[k]  <- 1/m}
        xi ~ dgamma(10,10); a.lam ~ dexp(1)
        pi[1:m] ~ ddirich(omeg[1:m])
for (s in 1:S) {beta[2*s-1] ~dnorm(0,0.01);   beta[2*s] ~ dnorm(0,0.01)}}
```

# References

Achcar, J., Martinez, E., Ruffino-Netto, A., Paulino, C. and Soares, P. (2008) A statistical model investigating the prevalence of tuberculosis in New York City using counting processes with two change-points. *Epidemiology and Infection*, **136**(12),1599–1605.

Akaike, H. (1986) The selection of smoothness priors for distributed lag estimation. In P. Goel and A. Zellner (eds), *Bayesian Inference and Decision Techniques*, pp. 109–118. Elsevier, Oxford, UK.

Albert, J. and Chib, S. (1993) Bayes inference via Gibbs sampling of autoregressive time series subject to Markov mean and variance shifts. *Journal of Business and Economic Statistics*, **11**, 1–15.

Ameen, J. and Harrison, P. (1985) Normal discount Bayesian models. In J. Bernardo, M. De Groot, D. Lindley and A. Smith (eds), *Bayesian Statistics 2*, pp. 271–298. North-Holland, Amsterdam.

Ardia, D. and Hoogerheide, L. (2010) Efficient Bayesian estimation and combination of GARCH-type models. *Rethinking Risk Measurement and Reporting: Examples and Applications from Finance*, 2.

Ardia, D. and Hoogerheide, L. (2012) Bayesian estimation of the GARCH(1,1) model with Student-t innovations. *R Journal*, **2**(2), 41–47.

Asai, M. (2006) Comparison of MCMC methods for estimating GARCH models. *Journal of the Japananese Statististical Society*, **36**(2), 199–212.

Baltagi, B. (2011) *Econometrics*, 5th edn. Springer, New York, NY.

Barnett, G., Kohn, R. and Sheather, S. (1996) Bayesian estimation of an autoregressive model using Markov Chain Monte Carlo. *Journal of Econometrics*, **74**(2), 237–254.

Barry, D. and Hartigan, J.A. (1993) A Bayesian analysis for change point problems. *Journal of the American Statistical Association*, **88**, 309–319.

Bauwens, L. and Lubrano, M. (1998) Bayesian inference on GARCH models using the Gibbs sampler. *The Econometrics Journal*, **1**, C23–C46.

Bauwens, L., Lubrano, M. and Richard, J. (2000) *Bayesian Inference in Dynamic Econometric Models*. Oxford University Press, New York, NY.

Bauwens, L., Dufays, A. and De Backer, B. (2011) Estimating and forecasting structural breaks in financial time series. CORE Discussion Paper 2011/55.

Berg, A., Meyer, R. and Yu, J. (2004) Deviance information criterion for comparing stochastic volatility models. *Journal of Business and Economic Statistics*, **22**(1), 107–120.

Bollerslev, T., Litvinova, J. and Tauchen, G. (2006) Leverage and volatility feedback effects in high-frequency data. *Journal of Financial Econometrics*, **4**(3), 353–384.

Box, G. and Jenkins, G. (1976) *Time Series Analysis: Forecasting and Control*. Holden-Day, San Francisco, CA.

Brandt, P. and Sandler, T. (2012) A Bayesian Poisson vector autoregression model. *Political Analysis*, **20**(3), 292–315.

Broemeling, L. and Cook, P. (1993) Bayesian estimation of the mean of an autoregressive process. *Journal of Applied Statistics*, **20**, 25–39.

Cameron, A. and Trivedi, P. (1999) *Regression Analysis of Count Data*. Oxford University Press, Oxford, UK.

Canova, F. (2007) *Methods for Applied Macroeconomic Research*. Princeton University Press, Princeton, NJ.

Cardinal, M., Roy, R. and Lambert, J. (1999) On the application of integer-valued time series models for the analysis of disease incidence. *Statistics in Medicine*, **18**, 2025–2039

Cargnoni, C., Müller, P. and West, M. (1997) Bayesian forecasting of multinomial time series through conditionally Gaussian dynamic models, *Journal of the American Statistical Association*, **92**, 640–647.

Caron, F., Doucet, A. and Gottardo R. (2012) On-line changepoint detection and parameter estimation with application to genomic data. *Statistics and Computing*, **22**(2), 579–595.

Carter, C. and Kohn, R. (1994) On Gibbs sampling for state space models. *Biometrika*, **81**, 541–553.

Chan, K. and Ledolter, J. (1995) Monte-Carlo EM Estimation for time series models involving counts. *Journal of the American Statistical Association*, **90**(429), 242–252.

Chen, C. and Lee, J. (1995) Bayesian inference of threshold autoregressive models. *Journal of Time Series Analysis*, **16**(5), 483–492.

Chib, S. (1996) Calculating posterior distributions and modal estimates in Markov mixture models. *Journal of Econometrics*, **75**(1), 79–97.

Chib, S. (1998) Estimation and comparison of multiple change-point models. *Journal of Econometrics*, **86**, 221–41.

Chib, S. and Greenberg, E. (1994) Bayes inference in regression models with ARMA(p,q) errors. *Journal of Econometrics*, **64**, 183–206.

Chiogna, M. and Gaetan, C. (2002) Dynamic generalized linear models with application to environmental epidemiology. *Journal of the Royal Statistical Society C*, **51**(4), 453–468.

Chu, P. and Xin, Z. (2007) A Bayesian regression approach for predicting seasonal tropical cyclone activity over the Central North Pacific. *Journal of Climate*, **20**, 4002–4013.

De Pooter, M., Segers, R. and van Dijk, H. (2006) On the practice of Bayesian inference in basic economic time series models using gibbs sampling. Tinbergen Institute Discussion Paper, 2006–076/4.

Durbin, J. and Koopman, S. (2000) Time series analysis of non-Gaussian observations based on state space models from both classical and Bayesian perspectives. *Journal of the Royal Statistical Society B*, **62**, 3–29.

Eckley, I., Fearnhead, P. and Killick, R. (2011) Analysis of changepoint models. In D. Barber, A.T. Cemgil and S. Chiappa (eds), *Bayesian Time Series Models*, pp. 203–224. Cambridge University Press, Cambridge, UK.

Elliott, M. and Shope, J. (2003) Use of a Bayesian changepoint model to estimate effects of a graduated driver's licensing program. *Journal of Data Science*, **1**, 43–63.

Enciso-Mora, V., Neal, P. and Subba Rao, T. (2009) Integer valued AR processes with explanatory variables. *Sankhya*, **71B**(2), 248–263.

Enders, W. (2004) *Applied Econometric Times Series*, 2nd edn. Wiley, Chichester, UK.

Engle, R. (1982) Autoregressive conditional heteroscedasticity with estimates of the variance of United Kingdom inflation. *Econometrica*, **50**, 987–1007.

Fahrmeir, L. and Lang, S. (2001) Bayesian inference for generalized additive mixed models based on Markov random field priors. *Journal of the Royal Statistical Society C*, **50**, 201.

Fahrmeir, L. and Kneib, T. (2011) *Bayesian Smoothing and Regression for Longitudinal, Spatial and Event History Data*. Oxford University Press, Oxford, UK.

Fahrmeir, L. and Tutz, G. (2001) *Multivariate Statistical Modelling based on Generalized Linear Models*, 2nd edn. Springer, New York, NY.

Farrell, P., MacGibbon, B. and Tomberlin, T. (2007) A hierarchical Bayes approach to estimation and prediction for time series of counts. *Brazilian Journal of Probability and Statistics*, **21**, 187–202.

Fearnhead, P. (2006) Exact and efficient Bayesian inference for multiple changepoint problems. *Statistics and Computing*, **16**, 203–213.

Ferreira, M. and Gamerman, D. (2000) Dynamic generalized linear models. In J. Ghosh, D. Dey and B. Mallick (eds.) *Generalized Linear Models: a Bayesian Perspective*. Marcel Dekker, New York, NY.

Fokianos, K. (2011) Some recent progress in count time series. *Statistics*, **45**, 49–58.

Fokianos, K. (2012) Count time series models. In T. Subba Rao, S. Subba Rao and C. R. Rao (eds), *Handbook of Statistics. Vol 30: Time Series – Methods and Applications*, pp. 315–347. Elsevier B.V., Amsterdam.

Fokianos, K., Rahbek, A. and Tjøstheim, D. (2009) Poisson autoregression. *Journal of the American Statistical Association*, **104**, 1430–1439.

Franke, J. and Seligmann, T. (1993) Conditional maximum likelihood estimates for INAR1 processes and their application to modelling epileptic seizure counts. In T. Subba Rao (ed), *Developments in Time Series Analysis*, pp. 310–330. Chapman & Hall, London, UK.

Freeland, R. and McCabe, B. (2004) Analysis of low count time series data by Poisson autoregression. *Journal of Time Series Analysis*, **25**(5), 701–722.

Fruhwirth-Schnatter, S. (1994) Data augmentation and dynamic linear models. *Journal of Time Series Analysis*, **15**, 183–202.

Frühwirth-Schnatter, S. (2006) *Finite Mixture and Markov Switching Models*. Springer, New York, NY.

Gao, M., Chang, X. and Wang, X. (2012) Bayesian parameter estimation in dynamic population model via particle Markov chain Monte Carlo. *Computational Ecology and Software*, **2**(4), 181–197.

Gelfand, A. E. and Dey, D. K. (1994). Bayesian model choice: asymptotics and exact calculations. *Journal of the Royal Statistical Society*, **56B**, 501–514.

Gelfand, A. E. and Ghosh, S. K. (1998). Model choice: A minimum posterior predictive loss approach. *Biometrika*, **85**(1), 1–11.

George, E. and McCulloch, R (1993) Variable selection via Gibbs sampling. *Journal of the American Statistical Association*, **88**, 881–889.

Geweke, J. and Amisano, G. (2010) Comparing and evaluating Bayesian predictive distributions of asset returns. *International Journal of Forecasting*, **26**(2), 216–230.

Geweke, J. and Jiang, Y. (2011) Inference and prediction in a multiple structural break model. *Journal of Econometrics*, **163**(2), 172–185.

Geweke, J. and Terui, N. (1993) Bayesian threshold auto-regressive models for nonlinear time series. *Journal of Time Series Analysis*, **14**(5), 441.

Geweke, J. and Whiteman, C. (2006). Bayesian forecasting. In G. Elliott, C.W.J. Granger and A. Timmermann (eds), *Handbook of Economic Forecasting*, pp. 4–80. Elsevier, Amsterdam.

Giakoumatos, S., Dellaportas, P. and Politis, D. (2005) Bayesian analysis of the Unobserved ARCH Model. *Statistics and Computing*, **15**, 103–111.

Giordani, P., Kohn, R. and van Dijk, D. (2007) A unified approach to nonlinearity, structural change, and outliers. *Journal of Economics*, **137**, 112–133.

Grunfeld, Y. and Griliches, Z. (1960) Is aggregation necessarily bad? *Review of Economics and Statistics*, **42**, 1–13.

Harvey, A., Ruiz, E. and Shepherd, N. (1994) Multivariate stochastic variance models. *Review of Economic Studies*, **61**, 247–264.

Harvey, A., Koopman, S. and Shephard, N. (2004) *State Space and Unobserved Component Models: Theory and Applications*. Cambridge University Press, Cambridge, UK.

Held, L., Hofmann, M., Höhle, M. and Schmid, V. (2006) A two component model for counts of infectious diseases. *Biostatistics*, **7**, 422–437.

Hoek, H., Lucas, A. and Vandijk, H. (1995) Classical and Bayesian aspects of robust unit-root inference. *Journal of Econometrics*, **69**, 27–59.

Johannes, M. and Polson, N. (2009) MCMC methods for financial econometrics. In Y. Ait-Sahalia and L. Hansen (eds), *Handbook of Financial Econometrics*, pp. 1–72. Elsevier, Oxford, UK.

Jørgensen, B., Lundbye-Christensen, S., Song, P. and Sun, L. (1999) A state space model for multivariate longitudinal count data. *Biometrika*, **86**, 169–181.

Jung, R. and Tremayne, A. (2011) Useful models of time series of counts or simply wrong ones? *Advances in Statistical Analysis*, **95**, 59–91.

Jung, R., Kukuk, M. and Liesenfeld, R. (2006) Time series of count data: modeling, estimation and diagnostics. *Computational Statistics and Data Analysis*, **51**, 2350–2364.

Kadiyala, K. and Karlsson, S. (1997) Numerical methods for estimation and inference in Bayesian VAR-models. *Journal of Applied Economics*, **12**, 99–132.

Kauppi, H. and Saikkonen, P. (2008) Predicting U.S. recessions with dynamic binary response models. *Review of Economics and Statistics*, **90**(4), 777–791.

Kedem, B. and Fokianos, K. (2002) *Regression Models for Time Series Analysis*. Wiley, Chichester, UK.

Keenan, D. (1982) A time series analysis of binary data. *Journal of the American Statistical Association*, **77**, 816–821.

Kim, S., Shephard, N. and Chib, S. (1998) Stochastic volatility: likelihood inference and comparison with ARCH models. *Review of Economic Studies*, **65**(3), 361–393.

Kitagawa, G. and Gersch, W. (1996) *Smoothness Priors Analysis of Time Series*. Springer, New York, NY.

Kleibergen, F. and Hoek, H. (2000) Bayesian analysis of ARMA Models. TI 2000-027/4, Tinbergen Institute Discussion Paper, Tinbergen Institute.

Koop, G. and Potter, S. (1999) Dynamic asymmetries in U.S. unemployment. *Journal of Business and Economic Statistics*, **17**, 298–312.

Koop, G. and Potter, S. (2009) Prior elicitation in multiple change-point models. *International Economic Review*, **50**(3), 751–772.

Koop, G. and Steel, M. (1994) A decision-theoretic analysis of the unit-root hypothesis using mixtures of elliptical models. *Journal of Business and Economic Statistics*, **12**, 95–107.

Koop, G., Osiewalski, J. and Steel, M. (1995) Bayesian long-run prediction in time series models. *Journal of Econometrics*, **69**, 61–80.

Kuo, L. and Mallick, B. (1998) Variable selection for regression models. *Sankhya*, **60B**, 65–81.

Lai, T. and Xing, H. (2011) A simple Bayesian approach to multiple change-points. *Statistica Sinica*, **21**, 539–569.

Lau, E., Cheng, C., Ip, D. and Cowling, B. (2012) Situational awareness of influenza activity based on multiple streams of surveillance data using multivariate dynamic linear model. *PLoS One*, **7**(5), e38346.

Leroux, B. and Puterman, M. (1992) Maximum penalized likelihood estimation for independent and Markov-dependent mixture models. *Biometrics*, **48**, 545–558.

Litterman, R. (1986) Forecasting with Bayesian vector autoregressions: five years of experience. *Journal of Business and Economic Statistics*, **4**, 25–38.

Lopes, H. and Polson, N. (2010) Bayesian inference for stochastic volatility modeling. In K. Böcker (ed.), *Rethinking Risk Measurement, Management and Reporting*, pp. 515–551. Risk Books, Zurich.

Lubrano, M. (1995a) Testing for unit roots in a Bayesian framework. *Journal of Economics*, **69**(1), 81–109.

Lubrano, M. (1995b) Bayesian tests for cointegration in the case of structural breaks. *Recherches Economiques de Louvain*, **61**, 479–507.

Lunn, D., Spiegelhalter, D., Thomas, A. and Best, N. (2009). The BUGS project: Evolution, critique and future directions. *Statistics in medicine*, **28**(25), 3049–3067.

Maddala, G. (1979) *Econometrics*. McGraw-Hill, New York, NY.

Maddala, G. and Kim, I. (1996) Structural change and unit roots. *Journal of Statistical Planning and Inference*, **49**(1), 73–103.

Maheu, J. and Song, Y. (2012) A new structural break model with application to Canadian inflation forecasting. Working Paper Series 27_12, The Rimini Centre for Economic Analysis.

Marriott, J. and Smith, A. (1992) Reparameterisation aspects of numerical Bayesian methodology for ARMA models. *Journal of Time Series Analysis*, **13**, 327–343.

Marriott, J., Ravishanker, N., Gelfand, A. and Pai, J. (1996) Bayesian analysis of ARMA processes: complete sampling-based inference under full likelihoods. In D. Berry *et al.* (eds), *Bayesian Analysis in Statistics and Econometrics*, pp. 243–256. Wiley, New York, NY.

Marriott, J., Naylor, J. and Tremayne, A. (2004) Bayesian graphical inference for economic time series that may have stochastic or deterministic trends. In R. Becker and S. Hurn (eds), *Contemporary Issues in Economics and Econometrics: Theory and Applications*, pp. 112–146. Edward Elgar, Cheltenham.

McCabe, B. and Martin, G. (2005) Bayesian predictions of low count time series. *International Journal of Forecasting*, **21**, 315–330.

McCulloch, R. and Tsay, R. (1994) Bayesian inference of trend and difference stationarity. *Econometric Theory*, **10**, 596–608.

McKenzie, E. (1986) Autoregressive moving-average processes with negative-binomial and geometric marginal distributions. *Advances in Applied Probability*, **18**, 679–705.

McKenzie, E. (2003) Discrete variate time series. In D. N. Shanbhag and C. R. Rao (eds), *Handbook of Statistics*, Vol. 21: *Stochastic Processes: Modelling and Simulation*, pp. 573–606. North-Holland, Amsterdam.

McLeod, A., Yu, H. and Mahdi, E. (2012) Time series analysis with R. In T. Rao, S. Rao and C. Rao (eds), *Handbook of Statistics*, Vol. 30: *Time Series Analysis: Methods and Applications*, pp. 661–712. North-Holland, Amsterdam.

Meyer, R. and Yu, J. (2000) BUGS for a Bayesian analysis of stochastic volatility models. *Econometrics Journal*, **3**, 198–215.

Migon, H. and Harrison, P. (1985) An application of non-linear Bayesian forecasting to television advertising. In J. Bernardo, M. De Groot, D. Lindley and A. Smith (eds), *Bayesian Statistics 2*, pp. 271–294. North-Holland, Amsterdam.

Migon, H., Gamerman, D., Lopes, H. and Ferreira, M. (2005) Dynamic models. In D. Dey and C.R. Rao (eds), *Handbook of Statistics, Volume 25: Bayesian Thinking, Modeling and Computation*, pp. 553–588. Elsevier, Amsterdam.

Mira, A. and Petrone, S. (1996) Bayesian hierarchical nonparametric inference for change-point problems. In J. Bernardo, J.O. Berger, A.P. Dawid and A.F.M. Smith (eds). *Bayesian Statistics 5*. Oxford University Press, Oxford, UK.

Monahan, J. (1984) A note on enforcing stationarity in autoregressive-moving average models. *Biometrika*, **71**, 403–404.

Nandram, B. and Petrucelli, J. (1997) A Bayesian analysis of autoregressive time series panel data. *Journal of Business and Economic Statistics*, **15**, 328–334.

Naylor, J. and Marriott, J. (1996) A Bayesian analysis of non-stationary AR series. In J. Bernardo, *et al.* (eds), *Bayesian Statistics 5*, pp. 705–712. Oxford University Press, Oxford, UK.

Nelson, C. and Plosser, C. (1982) Trends and random-walks in macroeconomic time-series – some evidence and implications. *Journal of Monetary Economics*, **10**(2), 139–162.

Nyberg, H. (2008) Testing an autoregressive structure in binary time series models. Discussion Paper No 243. Helsinki Center of Economic Research, Discussion Papers, ISSN 1795–0562.

Ord, K., Fernandes, C. and Harvey, A. (1993) Time series models for multivariate series of count data. In T. Subba Rao (ed.), *Developments in Time Series Analysis*, pp. 295–309. Chapman & Hall, London, UK.

Pang, X. (2010) Modeling heterogeneity and serial correlation in binary time-series cross-sectional data: a Bayesian multilevel model with AR(p) errors. *Political Analysis*, **18**, 470–498.

Pesaran, M., Pettenuzzo, D. and Timmermann, A. (2006) Forecasting time series subject to multiple structural breaks. *Review of Economic Studies*, **73**(4), 1057–1084.

Petris, G., Petrone, S. and Campagnoli, P. (2009) *Dynamic Linear Models with R*. Springer, New York, NY.

Piger, J. (2011) Econometrics: models of regime changes. In R. Meyers (ed.), *Complex Systems in Finance and Econometrics*, pp. 190–202. Springer, New York, NY.

Pitt, M. and Shephard, N. (1998) Time-varying covariances: A factor stochastic volatility approach. In J.M. Bernardo *et al.* (eds), *Bayesian Statistics 6*. Clarendon Press, Oxford, UK.

Pitt, M. and Shephard, N. (1999) Time varying covariances: A factor stochastic volatility approach. In: In J.M. Bernardo *et al.* (eds), *Bayesian Statistics 6*, pp. 547–570. OUP, London, UK.

Prado, R. and West, M. (2010) *Times Series: Modelling, Computation, and Inference*. CRC/Chapman & Hall, Boca Raton, FL.

Pruscha, H. (1993) Categorical time series with a recursive scheme and with covariates. *Statistics*, **24**, 43–57.

Quandt, R. (1958) The estimation of parameters of a linear regression system obeying two separate regimes. *Journal of the American Statistical Association*, **53**, 873–880.

Ravishanker, N. and Ray, B. (1997) Bayesian analysis of vector ARMA models using Gibbs sampling. *Journal of Forecasting*, **16**, 177–194.

Ray, B. and Tsay, R. (2002) Bayesian methods for change-point detection in long-range dependent processes. *Journal of Time Series Analysis*, **23**, 687–705.

Reis, E.A., Salazar, E. and Gamerman, D. (2006) Comparison of sampling schemes for dynamic linear models. *International Statistical Review*, **74**, 203–214.

Rosenberg, M. and Young, V. (1999) A Bayesian approach to understanding time series data. *North American Actuarial Journal*, **3**, 130–144.

Ruggeri, F. and Sivaganesan, S. (2005) On modeling change points in non-homogeneous Poisson processes. *Statistical Inference for Stochastic Processes*, **8**(3), 311–329.

Ruiz-Cárdenas, R., Krainski, E. and Håvard Rue, H. (2012) Direct fitting of dynamic models using integrated nested Laplace approximations-INLA. *Computational Statistics and Data Analysis*, **56**, 1808–1828.

Ruppert, D. (2011) *Statistics and Data Analysis for Financial Engineering*. Springer, Berlin.

Schotman, P. (1994). Priors for the AR(1) model, parameterisation issues and time series considerations. *Econometric Theory*, **10**, 579–595.

Sephton, D. (2009) Predicting recessions: a regression (probit) model approach. *Foresight: The International Journal of Applied Forecasting*, **12**, 26–32.

Shephard, N. (1996) Statistical aspects of ARCH and Stochastic Volatility. In D.R. Cox, O.E. Barndorff-Nielsen and D.V. Hinkley (eds), *Time Series Models in Econometrics, Finance and Other Fields*, pp. 1–67. Chapman and Hall, London, UK.

Shephard, N. (2008) Stochastic volatility models. In S.N. Durlauf and L.E. Blume (eds), *The New Palgrave Dictionary of Economics*, 2nd edn. Palgrave Macmillan, Basingstoke, UK.

Silva, I., Silva, M., Pereira, I. and Silva, N. (2005) Replicated INAR(1) processes. *Methodology and Computing in Applied Probability*, **7**, 517–542.

Spirling, A. (2007) Bayesian approaches for limited dependent variable change point problems. *Political Analysis*, **15**(4), 387–405.

Startz, R. (2008) Binomial autoregressive moving average models with an application to U.S. recessions. *Journal of Business and Economic Statistics*, **26**(1), 1–8.

Steel, M. (2008) Bayesian time series analysis. In S.N. Durlauf and L.E. Blume (eds), *The New Palgrave Dictionary of Economics*, 2nd edn. Palgrave Macmillan, Basingstoke, UK.

Tanizaki, H. and Mariano, R. (1998) Nonlinear and non-gaussian state-space modeling with Monte Carlo simulations. *Journal of Econometrics*, **83**(1,2), 263–290.

Thomas, R. (1997) *Modern Econometrics: an Introduction*. Addison-Wesley.

Thomson, J., Kimmerer, W., Brown, L., Newman, K., MacNally, R., Bennett, W., Feyrer, F. and Fleishman, E. (2010) Bayesian change point analysis of abundance trends for pelagic fishes in the upper San Francisco Estuary. *Ecological Applications*, **20**(5), 1431–1448.

Tong, H. (1983) *Threshold Models in Non-Linear Time Series Analysis*. Springer, New York, NY.

Tsay, R. (1988) Outliers, level shifts, and variance changes in time series. *International Journal of Forecasting*, **7**(1), 20.

Tsay, R. (1989) Testing and modeling threshold autoregressive processes. *Journal of the American Statistical Association*, **84**, 231–240.

Van Ophem, H. (1999) A general method to estimate correlated discrete random variables. *Econometric Theory*, **15**, 228–237.

Vrontos, I., Dellaportas, P. and Politis, D. (2000) Full Bayesian Inference for GARCH and EGARCH Models. *Journal of Business and Economic Statistics*, **18**, 187–198.

Wang, J. and Zivot, E. (2000) A Bayesian time series model of multiple structural changes in level, trend, and variance. *Journal of Business and Economic Statistics*, **18**, 374–386

West, M. (1996) Bayesian time series: models and computations for the analysis of time series in the physical sciences. In K. Hanson (ed.), *Maximum Entropy and Bayesian Methods*. Kluwer, Dordrecht.

West, M. (2013) Bayesian dynamic modelling. In P. Damien, P. Dellaportas, N. Polson and D. Stephens (eds), *Bayesian Theory and Applications*, pp. 145–166. Oxford University Press, Clarendon.

West, M. and Harrison, J. (1997) *Bayesian Forecasting and Dynamic Models*, 2nd edn. Springer-Verlag, Berlin.

Yelland, P.M. (2009). Bayesian forecasting for low-count time series using state-space models: An empirical evaluation for inventory management. *International Journal of Production Economics*, **118**, 95–103.

Yu, J. (2005) On leverage in a stochastic volatility model. *Journal of Econometrics*, **127**, 165–178.

Zeger, S. and Qaqish, B. (1988) Markov regression models for time series: A quasi-likelihood approach. *Biometrics*, **44**(4), 1019–1031.

Zellner, A. (1971) *An Introduction to Bayesian Inference in Econometrics*. Wiley, Chichester, UK.

Zellner, A. and Tiao, G. (1964) Bayesian analysis of the regression model with autocorrelated errors. *Journal of the American Statistical Association*, **59**, 763–778.

# 7

# Analysis of panel data

## 7.1 Introduction

Panel or longitudinal data sets occur when the continuous or discrete response $y_{it}$ for each subject i ($i = 1, \ldots, n$) is observed on several occasions $t = 1, \ldots, T_i$, where $T_i$ may differ between subjects. Times of observations $v_{it}$, and so spacings $\Delta_{it} = v_{it} - v_{i,t-1}$, may also differ between subjects. Time scales may include calendar time, age, work experience, and so on. The analysis of change in serial measurements over individuals or groups plays a major role in social and biomedical research, and is fundamental in understanding causal mechanisms of disease or testing behavioural hypotheses (Hsiao, 2007), and in the analysis of developmental and growth processes.

Relevant texts include Singer and Willett (2003), Frees (2004), Weiss (2005), Verbeke and Molenberghs (2000), Hedeker and Gibbons (2006), Arellano and Honoré (2001), and Baltagi (2008), with Bayesian overviews exemplified by Chib (2008), Walker *et al.* (2007), and Arellano and Bonhomme (2011). Options for Bayesian estimation include BUGS, R-INLA, MCMCglmm (Hadfield, 2010) and JMBayes.

Such data may be observational or follow an experimental design, as in clinical trials, typically with treatment and control groups, where repeated measures of outcomes and risk factors are obtained (Galbraith and Marschner, 2002). Observational longitudinal data are exemplified by panel data on economic units, such as patents and R&D spend by companies (Blundell *et al.*, 2002), or population survey data for socioeconomic groups or geographical areas.

The accumulation of information over both times and subjects increases the power of statistical methods to identify effects (e.g. treatment effects in medical applications), and permits the estimation of parameters (e.g. permanent effects or 'frailties' for subjects) not identifiable from cross-sectional analysis. In economic applications, a longitudinal model of (say) patent applications gives better scope to assess the role of persisting unobserved heterogeneity between firms, such as entrepreneurial and technical skills which affect patent applications but may be difficult to capture with observable variables (Hall *et al.*, 1986; Maddala, 1987; Winkelmann, 2000). Longitudinal designs also provide information to describe patterns of development across time variables, enabling out-of-sample predictions of development. For example, Lee and Hwang (2000) consider the best choice of prior for extended prediction beyond the observed time range of the sample.

*Applied Bayesian Modelling*, Second Edition. Peter Congdon.
© 2014 John Wiley & Sons, Ltd. Published 2014 by John Wiley & Sons, Ltd.

Major methodological questions in panel data include representing dependency within subjects, and distinguishing different sources of dependence (e.g. spurious vs. true state dependencies). Methods to represent these features include dynamic dependence (e.g. allowing lagged outcomes to affect current outcomes), random subject intercepts and predictor effects, and autocorrelated error schemes (Beck and Katz, 1996). Many methodological issues also derive from the application to categorical outcomes (binary, multinomial or count data) of methods originally developed for continuous outcomes (Chib and Carlin, 1999). Other modelling issues include detection of change points in longitudinal profiles (Yang and Gao, 2013), and variable selection methods for random effects (Fruhwirth-Schnatter and Wagner, 2011).

An additional major issue is the presence of missing data in panel studies (see Section 7.6), especially permanent loss or 'attrition' of subjects, where $T_i$ is less than the maximum span of the study in clinical trials or panel studies of economic interventions (Hausman and Wiseman, 1979). Intermittent missing data may occur in multi-site environmental studies (Caffo *et al.*, 2011). Bayesian perspectives on this issue include those of Daniels and Hogan (2008), and Little and Rubin (2002).

## 7.2 Hierarchical longitudinal models for metric data

The modelling of subject effects via univariate or multivariate random effects leads into a wide class of hierarchical models for growth data and other types of longitudinal observation (Chib, 2008). Thus in a simple linear growth curve model

$$y_{it} = b_{i1} + b_{i2}t + e_{it}$$

for $i = 1, \ldots, n$ and $t = 1, \ldots, T$, the subject level random effects $(b_{i1}, b_{i2})$ respectively describe differences in baseline levels of the outcome $(b_{i1})$, such as the underlying average attainment or morbidity, and differences in linear growth rates in attainment. If times are centred, as in

$$y_{it} = b_{i1} + b_{i2}(t - \bar{t}) + e_{it}$$

then $b_{i1}$ represents average outcome levels at time $\bar{t}$. The residuals $e_{it}$ may be taken as iid, and are generally assumed independent of the subject effects. The distribution of the random effects, whether parametric or non-parametric, and if parametric, whether normal or otherwise (Butler and Louis, 1992), constitutes the first stage of the prior density specification. The hyperparameters on the densities assumed for $\{b_{i1}, b_{i2}, e_{it}\}$ form the second stage of the prior specification.

In many studies the interest may especially be in identifying subject level effects from panel data with greater reliability than is possible with cross-sectional data (Horrace and Schmidt, 2000). A typical specification for continuous data is

$$y_{it} = b_i + X_{it}\beta + e_{it}$$

with subject effects $b_i$ and residuals $e_{it}$ both normal, and both random effects iid and independent of $X_{it}$. Control for unobserved heterogeneity is the basis for obtaining consistent estimates of the systematic (regression) part of the model (Hamerle and Ronning, 1995). Such heterogeneity induces correlation in the combined error $\eta_{it} = e_{it} + b_i$. Thus if subject level effects have variance $\sigma_b^2$, and equicorrelated measurement errors $e_{it}$ have variance $\sigma_e^2$, then the intra-subject (or intra-cluster) correlation between $\eta_{is}$ and $\eta_{it}$ at periods $s$ and $t$ is $\phi = \sigma_b^2/(\sigma_b^2 + \sigma_e^2)$.

The assumption that $b_i$ is independent of observed characteristics $X_{it}$ may be realised under randomization (e.g. in medical trials), but may be less likely in observational settings subject to selectivity effects, as in econometrics (Carro, 2007; Hsiao, 2007). Therefore fixed effects models may be less restrictive in not assuming the independence of $b_i$ and $X_{it}$, and robust in not needing to specify a density for $b_i$. On the other hand estimation may be problematic for large $n$ and small $T$, and there is heavier parameterisation compared to random effects approaches (Maddala, 1987). Alternatively one may link the means of random subject effects to known permanent subject level covariates (including averages of time varying predictors), as in

$$y_{it} = b_i + X_{it}\beta + e_{it},$$
$$b_i \sim N(\delta_1 \overline{X}_i + \delta_2 Z_i, \sigma_b^2).$$

For longer panels, variation over time in the impact of $b_i$ may be represented in factor type models (see Chapter 9) involving time specific loadings $\lambda_t$ on the subject effects (Arellano and Honoré, 2001; Dunson, 2007), as in

$$y_{it} = \lambda_t b_i + X_{it}\beta + e_{it}.$$

Either a particular loading is set to a known value (e.g. $\lambda_1 = 1$) to ensure identifiability, or with all $\lambda_t$ unknown, the variance of the $b_i$ is pre-defined (e.g. var($b_i$) = 1). The correlation between $\eta_{is}$ and $\eta_{it}$ at periods $s$ and $t$ becomes

$$\phi_{st} = \lambda_t \lambda_s \sigma_b^2 / [(\lambda_t^2 \sigma_b^2 + \sigma_e^2)^{0.5} (\lambda_s^2 \sigma_b^2 + \sigma_e^2)^{0.5}].$$

## 7.2.1 Autoregressive errors

If autocorrelation in the errors is present under the iid assumption on $e_{it}$, one may apply autoregressive or ARMA error schemes (Lee *et al.*, 2005; Cameron and Miller, 2011; Meligkotsidou *et al.*, 2012). The equicorrelated assumption $e_{it} \sim N(0, \sigma_e^2 I)$ is replaced by $e_{it} \sim N(0, \sigma_e^2 \Omega)$ where $\Omega$ represents the form of autocorrelation. This step might, however, be preceded by an investigation of dynamic models (see Section 7.2.2) to assess the existence of system dynamics (state dependence) as opposed to error dynamics (Maddala, 1987).

For example, suppose the errors follow a first order autoregression with parameter $\gamma$ so that corr $(e_{is}, e_{it}) = \gamma^{|t-s|}$. Then

$$\Omega = \begin{bmatrix} 1 & \gamma & \gamma^2 & \gamma^3 & .. \\ \gamma & 1 & \gamma & \gamma^2 & .. \\ \gamma^2 & \gamma & 1 & \gamma & .. \\ & & .. & .. & \\ .. & & \gamma^2 & \gamma & 1 \end{bmatrix}$$

with corresponding representation

$$y_{it} = b_i + X_{it}\beta + e_{it}$$
$$e_{it} = \gamma e_{i,t-1} + u_{it} \quad t > 1$$
$$u_{it} \sim N(0, \sigma_u^2)$$
$$e_{i1} \sim N(0, \sigma_u^2 / (1 - \gamma^2)).$$

Then rather than $b_i \sim N(0, \sigma_b^2)$, one may link initial conditions and $b_i$ via the prior

$$b_i \sim N(\psi e_{i1}, \sigma_b^2)$$

where $\psi$ can be positive or negative (e.g. Chamberlain and Hirano, 1999). This amounts to assuming a bivariate density for $b_i$ and $e_{i1}$ with independence when $\psi$ is zero.

### 7.2.2    Dynamic linear models

As well as allowing error dependence, panel models for continuous outcomes may also account for temporal autocorrelation through lags in the response (also called state dependence), leading to the general class of linear dynamic panel data models (e.g. Mander *et al.*, 1999; Bond 2002), with the AR($p$) case being

$$y_{it} = X_{it}\beta + \rho_1 y_{i,t-1} + \rho_2 y_{i,t-2} + .. + \rho_p y_{i,t-p} + e_{it}.$$

As for simple time series models, autoregression and trend may be incorporated together (Arellano and Bonhomme, 2011) leading to models such as

$$y_{it} = b_{i1} + b_{i2}t + \rho y_{i,t-1} + e_{it}.$$

Bayesian approaches to AR($p$) models for panel data observed for periods $t = 1, \dots, T$ include specifying a prior on latent data $(y_{i0}, \dots, y_{i,T-p})$ which may require knowledge of covariates $\{X_{i0}, \dots, X_{i,T-p}\}$ at earlier times (Nandram and Petruccelli, 1997; Juárez and Steele, 2010), or conditioning on the first $p$ observations (Cefis *et al.*, 2007). Another option is to specify a separate static model for the initial observations (Geweke and Keane, 2000), as for the case $p = 1$, with the model for time 1 data then being

$$y_{i1} = X_{i1}\beta^* + u_i$$

$$y_{it} = (1 - \rho)(X_{it}\beta + b_i + \zeta u_i) + \rho y_{i,t-1} + e_{it}, \quad t > 1.$$

In this scheme $\rho$ is confined to stationarity, $u_i$ is identically distributed with variance $\phi_1$, and $e_{it}$ may be identically distributed or autoregressive. The $\zeta$ coefficient represents any carry over effect of the first period error.

### 7.2.3    Extended time dependence

While panel data is often used to sharpen inferences about individual differences, in some circumstances a two stage model with random effects over time as well as over individuals at the first stage may be relevant. Thus time varying intercepts representing unmeasured influences at each time point as in

$$y_{it} = b_i + c_t + X_{it}\beta + e_{it},$$

where the $c_t$ may be fixed or random effects (e.g. a first or second order random walk). Population wide time effects may also be modelled using a smooth regression function $c(t)$ (Daniels and Hogan, 2008). For example, a degree $q$ polynomial truncated spline basis, and knots $\{\tau_1, \dots, \tau_K\}$, is $B = (t, t^2, \dots, t^q, (t - \tau_1)_+^q, \dots, (t - \tau_K)_+^q)$, with $q + K$ coefficients,

$$c(t) = \psi_1 t + \dots + \psi_q t^q + \zeta_1 (t - \tau_1)_+^q + \dots + \zeta_K (t - \tau_K)_+^q.$$

An additional shrinkage device might be a random effects penalty prior for the knot coefficients, $\zeta_k \sim N(0, \sigma_\zeta^2)$.

As compared to time series data, the borrowing of strength over subjects means that fewer time points are needed to model evolution in regression coefficients $\beta$, or in other parameters, such as lag coefficients on previous outcome values. Thus for continuous $y_{it}$, with predictors including lags in y, one might specify an autoregressive latent trait model (e.g. Bollen and Curran, 2004). An example is a time varying AR(1) parameter as in

$$y_{it} = b_{i1} + b_{i2}t + \rho_t y_{i,t-1} + \epsilon_{it}.$$

## Example 7.1    Returns to education

Consider data from years 1976–1982 ($T = 7$) of the Panel Study of Income Dynamics (PSID) on $n = 595$ subjects aged 18–65 in 1976. The data are considered by Cameron and Trivedi (2009) and Baltagi (2008), and originally by Cornwell and Rupert (1988); a longer PSID panel is considered by Geweke and Keane (2000). Of interest is the impact on log transformed wages of years of education (ED), after allowing for years of work experience (EXP), experience squared (EXP2), and weeks worked in a year (WKS).

Model 1 includes permanent subject effects to represent unmeasured attributes that affect wages:

$$y_{it} = \beta_0 + b_i + \beta_1 EXP_i + \beta_2 EXP2_i + \beta_3 WKS_i + \beta_4 ED_i + e_{it}$$

and assumes the $e_{it}$ are equicorrelated, with $e_{it} \sim N(0, \sigma_e^2)$, and $b_i \sim N(0, \sigma_b^2)$. Improved MCMC convergence in this and subsequent models is obtained by scaling (division by 10) of the experience and weeks worked predictors. A uniform prior on the ratio $S = \sigma_e^2/(\sigma_e^2 + \sigma_b^2)$ is adopted, namely

$$\sigma_e^2/(\sigma_e^2 + \sigma_b^2) \sim U(0, 1),$$

together with a gamma prior on $1/\sigma_e^2$, which has the benefit that $\phi = 1 - S$ is interpretable as the intra-cluster correlation. This model may be estimated in BUGS or using MCMCglmm[1].

The last 4000 of a 5000 iteration two chain run in BUGS provide posterior means (sd) for the beta coefficients ($\beta_0, \ldots, \beta_4$) of 2.98 (0.17), 1.08 (0.03), −0.05 (0.005), 0.0085 (0.006) and 0.14 (0.01). Variation in income between subjects considerably outweighs intra-career variation within subjects, with $\phi$ estimated at 0.97. As one possible composite discrepancy measure, predictive replicates are sampled, and a sum of squared predictive residuals (SSPR), $\sum_{i,t}(y_{it} - y_{rep,it})^2$, obtained as 197.

Departure from normality assumptions regarding the $b_i$ is apparent both in the posterior density of the Jacques–Bera (JB) statistic for the sampled $b_i$ (which has posterior mean 21.70), and also in a posterior predictive check (which has value 0.999) comparing the JB statistics for $b_i$ and replicate $b_i$ respectively. A Q-Q plot of the posterior mean $b_i$ also suggests (negative) skewness.

Such departures may be due to autocorrelation in the outcome or errors. The second model accordingly considers the dynamic representation (Geweke and Keane, 2000),

$$y_{i1} = X_{i1}\beta^* + u_i$$
$$y_{it} = (1 - \rho)(X_{it}\beta + b_i + \zeta u_i) + \rho y_{i,t-1} + e_{it}, \qquad t > 1,$$

where $u_i$ and $e_{it}$ ($t > 1$) are separate identically distributed error terms, and the predictors are now centred. $U(-1, 1)$ and $N(0, 1)$ priors are adopted on $\rho$ and $\zeta$ respectively. The second half of a 10 000 iteration two chain run in BUGS provide posterior means (sd) for the beta coefficients ($\beta_0, \ldots, \beta_4$) of 7.09 (0.03), 0.33 (0.07), $-0.05$ (0.02), 0.035 (0.03), and 0.085 (0.01). The coefficient on experience is much reduced, but returns to education remain significant. Posterior means (sd) on $\rho$ and $\zeta$ are respectively 0.79 (0.01), and 0.81 (0.05). The skewness in $b_i$ is reduced, but excess kurtosis remains. Overall model fit in fact deteriorates with the SSPR statistic increasing to 338.

A third model instead specifies AR1 errors namely

$$y_{it} = X_{it}\beta + b_i + e_{it} = \beta_0 + b_i + \beta_1 EXP_i + \beta_2 EXP2_i + \beta_3 WKS_i + \beta_4 ED_i + e_{it}$$

$$e_{it} = \gamma e_{i,t-1} + u_{it}, \qquad\qquad t > 1$$

$$u_{it} \sim N(0, \sigma_u^2),$$

$$e_{i1} \sim N(0, \sigma_u^2/(1 - \gamma^2)),$$

with $\gamma$ confined to stationarity. For periods $t > 1$, the regression may be re-expressed (e.g. Meligkotsidou *et al.*, 2012) as

$$y_{it} = (X_{it} - \gamma X_{i,t-1})\beta + b_i(1 - \gamma) + \gamma y_{i,t-1} + u_{it}.$$

The estimated $\gamma$ is clearly positive with 95% interval $(0.88, 0.91)$, and introducing the autocorrelated error, which to some degree supplants the permanent effects $b_i$, considerably reduces $\sigma_b^2$ and $\phi$. However, evidence of excess parameterisation in this model is provided by a larger SSPR of 459 than both earlier models, and increased DIC.

## 7.3    Normal linear panel models and normal linear growth curves

This section considers prior specification in the normal linear model for panel data, which sets the basis for panel models for discrete outcomes. Consider first a basic panel data model for observations $Y_i = (y_{i1}, y_{i2}, \ldots, y_{iT})$ on subjects $i = 1, \ldots, n$ and with predictor matrix $X_i$ of dimension $p$. Then the cross-sectional normal linear model generalises to

$$y_{it} = X_i\beta + e_{it}$$

where $e_i = (e_{i1}, e_{i2}, \ldots, e_{iT})$ is multivariate normal with mean zero and $T \times T$ dispersion matrix $\Sigma$. Assume $\psi = \Sigma^{-1}$ has a Wishart prior density with scale matrix $R$ and degrees of freedom $r$, and $\beta$ has a multivariate normal prior with mean $\beta_0$ and dispersion matrix $B_0$. Then the full conditional distribution of $\beta$ given $\Sigma^{-1}$ is multivariate normal

$$N_q(\beta^*, B^*), \tag{7.1}$$

where

$$\beta^* = B^*\left(B_0^{-1}\beta_0 + \sum_{i=1,\ldots,N} X_i\psi Y_i\right)$$

$$(B^*)^{-1} = B_0^{-1} + \sum_{i=1,\ldots,N} X_i\psi X_i.$$

The full conditional of $\psi$ is Wishart with $r + n$ degrees of freedom, and scale matrix $R^*$, where

$$(R^*)^{-1} = R^{-1} + \sum_{i=1,...,N} e_i e_i'.$$

This formulation extends to include unobserved heterogeneity between subjects and to time varying predictors $X_i = (X_{i1}, \ldots , X_{iT})$ of dimension $T \times p$. The normal linear mixed model is then

$$y_{it} = X_{it}\beta + W_{it}b_i + e_{it}$$

$$e_i \sim N(0, \Sigma)$$

or at subject (cluster) level

$$Y_i = X_i\beta + W_i b_i + e_i. \tag{7.2}$$

The $T \times q$ matrix $W_i$ is usually a subset of regression vector $X_i$, and $b_i$ a $q \times 1$ random effect with mean zero and covariance matrix $\Sigma_b$. If the density of $b_i$ is multivariate normal then the mean and variance of $Y_i$, unconditionally on $b_i$, are respectively

$$E(Y_i) = X_i\beta,$$

and

$$V(Y_i) = \Sigma + W_i \Sigma_b W_i.$$

There may be additional grouping variables, e.g. exam results over time for pupils $i$ within schools $j$, or clinical measures over time for patients $i$ within hospitals $j$. In this case the multi-level random effects $b_{ij}$ may be assumed, or $b_i$ may be related to fixed characteristics of the higher level grouping.

Suppose a Wishart prior is assumed on $\Sigma_b^{-1}$ with degrees of freedom $r_b$ and scale matrix $R_b$, and further that $\Sigma = \sigma^2 I_T$, where the precision $\tau = \sigma^{-2}$ is assigned a gamma prior with parameters $v_1$ and $v_2$. Again assume that $\beta$ has a multivariate normal prior with mean $\beta_0$ and dispersion matrix $B_0$. Then by rewriting 7.2 as

$$y_i - W_i b_i = X_i\beta + e_i$$

the density of $\beta$, conditional on $\{b_i, \sigma^2, \Sigma_b\}$, has the same form as 7.1 with dispersion matrix $\Sigma = \sigma^2 I_T$. The conditional density of $\tau$ is Gamma with parameters $v_1 + nT/2$ and $v_2 + 0.5\sum_{i=1,n}\sum_{t=1,T}e_{it}^2$, while that of $\Sigma_b^{-1}$ is Wishart with degrees of freedom $r_b + n$, and scale matrix $R_b^*$ where

$$(R_b^*)^{-1} = R_b^{-1} + \sum_{i=1,n} b_i b_i'.$$

Similarly the full conditional of $b_i$ conditions on $\beta$, whereby 7.2 can be rewritten as

$$H_{it} = y_{it} - X_{it}\beta = W_{it}b_i + e_{it}.$$

The $b_i$ then have variances $V_i$ given by

$$(V_i)^{-1} = \Sigma_b^{-1} + \tau W_i' W_i$$

and means

$$V_i \tau W_i H_i.$$

With suitable adaptations the linear mixed model may be applied with discrete outcomes. An alternative hierarchical parameterisation in models 7.2 may provide improved convergence and identifiability in Markov chain Monte Carlo sampling (Gelfand *et al.*, 1995). When $W_i$ is a subset of $X_i$, as in the case $W_i = X_{i1} = 1$ (with $q = 1$) under intercept variation, it may be preferable to merge the fixed effect $\beta_0$, with the random effect $b_i$, such that $b_i$ has a non-zero mean, namely $b_i \sim N(\beta_0, \sigma_b^2)$. More generally, assuming $W_i = X_i$ (i.e. that all predictors have randomly varying effects), one has

$$Y_i = W_i b_i + e_i,$$

$$b_i \sim N(\beta, \Sigma_b).$$

## 7.3.1   Growth curves

In a growth curve analysis, the design matrix $X_{it}$ consists of time functions with equal values over subjects, for example, powers or orthogonal polynomials of time. So for a linear growth model $X_{it} = (1, t)$, the coefficients $\{\beta_1, \beta_2\}$ on the time variables represent the population level relationship between the mean outcome and time or age $t$. However, average growth curves will often conceal substantial variability in growth, and moreover such variation may be correlated with intercept variability.

To reflect such variability, the normal mixed linear growth curve involves subject-specific intercepts and growth rates,

$$y_{it} = b_{i1} + b_{i2}t + e_{it},$$

with $e_i \sim N(0, \Sigma)$, and bivariate normal random effects $b_i = (b_{i1}, b_{i2})$

$$b_i \sim N_2(\beta, \Sigma_b),$$

with population level parameters being the intercept $\beta_1$ and average linear growth rate $\beta_2$. Questions of interest might include establishing whether variations in growth rates $b_{i2}$ can be explained by attributes $Z_i$ of individuals: for example, whether differential declines in marital quality are related to initial spouse age, or to spouse education (Karney and Bradbury, 1995).

Generalisations of the linear growth curve are commonly needed. For example, quadratic dependence, varying by subject, leads to

$$y_{it} = b_{i1} + b_{i2}(t - \bar{t}) + b_{i3}(t - \bar{t})^2 + e_{it}$$

with $b_{i3}$ representing acceleration or deceleration in growth, such that the growth rate $\delta_i$ depends on time, namely $\delta_i = b_{i2} + 2b_{i3}(t - \bar{t})$. Similarly the smooth curve approach (Section 7.2.3) can be made subject specific, as in a linear spline ($q = 1$) (e.g. Howe *et al.*, 2013), leading to

$$y_{it} = b_{i1} + c_{it} + e_{it}$$

$$c_{it} = \psi_{1i}t + \zeta_{1i}(t - \tau_1)_+ + \ldots + \zeta_{Ki}(t - \tau_K)_+.$$

If individuals $i$ have different observation times, or are nested hierarchically within groups $j$, then more complex growth curve models have been suggested, for example, when observation times $v_{it}$ vary between subjects (Diggle, 1988). Then the series for individual $i$ may be modelled as

$$y_i(v_{it}) = \mu_i(v_{it}) + W_i(v_{it}) + e_{it} + b_i. \tag{7.3}$$

This representation contains an iid measurement error $e_{it}$, as well as autoregressive errors $W_i(v_{it})$ (cf. Chamberlain and Hirano, 1999). The prior for the latter would incorporate a model for correlation $\rho(\Delta)$ between successive observations according to the time difference $\Delta_t = v_{t+1} - v_t$ between readings. The error association typically decreases in $\Delta$, since measurements closer in time tend to be more strongly associated. The model includes subject level errors $b_i$ which may depend on covariates. These stable effects may also pre-multiply covariates, including the times $v_{it}$ themselves, in which case they become variable growth rates.

Suppose individuals $i$ are classified by group $j = 1, \ldots, J$. Assume for simplicity equally spaced observation times for all subjects. Then the corresponding model to 7.2 contains measurement error, as well as autoregressive dependence, at observation level, permanent effects $b_{ij}$ specific to subject $i$ and group $j$, and growth curve parameters varying over group or over individuals. For example, a group varying linear growth model (see Example 7.2) might take the form

$$y_{ijt} = A_j + B_j t + b_{ij} + w_{ijt} + e_{ijt}$$

where the $w_{ijt}$ are autoregressive with

$$w_{ijt} = \gamma_1 w_{ij,t-1} + u_{ijt}$$

and both $e_{ijt}$ and $u_{ijt}$ are iid errors.

## Example 7.2    Hypertension trial

As an example of hierarchical data (observation sequences for individuals themselves nested within groups or institutions), consider data on $i = 1, \ldots, 288$ patients randomized to receive one of three drug treatments for hypertension (Brown and Prescott, 1999). The drugs are denoted A = Carvedilol, B = Nifedipine, and C = Atenolol, with drug A being the new drug, and the other two being conventional treatments. Patients are allocated to one of $j = 1, \ldots, 29$ clinics.

There are five readings on each patient: a pre-treatment (week 1) baseline reading $P_{0i}$ of diastolic blood pressure (DBP), and $T = 4$ post-treatment readings $y_{ijt}$ at two weekly intervals (weeks 3, 5, 7 and 9 after treatment). Including a baseline measure as a covariate may assist in adjusting for any imbalances remaining after randomization, and in increasing precision of the estimated treatment effect (Schmidt et al., 2011). Treatment success is judged in terms of reducing blood pressure.

A first analysis of these data (model 1) is a fixed effects model without random effects over patients or patient-clinic combinations. It involves baseline and treatment effects, with the new treatment as reference category (namely $\delta_A = 0$, with $\delta_B$ and $\delta_C$ then unknown)

$$y_{ijt} = \beta_1 + \beta_2 P_{0i} + \delta_B + \delta_C + e_{ijt},$$

with $e_{ijt}$ identically distributed. A single baseline measure is missing, and an imputation is made based on a missing at random (MAR) assumption (see Section 7.6) such that the response mechanism itself does not need to be modelled. The response data themselves are also subject to both early dropout and intermittent missingness (for example, subject 202 is missing for visit 1, but observed for subsequent visits). Means for all (288 × 4) combinations of patient and visit are estimated under a MAR assumption for such missing $y$ values.

Centred treatment effects (delta.c[j] in the code) are obtained by subtracting $\delta_A$, $\delta_B$ and $\delta_C$ from the mean of the three effects. Estimates from the last 4000 of a two chain run of 5000 iterations show the lowest DBP readings for drug C, with centred effect (mean and 95% CRI) of $-1.55$ ($-2.29$, $-0.82$), whereas the new drug has a significant positive effect. A low CPO estimate is apparent for the third visit of patient 249, with DBP reading of 140, compared to previous readings of 120 and 118. Predictions $y_{rep,ijt}$ are used to obtain a predictive discrepancy criterion $\sum_{i,j,t}(y_{ijt} - y_{rep,ijt})^2$. This criterion, the LPML, and the DIC, are respectively 168246, $-3925$, and 9569.

A second model introduces a normal subject level random intercept

$$y_{ijt} = \beta_1 + b_i + \beta_2 P_{0i} + \delta_B + \delta_C + e_{ijt}.$$

As in Example 7.1, a uniform prior is adopted on the ratio of the variance of $e_{ijt}$ to the variance of $\eta_{ijt} = b_i + e_{ijt}$. Estimates for this model show slightly enhanced treatment parameters in absolute terms, but also reduced precision, with the 95% interval for centred $\delta_C$ now being $(-2.80, -0.52)$. The posterior mean for $\phi = \sigma_b^2/(\sigma_b^2 + \sigma_e^2)$ is 0.51. The model assessment criteria show major gain through introducing patient heterogeneity: the predictive discrepancy falls to 83646, the LPML increases to $-3671$, and the DIC falls to 9030.

The assumption of normality regarding the permanent effects may be questioned: for example, a D'Agostino-Pearson test applied to posterior mean $b_i$ suggests non-normality. Instead one may consider a non-parametric mixture approach to representing the $b_i$ (e.g. Butler and Louis, 1992; Chib, 2008). In particular, parametric assumptions about the permanent subject effects are avoided by a Dirichlet process mixture (e.g. Escobar and West, 1998; Hirano, 2002) whereby

$$b_i \sim N(v_i, \phi_i)$$

$$(v_i, \phi_i) \sim G,$$

$$G \sim D(G_0, \alpha).$$

In practice there will be clustering of $v_i$ and $\phi_i$ values, and a maximum of $M = 20$ clusters is assumed. The baseline prior $G_0$ has form

$$v_m \sim N(0, f_m \phi_m)$$

$$\phi_m \sim Ga(1, 0.001); \qquad m = 1, \dots, M$$

where the $f_m$ determine the relative spread of cluster means and patient specific random effects, and are assigned a Ga(1, 1) prior. The category $L_i$ to which patient $i$ is assigned has prior

$$L_i \sim \text{Categorical}(p_1, p_2, \dots, p_M).$$

Only $M^*$ clusters (between 1 and $M$) are selected at any iteration as appropriate for one or more observations. The mixture weights $p_1, p_2, \dots, p_M$ are determined by the stick-breaking method, whereby $r_1, r_2, \dots, r_{M-1}$ are a sequence of Beta(1, $\alpha$) random variables (with $r_M = 1$), and

$$p_1 = r_1$$

$$p_2 = r_2(1 - r_1)$$

$$p_3 = r_3(1 - r_2)(1 - r_1)$$

$$\dots$$

The precision parameter $\alpha$ is assigned an exponential prior with mean $a_\alpha = 1$. Sensitivity to prior settings may be assessed by setting alternative values for $a_\alpha$; larger values of $\alpha$ (assuming smaller $a_\alpha$) imply more clusters $M^*$ and greater differentiation between the $b_i$.

The second half of a two chain run of 10 000 iterations in BUGS provide a posterior mean $M^*$ of 5.1. The predictive discrepancy criterion now stands at 83 950, with the LPML unchanged at $-3671$. However, substantive inferences may be regarded as robustified: thus treatment effects are absolutely smaller than in model 2, with the centred $\delta_C$ now having a 95% CRI of $(-2.61, -0.39)$, while the posterior mean baseline effect $\beta_2$ is smaller (0.43 vs. 0.48) than under the second model. There is a more peaked density of patient effects $b_i$ than under model 2 (kurtosis of 4.6 as against 3.8).

To introduce the information on clinics into the analysis one may adopt a form of the multi-level growth curve model (model 4), which has BUGS data input in a different form. Thus patient-time data is sorted within clinics, and DBP baseline data is input in patients within clinics order. Since this form of input requires known data, the posterior mean baseline reading from model 1 is substituted for the missing baseline value. Corresponding to the broad decline over time in DBP readings, this model introduces a linear growth effect at clinic level. Intercepts and baseline effects also vary at clinic level and there is now a permanent error at patient-clinic level. Thus with patients $i = 1, \ldots, m_j$ nested within clinics $j$, the model has the form

$$y_{ijt} = \beta_{1j} + \beta_{2j}t + \beta_{3j}P_{0ij} + \delta_B + \delta_C + b_{ij} + w_{ijt} + e_{ijt}$$

with

$$\beta_{kj} \sim N(\mu_k, \phi_k), \qquad k = 1, \ldots, 3,$$
$$b_{ij} \sim N(0, \sigma_b^2),$$
$$e_{ijt} \sim N(0, \sigma_e^2), \text{ and}$$
$$w_{ijt} \sim N(\gamma w_{ij,t-1}, \sigma_w^2), \qquad t > 1.$$

The initial conditions $w_{ij1}$ have a distinct variance term, $\sigma_w^2/(1 - \gamma^2)$. Because of the different sources of random variation, there may be identifiability issues. Convergence in the variance terms $\sigma_e^2, \sigma_b^2$, and $\sigma_w^2$, is assisted by setting a prior on the inverse of the total of these three variances $\sigma_T^2 = \sigma_e^2 + \sigma_b^2 + \sigma_w^2$, and using a Dirichlet prior to apportion the total precision between the three sources.

The second half of a two chain run of 25 000 iterations gives a reduced predictive discrepancy of 37 460, and LPML of $-3541$. This model confirms a significant linear decline in the blood pressure readings with 95% interval for $\mu_2$ (the average linear growth coefficient) between $-1.34$ and $-0.52$, and also confirms the beneficial effect of drug C (see Table 7.1, which includes centred treatment effects). The baseline effect remains important (with posterior mean for $\mu_2$ of 0.49), and this form of dependence coexists with autoregressive dependence in the errors, with $\gamma$ averaging 0.46.

## 7.3.2   Subject level autoregressive parameters

While random coefficients attached to powers of time or spline functions may represent diversity in growth curves, another option is to incorporate elements of an autoregressive approach. One possibility is the autoregressive latent trait model mentioned above.

**Table 7.1**  Multilevel growth model (hypertension trial).

| Parameter | Mean | St devn | MC error | 2.5% | 97.5% |
|---|---|---|---|---|---|
| $\delta_A$ | 1.15 | 0.48 | 0.03 | 0.21 | 2.08 |
| $\delta_B$ | 0.34 | 0.55 | 0.02 | −0.73 | 1.44 |
| $\delta_C$ | −1.49 | 0.55 | 0.02 | −2.60 | −0.45 |
| $\mu_1$ | 95.3 | 0.6 | 0.04 | 94.3 | 96.7 |
| $\mu_2$ | −0.94 | 0.21 | 0.01 | −1.34 | −0.52 |
| $\mu_3$ | 0.49 | 0.10 | 0.00 | 0.30 | 0.69 |
| $\gamma$ | 0.46 | 0.21 | 0.02 | 0.12 | 0.86 |
| $\sigma_e^2$ | 17.6 | 8.7 | 0.6 | 2.7 | 31.8 |
| $\sigma_b^2$ | 27.4 | 7.4 | 0.4 | 10.2 | 39.6 |
| $\sigma_w^2$ | 20.7 | 9.5 | 0.6 | 5.0 | 37.4 |

Another is to allow AR coefficients varying over subjects (Nandram and Petruccelli, 1997; Rahiala, 1999; Juárez and Steel, 2010). Thus a growth model in $p$ lags might take the form

$$y_{it} = \rho_{0i} + \rho_{1i}y_{i,t-1} + \; \cdots \; + \rho_{pi}y_{i,t-p} + X_{it}\beta + e_{it},$$

where $\rho_{0i}$ represent different levels of the outcome for individuals. Then $\{\rho_{0i}, \rho_{1i}, \; \cdots \; , \rho_{pi}\}$ are taken to be multivariate normal or Student $t$ with means $\{\rho_0, \rho_1, \; \cdots \; , \rho_p\}$. This may allow greater flexibility than varying linear or spline growth rates in representing stationary growth paths, or if shapes of individual trajectories vary considerably (Rahiala, 1999).

One may also introduce subject level variation in autoregressive error structures (e.g. Chamberlain and Hirano, 1999). For example, AR(1) dependence in the over-dispersion errors in a log link regression for a Poisson outcome might be taken to vary over subjects:

$$\log \mu_{it} = X_{it}\beta + b_i + e_{it},$$

$$e_{it} = \gamma_i e_{it-1} + u_{it}.$$

In a study of suicide incidence trends, Congdon (2001) compares this 'differential persistence' model with a model allowing variable growth rates in incidence.

### Example 7.3   Protein content in milk

This example demonstrates options for representing varying autoregressive dependence in data on the percentage protein content $y_{it}(t = 1, \; \cdots \; , T_i)$ of cow milk according to diet regime. This may be an alternative to polynomial growth models in the face of diversity in the shapes of individual trajectories. 52 cows were observed for a maximum of $T_i = 19$ weeks, though some were observed for less than this – the numbers of weeks vary from 14 to 19. There are also a few missing cases in intervening weeks, before the end of the observation period on particular cows. These are assumed to be missing at random.

Following Rahiala (1999), a lag 5 model is fitted which conditions on the first five observations, with lags assumed at 1, 2, 3 and 5 weeks, varying coefficients at lag 2 and 5, and with the mean lag at 5 taken as zero. Thus

$$y_{it} \sim N(v_{it}, s_1^2),$$

$$v_{it} = \rho_{0i} + \rho_1 y_{i,t-1} + \rho_{2i} y_{i,t-2} + \rho_3 y_{i,t-3} + \rho_{5i} y_{i,t-5} + X_i \beta,$$

$$\{\rho_{0i}, \rho_{2i}, \rho_{5i}\} \sim N_3 \left(\mu_\rho, \Sigma_\rho\right).$$

where $\mu_\rho = (\rho_0, \rho_2, 0)$. There is a single covariate $X_i$ for diet type (barley $= 1, 0 =$ lupins). The homogenous coefficients $\{\rho_{1,}\rho_2, \rho_3\}$ are not confined to stationarity a priori, and are assigned $N(0, 1)$ priors. Note that in BUGS, the data are input in a $52 \times 19$ format though some cows are observed for less than 19 weeks. A Wishart prior is assumed for $\Sigma_\rho^{-1}$ with scale matrix $Q$ and $k = 3$ degrees of freedom.

There may be sensitivity in inferences to the setting of this scale matrix, and substantively based information, prior elicitation, or information from non-Bayesian estimates may be relevant. For example, consider ranked values for varying lag 2 and lag 5 coefficients $\{\rho_{[2i]}, \rho_{[5i]}\}$. As considered in Chapter 2, a prior expectation may be that the gap between 95th and 5th percentiles of these varying lag coefficients is 0.5, so that the prior standard deviation is 0.152 ($= 0.5/3.29$), with corresponding prior variance 0.023. In the Wishart prior (with $k = 3$ degrees of freedom) this implies diagonal elements $Q_{22}$ and $Q_{33}$ of 0.069. For setting $Q_{11}$ one might consider an admittedly ad hoc data based procedure, using the variance, 0.0506, of the 52 observed within-cow averages. Multiplying by 10 provides a down-weighting to 0.51, implying $Q_{11} = 1.53$.

Because not all the first five values in each animal's series is observed (e.g. $y_{20,2}$ is missing) the likelihood for the first five points must still be represented in some way under the conditional approach, even though they are not included in the differential autoregression scheme. A random effects model is accordingly adopted for the first five periods, with the observations $y_{it}$ ($t = 1, \ldots, 5$) for animal $i$ drawn from a normal density with variance $\sigma_2^2$ and means $m_i$, and with the means themselves drawn randomly from a normal density with mean $M$ and variance $\sigma_3^2$.

The second half of a two chain run of 10 000 iterations shows significant positive lags at 1 and 3 weeks, with posterior means (sd) for $\rho_1$ and $\rho_3$ of respectively 0.41 (0.04) and 0.16 (0.04). There is also a negative correlation, with mean (sd) of $-0.60$ (0.24) between the random intercepts $\rho_{0i}$, and lag 5 growth coefficients $\rho_{5i}$. The effect of the lupin diet is inconclusive, albeit mostly positive, with 95% CRI ($-0.05, 0.28$). A stationarity check (Section 6.2.2) on the collective five autoregressive coefficients (including $\rho_4 = 0$) indicates that the overall model is in fact stationary (as can be seen by monitoring tot.NS in the BUGS code).

## 7.4 Longitudinal discrete data: Binary, categorical and Poisson panel data

Bayesian estimation approaches are often advantageous in the analysis of longitudinal data for binary and categorical responses (multinomial or ordinal) and for count response data (e.g. Qiu *et al.*, 2002; Kaciroti *et al.*, 2008; Wang *et al.*, 2008; Varin and Czado, 2010; Giardina *et al.*, 2011; Dagne and Huang, 2012).

### 7.4.1 Binary panel data

Panel analysis of count data raises distinctive new issues (see Section 7.4.3) whereas, as discussed in Chapter 3, binary and categorical outcomes may be modelled in terms of latent continuous variables, so that analysis of binary panels continues the themes of Section 7.2. Use of the latent responses facilitates MCMC sampling and may have inferential benefits.

Thus for binary outcomes, a continuous latent variable or underlying propensity $Z_{it}$ exists such that $y_{it} = 1$ if $Z_{it} > 0$, and $y_{it} = 0$ if $Z_{it} \leq 0$, with $Z_{it}$ obtained by truncated sampling within ranges determined by the observed $y_{it}$ (Albert and Chib, 1993). Consider a panel model without heterogeneity, and distinguish between regression effects $\eta_{it} = X_{it}\beta$ where $X_{it}$ are known predictors, and a stochastic error $e_{it}$, so that

$$Z_{it} = \eta_{it} + e_{it}.$$

It is assumed that the probability of success $\Pr(y_{it} = 1)$ is expressed as $\pi_{it} = F()$, where $F(.)$ is a distribution function and so lies between 0 and 1. So a success occurs according to

$$\Pr(y_{it} = 1) = \Pr(Z_{it} > 0) = \Pr(e_{it} > -\eta_{it}) = 1 - F(-\eta_{it}).$$

For forms of $F$ that are symmetric about zero, such as the normal cumulative distribution function, $F = \Phi$, the last element of this expression equals $F(\eta_{it})$. If $F = \Phi$, $Z_{it}$ may be sampled from a truncated normal, truncated is to the right (with ceiling zero) if the observation is $y_{it} = 0$, and to the left by zero if $y_{it} = 1$.

Restrictions are needed for identifiability of such latent variable regressions, such as fixing the error variance or regression intercept. For example, if a constant variance var($e_{it}$) = $\phi$ is assumed for the errors, it is necessary that $\phi = 1$ (or some other preset value). Note, however, that time varying variances $\phi_t$ may be identifiable provided a particular variance (e.g. $\phi_1$) is set to a pre-specified value. Note that if $\phi$ is taken as unknown (or different fixed values of $\phi$ are adopted in a sensitivity analysis) then only ratios $\gamma_j = \beta_j/\phi^{0.5}$ are identified (Giardina et al., 2011).

To approximate a logit link, $Z_{it}$ can be sampled from a Student t density with 8 degrees of freedom, since, following Albert and Chib (1993), a $t(8)$ variable is approximately 0.634 times a logistic variable. This sampling based approach to the logit link additionally allows for outlier detection if the scale mixture version of the Student $t$ density is used. The scale mixture option retains truncated Normal sampling but adds a mixture variables $\lambda_i$, such that

$$Z_{it} \sim N(X_{it}\beta, \phi\lambda_i^{-1}) \quad I(L, U)$$

with $\lambda_i$ sampled from a Gamma density $Ga(4, 4)$. The resulting regression coefficients need to be scaled from the $t$ (8) to the logistic. The sampling limits are $\{L = 0, U = \infty\}$ when $y_{it} = 1$ and $\{L = -\infty, U = 0\}$ when $y_{it} = 0$.

The latent variable regression may be extended to include subject and period effects as in

$$Z_{it} = X_{it}\beta + b_i + c_t + \epsilon_{it},$$

where the $c_t$ are period effects invariant over individuals (e.g. reflecting changes in national economic conditions), and $b_i$ are permanent subject effects interpretable as propensities to experience the event, or as utilities of a certain choice. This representation may include time varying loadings on the $b_i$, as in the one factor model (Heckman, 1981)

$$Z_{it} = X_{it}\beta + \lambda_t b_i + c_t + \epsilon_{it}$$

where the $\lambda_t$ may be used to describe the correlation between time points $t$ and $s$.

Permanent subject effects $b_i$ that induce correlation over time may be considered as producing spurious state dependence in binary panel or multinomial data (Chib and Jeliazkov, 2006); examples are labour participation histories (see Example 7.4), and panel data on multinomial choices, such as product brand choices (Dubé et al., 2010). By contrast, true state dependence in a model for a binary panel outcome would lead to a model including autoregression on $y_{it}$ (e.g. Sutradhar $et$ $al.$, 2010), as in the AR1 case

$$Z_{it} = X_{it}\beta + \rho y_{i,t-1} + b_i + e_{it}.$$

Here $\rho y_{i,t-1}$ represents the impact of actual behaviour in the preceding period on utility or preference in the current period. If $\rho$ is not significantly different from zero, but $\sigma_b^2 > 0$ or most $b_i$ are significant (see Section 7.5), then spurious dependence is predominant. Dependence on lagged $Z_{it}$ rather than $y_{it}$ may be included instead, but characterised as preference persistence (Giardina $et$ $al.$, 2011; Stegmueller, 2013).

Autoregression on lagged $y$ or $Z$ values raises questions about the modelling of the initial data (for example, the first observation in the case of AR1 dependence). One option (Giardina $et$ $al.$, 2011) in schemes with preference persistence and excluding permanent subject effects is

$$Z_{it} = X_{it}\beta + \rho Z_{i,t-1} + e_{it} \quad t > 1$$
$$Z_{i1} = \beta_0 + e_{i1}$$

with var$(e_{it}) = \phi$ for $t = 1, \dots, T$. Identification is achieved either by setting $\phi$ and leaving $\beta_0$ unknown, or by setting $\beta_0$ and leaving $\phi$ unknown. In a scheme including permanent subject effects, one possibility is a separate static model for the initial observations, as in

$$Z_{i1} = X_{i1}\beta^* + u_i$$
$$Z_{it} = X_{it}\beta + b_i + \zeta u_i + \rho Z_{i,t-1} + e_{it}, \quad t > 1,$$

where, by virtue of the model form, $\rho$ is not necessarily constrained to stationarity, the $u_i$ are identically distributed with known variance $\phi_1$, and the $e_{it}$ also have known variance $\phi_e$.

## Example 7.4    Employment transitions

Consider data on monthly employment transitions of recently married women as (Skrondal and Rabe-Hesketh, 2004). There are $n = 155$ women with varying lengths of employment history, with $N = \sum_{i=1}^{Ti} = 1580$ observations. The response is $y_{it} = 1$ if a woman is in paid employment and $y_{it} = 0$ otherwise, though the models are framed in terms of latent normal augmented data $Z_{it}$ (equivalent to a probit link). Predictors are husband unemployment status ($x_{2i} = 1$ if unemployed, 0 otherwise), time in months from start of study ($x_{3i}$), whether parent to child under 1 ($x_{4i} = 1$ if parent to child under 1, $x_{4i} = 0$ otherwise), child under 5 ($x_{5i} = 1$ if parent to child under 5, $x_{5i} = 0$ otherwise), and age of woman ($x_{6i}$).

Different dynamic specifications are considered, with the first (model 1) being the separate static model for period 1, as mentioned above. Thus priors $\rho \sim N(0, 1), \zeta \sim N(0, 1), \{b_i \sim N(0, \sigma_b^2), \ i = 1, \dots, n\}, \ \sigma_b \sim U(0, 100)$, and $\beta_j \sim N(0, 1000), \ \beta_j^* \sim N(0, 1000)$ for covariates $j$, are adopted. Additionally it is assumed that residual variances $\phi_e = \phi_1 = 1$. Convergence is obtained after 2500 iterations, and inferences are based on the second half of a run of 5000 iterations. Predictive assessment is based on sampling replicates $Z_{new,it}$ without constraint, so that the corresponding $y_{new,it}$ is obtained according to whether $Z_{new,it}$ is positive or negative. A predictive assessment is based on the proportion of correctly identified transitions ($y_{new} = y$), denoted SPRD in the BUGS code[2].

The $\beta$ and $\beta^*$ estimates both show that a husband being unemployed also reduces the woman's chance of entering or retaining employment, perhaps reflecting local labour market conditions, while a child under 1 also reduces the probability of being employed. Specifically, the $\beta_4$ coefficient has posterior mean and a 95% CRI $-1.57$ ($-2.03, -1.14$). The proportion of correctly identified transitions is 0.559 and the DIC is 1000 (with $d_e = 169$).

The second model assumes pre-period latent responses $Z_{i0} \sim N(\mu_0, 1)$ so that for all periods $t = 1, \ldots, T$ of observed data the same model, namely

$$Z_{it} = X_{it}\beta + \rho Z_{i,t-1} + b_i + e_{it}$$

may be estimated. It is assumed that $\mu_0 \sim N(0, 1000)$. From the second half of a two chain run of 5000 iterations, similar results to model 1 are obtained for the $\beta$ coefficients. The posterior mean (and 95% CRI) for $\mu_0$ is 0.15 ($-0.24, 0.53$), and both this parameter and individual $Z_{i0}$ show no problems with convergence. The DIC is reduced to 990 (with $d_e = 207$), and the proportion of correctly identified transitions is increased to 0.564.

A third model allows variability of the permanent effects $b_i$ to vary according to previous observed status $y_{i,t-1}$. Thus

$$Z_{it}|y_{i,t-1} = 1 = X_{it}\beta + \rho Z_{i,t-1} + \lambda b_i + e_{it}$$
$$Z_{it}|y_{i,t-1} = 0 = X_{it}\beta + \rho Z_{i,t-1} + b_i + e_{it}$$

with $\lambda > 0$, so expressing potential differences in unobserved variability for currently unemployed subjects as against employed subjects. As in the second model, $Z_{i0} \sim N(\mu_0, 1)$ for the period preceding the first observations, and extending the differential heterogeneity principle one has

$$Z_{i1}|Z_{i0} > 0 = X_{i1}\beta + \rho Z_{i0} + \lambda b_i + e_{i1}$$
$$Z_{i1}|Z_{i0} \leq 0 = X_{i1}\beta + \rho Z_{i0} + b_i + e_{i1.}$$

An exponential prior is adopted for $\lambda$, namely $\lambda \sim E(1)$. This model reduces the DIC to 956 (from the second half of a two chain run of 5000 iterations), with the proportion of correctly identified transitions raised to 0.569. In contrast to the two earlier models, parenthood of a child under 5 now significantly also reduces chances of employment, in addition to husband unemployment and parenthood of a child under 1. The posterior mean of $\lambda$ and 95% CRI for is 0.19 (0.02, 0.43), so that heterogeneity is smaller for women already employed.

## 7.4.2   Ordinal panel data

Models for ordinal responses over time are important because in many settings involving human subjects, classifications are on a graded scale, with precise quantification not being possible. Examples include pre- and post-treatment observations on rankings of illness symptoms (e.g. no symptoms, mild, definite) or changed illness states, as well as survey questions on changing views on controversial topics. Methodological aspects considered in the literature on ordinal longitudinal data include translation between different ordinal scales (Parmigiani et al., 2003), missing data modelling (Kaciroti et al., 2006), and Markov transition approaches to ordinal data (Noorian and Ganjali, 2012), whereby the focus is on probabilities $\Pr(y_{it} = c|y_{i,t-1} = d)$.

For ordered choices $c = 1, \ldots, C$, subjects $i$ and periods $t$, observed categorical data may be considered as generated by underlying utilities or propensities

$$Z_{it} = X_{it}\beta + W_{it}b_i + e_{it},$$

in relation to ordered thresholds $\tau_c$, under a specific cumulative distribution function $F(e)$. For example, if $e$ follows a standard logistic density with mean 0 and variance $\pi^2\sigma^2/3$, then $F(e) = 1/[1 + \exp(-e/\sigma)]$. Then $y_{it} = 1$ if $Z_{it} \leq \tau_1, y_{it} = 2$ if $\tau_1 < Z_{it} \leq \tau_2$, etc. until $y_{it} = C$ if $Z_{it} > \tau_{C-1}$ under the chosen cumulative density. Specifically one has

$$\Pr(y_{it} = c) = \Pr(\tau_{c-1} < Z_{it} \leq \tau_c) = F[(\tau_c - Z_{it})/\sigma] - F[(\tau_{c-1} - Z_{it})/\sigma].$$

If all $C - 1$ thresholds are free parameters then the regression term excludes an intercept for identifiability.

The regression specifications for $Z_{it}$ might be modified. Instead of the proportional odds assumption $\beta_c = \beta$, covariate effects $\beta$ and random effects $b_i$ may differ by category. In a panel data setting, one might also consider shifts in the location of thresholds between periods – for example, if the analysis was intended to assess whether there had been a shift in attitudes.

Setting $\eta_{it} = X_{it}\beta + W_{it}b_i$, choice of category is determined by cumulative probabilities

$$\Pr(y_{it} \leq c) = \Pr(Z_{it} \leq \tau_c) = \gamma_{itc} = \pi_{it1} + \ldots + \pi_{itc}.$$

Under a standard logistic $F$,

$$\Pr(Z_{it} \leq \tau_c) = \Pr(Z_{it} - \eta_{it} \leq \tau_c - \eta_{it})$$
$$= 1/[1 + \exp(\eta_{it} - \tau_c)]$$
$$= \exp(\tau_c - \eta_{it})/[1 + \exp(\tau_c - \eta_{it})].$$

Equivalently

$$y_{it} \sim \text{Categoric}(\pi_{it,1:C})$$

$$\pi_{it1} = \gamma_{it1}$$

$$\pi_{itc} = \gamma_{itc} - \gamma_{it,c-1} \qquad c = 2, \ldots, C$$

$$\pi_{itC} = 1 - \gamma_{it,C-1},$$

and

$$\text{logit}(\gamma_{itc}) = \tau_c - X_{it}\beta - W_{it}b_i.$$

For an ordinal probit model one has

$$\gamma_{itc} = \Phi(\tau_c - X_{it}\beta - W_{it}b_i).$$

An ordinal longitudinal model can be estimated directly with a categorical likelihood or using the augmented data approach with appropriate density for the errors, in which case the variance of the error term $e$ has to be set. For example taking the errors to be Normal with variance 1 leads to an ordinal probit model, whereas taking e to be Student $t\,(8)$ with variance 1 leads (to a close approximation) to the ordinal logit model.

## Example 7.5   Ratings of schizophrenia

To illustrate longitudinal approaches for ordinal responses, consider data on the impacts of drug treatments on symptom severity among $n = 324$ schizophrenic patients, collected as part of the NIMH Schizophrenic Collaborative Study (Hedecker and Gibbons, 1994). There were originally four treatments: chlorpromazine, fluphenazine, thioridazine, and placebo. Since previous analysis revealed similar effects of the three anti-psychotic drugs, the treatment is reduced to a binary comparison of any drug vs. placebo. The severity rating, derived from item 79 of the Inpatient Multidimensional Psychiatric Scale, has $C = 7$ categories: $1 =$ normal, $2 =$ borderline, $3 =$ mildly ill, $4 =$ moderately ill, $5 =$ markedly ill, $6 =$ severely ill and $7 =$ extremely ill. There are $T = 4$ waves of observation including the first wave (at week 0) coincident with treatment, giving $N = 1296$ observations. For the majority of subjects follow-up readings are at weeks $v = 1$, $v = 3$ and $v = 6$. For a few subjects, their third readings are at $v = 2$ or $v = 4$ rather than $v = 3$, while for some other subjects final readings are at $v = 4$ or $v = 5$.

Two models are compared, the first (model 1) including bivariate random intercepts and slopes on time, but with no dynamic dependence. Thus

$$y_{it} \sim \text{Categorical}(\pi_{it,1:C}),$$

$$\text{logit}(\gamma_{itc}) = \tau_c - \eta_{it},$$

$$\eta_{it} = X_{it}\beta + W_{it}b_i.$$

The predictors $W_{it}$ are the intercept and a time variable, namely the square root of weeks $v_{it}$ at observation $t$, whereas $X_{it}$ includes a treatment effect $(X_1)$, patient gender $(X_2 = 1$ for males, $0$ for female), and a treatment by time interaction, so that

$$\eta_{it} = b_{i1} + b_{i2}v_{it}^{0.5} + \beta_1 X_{1i} + \beta_2 X_{2i} + \beta_3 X_{1i}v_{it}^{0.5}$$

$$b_{i,1:2} \sim N_2\left(\mu_{1:2}, \Sigma_b\right)$$

where the means of $b_{i1}$ and $b_{i2}$ are the overall intercept and average change in severity in the root week scale. Because the regression includes an intercept, there are only $C - 2 = 5$ unknown thresholds $\{\tau_2, \tau_3, \ldots, \tau_6\}$ with $\tau_1$ (governing the transition from normality to borderline illness) set to zero.

A second model (model 2) includes dynamic dependence, and involves a form in lag in $y_{it}$ reflecting the ordinal nature of the predictor. Thus

$$\eta_{it} = b_{i1} + b_{i2}v_{it}^{0.5} + \rho S_{it} + \beta_1 X_{1i} + \beta_2 X_{2i} + \beta_3 X_{1i}v_{it}^{0.5}$$

where $S_{it} = 1$ if $|y_{it} - y_{i,t-1}| \leq 1$. Thus $S_{it} = 1$ if the absolute gap between successive ordinal observations is 1 or 0; in other words, if the successive observations are similar on the ordinal scale. For example, if $y_{i1} = 6$ and $y_{i2} = 7$, then $S_{i2} = 1$, but if $y_{i1} = 5$ and $y_{i2} = 7$, then $S_{i2} = 0$. This model involves an additional step to generate latent severity scores $y_{i0}$ at wave 0, allowing wave 1 observations to be included in the likelihood. The $y_{i0}$ are sampled from a categorical density with unknown probability vector $\pi_0$. The lag coefficient $\rho$ is assigned a $N(0, 1)$ prior.

The predictive success of the two models can be assessed in a $C \times C$ cross-tabulation $M_{cd}$ comparing (over all $N = \sum_{i=1}^{T_i}$ observations) actual $y$ categories, $y_{it} = c$, with the categories $y_{\text{new},i,t} = d$ of replicate data. Thus predictive discrepancies will be apparent if replicate categories are widely separated from actual categories. The relevant BUGS code is

```
for (t in 1:T) { y[i,t] ~ dcat(pi[i,t,1:C]);
ynew[i,t] ~ dcat(pi[i,t,1:C])
for (c in 1:C) {for (d in 1:C){prmatch[c,d,i,t]
     <- equals(y[i,t],c)*equals(ynew[i,t],d)}}
with a subsequent summation
for (c in 1:C){for (d in 1:C) {M[c,d] <- sum(prmatch[c,d,,])}}
```

Inferences for both models are based on the second halves of runs of 10 000 iterations in BUGS. Of particular substantive importance is the estimate of $\beta_3$ (the treatment by time interaction) which in the first model has mean and 95% CRI of $-1.22$ ($-1.38$, $-1.06$). There is also a higher average severity score for males under this model with posterior mean and 95% CRI for $\beta_2$ of 1.01 (0.80, 1.23).

Table 7.2 shows the predictive cross-tabulation mentioned above (posterior means of $M_{cd}$), including proportions in each row of exact matches $M_{cc}/\sum_{d=1,...,C}M_{cd}$, and matches that are within a single ordinal category, for example $(M_{21} + M_{22} + M_{23})/\sum_{d=1,...,C}M_{2d}$. It can be seen that model 1 tends to overpredict lower severity categories, especially category 1. The model including the lag effect[3] has better predictive concordance, with estimate for $\rho$ of 2.58. The

**Table 7.2**  Ordinal predictive concordance: Cross-tabulation of actual and predicted categories.

| (A) Model 1 Without Lag | | | | | | | | | |
|---|---|---|---|---|---|---|---|---|---|
| Actual data category | Replicate data category | | | | | | | Total | Proportion with exact predictive match | Proportion with close predictive match |
| | 1 | 2 | 3 | 4 | 5 | 6 | 7 | | | |
| 1 | 22 | 7 | 9 | 10 | 9 | 4 | 0 | 61 | 0.36 | 0.47 |
| 2 | 32 | 13 | 17 | 22 | 21 | 10 | 1 | 116 | 0.11 | 0.54 |
| 3 | 35 | 16 | 23 | 33 | 37 | 20 | 2 | 166 | 0.14 | 0.43 |
| 4 | 44 | 21 | 32 | 52 | 68 | 47 | 6 | 270 | 0.19 | 0.56 |
| 5 | 46 | 24 | 37 | 67 | 102 | 85 | 12 | 373 | 0.27 | 0.68 |
| 6 | 28 | 14 | 23 | 44 | 76 | 77 | 12 | 274 | 0.28 | 0.60 |
| 7 | 3 | 2 | 2 | 5 | 10 | 12 | 2 | 36 | 0.06 | 0.39 |
| Total | 211 | 99 | 146 | 237 | 328 | 259 | 43 | 1296 | | |

| (B) Model 2 With Lag | | | | | | | | | |
|---|---|---|---|---|---|---|---|---|---|
| Actual data category | Replicate data category | | | | | | | Total | Proportion with exact predictive match | Proportion with close predictive match |
| | 1 | 2 | 3 | 4 | 5 | 6 | 7 | | | |
| 1 | 22 | 9 | 10 | 11 | 7 | 2 | 0 | 61 | 0.36 | 0.51 |
| 2 | 35 | 17 | 20 | 23 | 16 | 5 | 0 | 116 | 0.14 | 0.62 |
| 3 | 33 | 20 | 28 | 37 | 33 | 14 | 1 | 166 | 0.17 | 0.51 |
| 4 | 31 | 22 | 37 | 65 | 74 | 37 | 4 | 270 | 0.24 | 0.65 |
| 5 | 16 | 15 | 32 | 76 | 125 | 97 | 14 | 373 | 0.33 | 0.80 |
| 6 | 9 | 8 | 18 | 46 | 87 | 90 | 15 | 274 | 0.33 | 0.70 |
| 7 | 3 | 2 | 3 | 5 | 9 | 12 | 3 | 36 | 0.07 | 0.42 |
| Total | 149 | 94 | 152 | 267 | 356 | 262 | 44 | 1296 | | |

treatment-time parameter $\beta_3$ is amplified under this model, with a posterior mean and 95% CRI of $-1.46$ ($-1.64$, $-1.29$), while the male severity excess is reduced to 0.59 (0.38, 0.79).

### 7.4.3   Panel data for counts

Assume counts $y_{it}$ observed at times $t = 1, \ldots, T$, and taken as Poisson $y_{it} \sim \text{Po}(\mu_{it})$, or variations such as zero inflated Poisson or negative binomial (Boucher *et al.*, 2008; Dagne, 2010; Neelon *et al.*, 2010). Modelling issues with such data include the most suitable form for allowing dynamic dependence on previous observations, the modelling of initial waves in dynamic models (Fotouhi, 2007), and the choice for the 'frailty' model assumed for permanent effects (Achcar *et al.*, 2008).

Thus, as mentioned by Blundell *et al.* (2002), taking a linear impact of the first lag outcome $y_{i,t-1}$ in a log-link regression, as in

$$\mu_{it} = \exp(X_{it}\beta + \rho y_{i,t-1} + W_{it}b_i), \quad t = 2, .., T$$

can lead to explosive series or to problems with transforming zero values. Instead one may consider transformations of the lagged outcome

$$\mu_{it} = \exp(X_{it}\beta + \rho g(y_{i,t-1}) + W_{it}b_i), \quad t = 2, .., T$$

such as $g(y) = \log(y + 1)$. Other options include a linear feedback model, based on the INAR time series specification considered in Chapter 6 (e.g. Sutradhar, 2011) whereby

$$\mu_{it} = \rho y_{i,t-1} + \exp(X_{it}\beta + W_{it}b_i),$$

while a first order version of the full autoregressive conditional Poisson specification (Jung *et al.*, 2006) leads to

$$\mu_{it} = \rho y_{i,t-1} + \eta \mu_{i,t-1} + \exp(X_{it}\beta + W_{it}b_i).$$

As to representing initial conditions in models with dynamic dependence, consider models with only random intercepts at subject level (i.e. $W_{it} = 1$). Then the model for the initial period data can specify a unique regression such as

$$\mu_{i1} = \exp(X_{i1}\beta^* + c_i),$$

with carry over of the initial period residual to subsequent periods

$$\mu_{it} = \exp(X_{it}\beta + \zeta c_i + b_i + \rho g(y_{i,t-1})) \qquad t > 1.$$

Alternatively the initial period and subsequent period random effects ($c_i$ and $b_i$) can be taken to be bivariate random effects, as in the specification

$$\mu_{i1} = \exp(X_{i1}\beta^* + c_i),$$

$$\mu_{it} = \exp(X_{it}\beta + b_i + \rho g(y_{i,t-1})), \qquad t > 1,$$

$$(b_i, c_i) \sim N_2(v, \Omega).$$

It is also possible to condition on the initial period, or specify a minimal model such as $y_{i1} \sim \text{Po}(\mu_1)$, (Arellano and Carrasco, 2003; Wooldridge, 2005) with the interrelation between permanent effects and initial conditions specified in a prior such as

$$b_i | y_{i1} \sim N(\phi y_{i1}, \sigma_b^2).$$

## Example 7.6   Burglary rates in England districts

Consider data (available from http://www.ons.gov.uk/ons/datasets-and-tables/index.html) on totals of domestic burglary offences $y_{it}$ for $n = 345$ English local authority districts from 2002/03 through to 2011/12 ($T = 10$ financial years). Districts are ordered alphabetically (Adur is area 1 through to York). Differences in crime rate may be partly explained in terms of observed characteristics such as area deprivation and urban-rural status. However, unobserved variation is likely to remain in terms of unmeasured aspects of the urban socioeconomic and physical environment – suggesting the need for area level random effects. The general trend in the data is for reduced crime, but with varying trends between areas that can also be taken as random effects. Differences in trends (assumed linear) between areas may be linked to observed characteristics of areas. Offsets are provided by populations $P_{it}$.

Let $X_{it}$ denote area attributes, $b_{i1}$ denote varying crime levels and $b_{i2}$ denote varying linear trends. Also let $Z_i$ denote area attributes influencing trends. To account for remaining overdispersion, a multiplicative conjugate gamma form for area-time heterogeneity is adopted. This facilitates sampling of replicate data as compared to a specific negative binomial likelihood. Thus

$$y_{it} \sim Po(\mu_{it}c_{it})$$

$$\mu_{it} = P_{it}\rho_{it}$$

$$c_{it} \sim Ga(\alpha, \alpha)$$

$$\log(\rho_{it}) = X_{it}\beta + W_{it}b_i = b_{i1} + b_{i2}t + X_{it}\beta,$$

$$b_{i,1:2} \sim N\left(\delta_{i,1:2}, \Sigma_b\right),$$

$$\delta_{i1} = 0$$

$$\delta_{i2} = \gamma_0 + Z_i\gamma.$$

There are two predictors of varying crime levels, a measure of social deprivation in 2001 (Carstairs score), $x_{i1}$, and a binary measure ($x_{i2}$) of urban status for each district, based on a broader six-fold categorisation: major urban (category 1), large urban (category 2), other urban (category 3), significant rural (category 4, over 26% of population in rural settlements and market towns) rural-50 (category 5, between 50-80% in rural settlements and market towns), and rural-80 (category 6, at least 80% in rural settlements and market towns). While there may be alternatives, a binary urban-rural split is based on categories taking $x_{i2} = 1$ for categories 1–3, and $x_{i2} = 0$ for categories 4–6. For predicting changing burglary rates, we take $z_{i1} = x_{i1}$, while $z_{i2} = 1$ for major urban areas, and $z_{i2} = 0$ otherwise.

MCMC convergence of the $\beta$ and $\gamma$ coefficients is assisted by a fully centred parameterisation using the area random effects. This is possible because both $x$-predictors are not time varying (i.e. $X_{it} = X_i$ ). Specifically the above model is re-expressed as

$$\log(\rho_{it}) = b_{i1} + b_{i2}t,$$

$$b_{i,1:2} \sim N\left(\delta_{i,1:2}, \Sigma_b\right),$$

$$\delta_{i1} = \beta_0 + \beta_1 x_{i1} + \beta_2 x_{i2},$$

$$\delta_{i2} = \gamma_0 + \gamma_1 z_{i1} + \gamma_2 z_{i2}$$

so that the random effects depend on $X_i$ rather than being independent of them (Maddala, 1987).

To assess model performance a mixed predictive check (see Section 2.2.6) is carried out which involves sampling replicates of both sets of random effects $b_{i,1:2}$ (bivariate normal) and $c_{it}$ (gamma). Denoting these as $b_{new,i,1:2}$ and $c_{new,it}$, replicate values of the response $y_{new,it}$ are thus obtained as

$$y_{new,it} \sim Po(\mu_{new,it}c_{new,it}); \mu_{new,it} = P_{it}\rho_{new,it}; c_{new,it} \sim Ga(\alpha, \alpha);$$

$$\log(\rho_{new,it}) = b_{new,i1} + b_{new,i2}t,$$

and then compared with actual values. Thus at iteration $s$, the comparisons

$$R_{it}^{(s)} = I(y_{new,i,t}^{(s)} > y_{it}|\theta^{(s)}) + 0.5I(y_{new,it}^{(s)} = y_{it}|\theta^{(s)}),$$

are carried out, in order to assess possible over- or under-prediction.

From the second half of a two chain run of 10 000 iterations, we have that there are 152 area-time combinations (out of 3450) which are under-predicted, with $Pr(R_{it} = 1) < 0.05$, and 173 area-time observations with over-prediction, with $Pr(R_{it} = 1) > 0.95$. So 325 area-time combinations (i.e. 9.4% of all observations) are in the two 5% tail areas, and predictions of the model are therefore satisfactory.

In terms of substantive findings, there are positive estimates for $\beta_1$ and $\beta_2$, so that deprived and urban areas have higher burglary rates at the start of the 10 year period. As to influences on changing crime rates, there is a posterior mean (95% CRI) for $\gamma_2$ of 0.035 (0.023, 0.046), while the posterior mean for (95% CRI) for $\gamma_0$ is $-0.077$ ($-0.083$, $-0.072$). Thus there is a general decline in burglary rates, but the decline is less in the most urban areas as compared to other types of area.

Denoting $v_{it} = \mu_{it}c_{it}$, correlation in the errors is assessed by first order dependence in the standard residuals (Sutradhar, 2011)

$$e_{it} = (y_{it} - v_{it})/v_{it}^{0.5}.$$

There is no evidence of marked serial correlation in the first model, with the 95% interval for first order temporal autocorrelation (ACF1 in the BUGS code) being ($-0.008$, 0.055). However, to correct for autocorrelation, one may model dynamic dependence via impacts of the lagged outcome. In this regard, it has been argued that it is 'best to model dynamics via a lagged dependent variable rather than via serially correlated errors' (Beck and Katz, 1996).

A second model therefore includes a lag in the response. Since there is a population offset the lag has the form

$$g_{it} = \log\left(\frac{y_{i,t-1} + 0.5}{P_{i,t-1} + 0.5}\right).$$

Additionally the second model assumes log-normal area-time errors $u_{it}$ (rather than multiplicative gamma errors), as this permits centred sampling about the lag effect, which in turn assists in MCMC convergence. The model for times $t = 2, \ldots, T$ is thus

$$y_{it} \sim Po(P_{it}\rho_{it})$$

$$\log(\rho_{it}) = b_{i1} + b_{i2}t + u_{it},$$

$$b_{i,1:2} \sim N\left(\delta_{i,1:2}, \Sigma_b\right),$$

$$\delta_{i1} = \beta_0 + \beta_1 x_{i1} + \beta_2 x_{i2},$$

$$\delta_{i2} = \gamma_0 + \gamma_1 z_{i1} + \gamma_2 z_{i2},$$

$$u_{it} \sim N(\beta_3 g_{it}, \sigma_u^2).$$

For the first period, a model with an intercept and random effects is adopted, namely $y_{i1} \sim$ Po$(P_{i1}\rho_{i1})$,

$$\log(\rho_{i1}) = \alpha_0 + u_{i1}.$$

The last 5000 iterations of a two chain run of 20 000 iterations show a significant lag effect with $\beta_3$ having a posterior mean (95% CRI) of 0.413 (0.386, 0.438). The DIC is slightly higher under this model than the first (34355 vs. 34353), but the proportion of area-time observations that are over- or under-predicted falls to 8.4%. The estimate for first order error autocorrelation now has a 95% interval $(-0.038, 0.023)$.

## 7.5   Random effects selection

As discussed above, the assumption of normality of subject level random effects $b_i$ may be modified to account for features such as extreme observations, excess skew or kurtosis. As well as the Dirichlet process prior considered in Example 7.2, robust modelling of random effects at both observation and subject level may be achieved by student-mixture models (Chib, 2008), as in the scheme

$$y_i = X_i \beta + W_i b_i + e_i,$$

$$e_i \sim N(0, \lambda_i^{-1} \sigma^2 I_T),$$

$$b_i \sim N\left(0, \kappa_i^{-1} \Sigma_b\right),$$

$$\lambda_i \sim \text{Ga}\left(\frac{v_e}{2}, \frac{v_e}{2}\right),$$

$$\kappa_i \sim \text{Ga}\left(\frac{v_b}{2}, \frac{v_b}{2}\right),$$

with low values for $\lambda_i$ and $\kappa_i$ indicating discrepant observations and outlier subjects respectively.

Another possibility for the subject level effects is a discrete mixture of Gaussian densities, such as a spike and slab prior with one component of the mixture having considerably lower variance than the other,

$$b_i \sim (1 - \delta)N\left(0, r\Sigma_b\right) + \delta N\left(0, \Sigma_b\right)$$

where $r \ll 1$. Lasso random effect models (Fruhwirth-Schnatter and Wagner, 2011) extend this principle to allow case specific component indicators $\delta_i$ and include a hierarchical prior on the variances.

For example, a mixture of Laplace densities is obtained under

$$b_i \sim (1 - \delta_i)N(0, \psi_{1i}) + \delta_i N(0, \psi_{2i}),$$

$$\psi_{1i} \sim E(1/(2rQ)),$$

$$\psi_{2i} \sim E(1/(2Q)),$$

with $r$ set small, so that $\psi_{1i} \simeq 0$. The $\delta_i$ are binary with unknown probability $\omega$, the prior proportion of subjects with non-zero random effects. If $Q$ is also unknown, there may be identification issues under independent priors, as different combinations of $\omega$ and $Q$ can give similar $b_i$. This scheme can be applied separately to random slopes and intercepts, though joint selection schemes can also be used, for example, based on Cholesky decomposition of $\Sigma_b$ (Kinney and Dunson, 2007).

### Example 7.7    Random effects selection for simulated data

This example considers random effect selection for simulated data with $n = 100$ cases and $T = 10$ periods, namely

$$y_{it} \sim N(\mu + b_i + \alpha_1 x_{1it} + \alpha_2 x_{2it} + \alpha_3 x_{3it} + \alpha_4 x_{4it}, \sigma^2)$$

with the $x$ variables standard normal, $\alpha = (0.5, -0.5, 0.7, -0.7)$, $\mu = 1$, $\sigma^2 = 0.5$, and half the $b_i$ set to zero. Specifically we take $b_i = -4$ for $i = 1, \dots, 5$; $b_i = -1$ for $i = 6, \dots, 25$; $b_i = 0$ for $i = 26, \dots, 75$; $b_i = 1$ for $i = 76, \dots, 95$; and $b_i = 4$ for $i = 96, \dots, 100$ (cf. Fruhwirth-Schnatter and Wagner, 2011).

Now regarding the sampled data as observations with unknown DGP, the random effect selection scheme above is adopted:

$$y_{it} \sim N(\mu + b_i + \alpha_1 x_{1it} + \alpha_2 x_{2it} + \alpha_3 x_{3it} + \alpha_4 x_{4it}, \sigma^2)$$

$$b_i \sim (1 - \delta_i)N(0, \psi_{1i}) + \delta_i N(0, \psi_{2i}),$$

$$\delta_i \sim \text{Bern}(\omega),$$

$$\psi_{1i} \sim E(1/(2rQ)),$$

$$\psi_{2i} \sim E(1/(2Q))$$

is adopted, with $N(0, 1000)$ priors on fixed effects coefficients, and $r = 0.000025$.

An initial prior setting involved separate priors on $Q$ and $\omega$, namely

$$\omega \sim \text{Beta}(1, 1)$$

$$1/Q \sim \text{Ga}(0.5, 0.2275)$$

where the latter prior is relatively diffuse, and provides a prior median of 1 for $Q$. This prior, however, demonstrates identification and convergence problems (in OpenBUGS 3.2.2) with a two chain run showing non-convergent sets of samples for $\omega$ and $Q$.

An alternative bivariate normal prior scheme[4] is then adopted on $h_1 = \log(1/Q)$ and $h_2 = \text{logit}(\omega)$ jointly, with zero prior mean, and prior identity covariance for $\Sigma_h$. This prior has no convergence issues, and the second half of a two chain run of 20 000 iterations provides a posterior mean (95% CRI) for $\omega$ of 0.59 (0.47, 0.71). The selection indicators $\delta_i$ for cases 26, $\dots$, 75 have posterior means ranging from 0.12 to 0.52 with average 0.22. The remaining selection indicators have posterior means exceeding 0.95, except for cases 82, 83 and 90 (with posterior mean $\delta_i$ of 0.67, 0.80 and 0.70), possibly due to vagaries of the random data generation.

# 7.6    Missing data in longitudinal studies

A practical feature of many surveys or trials is attrition or intermittent non-response, with implications for biased inferences due to missing data (Winer, 1983; Ibrahim *et al.*, 2005; Mason *et al.*, 2012). Such problems occur in periodic longitudinal surveys where a separate sample of the population is chosen each time, and in clinical trials or other studies with a cohort design, where patients may drop out because of deteriorating health, or adverse treatment effects (Carpenter *et al.*, 2002; Goldstein, 2009; Gustavson *et al.*, 2012 ). Such drop-out may be 'informative', inducing dependence between missing values of the variable being measured and the drop-out mechanism.

Among approaches to this problem are weighting adjustments for non-response (Rizzo *et al.*, 1996) and the use of model based imputation methods. Alternatively attrition may be ignored, with only observed cases used for analysis (e.g. as under listwise deletion), which amounts to assuming that unobserved data is missing completely at random (i.e. missingness is unrelated to both observed and unobserved data).

Explicit models for non-response introduce indicators for response/non-response as well as the main study outcomes. Consider response on an outcome $Y$ in particular, and let $R = \{R_{it}\}$ indicate whether response was made ($R_{it} = 1$), or missing ($R_{it} = 0$), for subjects $i = 1, \ldots, n$ at times $t = 1, \ldots, T$. Also assume that certain covariates $X$ are fully measured for all respondents (e.g. demographic attributes or survey design variables such as geographic area of sampling).

One may define (Daniels and Hogan, 2008) the full data model $p(R, Y | \theta_R, \theta_Y, X)$, namely the joint density for the response $R$ and the main study outcome $Y$, and where $Y = \{Y_{obs}, Y_{mis}\}$ includes both observed and missing data. Then the full data model may be expressed as

$$p(Y_{mis} | Y_{obs}, R, X, \theta_E) p(Y_{obs}, R | X, \theta_O)$$

where the first density governs extrapolation, and $\theta_E$ can be identified only subject to modelling assumptions or constraints. Auxiliary outcomes $Z$ (e.g. additional biomarkers collected repeatedly in a trial along with the outcome of interest) may be used to increase the level of information used in extrapolation.

The full data model may be factored in different ways. The selection model specification has the form

$$p(R, Y | \theta_R, \theta_Y, X) = p(Y | X, \theta_Y) p(R | Y, X, \theta_R)$$

allowing the missing data mechanism (MDM) to be influenced by the outcomes $Y$, whether observed or missing. However, identification of the missing data mechanism depends on (i.e. may be sensitive to) the likelihood assumption regarding $Y$ (e.g. Kenward, 1998), and also on how the missing data model $p(R | Y, X, \theta_R)$ expresses dependence on $Y$, for example by assuming linearity or otherwise (Daniels and Hogan, 2008, Section 8.3).

In panel studies subject to monotone dropout, let $H_t$ denote observations on the outcome at times up to an including $t$, and assume the data are observed up to $t - 1$ but may be missing at $t$. Then dropout is random if $p(R_t | H_t, X, \theta_R) = p(R_t | y_1, \ldots, y_{t-1}, X, \theta_R)$, but dropout is non-random if it is related to the current, potentially missing, outcome. In practice, any such dependence can be assessed by survival methods or by binary regression, relating $\Pr(R_{it} = 1)$ to previous and current outcomes $y_{i1}, \ldots, y_{i,t-1}, y_{i,t}$ (e.g. Carpenter *et al.*, 2002). With intermittent missingness, dependence on potentially missing lagged values will also indicate non-random missingness.

Pattern mixture models (Little, 2009) express the joint distribution by conditioning on observed response patterns, so that

$$p(R, Y | \phi_R, \phi_Y, X) = p(Y | X, R, \phi_Y) p(R | X, \phi_R).$$

Often the response pattern is assumed to be multinomial, and given a particular pattern $R = r$, the data likelihood can be expressed as

$$Y | X, R = r \sim P(Y | X, \phi_Y^{(r)}).$$

For example, for data with $T = 3$ waves, and assuming all subjects are observed at wave 1, but with monotone attrition thereafter, there are $M = 3$ possible response patterns: observed for all waves ($r = 1$), observed at waves 1 and 2 ($r = 2$), and observed at wave 1 only ($r = 3$). With 3 waves and intermittent missingness there are 8 possible missingness patterns. Then for continuous data one might assume

$$Y_i | X_i, R = r \sim N(X_i \beta^{(r)} + W_i b_i, \tau^{(r)}),$$

$$b_i \sim N\left(0, \Sigma_b^{(r)}\right),$$

though in general some parameters are non-identified. In practice different patterns may be amalgamated (Hedeker and Gibbons, 1997), identifying restrictions applied (Little, 1993), or informative priors used to link unidentified to identified parameters (Kaciroti et al., 2008).

A further main strategy involves shared random parameters, denoted $b_i$, typically at subject level. It may be assumed that response and missingness indicators are independent given the shared parameters, so that the full data model (e.g. Creemers et al., 2011) becomes

$$p(R_i, Y_i | b_i, \theta_R, \theta_Y, X_i) = p(Y_i | b_i, X_i, \theta_Y) p(R_i | b_i, X_i, \theta_R)$$

and on integration

$$p(R_i, Y_i | \theta_R, \theta_Y, X_i) = \int p(Y_i | b_i, X_i, \theta_Y) p(R_i | b_i, X_i, \theta_R) p(b_i) db_i.$$

Shared parameter schemes may also be combined with a pattern mixture strategy with the distribution of the shared random effects differing between response patterns (Hogan and Laird, 1997).

A baseline assumption in examining sensitivity in inferences and fit is that of missingness at random (MAR) whereby the selection missing data model reduces to

$$p(R | Y_{\text{obs}}, Y_{\text{mis}}, X, Z_{\text{obs}}, Z_{\text{mis}}, \theta_R) = p(R | Y_{\text{obs}}, X, Z_{\text{obs}}, \theta_R),$$

so that non-response can depend on observed outcomes (e.g. on earlier observed $Y$ responses, or observed responses on other items $Z$), or possibly on observed responses by other subjects, but given these, it will not depend on the missing item responses themselves. However, if missingness depends on unknown observations it may be typified as missingness not at random (MNAR). If the MAR assumption holds and models for the outcome and missing data are separate (i.e. $\theta_Y$ and $\theta_R$ are a priori independent) then the missingness pattern is called ignorable (Rubin, 1976; Daniels and Hogan, 2008, p. 99), and inferences about $\theta_Y$ may be made without modelling the missing data mechanism.

## Example 7.8   Smoking cessation

Consider the Commit to Quit (CTQII) trial data (Marcus *et al.*, 2005; Daniels and Hogan, 2008, Section 7.5) on the efficacy of moderate-intensity exercise as an aid for smoking cessation. The study involved $n = 217$ subjects and $T = 8$ time points, and a binary outcome $Y = 1$ if smoking cessation was initiated or maintained, and $Y = 0$ otherwise. There is an auxiliary metric outcome (weight gain) $z_{it}$ at each time point. Both outcomes are subject to intermittent missingness, with overall non-response being 30.9% for cessation and 33.5% for weight change.

The likelihood may potentially be modelled as bivariate normal (with augmented continuous latent data replacing the observed $Y$ values), but here the potential interdependence is handled by taking the probability $\omega_{it} = \Pr(y_{it} = 1)$ to be conditional on $Z$, specifically via a one period lag in $Z$ as well as in $Y$. Thus with the treatment variable $X$ ($x_i = 0$ for 'wellness' and $x_i = 1$ for 'exercise') also included, and a time trend $c_{1t}$ (modelled using fixed effects), one has for times $t = 3, \ldots, T$

$$\text{logit}(\omega_{it}) = c_{1t} + \delta_1 x_i + \alpha_1 z_{i,t-1} + \alpha_2 y_{i,t-1}.$$

The normal likelihood for the weight gain variable $z_{it} \sim N(\mu_{it}, 1/\tau_z)$ (for times $t = 3, \ldots, T$) uses the following representation for the means

$$\mu_{it} = c_{2t} + \delta_2 x_i + \alpha_3 z_{i,t-1}.$$

The period 1 and 2 data precede the intervention, and for both responses are modelled separately, and in pooled regressions (aggregating both periods) involving only treatment received as a predictor.

The substantive interest is whether the treatment leads to higher cessation rates for times $t = 3, \ldots, T$. This can be assessed by totalling the $y_{it}$ (including imputed values for missing data) over subjects and periods, and within the two treatment groups, and then obtaining the overall cessation rates $C_{r,t}$ according to treatment ($r = 1$ for 'wellness', and $r = 2$ for 'exercise'). In the program code[5] this is handled using the commands

```
# overall cessation rates
for (t in 3:T) {C[1,t] <- sum(cesstr[1,t,1:n])/Ttrt[1]
               C[2,t] <- sum(cesstr[2,t,1:n])/Ttrt[2]
# probability of treatment differences in cessation
# 1 for "wellness", 2 for "exercise"
               trtdiff[t] <- step(C[1,t]-C[2,t])
for (i in 1:n) { cesstr[1,t,i] <- cess[i,t]*equals(trt[i],1)
                cesstr[2,t,i] <- cess[i,t]*equals(trt[i],2)}}
```

where cess[i,t] is the cessation response, and Ttrt[1:2] contains the numbers in each treatment group.

A selection approach is adopted with a missingness data mechanism in both responses. Thus indicators $R_{1it}$ ($R_{1it} = 1$ if $Y$ observed, $R_{1it} = 0$ otherwise) and $R_{2it}$ (for $Z$) are modelled using binary regression with predicted probabilities $\eta_{1it} = \Pr(R_{1it} = 1)$ and $\eta_{2it} = \Pr(R_{2it} = 1)$. In a baseline model these missingness probabilities are related to treatment and a time trend only (i.e. a MAR assumption), whereas in an extended model, they are related additionally to one period lagged values in $Y$ and $Z$. Because of intermittent missingness these lagged values are potentially missing, so a significant impact would be consistent with non-random missingness. Thus, in the baseline model, the missing data logit regression for cessation is

$$\text{logit}(\eta_{1it}) = d_{1t} + \gamma_1 x_i,$$

while under the extended model

$$\text{logit}(\eta_{1it}) = d_{1t} + \gamma_{11}x_i + \gamma_{12}z_{i,t-1} + \gamma_{13}y_{i,t-1},$$

with parallel regressions for weight gain.

As a measure of fit, a posterior predictive loss (PPL) criterion is used based on an indicator of the type $H(R, R \times y)$ (Daniels and Hogan, 2008) where $(R \times y) = (R_1 y_1, R_2 y_2, \ldots, R_T y_T)$. The PPL criterion is based on replicates of both $R$ and $y$, with $H_{\text{rep}} = H(R_{\text{rep}}, R_{\text{rep}} \times y_{\text{rep}})$, and has the form

$$\text{PPL}(k) = \text{Var}(H_{\text{rep}}) + \frac{k}{(k+1)} \ (H - E(H_{\text{rep}}))^2.$$

The indicator used (among a range of possible options) is $H_i = R_{1iT}y_{iT} - R_{1i1}y_{i1}$, namely an indicator of change since baseline. Inferences below are in all cases based on the second half of two chain runs of 10 000 iterations.

Estimates for the baseline model show a significant dynamic dependence in both outcomes, with $\alpha_2$ and $\alpha_3$ having respective posterior means (sd) of 2.83 (0.21) and 0.91 (0.02). The 95% CRI for $\alpha_1$, namely $(-0.07, 0.02)$ is biased to negative values, albeit non-significant, in line with weight gain diminishing the chances of maintaining or initiating smoking cessation. Cessation appears lower (albeit not significantly so) under the 'exercise' treatment, with $\delta_1$ having a 95% CRI of $(-0.48, 0.15)$.

However, the aggregated cessation rates show a clearly significant difference with the probability that $C_{1t} > C_{2t}$ exceeding 0.98 for periods 4 through to 6 (Daniels and Hogan, 2008, p. 159). The estimates for the missingness regressions show no treatment differences, but do show a trend towards increased missingness at later periods. The PPL criterion has values 50.8 for $k = 1$, and 64.9 for $k = 1000$, with the setting $k = 1000$ putting more emphasis on fit.

Under the extended model there are significant impacts of lagged $Y$ and $Z$ values (which may be missing) on the probabilities $(\eta_1, \eta_2)$ of missingness. So there is support for missingness being non-random. Thus the posterior means (sd) for $\gamma_{12}$ and $\gamma_{13}$ are respectively 0.21 (0.02) and 1.17 (0.21). The PPL measure indicates better fit with values 37.5 for $k = 1$, and 48.0 for $k = 1000$.

As an indication of sensitivity in inferences, consider the overall cessation rates under treatment 1 ('wellness') for periods 3 to 8. Under the extended model these are respectively 33.6%, 37.9%, 44.9%, 39.8%, 37.2% and 35.5%, whereas under the baseline model they are 39.3%, 46%, 55.8%, 52.9%, 51.5% and 46.4%. However, cessation rates are still significantly higher for treatment 1 than treatment 2 under the extended model.

As a simple example of a shared parameter selection model, the following stipulation (model 3) is used for $t > 2$:

Cessation:

$$y_{it} \sim \text{Bern}(\omega_{it})$$

$$\text{logit}(\omega_{it}) = c_{1t} + \delta_1 x_i + \alpha_1 y_{i,t-1} + b_i,$$

$$b_i \sim N(0, 1/\tau_b)$$

Weight change:

$$z_{it} \sim N(\mu_{it}, 1/\tau_z)$$

$$\mu_{it} = c_{2t} + \delta_2 x_i + \alpha_2 z_{i,t-1} + \lambda_1 b_i$$

Missingness:

$$R_{1it} \sim \text{Bern}(\eta_{1it}); R_{2it} \sim \text{Bern}(\eta_{2it})$$

$$\text{logit}(\eta_{1it}) = d_{1t} + \gamma_1 x_i + \lambda_2 b_i$$

$$\text{logit}(\eta_{2it}) = d_{2t} + \gamma_2 x_i + \lambda_3 b_i.$$

Thus a shared random effect is used to explain unobserved heterogeneity in both the outcomes and the response indicators. This model form typically raises identifiability issues (see Chapter 9), and a positivity constraint $\lambda_j \sim E(1)$ is accordingly used for the loadings, as results from initial exploratory analysis without constraint supported this informative prior.

This model in fact provides improved fit: the PPL measure has values 34.5 for $k = 1$, and 44.1 for $k = 1000$. Subjects with higher estimated $b_i$ are more likely to be quitters and to have weight gain. Such subjects also tend to have higher chances of responding, with $\lambda_2$ and $\lambda_3$ having posterior means (sd) of 6.56 (1.26) and 5.97 (1.15). Cessation rates under wellness are intermediate between those under the two earlier models.

### Example 7.9    Mastitis data

To illustrate contrasting inferences under selection and pattern mixture models consider data on milk yields (thousands of litres) for $n = 107$ dairy cows and $T = 2$ years (Kenward, 1998). In the first year there is complete response, but in the second year, when some cows became infected with mastitis, there are 26 cows with missing yields. Of substantive interest is the change in average yield between the two years as this may be affected by the mastitis infection. Inferences are in all examples based on the second half of two chain runs of 20 000 iterations.

Initially a selection model is adopted with a bivariate normal likelihood assumed for the two readings, $y_i = (y_{i1}, y_{i2})$

$$y_i \sim N_2 (\mu, \Sigma).$$

The missing data indicators are $R_i = 1$ if year 2 yields are observed, and $R_i = 0$ otherwise. The missing data model

$$R_i \sim \text{Bern}(\eta_i)$$

could allow dependence on both the known first year yield alone (i.e. in line with MAR), or on both the first year yield and the possibly missing second year yield:

$$\text{logit}(\eta_i) = \beta_0 + \beta_1 y_{i1} + \beta_2 y_{i2}.$$

Of substantive interest is the population difference $\Delta = \mu_2 - \mu_1$ between yearly yields, under different assumptions about the MDM. Also considered as a sensitivity indicator is the correlation $\rho$ between year 1 and year 2 yields. A relatively informative $N(0, 10)$ prior is adopted for the $\beta$ coefficients, since occasional numeric errors may be produced by combinations of extreme $\beta_2$ and $y_{i2}$ sampled values (when $y_2$ is missing).

If the MDM has the MAR form $\text{logit}(\eta_i) = \beta_0 + \beta_1 y_{i1}$, then posterior means (sd) for $\Delta$ and $\rho$ are respectively 0.71 (0.11) and 0.57 (0.07), with $\beta_1$ having mean (sd) of $-0.39$ (0.25). By contrast, when the MDM has the MNAR form, namely $\text{logit}(\eta_i) = \beta_0 + \beta_1 y_{i1} + \beta_2 y_{i2}$, then posterior means (sd) for $\Delta$ and $\rho$ are respectively 0.32 (0.14) and 0.44 (0.09) (cf. Kenward, 1998, Table 1). The MDM regressor effects are amplified considerably, with $\beta_1$ and $\beta_2$ having respective means (sd) of $-2.62$ (0.83) and $-2.83$ (0.91).

For the pattern mixture approach applied to these data there are just two patterns to consider, $r = 1$ if both years are observed, and $r = 2$ if only the first wave is observed. If the bivariate normal likelihood is retained, one has

$$y_i \sim N_2\left(\mu^{(r)}, \Sigma^{(r)}\right), \quad r = 1, 2,$$

where only the first element in the pattern 2 means

$$\mu^{(2)} = (\mu_1^{(2)}, \ \mu_2^{(2)})$$

is identified. Similarly only the top left cell in $\Sigma$ (the variance of year 1 yields) is identified for the pattern $r = 2$.

We set an informative prior centred on the identified mean parameter, involving a log-gamma differential prior between the parameters with variance 0.2 (cf. Kaciroti *et al.*, 2008)

$$\mu_2^{(2)} = \mu_2^{(1)} + \log(\lambda),$$

$$\lambda \sim Ga(5, 5).$$

The prior variance of $\lambda$ can be varied to assess sensitivity, as can the density assumed for $\lambda$ (e.g. log-normal rather than gamma). For the covariance it is assumed that the ratio of year 2 to year 1 variances in the pattern 2 covariance matrix is the same as that ratio in the pattern 1 covariance matrix. The correlation parameter $\rho$ in the pattern 2 matrix is assumed to equal the correlation in the pattern 1 matrix, though one could invoke a differential prior for this parameter too. A Dirichlet prior is adopted for the proportions $\pi^{(r)}$ in each response pattern.

The estimate for $\Delta$ is obtained as a weighted average of the $\Delta^{(r)}$ for each pattern. The posterior mean $\Delta$ under the pattern mixture approach is then intermediate between that obtained under the MAR and MNAR missing data selection mechanisms, with posterior mean (sd) of 0.52 (0.15). The parameter $\mu_2^{(2)}$ has posterior mean and 95% credible interval of 5.89 (4.82, 6.76).

## Exercises

**7.1.** In Example 7.2 include a posterior predictive check (e.g. using the Jarque–Bera criterion) of the normality of the permanent effects $(b_i)$ in model 2.

**7.2.** In Example 7.3 investigate whether an improved fit results from making all the lag coefficients random (including lags 1 and 3), and with all mean lag coefficients unknown (including the mean of $\rho_{5i}$).

**7.3.** In Example 7.3, assess sensitivity of predictive fit (the sum of squared deviations between observations and predictive replicates) and inferences regarding the diet coefficient $\beta$ to alternative settings $Q = kI$ of the prior scale matrix for $\Sigma_\rho^{-1}$ under a Wishart prior. For example, a suggestion is to compare the settings $k = 0.001$, $k = 0.1$ and $k = 1$.

**7.4.** In Example 7.4 on employment histories, estimate model 2 using a lag in $y$ rather than $Z$, and also apply the random effects selection of Section 7.5, to assess whether there is exclusive true state dependence (and no spurious state dependence) for a subset of subjects.

**7.5.** In model 1 of Example 7.5, try a scale mixture version for the random effects $b_{i,1:2}$ with degrees of freedom an unknown (equivalent to a bivariate Student $t$). The analysis may be sensitive to the prior adopted for the degrees of freedom (e.g. a uniform between 4 and 100, an exponential with unknown mean, etc.). Are there any clear outliers (scale factors clearly under one)?

**7.6.** In Example 7.8, extend the likelihood model for cessation to include an interaction between lagged cessation and lagged weight gain, so that the regressions involved in the $(Y, Z)$ likelihood are

$$\text{logit}(\omega_{it}) = c_{1t} + \delta_1 x_i + \alpha_1 z_{i,t-1} + \alpha_2 y_{i,t-1} + \alpha_3 z_{i,t-1} y_{i,t-1},$$

$$\mu_{it} = c_{2t} + \delta_2 x_i + \alpha_4 z_{i,t-1}.$$

Also extend the PPL criterion to include both the cessation and weight change outcomes (using criterion $H_{2i} = R_{2iT} z_{iT} - R_{2i1} z_{i1}$), and compare the fit of this model against a model without $z_{i,t-1} y_{i,t-1}$ in the logit regression for $\omega_{it}$.

**7.7.** In Example 7.9, include a predictor selection mechanism (see Chapter 3) in the MNAR version of the missing data logit regression for $\eta_i$ (under a selection model approach). How does this affect inferences about $\Delta$?

**7.8.** In Example 7.9, include code to derive the posterior predictive loss under the two selection models (MAR and MNAR missing data options) using the criterion $H_i = R_i(y_{i2} - y_{i1})$.

# Notes

1. An illustrative code for MCMCglmm applied to the data in Example 7.1 is

```
install.packages("MCMCglmm"); library(MCMCglmm)
D <- read.table("psid.txt",header=T) # id is subject identifier
M <-MCMCglmm(lwage~1+exp+exp2+wks+ed,random=~id, family="gaussian",
data=D, verbose=FALSE, nitt=20000, thin=5, burnin=1000, DIC=TRUE)
# posterior plots, modes and HPD intervals for fixed effects
plot(M$Sol)
posterior.mode(model1$Sol)
HPDinterval(M$Sol[,1], 0.95)
# autocorrelation in sampled values of fixed effects
autocorr(M$Sol)
# posterior plots for random effect variances
plot(M$VCV)
#HPD for subject and unit level variances, and intra-subject
 correlation
HPDinterval(M$VCV[,1], 0.95)
HPDinterval(M$VCV[,2], 0.95)
HPDinterval(M$VCV[,"id"]/(M$VCV[,"id"]+M$VCV[, "units"]))
# DIC fit
M$DIC
```

2. The code for the first model in Example 7.4 is

```
model {# id and obs are subject and period identifiers
for (i in 1:N) {y[id[i],obs[i]]<- Y[i]
```

```
x1[id[i],obs[i]] <- hunemp[i];     x2[id[i],obs[i]] <- time[i]
x3[id[i],obs[i]] <- child1[i];
x4[id[i],obs[i]] <- child5[i]
x5[id[i],obs[i]] <- age[i]}
# predictive match over profile of individuals
for (i in 1:n) { sprm[i] <- sum(prmatch[i,1:T[i]])
# sampling limits
for (t in 1:T[i]) {L[i,t] <- -20*equals(y[i,t],0);
                   U[i,t] <- 20*equals(y[i,t],1)
# individual predictive match
      prmatch[i,t] <- equals(y[i,t],1)*step(Znew[i,t])}
# Permanent subject effects
      b[i] ~ dnorm(0,inv.sig2.b)
# Period 1
      Z[i,1] ~ dnorm(mu[i,1],1) I(L[i,1],U[i,1])
# Replicate sampling
      Znew[i,1] ~ dnorm(mu[i,1],1)
      u[i] <- Z[i,1]-mu[i,1]
      mu[i,1] <- beta.star[1] +beta.star[2]*x1[i,1]
                            +beta.star[3]*x2[i,1]
+beta.star[4]*x3[i,1] +beta.star[5]*x4[i,1] +beta.star[6]*x5[i,1]
# Periods 2,...,T[i]
for (t in 2:T[i]) {
Z[i,t] ~ dnorm(mu[i,t],1) I(L[i,t],U[i,t])
Znew[i,t] ~ dnorm(mu[i,t],1)
        mu[i,t] <- rho*Z[i,t-1] +beta[1] +b[i] +beta[2]*x1[i,t]
+beta[3]*x2[i,t] +beta[4]*x3[i,t] +beta[5]*x4[i,t] +beta[6]*x5[i,t]
                        +zeta*u[i]}}
# priors
for (j in 1:6) {beta[j] ~ dnorm(0,0.0001); beta.star[j]
                              ~ dnorm(0,0.0001)}
rho ~ dnorm(0,1);zeta ~ dnorm(0,1);sig.b ~ dunif(0,100);
inv.sig2.b <- 1/pow(sig.b,2);
# total predictive assessment
SPRD <- sum(sprm[])/sum(T[])}
```

3. The code for the dynamic dependence model in Example 7.5 is

```
model { for (i in 1:N) {
y[id[i],time[i]] <- svr[i]
v[id[i],time[i]] <- sqrt(Week[i])}
# latent wave 0 data
for (i in 1:n) { y0[i] ~ dcat(pi0[1:C])
for (t in 1:T) { y[i,t] ~ dcat(pi[i,t,1:C])
                        ynew[i,t] ~ dcat(pi[i,t,1:C])
  for (c in 1:C) {for (d in 1:C) {prmatch[c,d,i,t] <-
                                  equals(y[i,t],c)*
                        equals(ynew[i,t],d) }}
                        pi[i,t,1] <- gamma[i,t,1];
   for (c in 1:C-1) { logit(gamma[i,t,c]) <- tau[c] - eta[i,t]}
   for (c in 2:C-1) { pi[i,t,c] <- gamma[i,t,c] - gamma[i,t,c-1]}
                        pi[i,t,C] <- 1-gamma[i,t,C-1]}}
for (i in 1:n) { eta[i,1] <- b[i,1] +b[i,2]*v[i,1]
                        +rho*step(1-abs(y[i,1]-y0[i]))
+beta[1]*Tr[i] +beta[2]*G[i] +beta[3]*Tr[i]*v[i,1]
for (t in 2:T) { eta[i,t] <- b[i,1] +b[i,2]*v[i,t]
```

```
                    + rho*step(1-abs(y[i,t]-y[i,t-1]))
+beta[1]*Tr[i] +beta[2]*G[i] +beta[3]*Tr[i]*v[i,t]}}
# permanent effects
for (i in 1:n) { b[i,1:2] ~ dmnorm(mu[1:2],P.b[,])}
# precision matrix
                P.b[1:2,1:2] ~ dwish(S.b[,],2)
                Sigma.b[1:2,1:2] <- inverse(P.b[,])
for (i in 1:2) {mu[i] ~ dnorm(0,0.01)
for (j in 1:2) {Corr.b[i,j] <- Sigma.b[i,j] /
                sqrt(Sigma.b[i,i]*Sigma.b[j,j])
                S.b[i,j] <- equals(i,j)}}
# thresholds
                tau[1] <- 0
                tau[C-1] ~ dnorm(0, 0.1) I(tau[C-2],);
                for (k in 2:C-2) {tau[k] ~ dnorm(0, 0.1)
                                I(tau[k-1],tau[k+1])}
# covariate effects
for (j in 1:3) {beta[j] ~ dnorm(0,0.01)}
rho ~ dnorm(0,1)
for (c in 1:C){alph0[c] <- 1
for (d in 1:C) {M[c,d] <- sum(prmatch[c,d,,])}}
# categorical probability for latent wave 0 data
pi0[1:C] ~ ddirch(alph0[1:C])}
```

4. The code for this approach in Example 7.7 is

```
model { for (i in 1:n) {b[i] ~ dnorm(0,inv.psi[i]); inv.psi[i]
                                <- 1/psi[i]
                    delta[i] ~ dbern(omega)
                    psi[i] <- equals(delta[i],0)
                *psi.spk + equals(delta[i],1)*psi.slb
for (t in 1:T) {y[i,t] ~ dnorm(nu[i,t],tau)
nu[i,t] <-  mu +alph[1]*x[i,t,1] +alph[2]*x[i,t,2]
+alph[3]*x[i,t,3] +alph[4]*x[i,t,4] +b[i]}}
tau ~ dgamma(1,0.001)
for (j in 1:4) {alph[j] ~ dnorm(0,0.001)}
mu ~ dnorm(0,0.001)
log(invQ) <- h[1];  logit(omega) <- h[2]
h[1:2] ~ dmnorm(nought[],tau.h[,])
Q <- 1/invQ
invQ.slb <- 0.5*invQ; invQ.spk <- 0.5*invQ/r
psi.slb ~ dexp(invQ.slb); psi.spk ~ dexp(invQ.spk)}
```

5. The code for the baseline model in Example 7.8 is

```
model {# baseline periods 1 and 2 pooled regression
                (cess=1 if subject not smoking)
for (i in 1:n) {  x[i] <- equals(trt[i],2)
                logit(p.bs[i]) <- lam1[trt[i]]; mu.bs[i]
                                <- lam2[trt[i]]
for (t in 1:2) {  cess[i,t] ~ dbern(p.bs[i]); wtgn[i,t]
                        ~ dnorm(mu.bs[i],tau.bs);
                R1[i,t] ~ dbern(p.bs.R[1]);
                R2[i,t] ~ dbern(p.bs.R[2])}
# periods 3,...,T
```

```
for (t in 3:T) {  cess[i,t] ~ dbern(omega[i,t])
                        twtgn[i,t-1] <- sum(wtgn[i,1:t-1]);
                         tcess[i,t-1] <- sum(cess[i,1:t-1])
                        logit(omega[i,t]) <- c1[t]+delta[1]*x[i]
                        +alph[1]*wtgn[i,t-1]+alph[2]*cess[i,t-1]
                        wtgn[i,t] ~ dnorm(mu[i,t],tau)
                        mu[i,t] <-    c2[t]+delta[2]*x[i]
                                      +alph[3]*wtgn[i,t-1]
                        R1[i,t] ~ dbern(eta1[i,t])
                        R2[i,t] ~ dbern(eta2[i,t])
                        logit(eta1[i,t]) <- d1[t]+gam[1]*x[i]
                        logit(eta2[i,t]) <- d2[t]+gam[2]*x[i]}}
# Cessation Rates (trt=1 "Wellness", trt=2 Exercise)
for (t in 3:T) {C[1,t] <- sum(cesstr[1,t,1:n])/Ttrt[1]
                        C[2,t] <- sum(cesstr[2,t,1:n])/Ttrt[2]
# probability of treatment differences in cessation
                        trtdiff[t] <- step(C[1,t]-C[2,t])
for (i in 1:n) { cesstr[1,t,i] <- cess[i,t]*equals(trt[i],1)
                        cesstr[2,t,i] <- cess[i,t]
                                   *equals(trt[i],2)}}
# PPL (for cessation)
for (i in 1:n) {for (t in 1:2) {
cess.rep[i,t] ~ dbern(p.bs[i]);
R1.rep[i,t] ~ dbern(p.bs.R[1])}
for (t in 3:T) {cess.rep[i,t] ~  dbern(omega[i,t])
                        R1.rep[i,t] ~ dbern(eta1[i,t])}
for (t in 1:T) {
cess.R[i,t] <- R1[i,t]*cess[i,t]
cess.rep.R[i,t] <- R1.rep[i,t]*cess.rep[i,t]}
                        H.rep[i] <- cess.rep.R[i,T]-cess.rep.R[i,1]
# H[i] are known
                        H[i] <-        cess.R[i,T]-cess.R[i,1]}
# priors
for (j in 1:2) {gam[j] ~ dnorm(0,0.001); lam1[j] ~ dnorm(0,0.001);
                        lam2[j] ~ dnorm(0,0.001); delta[j]
                                          ~ dnorm(0,0.001);
                        p.bs.R[j] ~ dbeta(1,1)}
for (j in 1:3) {alph[j] ~ dnorm(0,0.001)}
for (t in 3:T) {c1[t] ~ dnorm(0,0.001); c2[t] ~ dnorm(0,0.001)
                        d1[t] ~ dnorm(0,0.001); d2[t]
                                          ~ dnorm(0,0.001)}
                        tau.bs ~ dgamma(1,0.001);   tau
                                          ~ dgamma(1,0.001)}
```

# References

Achcar, J., Coelho-Barros, E. and Martinez, E. (2008) Statistical analysis for longitudinal counting data in the presence of a covariate considering different "frailty" models. *Brazilian Journal of Probability and Statistics*, **22**(2), 183–205.

Albert, J. and Chib, S. (1993) Bayesian analysis of binary and polychotomous response data. *Journal of the American Statistical Association*, **88**(422), 669–679.

Arellano, M. and Bonhomme, S. (2011) Nonlinear panel data analysis. *Annual Review of Economics*, **3**, 395–424.

Arellano, M. and Carrasco, R. (2003). Binary choice panel data models with predetermined variables. *Journal of Econometrics*, **115**(1), 125–157.

Arellano, M. and Honoré, B. (2001) Panel data models: some recent developments. In J. Heckman and E. Leamer (eds), *Handbook of Econometrics*, Volume 5, pp. 3229–3296. North-Holland, Amsterdam.

Baltagi, B. (2008) *Econometric Analysis of Panel Data*, 4th edn. Wiley, Chichester, UK.

Beck, N. and Katz, J. (1996) Nuisance vs. substance: specifying and estimating time-series-cross-section models. *Political Analysis*, **6**(1), 1–36.

Blundell, G.R. and Windmeijer, F. (2002) Individual effects and dynamics in count data models. *Journal of Econometrics*, **108**(1), 113–131.

Blundell, G.R., Griffith R. and Windmeijer F. (2002) Individual Effects and Dynamics in Count Data Models, *Journal of Econometrics*, **108**(1): 113-131.

Bollen, K. and Curran, P. (2004) Autoregressive Latent Trajectory (ALT) models: a synthesis of two traditions. *Sociological Methods and Research*, **32**, 336–383.

Bond, S. (2002) Dynamic panel data models: a guide to micro data methods and practice. *Portuguese Economic Journal*, **1**, 141–162.

Boucher, J.-P., Michel Denuit, M. and Guillen, M. (2008) Models of insurance claim counts with time dependence based on generalization of poisson and negative binomial distributions. *Variance Journal*, **2**(1), 135–162.

Brown, H. and Prescott, R. (1999) *Applied Mixed Models in Medicine*. Wiley, Chichester, UK.

Butler, S. and Louis, T. (1992) Random effects models with non-parametric priors. *Statistics in Medicine*, **11**, 1981–2000.

Caffo, B., Peng, R., Dominici, F., Louis, T. and Zeger, S. (2011) Parallel Bayesian MCMC imputation for multiple distributed lag models: A case study in environmental epidemiology. In S. Brooks, A. Gelman, G. Jones and X. Meng (eds), *Handbook of Markov Chain Monte Carlo*, pp. 493–512. Chapman & Hall/CRC, Boca Raton, FL.

Cameron, C. and Miller, D. (2011) Robust inference with clustered data. In A. Ullah and D. Giles (eds), *Handbook of Empirical. Economics and Finance*. CRC, Boca Raton, FL.

Cameron, C. and Trivedi, P. (2009) *Microeconometrics Using Stata*. Stata Press, College Station, TX.

Carpenter, J., Pocock, S. and Johan Lamm, C. (2002) Coping with missing data in clinical trials: A model-based approach applied to asthma trials. *Statistics in Medicine*, **21**(8), 1043–1066.

Carro, J.M. (2007) Estimating dynamic panel data discrete choice models with fixed effects. *Journal of Econometrics*, **140**(2), 503–528.

Cefis, E., Ciccarelli, M. and Orsenigo, L. (2007). Testing Gibrat's legacy: A Bayesian approach to study the growth of firms. *Structural Change and Economic Dynamics*, **18**(3), 348–369.

Chamberlain, G. and Hirano, K. (1999) Predictive distributions based on longitudinal earnings data. *Annales d'Economie et de Statistique*, **55**, 211–242.

Chib, S. (2008) Panel data modeling and inference: a Bayesian primer. In L. Matyas and P. Sevestre (eds), *The Econometrics of Panel Data*, pp. 479–515. Springer-Verlag, Berlin.

Chib, S. and Carlin, B (1999) On MCMC sampling in hierarchical longitudinal models. *Statistics and Computing*, **9**(1), 17–26.

Chib, S. and Jeliazkov, I. (2006) Inference in semiparametric dynamic models for binary longitudinal data. *Journal of the American Statistical Association*, **101**(474), 685–700.

Chib, S., Greenberg, E. and Winkelmann, R. (1998) Posterior simulation and Bayes factors in panel count data models. *Journal of Econometrics*, **86**(1), 33–54.

Congdon, P. (2001) Bayesian models for suicide monitoring. *European Journal of Population*, **15**(3), 1–34.

Cornwell, C. and Rupert, P. (1988) Efficient estimation with panel data: An empirical comparison of instrumental variables estimators. *Journal of Applied Econometrics*, **3**(2), 149–155.

Creemers, A., Hens, N., Aerts, M., Molenberghs, G., Verbeke, G. and Kenward, M.G. (2011). Generalized shared-parameter models and missingness at random. *Statistical Modelling*, **11**(4), 279–310.

Dagne, G. (1999) Bayesian analysis of hierarchical Poisson models with latent variables. *Communications in Statistics-Theory and Methods*, **28**(1), 119–136.

Dagne, G. (2010) Bayesian semiparametric zero-inflated Poisson model for longitudinal count data. *Mathematical Biosciences*, **224**(2), 126–130.

Dagne, G. and Huang, Y. (2012) Mixed-effects Tobit joint models for longitudinal data with skewness, detection limits, and measurement errors. *Journal of Probability and Statistics*, **2012**, Article ID 614102, http://dx.doi.org/10.1155/2012/614102.

Daniels, M. and Hogan, J. (2008) *Missing Data in Longitudinal Studies; Strategies for Bayesian Modeling and Sensitivity Analysis*. Chapman & Hall, London, UK.

Diggle, P. (1988) An approach to the analysis of repeated measurements. *Biometrics*, **44**, 959–971.

Dubé, J.-P., Hitsch, G. and Rossi, P. (2010) State dependence and alternative explanations for consumer inertia. *RAND Journal of Economics*, **41**(3), 417–445.

Dunson, D. (2007) Bayesian methods for latent trait modelling of longitudinal data. *Statistical Methods in Medical Research*, **16**(5), 399–415.

Escobar, M. and West, M. (1998) Computing nonparametric hierarchical models. In D. Dey *et al.* (ed.), *Practical Nonparametric and Semiparametric Bayesian Statistics*. Springer, Berlin.

Fotouhi, A.R. (2007). The initial conditions problem in longitudinal count process: A simulation study. *Simulation Modelling Practice and Theory*, **15**(5), 589–604.

Frees, E. (2004) *Longitudinal and Panel Data*. Cambridge University Press, Cambridge, UK.

Frühwirth-Schnatter, S. and Wagner, H. (2011) Bayesian variable selection for random intercept modeling of Gaussian and non-Gaussian data. In J. Bernardo, J. Bayarri, J. Berger, A. Dawid, D. Heckerman, A. Smith and M. West (eds), *Bayesian Statistics 9*, **165–200**. Clarendon Press, Oxford, UK.

Galbraith, S. and Marschner, I. (2002) Guidelines for the design of clinical trials with longitudinal outcomes. *Control Clinical Trials*, **23**(3), 257–273.

Gelfand, A.E., Sahu, S.K. and Carlin, B.P. (1995) Efficient parametrisations for normal linear mixed models. *Biometrika*, **82**(3), 479–488.

Geweke, J. and Keane, M. (2000) An empirical analysis of earnings dynamics among men in the PSID: 1968–1989. *Journal of Econometrics*, **96**, 293–356.

Giardina, F., Guglielmi, A., Quintana, F. and Ruggeri, F. (2011) Bayesian first order auto-regressive latent variable models for multiple binary sequences. *Statistical Modelling*, **11**(6), 471–488.

Goldstein, H. (2009) Handling attrition and non-response in longitudinal data. *Longitudinal and Life Course Studies*, **1**(1), 63–72.

Gustavson, K., von Soest, T., Karevold, E. and Røysamb, E. (2012) Attrition and generalizability in longitudinal studies: findings from a 15-year population-based study and a Monte Carlo simulation study. *BMC Public Health*, **12**, 918.

Hadfield, J. (2010) MCMC Methods for multi-response generalized linear mixed models: the MCMCglmm R package. *Journal of Statistical Software*, **33**, 1–22.

Hall, B., Griliches, Z. and Hausman, J. (1986) Patents and R&D: Is there a lag? *International Economic Review*, **27**, 265–283.

Hamerle, A. and Ronning, G. (1995) Panel analysis for qualitative variables. In G. Arminger, C. Clogg and M. Sobel (eds), *Handbook of Statistical Modeling for the Social and Behavioral Sciences*, pp. 401–452. Plenum, New York, NY.

Hausman, J. and Wiseman, A. (1979) Attrition bias in experimental and panel data: the Gary Income Maintenance Experiment. *Econometrica*, **47**, 455–473.

Heckman, J. (1981) Statistical models for discrete panel data. In C. Manski and D. McFadden (eds), *Structural Analysis of Discrete Data with Econometric Applications*, pp 114–178. The MIT Press, Cambridge, MA.

Hedeker, D. and Gibbons, R. (1994) A random-effects ordinal regression model for multilevel analysis. *Biometrics*, **933–944**.

Hedeker, D. and Gibbons, R. (1997) Application of random-effects pattern-mixture models for missing data in longitudinal studies. *Psychological Methods*, **2**(1), 64–78.

Hedeker, D. and Gibbons, R. (2006) *Longitudinal Data Analysis*. Wiley, Hoboken, NJ.

Hirano, K. (2002) Semiparametric Bayesian inference in autoregressive panel data models. *Econometrica*, **70**, 781–799.

Hogan, J. and Laird, N. (1997) Mixture models for the joint distribution of repeated measures and event times. *Statistics in Medicine*, **16**(3), 239–257.

Horrace, C. and Schmidt, P. (2000) Multiple comparisons with the best, with economic applications, *Journal of Applied Econometrics*, **15**, 1–26.

Howe, L., Tilling K., Matijasevich A., et al. (2013) Linear spline multilevel models for summarising childhood growth trajectories: a guide to their application using examples from five birth cohorts. *Statistical Methods in Medical Research* [Epub ahead of print].

Hsiao, C. (2007) Panel data analysis – advantages and challenges. *Test*, **16**(1), 1–22.

Ibrahim, J., Chen, M.-H., Lipsitz, S. and Herring, A. (2005) Missing data methods for generalized linear models: a comparative review. *Journal of the American Statistical Association*, **100**, 332–346.

Juárez, M.A. and Steel, M.F. (2010) Non-Gaussian dynamic Bayesian modelling for panel data. *Journal of Applied Econometrics*, **25**(7), 1128–1154.

Jung, R., Kukuk, M. and Liesenfeld, R. (2006) Time series of count data: modeling, estimation and diagnostics. *Computational Statistics and Data Analysis*, **51**, 2350–2364.

Kaciroti, N., Raghunathan, T., Schork, M., Clark, N. and Gong, M. (2006) A Bayesian approach for clustered longitudinal ordinal outcome with nonignorable missing data: evaluation of an asthma education program. *Journal of the American Statistical Association*, **101**(474), 435–446.

Kaciroti, N., Raghunathan, T., Schork, M. and Clark, N. (2008) A Bayesian model for longitudinal count data with non-ignorable dropout. *Journal of the Royal Statistical Society C*, **57**(5), 521–534.

Karney, B. and Bradbury, T. (1995) The longitudinal course of marital quality and stability: a review of theory, methods, and research. *Psychological Bulletin*, **118**(1), 3–34.

Kenward, M. (1998) Selection models for repeated measurements with non-random dropout: an illustration of sensitivity. *Statistics in Medicine*, **17**(23), 2723–2732.

Kinney, S. and Dunson, D. (2007) Fixed and random effects selection in linear and logistic models. *Biometrics*, **63**, 690–698.

Lee, J. and Hwang, R. (2000) On estimation and prediction for temporally correlated longitudinal data. *Journal of Statistical Planning and Inference*, **87**(1), 87–104.

Lee, J., Lin, T., Lee, K. and Hsu, Y. (2005) Bayesian analysis of Box-Cox transformed linear mixed models with ARMA(p,q) dependence. *Journal of Statistical Planning and Inference*, **133**(2), 435–451

Little, R. (1993) Pattern-mixture models for multivariate incomplete data. *Journal of the American Statistical Association*, **88**(421), 125–134.

Little, R. (2009) Selection and pattern-mixture models. In G. Fitzmaurice, M. Davidian, G. Verbeke and G. Molenberghs (eds), *Longitudinal Data Analysis*, pp. 409–432. Chapman & Hall, London, UK.

Little, R. and Rubin, D. (2002) *Statistical Analysis with Missing Data*, 2nd edn. Wiley, New York, NY.

Maddala, G. (1987) Limited dependent variable models using panel data. *The Journal of Human Resources*, **22**(3), 307–338.

Mander, A., Hughes, M., Sharp, S. and Lamm, C. (1999). Autoregressive models for describing non-linear changes in biological parameters fitted using BUGS. *Statistics in Medicine*, **18**(20), 2709–2722.

Marcus, B., Lewis, B., Hogan, J., King, T., Albrecht, A., Bock, B., Parisi, A., Niaura, R. and Abrams, D. (2005) The efficacy of moderate-intensity exercise as an aid for smoking cessation in women: a randomized controlled trial. *Nicotine and Tobacco Research*, **7**(6), 871–880.

Mason, A., Richardson, S., Plewis, I. and Best, N. (2012) Strategy for modelling nonrandom missing data mechanisms in observational studies using Bayesian methods. *Journal of Official Statistics*, **28**(2), 279–302.

Meligkotsidou, L., Tzavalis, E. and Vrontos, I. (2012) A Bayesian panel data framework for examining the economic growth convergence hypothesis: do the G7 countries converge? *Journal of Applied Statistics*, **39**(9), 1975–1990.

Nandram, B. and Petruccelli, J.D. (1997) A Bayesian analysis of autoregressive time series panel data. *Journal of Business and Economic Statistics*, **15**(3), 328–334.

Neelon, B., O'Malley, A. and Normand, S. (2010) A Bayesian model for repeated measures zero-inflated count data with application to outpatient psychiatric service use. *Statistical Modelling*, **10**(4), 421–439.

Noorian, S. and Ganjali, M. (2012) Bayesian analysis of transition model for longitudinal ordinal response data: application to insomnia data. *International Journal of Statistics in Medical Research*, **1**, 148–161.

Parmigiani, G., Ashih, H., Samsa, G., Duncan, P., Lai, S. and Matchar, D. (2003) Cross-calibration of stroke disability measures: Bayesian analysis of longitudinal ordinal categorical data using negative dependence. *Journal of the American Statistical Association*, **98**, 273–281.

Qiu, Z., Song, P. and Tan, M. (2002) Bayesian hierarchical models for multi-level repeated ordinal data using WinBUGS. *Journal of Biopharmaceutical Statistics*, **12**(2), 121–135.

Rahiala, M. (1999) Random coefficient autoregressive models for longitudinal data. *Biometrika*, **86**(3), 718–722.

Rizzo, L., Kalton, G. and Brick, M. (1996) A comparison of some weighting adjustment methods for panel nonresponse. *Survey Methodology*, **22**, 43–53.

Rubin, D. (1976) Inference and missing data. *Biometrika*, **63**, 581–592.

Schmidt, W., Arnold, B., Boisson, S., Genser, B., Luby, S., Barreto, M., Clasen, T. and Cairncross, S. (2011) Epidemiological methods in diarrhoea studies – an update. *International Journal of Epidemiology*, **40**(6), 1678–1692.

Singer, J. and Willett, J. (2003) *Applied Longitudinal Data Analysis: Modeling Change and Event Occurrence*. Oxford University Press, New York, NY.

Skrondal, A. and Rabe-Hesketh, S. (2004) *Generalized Latent Variable Modeling: Multilevel, Longitudinal and Structural Equation Models*. Chapman & Hall/CRC, Boca Raton, FL.

Stegmueller, D. (2013) Modeling changing preferences: a Bayesian robust dynamic latent ordered probit model. *Political Analysis*. doi: 10.1093/pan/mpt001 [Epub ahead of print].

Sutradhar, B. (2011) Longitudinal models for count data. In *Dynamic Mixed Models for Familial Longitudinal Data, Springer Series in Statistics*, pp. 181–240. Springer, New York, NY.

Sutradhar, B., Bari, W. and Das, K. (2010) On probit versus logit dynamic mixed models for binary panel data. *Journal of Statistical Computation and Simulation*, **80**(4), 421–441.

Varin, C. and Czado, C. (2010) A mixed autoregressive probit model for ordinal longitudinal data. *Biostatistics*, **11**(1), 127–138.

Verbeke, G. and Molenberghs, G. (2000) *Linear Mixed Models for Longitudinal Data*. Springer, New York, NY.

Walker, L., Gustafson, P. and Frimer, J. (2007) The application of Bayesian analysis to issues in developmental research. *International Journal of Behavioral Development*, **31**(4), 366–373.

Wang, L., Zhang, Z., McArdle, J. and Salthouse, T. (2008) Investigating ceiling effects in longitudinal data analysis. *Multivariate Behavioral Research*, **43**(3), 476–496.

Weiss, R. (2005) *Modelling Longitudinal Data*. Springer, New York.

Winer, R. (1983) Attrition bias in econometric models estimated with panel data. *Journal of Marketing Research*, **20**, 177–186.

Winkelmann, R. (2000) *Econometric Analysis of Count Data*, 3rd edn. Springer, Berlin.

Wooldridge, J. (2005) Simple solutions to the initial conditions problem in dynamic, nonlinear panel data models with unobserved heterogeneity. *Journal of Applied Econometrics*, **20**, 39–54.

Yang, L. and Gao, S. (2013) Bivariate random change point models for longitudinal outcomes. *Statistics in Medicine*, **32**, 1038–1053.

Zhang, D. (2004) Generalized linear mixed models with varying coefficients for longitudinal data. *Biometrics*, **60**(1), 8–15.

# 8

# Models for spatial outcomes and geographical association

## 8.1 Introduction

Advances in spatial data analysis refer to a central core of knowledge but show many distinct features in the specialisms involved. Thus many Bayesian applications have occurred in spatial epidemiology, with methodologically oriented overviews including Pfeiffer *et al.* (2008), Waller and Gotway (2004), Beale *et al.* (2008), Graham *et al.* (2004), Schrödle and Held (2011), Auchincloss *et al.* (2012) and Jerrett *et al.* (2010). Here a major element is the assessment of spatial clustering of relative disease risk, often for irregular lattice systems (e.g. administrative areas). A more long-standing tradition of spatial modelling has occurred in spatial econometrics with Anselin (2006, 2010), Pace and LeSage (2010), Getis *et al.* (2004), Arbia and Baltagi (2009) and LeSage (2008) providing recent overviews, and with LeSage and Pace (2009) reviewing Bayesian principles in this area. Here the major emphasis lies in describing behavioural relationships by regression models, whether the data are defined over areas, or for individual actors (house purchasers, firms, etc.) involved in spatially defined behaviours. A third major specialism occurs in geostatistics, where a continuous spatial framework is adopted, and the goal is often to smooth or interpolate between observed readings (e.g. of mineral concentrations) at sampled locations (Diggle and Ribeiro, 2007; Gaetan and Guyon, 2009). Providing a common thread is a central core of spatial statistics, exemplified by the works of Banerjee *et al.* (2004), Gelfand *et al.* (2010), Anselin and Rey (2010), Ripley (2004), Haining (2003), Shi (2009), Cressie and Wikle (2011), and Cressie (1993). Bayesian approaches to spatial data can be implemented in WinBUGS/OpenBUGS, R-INLA, R2BayesX, geoRglm (http://gbi.agrsci.dk/~ofch/geoRglm/), spTimer (Bakar and Sahu, 2013), spatialprobit (Wilhelm and Matos, 2013), and CARBayes.

As for panel data, spatial lags in observed outcomes or predictors may represent substantive processes, for example, in the spatial pattern of democratic governments (Ward and Gleditsch, 2008). However, in many regression applications, spatial dependence occurs because of omitted or unmeasured spatially correlated predictors, and so is reflected in regression errors, causing departures from the independent errors assumption of the conventional regression.

*Applied Bayesian Modelling*, Second Edition. Peter Congdon.
© 2014 John Wiley & Sons, Ltd. Published 2014 by John Wiley & Sons, Ltd.

In problems involving both space and time dimensions (Section 8.6), errors may be correlated in both time and space simultaneously (Lagazio et al., 2001). Spatial correlation reduces the independent evidence available to model the process under investigation, and may, if not allowed for, lead to overestimation of the significance of regression relationships (Richardson and Monfort, 2000, p. 211).

Allowance for spatial covariance in errors is illustrated by normal linear geostatistical regression (Kyriakidis and Nagle, 2010) : for $n \times 1$ vectors $y(s)$ and $\epsilon(s)$ defined at points $s_i = (s_{1i}, s_{2i})$, and $X(s)$ of dimension $n \times P$,

$$y(s) = X(s)\beta + \epsilon(s) + u(s),$$

$$\epsilon(s) \sim N_n(0, \Sigma(s)),$$

$$u(s) \sim N(0, \tau^2),$$

where the $n \times n$ matrix $\Sigma_{ij}(s) = \sigma_\epsilon^2 r_{ij}(s)$ expresses spatial covariance in errors as a function of location or inter-point distances. Thus for a distance metric $d_{ij}$, one could specify off-diagonal correlations as

$$r_{ij}(s_i, s_j) = \sigma_\epsilon^2 \exp(-\gamma d_{ij}),$$

where $\gamma > 0$ is an unknown parameter. Such covariance modelling underlies geostatistical spatial interpolation (Section 8.4), namely predicting $y(s_{new})$ at an unobserved location.

By contrast for area (lattice) data, it is common to adopt a known spatial interaction scheme (e.g. Bailey and Gatrell, 1995; Paez and Scott, 2004), often based on area contiguities, providing simpler identifiability of other model aspects such as regression or risk identification (Anselin, 2001). Simultaneous autoregressive models (Section 8.2) involve spatial lags in outcomes or in regression errors, and are also usually expressed using known interaction matrices. In regression modelling of discrete area data, conditional priors are often more convenient, for example, in specifying spatially correlated errors for area disease risks (Section 8.3) (e.g. Mollié, 2001). There may be benefit in conceptual interchange, such as using geostatistical ideas in ecological regression of spatial disease patterns (Goovaerts and Gebreab, 2008).

An additional issue raised clearly by writers such as Fotheringham et al. (2000) and Lesage (1998) is that of spatial heterogeneity, for example, in terms of regression coefficients varying over space, or spatially varying heteroscedasticity in an iid error term (Sections 8.3 and 8.5). There may be identifiability problems in separating spatial dependence (e.g. correlation) from spatial heterogeneity (de Graaff et al., 2001; Anselin, 2001).

## 8.2    Spatial regressions and simultaneous dependence

Consider regression models for continuous outcomes, where the form of interaction between areas is taken as known, and the focus is on making inferences about regression impacts that allow for spatial dependence, on estimating different forms of spatial correlation (e.g. in the outcomes themselves or in regression errors), or on allowing for spatial heterogeneity. This framework encompasses latent continuous variable models, such as for binary outcomes (Anselin, 2001), using for instance the sampling methods of Albert and Chib (1993).

Thus consider an $n \times n$ matrix $C$ of binary contiguity indicators, with $c_{ij} = 1$ if areas $i$ and $j$ are adjacent, and $c_{ij} = 0$ otherwise (with $c_{ii} = 0$ for all $i$). Alternatively a distance based interaction scheme might involve elements such as $c_{ij} = 1/d_{ij}$ $(i \neq j)$, again with $c_{ii} = 0$. Specialized interaction schemes are possible for dyadic data, such as free trade agreements between pairs of countries), where the distance is defined in terms of the dyad $p$ formed by country-pair $i - j$ and dyad $q$ by country-pair $k - l$ (Neumayer and Pluemper, 2009). It is common practice to scale the interactions to sum to unity in rows, with $W$ as the row-standardised matrix,

$$W = [w_{ij}] = [c_{ij}/\sum_j c_{ij}].$$

A widely used model for representing spatial error dependence for a metric outcome $y$ is known as the spatial error model (e.g. Anselin, 2006; Kissling and Carl, 2008) or simultaneous autoregressive model (Cressie and Wikle, 2011, Ch 4; Wall, 2004),

$$y = X\beta + \epsilon,$$

with autocorrelated errors,

$$\epsilon = \rho C\epsilon + u,$$

or

$$\epsilon = \rho W\epsilon + u.$$

where $\rho$ is an unknown correlation parameter, and the $u$ are iid normal.

This model expresses spatial dependence not accounted for by the included predictor variables, measurement errors, and possibly also mismatch between the spatial units used to measure the variables and the scale at which the process occurs (Anselin and Bera, 1998). The $u$ denote spatially unstructured iid errors, which initially at least may be taken to have a constant variance $u_i \sim N(0, \sigma_u^2)$. The spatial error coefficient $\rho$ has bounds $\{1/\omega_{min}, 1/\omega_{max}\}$ where $\omega_{min}$ and $\omega_{max}$ are the minimum and maximum eigenvalues of $C$ or $W$. If the interactions are scaled within rows $\omega_{max}$ is 1 (Anselin, 2001; Voss et al., 2006), and since spatial correlation is positive in the great majority of econometric or health applications, a prior on $\rho$ constrained to [0, 1] is often appropriate (Stern and Cressie, 2000), though a prior allowing negative values may be useful for establishing whether $\rho$ is significantly positive. Though feasible values for $\lambda$ or $\rho$ depend on the eigenvalues for $W$, unconstrained uniform priors for these parameters may also be adopted, as non-feasible values will be rejected under sampling (Smith and LeSage, 2004).

The log-likelihood under the spatial error model is

$$L(\beta, \sigma^2, \rho|y) = -0.5n \log(\sigma_u^2) + \log|I - \rho W| - \frac{1}{2\sigma_u^2}[(y - X\beta)'(I - \rho W)'(I - \rho W)(y - X\beta)].$$

Denoting

$$y^* = y - \rho Wy,$$
$$X^* = X - \rho WX,$$
$$e = y^* - X^*\beta,$$

one has

$$L(\beta, \sigma^2, \rho|y) = -0.5n \log(\sigma_u^2) + \log|I - \rho W| - \frac{1}{2\sigma_u^2}e'e.$$

MCMC sampling of $\rho$ may be faciliated (LeSage and Pace, 2009) by using a grid-prior (implying griddy Gibbs sampling rather than Metropolis–Hastings) over a range of feasible values $\{\rho_1, \ldots, \rho_R\}$, typically with equal prior probabilities $\{\pi_1, \ldots, \pi_R\}$, though unequal weighting is possible (LeSage, 1997). A grid prior can be combined with precalculation of the log determinants of $I - \rho_r W$ at each feasible value, avoiding repeated computations during later sampling. For the interaction matrix $W$ in row standardised form, a set of values between 0 and 1 may be used, and in BUGS selection from a grid prior can be coded using

```
for (r in 1:R) {rho.grid[r] <- (R-r)/(R-1); pi[r] <- 1/R}
k.rho ~dcat(pi[1:R]); rho <- rho.grid[k.rho].
```

If the non-standard likelihood option is used (e.g. dloglik in OpenBUGS) then the log determinant value is equally apportioned between cases.

The spatial error model may be restated as

$$y = X\beta + (I - \rho W)^{-1}u,$$

and since it may be unrealistic to assume $X$ and $u$ are uncorrelated, an alternative is

$$u = X\gamma + v,$$

where the $v$ are iid. Then

$$y = X\beta + (I - \rho W)^{-1}X\gamma + (I - \rho W)^{-1}v,$$

and re-expression with iid errors (LeSage and Pace, 2009) leads to what is termed the spatial Durbin model (Anselin, 1988; Seya *et al.*, 2012)

$$y = \rho Wy + X(\beta + \gamma) - \rho WX\beta + v,$$

or re-parameterised

$$y = \rho Wy + X\theta_1 + WX\theta_2 + v.$$

Thus crime rates in an area may be influenced by crime rates in neighbouring areas, by policing intensity in the area itself, and by policing intensity in neighbouring areas.

Spatially lagged effects can be confined to dependent variables. The spatial lag model for $y$ continuous has the form

$$y = \lambda Wy + X\beta + u,$$

where the $u$ are iid normal. This model is often intended to represent neighborhood agglomeration effects (Anselin, 2006), for example, in modelling the spread of innovations, forms of government (Ward and Gleditsch 2008), spillovers in economic growth, or geographic variation in house prices, where the spatially weighted sum of house prices in neighbouring areas (the spatial lag) enters as an additional explanatory variable in the regression for house price formation. Metropolis–Hastings sampling for this model is considered by Holloway *et al.* (2002) in an application considering adoption of new agricultural practices.

The spatial errors model can be combined with spatial lag effects (e.g. Elhorst, 2010) as in

$$y = \lambda W_1 y + X\beta + \epsilon,$$

$$\epsilon = \rho W_2 \epsilon + u,$$

$$u \sim N(0, \sigma_u^2).$$

The log-likelihood for the combined lag and error model has the form

$$L(\beta, \lambda, \sigma^2, \rho | y) = -0.5n \log(\sigma_u^2) + \log |I - \lambda W_1| + \log |I - \rho W_2| - \frac{1}{2\sigma_u^2} e'e.$$

where

$$e = B(Ay - X\beta),$$

$$A = (I - \lambda W_1),$$

$$B = (I - \rho W_2).$$

Whatever form of spatial regression is used, there are possible issues of identifiability, since spatial clustering of regression errors may be produced by spatial heterogeneity, due to heteroscedasticity or spatially varying regression coefficients (Messner and Anselin, 2004). Then in a single cross-section, spatial autocorrelation and spatial heteroscedasticity are observationally equivalent (Anselin and Bera, 1998; Abreu *et al.*, 2005). Spatial heteroscedasticity may be parameterised in various ways: either one may suppose all areas to have distinct variances, or there may be area categories $G_i$, with $\mathrm{var}(u) = \sigma_{ug}^2$ if $G_i = g$. LeSage (2000) mentions scale mixtures (with each area having its own variance) to robustify inferences against outlier data points. Thus a normal scale mixture (equivalent to Student t) in the spatial errors model specifies

$$y = X\beta + \epsilon,$$

$$\epsilon = \rho W \epsilon + u,$$

$$u_i \sim N(0, \sigma_u^2/\kappa_i),$$

$$\kappa_i \sim Ga(0.5v, 0.5v),$$

where $v$ is the degrees of freedom. The log-likelihood is

$$L(\beta, \sigma^2, \rho|y) = -0.5n \log(\sigma_u^2) + 0.5 \sum_i \log(\kappa_i) + \log|I - \rho W| - \frac{1}{2\sigma_u^2} e'e,$$

where $e_i = \kappa_i^{0.5}(y_i^* - X_i^*\beta)$, $y^* = y - \rho Wy$, and $X^* = X - \rho WX$.

## 8.2.1    Regression with localised dependence

Indices to summarise spatial dependence depend on the form of indicator for each area (e.g. whether continuous or binary), and global indicators include Moran and join-count indicators. For binary measures $b_i$, a measure of spatial concordance is the total of area joins where both $b_i$ and $b_j$ are 1, namely

$$J_{11} = \sum_{i=1}^{n} \sum_{j=1}^{n} c_{ij} b_i b_j,$$

while for deviations $z_i = x_i - \bar{x}$ in a continuous measure, the Moran index is

$$I = \frac{n}{\displaystyle\sum_{i=1}^{n}\sum_{j=1}^{n} c_{ij}} \frac{\displaystyle\sum_{i=1}^{n}\sum_{j=1}^{n} c_{ij} z_i z_j}{\displaystyle\sum_{i=1}^{n} z_i^2}.$$

Global autocorrelation indices may not capture more localised spatial dependencies. Getis (1989) proposed a measure to detect local pockets of dependence, specified between area or point $i$ and all other observations within distance $d$ of $i$ as

$$L_i(d) = \left[ \sum_{i \neq j} w_{ij}(d) y_i y_j / \sum_{i \neq j} y_i y_j \right]^{0.5},$$

where $w_{ij}(d)$ equals 1 if point or area $j$ is within distance $d$ of $i$, and $w_{ij}(d) = 0$ otherwise. This measure assumes positive $y_i$, and may necessitate transformation of the actual observations. A variation of this approach (e.g. Ord and Getis, 1995; Paez and Scott, 2004) takes

$$G_i(d) = \left[ \sum_{i \neq j} w_{ij}(d) y_j / \sum_{i \neq j} y_j \right].$$

The indices $L_i(d)$ and $G_i(d)$ measure the proportion of the total association between place $i$ and all other places which occurs within distance $d$ of $i$. Define $W_i(d) = \sum_{i \neq j} w_{ij}(d)$. If there is no spatial patterning, then $E(y_j)$ is a constant independent of $d$, leading to null association values (or expected values under no autocorrelation) $L_{iN}(d) = [W_i(d)/(n-1)]^{0.5}$ and $G_{iN}(d) = [W_i(d)/(n-1)]$.

Considering $G_i$ statistics in particular, if values of $y$ near area $i$ tend to be high, the ratios $G_{iN}(d)/G_i(d)$ will be less than 1. Ord and Getis (1995), Getis (2010) and Getis and Griffith (2002) propose partitioning observed values of responses $y$ and predictors $X$ into spatially filtered components $\{y^*, X^*\}$, and remainders $\{y - y^*, X - X^*\}$ that encapsulate spatial association with other areas. For a predictor $x$, its filtered value in area $i$ is

$$x_i^* = x_i \frac{G_{iN}(d)}{G_i(d)},$$

so that when $G_i(d)$ is high relative to the null value, the remainder $x_i - x_i^*$ will be positive, reflecting spatial autocorrelation among high $x$-values. The set of values $L_x = x - x^*$ for all $i$ represents the spatial component of $x$. To allow for both 'own area' predictor effects and predictor effects spilling over from neighbouring areas (in crime rate applications exemplified by the effect of local poverty and poverty in nearby areas), the corresponding spatial regression is on both filtered and spatial components of $x$ variables, and on the spatial lag $L_y = y - y^*$ of the response. For example, in the Columbus crime data application (Example 8.2), with response CR and predictors income (INC) and housing values (HV), the potential predictors can be $INC^*$, $L_{INC}$, $HV^*$, $L_{HV}$ and $L_{CR}$.

One would select that distance $d$ at which spatial dependence shown by the $G_i(d)$ is greatest (Getis, 1995), for example by discretizing the range of $d$ into a small number, $D$, of bands. The optimal distance band $d_{opt}$ can be determined separately for each predictor and for the response, for example, using the global $G(d)$ statistic of Getis and Ord (1992). Alternatively one may assume a common distance band across all predictors and the response. In a Bayesian framework choice among the bands may be handled via a categorical indicator subject to probabilities following a Dirichlet prior, or using a profile fit over differing potential values for $d_{opt}$. Bayesian predictor selection can be applied in combination with selection of the optimal distance band.

## 8.2.2   Binary outcomes

For binary data defined over areas $i = 1, \ldots, n$

$$y_i \sim \text{Bern}(\pi_i),$$

one might model residual spatial errors under a generalized linear scheme with logit or probit links on $\pi_i$. One might for instance include a spatial error term under a logit link with conditional priors for spatial errors (e.g. Liao *et al.*, 2010),

$$\text{logit}(\pi_i) = x_i \beta + \epsilon_i,$$

where the $\epsilon_i$ follow a conditional autoregressive scheme (see Section 8.3).

However, much research (e.g. Holloway, 2002; Smith and LeSage, 2004; Jaimovich, 2012) has focused on the latent variable representation in tandem with joint error priors under spatial error or spatial lag models. The probit is often applied to choice behaviour of individual units, with spatial correlation expected in choices made by individuals located in similar places (Smith and LeSage, 2004). Thus

$$y_i = 1 \quad if \quad z_i > 0,$$
$$y_i = 0 \quad if \quad z_i \leq 0,$$

where the latent variable $z_i$ can be interpreted as a difference in utilities $U_{1i} - U_{0i}$ between binary options, with $\Pr(y_i = 1) = \Pr(U_{1i} > U_{0i}) = \Pr(z_i > 0) = \Phi(X_i \beta)$.

Then the spatial lag model for binary data based on the latent responses is

$$z = \lambda W z + X \beta + u, \qquad u \sim N(0, I),$$

where the variance of the $u_i$ terms is known for identifiability (Anselin, 2006), so that

$$z = (I - \lambda W)^{-1} X \beta + v,$$
$$v = (I - \lambda W)^{-1} u \sim N_n(0, [(I - \lambda W)'(I - \lambda W)]^{-1}).$$

Spatial heteroscedasticity can be modelled by taking $u_i \sim N(0, 1/\kappa_i)$ where $\kappa_i \sim Ga(0.5v, 0.5v)$, and $v$ is the Student $t$ degrees of freedom. The spatial errors model is

$$z = X \beta + \epsilon,$$
$$\epsilon = \rho W \epsilon + u,$$
$$u \sim N(0, I),$$
$$\epsilon = (I - \rho W)^{-1} u \sim N_n(0, [(I - \rho W)'(I - \rho W)]^{-1}).$$

**Example 8.1    Tuberculosis notification rates in London districts**

Infectious disease patterns may reflect spatial agglomeration, and so a spatial lag model is plausible. The data considered here are TB infection rates for $n = 32$ London health districts and based on a three year (2007-09) average of TB notification totals; the rates $y_i$ are obtained as this average times 100 000 divided by 2005 area populations. Two models are compared, a first-order pure spatial autoregression (LeSage, 1997) without covariates, namely

$$y = \beta_1 + \lambda W y + u,$$

and a mixed-spatial autoregressive model

$$y = \beta_1 + \lambda W y + X \beta + u,$$

including two predictors in addition to a constant term, namely percent of population non-white $(x_2)$ and a deprivation index $(x_3)$. The spatial interaction matrix $C$ is based on contiguity with row standardisation to form $W$.

The log likelihood kernel for the pure spatial autoregression model is

$$L(\beta, \sigma^2, \lambda | y) = -0.5n \log(\sigma_u^2) + \log |I - \lambda W| - \frac{1}{2\sigma_u^2} e' e.$$

where $e = y^* - \beta_1$, and $y^* = y - \lambda W y$. The analysis is complicated by the presence in the likelihood of the determinant $|I - \lambda W|$, and by the non-standard posterior for $\lambda$. The model is estimated here in WinBUGS using the non-standard likelihood device, with the log determinant apportioned equally over cases; it can also be estimated in OpenBUGS using the dloglik option. Additionally, an equal probability grid prior of $R = 100$ potential values of $\lambda_r$ from 0 to 0.99 is used, combined with precalculated determinant values $|I - \lambda_r W|$, while regression coefficients are assigned relatively diffuse normal priors, and $\sigma_u$ is assigned a $U(0, 100)$ prior.

A model without any predictors has a LPML of $-116.6$ (from a 50 000 iteration run with two chains), with a mean (95% interval) for $\lambda$ of 0.45 (0.06,0.81). This pure spatial lag model appears to control for spatial correlation in regression errors: the 95% interval of $(-0.07,0.26)$ on a Moran coefficient for the residuals $e = y^* - \beta_1$ spans the zero value.

A second model including the two predictors shows the LPML rise to $-94$, with $\beta_2$ and $\beta_3$ having means (95% intervals) of 1.64 (1.20,2.08), and 0.26 $(-0.37,0.88)$. The central role of ethnic mix in explaining TB infection rates for different areas confirms work by Beckhurst et al. (2000). The necessity for a spatial lag effect under this extended model seems in doubt with the mean (95% interval) for $\lambda$ reduced to 0.17 (0, 0.44).

A subsidiary analysis investigates the impact of converting the originally metric outcome to a binary one: $y = 1$ if the TB incidence rate exceeds 50 per 100 thousand, and $y = 0$ otherwise. The models compared are the same, namely a first-order spatial autoregression without covariates, and a mixed-spatial autoregressive model including predictors, population non-white $(x_2)$ and deprivation $(x_3)$. Thus

$$y_i = 1 \quad \text{if} \quad z_i > 0; \quad y_i = 0 \quad \text{if} \quad z_i \le 0,$$

where the regression model is

$$z = \lambda W z + X \beta + u, \quad u \sim N(0, I).$$

The log likelihood kernel is

$$L(\beta, \lambda | y) = \log |I - \lambda W| - 0.5 e' e.$$

where $e = z^* - X\beta$, and $z^* = z - \lambda W z$.

As above, a non-standard likelihood is used, with a grid prior on $\lambda$, and corresponding values for $\log |I - \lambda W|$ obtained beforehand. Initial values for the $\lambda$ grid point and the $\beta$ coefficients are provided, while initial $z$ values are generated (using 'gen inits') from the priors. For a pure lag model without predictors, the last 15 000 iterations of a two chain run of 25 000 iterations provide a posterior mean (95% interval) for $\lambda$ of 0.47 (0.05, 0.86) with the LPML estimated at $-12.3$. The Moran coefficient has 95% interval $(-0.14, 0.32)$.

A model including covariates is run for 50 000 iterations with relatively slow convergence apparent in the intercept $\beta_1$. The last 25 000 iterations show the LPML now at $-11.5$, with posterior mean (95% interval) for $\lambda$ of 0.34 (0.02, 0.72), and a significant coefficient for $x_2$, with mean (95% interval) of 0.10 (0.07, 0.15). The lesser improvement in fit as predictors are added may reflect the loss of information following the conversion of the response variable to binary form.

### Example 8.2    Columbus crime

Anselin (1988) considers data on crime rates $y_i$ (burglaries and vehicle thefts per 1000 households) in 49 neighbourhoods of Columbus, Ohio, with predictors income $(x_1)$ and housing values $(x_2)$ in thousands of dollars. There are unusual data features such as rates near zero in some neighbourhoods (4,16) compared with an average rate of 35/1000 and rates over 60/1000 in four neighbourhoods. Classical linear regression gives coefficients (s.e.) on $x_1$ and $x_2$ of $-1.6$ (0.33) and $-0.27$ (0.10), but analysis of residuals shows spatial correlation. Maximum likelihood estimation of a spatial error model gives $\rho = 0.56$ (0.13), with the coefficient on income reduced to $\beta_1 = -0.94$ (0.33), while that on housing value is enhanced, with $\beta_2 = -0.30$ (0.09).

Here a spatial error model (model 1) and a combined spatial lag and error model (model 2) are estimated, with the same $W$ (based on contiguity) throughout. These models are respectively

$$y = X\beta + \epsilon,$$

$$\epsilon = \rho W\epsilon + u,$$

and

$$y = \lambda Wy + X\beta + \epsilon,$$

$$\epsilon = \rho W\epsilon + u,$$

with $u$ iid normal. As in Example 8.1, grid priors are adopted on the spatial correlation parameters together with a pre-calculated set of log-determinants. Thus in model 1 the log determinants $B_r = \log |I - \rho_r W|$ are pre-evaluated at the selected grid values $\{\rho_1, \dots, \rho_R\}$ in the spatial error model. A regular grid of $R = 100$ values $\rho_r$ from 0 to 0.99 is used. This avoids repeated determinant evaluations in the log likelihood

$$L(\beta, \sigma^2, \rho | y) = -0.5n \log(\sigma_u^2) + \log |I - \rho W| - \frac{1}{2\sigma_u^2} e'e,$$

where $e = y^* - X^*\beta$, $y^* = y - \rho Wy$, and $X^* = X - \rho WX$. A two chain run of 50 000 iterations in OpenBUGS gives posterior means (posterior sd) on the regression coefficients for income and housing value of $-1.07$ (0.42) and $-0.30$ (0.11) respectively. The $\rho$ coefficient has mean 0.45 (sd=0.17) with 95% interval from 0.10 to 0.76. The LPML is obtained as $-148.3$ with the worst log(CPO) for neighbourhood 4 at $-10.8$.

The log-likelihood for the combined lag and spatial errors model

$$L(\beta, \sigma^2, \lambda, \rho | y) = -0.5n \log(\sigma_u^2) + \log |I - \lambda W| + \log |I - \rho W| - \frac{1}{2\sigma_u^2} e'e,$$

involves a double lagged transform of the $y$, namely

$$y^* = y - \rho Wy,$$

$$y^{**} = y^* - \lambda Wy^*,$$

$$e = y^{**} - X^*\beta.$$

A regular grid of 100 values from 0 to 0.99 for both $\rho_r$ and $\lambda_r$ are used. This model reduces the posterior mean of $\rho$ to 0.26 (with 95% interval 0.01 to 0.64), while $\lambda$ has mean 0.23. Posterior means (posterior sd) on income and housing value are virually unchanged at $-1.07$ (0.38) and $-0.29$ (0.11). The LPML is unchanged at $-148.3$, though the effect of neighbourhood 4 is apparent: its LPML falls to $-12.3$ implying a slightly better fit for model 2 to remaining neighbourhoods.

A regression allowing for localised dependence using the $G_i(d)$ statistic (e.g. Getis, 1995) is applied using $D = 10$ distance bands based on the 5th, 10th, ... 45th, and 50th percentiles of the inter-neighbourhood distances. The regression is on filtered and spatial components, $x^*$ and $L_x = x - x^*$, respectively, of the two predictors (with notation as in Section 8.2.1), and on the spatial component $L_y = y - y^*$ of the response. To avoid division by zero in $x_i^* = x_i \frac{G_{iN}(d)}{G_i(d)}$ (since a few neighbourhoods have no adajacent areas within the first distance band), the alternative $x_i^* = x_i \frac{G_{iN}(d)}{(G_i(d)+0.000001)}$ is used. A profile over $d_{opt} = 1$, $d_{opt} = 2$ etc. shows the highest LPML, namely $-132.4$ for $d_{opt} = 1$. Moran statistics applied to the regression residuals show no spatial correlation. The posterior means (posterior sd) for the coefficients on $x_1^*, L_{x_1}, x_2^*, L_{x_2}$, and $L_y$ are $-0.94$ (0.26), $-0.86$ (0.27), $-0.01$ (0.09), $-0.35$ (0.08) and 0.53 (0.08). So the effect of income applies both to filtered and spatial components, but the effect of housing value is confined to its spatial component.

## 8.3    Conditional prior models

Simultaneous spatial models may involve slow MCMC computation in large datasets, requiring inverse or determinant calculations for large matrices. Additionally simultaneous autoregressive schemes are primarily designed for continuous univariate responses. Conditional spatial priors and associated regression schemes, allowing discrete outcomes and generalizing relatively simply to multivariate outcomes, are an alternative, provided they are consistent with valid joint priors. As mentioned by Wakefield *et al.* (2000), modelling of spatially correlated errors may proceed by initially specifying either the joint multivariate distribution of the vector $\epsilon = (\epsilon_1, \dots, \epsilon_n)$ for $n$ areas, or the univariate density of each areas error, $\epsilon_i$, conditional on errors in other areas $\epsilon_{[i]} = \{\epsilon_j, j \neq i\}$. Conditions that ensure the joint density is proper when the model specification starts with a conditional rather than the joint prior are discussed by Wakefield *et al.* (2000), and Besag and Kooperberg (1995).

Let $\mu$ denote the vector mean response in a general linear model for $y$ continuous or discrete, with link $g$. One possible conditional prior, the conditional autoregressive or CAR($\rho$) prior (Bell and Broemeling, 2000; Assunção and Krainski, 2009) expresses each $\epsilon_i$ in

$$g(\mu_i) = X_i\beta + \epsilon_i,$$

as a univariate normal

$$\epsilon_i|\epsilon_{[i]} \sim N(\rho M_{1i}, \sigma_\epsilon^2),$$

where the conditional mean

$$M_{1i} = \sum_{j \neq i} c_{ij}\epsilon_j,$$

is a weighted average of errors in other areas, the spatial interaction matrix $C = (c_{ij})$ is symmetric, and $\rho$ is bounded by the inverses of the minimum and maximum eigenvalues of $C$. The covariance of the vector $\epsilon$ in the corresponding joint prior is then $\Sigma = \sigma_\epsilon^2(I - \rho C)^{-1}$ (Richardson *et al.*, 1992, p 541; Wakefield *et al.*, 2000; De Oliveira, 2011). If the $\epsilon_i$ have non-zero means $\mu_i$, then this prior is expressed

$$\epsilon_i | \epsilon_{[i]} \sim N(\mu_i + \rho \sum_{j \neq i} c_{ij}(\epsilon_j - \mu_j), \sigma_\epsilon^2).$$

A related scheme is the intrinsic conditionally autoregressive or ICAR($\rho$) prior (Stern and Cressie, 2000), implemented as the car.proper option in BUGS, with

$$\epsilon_i | \epsilon_{[i]} \sim N\left(\rho M_{2i}, \frac{\sigma_\epsilon^2}{\sum_{j \neq i} c_{ij}}\right)$$

where $M_{2i}$ is the standardised weighted average of $\epsilon_j (j \neq i)$, namely

$$M_{2i} = \sum_{j \neq i} w_{ij}\epsilon_j = \frac{\sum_{j \neq i} c_{ij}\epsilon_j}{\sum_{j \neq i} c_{ij}}.$$

The corresponding joint prior on $(\epsilon_1, \ldots, \epsilon_n)$ is proper only when $|\rho| < 1$.

A widely adopted scheme, sometimes denoted the ICAR(1) prior, and implemented by the car.normal option in BUGS, assumes $\rho = 1$, and usually takes $c_{ij} = 1$ for adjacent areas, $c_{ij} = 0$ otherwise. Let $d_i = \sum_{j \neq i} c_{ij}$ be the number of areas (excluding area $i$ itself) which are adjacent to area $i$, and let this collection of areas define the neighbourhood $\partial_i$ of areas adjacent to area $i$. Then

$$\epsilon_i | \epsilon_{[i]} \sim N\left(\frac{\sum_{j \in \partial_i} \epsilon_j}{d_i}, \frac{\sigma_\epsilon^2}{d_i}\right).$$

This scheme is technically improper (Sun et al., 1999), but propriety is achieved in effective terms (Rodrigues and Assuncao, 2008) under MCMC sampling by re-centring the sampled $\epsilon_i$ to sum to zero at each iteration.

Another scheme (Leroux et al., 1999; Lee, 2011), which reduces to a normal iid prior when $\rho = 0$, sets

$$\epsilon_i | \epsilon_{[i]} \sim N\left(\frac{\rho}{1 - \rho + \rho \sum_{j \neq i} c_{ij}} M_{1i}, \frac{\sigma_\epsilon^2}{1 - \rho + \rho \sum_{j \neq i} c_{ij}}\right),$$

with $0 \leq \rho \leq 1$. When the $c_{ij}$ are defined by contiguity, with $d_i = \sum_{j \neq i} c_{ij}$, one obtains

$$\epsilon_i | \epsilon_{[i]} \sim N\left(\frac{\rho}{1 - \rho + \rho d_i} \sum_{j \in \partial_i} \epsilon_j, \frac{\sigma_\epsilon^2}{1 - \rho + \rho d_i}\right).$$

Setting $\rho = 1$ leads to the ICAR(1) model.

Predictive assessment of CAR models may be based on leave one out predictive checks (Stern and Cressie, 2000). The model is estimated for data $y_{[i]}$ omitting area $i$, and the posterior predictive distribution $p(y_{rep,i}|y_{[i]})$ obtained. Then for count data $y_i$, outlier status may be assessed using indicators

$$C_i^{(t)} = I(y_{rep,i}^{(t)} > y_i) + 0.5I(y_{rep,i}^{(t)} = y_i),$$

which are monitored through MCMC iterations (e.g. using the step function in BUGS), and the probabilities

$$p_i = \Pr(y_{rep,i} > y_i|y_{[i]}) + 0.5\Pr(y_{rep,i} = y_i|y_{[i]}),$$

obtained as the posterior means of $C_i$. An alternative is the approximate mixed predictive scheme of Marshall and Spiegelhalter (2007), which is less computationally intensive, and would involve sampling new spatial random effects $\epsilon_{rep,i}$ to defining corresponding means $\mu_{rep,i}$, and then sampling $y_{rep,i}$ from these means. However, while outlying areas (exceptionally high or low relative risks) may in abstract terms represent model failures, they may also be due to genuine variations in risk (Catelan and Biggeri, 2011).

These conditional schemes extend to multivariate effects; for applications see Catelan and Biggeri (2008), Congdon (2009) and Song $et\ al.$ (2006). Suppose there are $P$ sets of spatial effects $\epsilon_{pi}$ for each area $i$, and define $D = \text{Diag}(d_1, \dots, d_n)$. Then the proper version of the multivariate conditional autoregressive or MCAR prior (e.g. Mardia, 1988) takes the $nP$ vector of effects $\epsilon = (\epsilon_{11}, \epsilon_{21}, \dots, \epsilon_{P1}, \epsilon_{12}, \epsilon_{22}, \dots, \epsilon_{P2}, \dots, \epsilon_{1n}, \dots, \epsilon_{Pn})$ as multivariate normal with zero mean and precision matrix, $Q = (D - \rho C) \otimes \Phi$, namely

$$p(\epsilon|\Phi, \rho) = \left(\frac{1}{2\pi}\right)^{nP/2} |D - \rho C|^{P/2} |\Phi|^{n/2} \exp[-\frac{1}{2}\epsilon'Q\epsilon],$$

where $\Phi^{-1}$ is a $P \times P$ matrix describing covariation between effects $\epsilon_{pi}$ within areas. The conditional prior for $\epsilon_i = (\epsilon_{1i}, \epsilon_{2i}, \dots, \epsilon_{Pi})$ given the remaining effects $\epsilon_{[i]} = (\epsilon_1, \dots \epsilon_{i-1}, \epsilon_{i+1}, \dots, \epsilon_n)$ is multivariate normal of dimension $P$ with means $M_i = (M_{1i}, \dots, M_{Pi})$, given by

$$E(\epsilon_{pi}|\epsilon_{[i]}) = M_{pi} = \rho \sum_{j\neq i} c_{ij}\epsilon_{pj} / \sum_{j\neq i} c_{ij}$$

and precisions

$$\text{Prec}(\epsilon_i|\epsilon_{[i]}) = d_i\Phi.$$

For contiguity weights, the $M_{pi} = \rho\sum_{j\in\partial_i} \epsilon_{pj}/d_i$ are locality averages of the $p^{\text{th}}$ spatial effect. With $\rho = 1$ the multivariate version of the intrinsic CAR prior is obtained (e.g. Rue and Held, 2005), and a multivariate version of the convolution prior combines $P$-variate ICAR spatial effects with a $P$-variate vector of iid effects. The mv.car option in BUGS implements the multivariate ICAR(1) prior.

A multivariate extension of the prior of Leroux $et\ al.$ (1999) has the advantage of allowing the data to determine the appropriate mix between spatial and iid dependence with a single set of random effects rather than the two sets of the multivariate convolution prior. Thus

$$E(\epsilon_i|\epsilon_{[i]}) = [M_{1i}, \dots, M_{Pi}] = \rho \sum_{j\neq i} c_{ij} I_P \epsilon_j / [1 - \rho + \rho \sum_{j\neq i} c_{ij}]$$

$$\text{Prec}(\epsilon_i|\epsilon_{[i]}) = [1 - \rho + \rho \sum_{j\neq i} c_{ij}]\Phi$$

where $0 \le \rho \le 1$, and $\Phi$ is of dimension $P \times P$. When the $c_{ij}$ are binary adjacency indicators, the conditional expectations become

$$E(\epsilon_{pi}|\epsilon_{[i]}) = M_{pi} = \frac{\rho \sum\limits_{j \in \partial_i} \epsilon_{pj}}{[1 - \rho + \rho d_i]}.$$

## 8.3.1   Ecological analysis involving count data

Spatial dependence figures strongly in regression analysis of area disease counts with focus on detecting the underlying pattern of relative risk in the face of overdispersion and spatially correlated errors, in turn reflecting omitted predictors (e.g. environmental, social) which are spatially clustered. For example, suppose $y_i$ denotes disease totals in a set of small areas, with expected events $E_i$ (in the demographic sense) derived from a standard schedule of disease rates (Cressie and Wikle, 2011, p. 195). The outcomes may, subject to the necessity to take account of overdispersion, be taken as Poisson,

$$y_i \sim Po(E_i\theta_i),$$

where $\theta_i$ denote relative disease risks by area, averaging 1 when $\sum\limits_i y_i = \sum\limits_i E_i$. This form of age-independent sampling may be justified by assuming area-age disease rates $\pi_{ij}$ result from proportional relative risks and age rates, namely $\pi_{ij} = \theta_i\pi_j$ (Wakefield et al., 2000).

Maximum likelihood fixed effects estimates of relative risk (e.g. standard mortality or incidence ratios) are $\theta_i = y_i/E_i$ with standard error $\sqrt{y_i}/E_i$. The terminology fixed effects means that there is no pooling of strength mechanism, as in Marshall and Spiegelhalter (2007, p. 417) and Plummer (2008, p 528). Equivalently, the fixed effects assumption is that other effects $\theta_{[i]} = \{\theta_j, j \ne i\}$ in areas $j \ne i$ provide no information about $\theta_i$. Fixed effects estimates may be unreliable as estimates of relative risk (see e.g. Bernardinelli and Montomoli, 1992; Assuncao et al., 2002; Lee, 2011), especially if the resulting maps are distorted by low event counts or populations at risk, such that small changes in event totals produce major shifts in the estimated RR. Mollié (2001, p. 269) describes how apparently extreme $\theta_i$, especially when based on small populations, are not significantly different from unity. In devising models to 'pool strength' and reduce such anomalies, one may envisage the total variability in relative risks to have three seperate components: the influence of known predictors $X_i$, within area sampling variation around the true underlying rate, and between area variation in the true rates due to unknown risk factors, likely to show spatial correlation to some degree.

Let $\epsilon_i$ denote a relatively smooth underlying spatial signal following an ICAR(1) prior, and $u_i$ represent unstructured 'white noise' variability. Then a substantive basis is provided for the mixed regressive-convolution or BYM scheme (Besag et al., 1991),

$$y_i \sim Po(E_i\theta_i),$$
$$\log(\theta_i) = X_i\beta + u_i + \epsilon_i,$$

where the $u_i$ are iid normal, $u_i \sim N(0, \sigma_u^2)$. To assess the relative strength of spatial and unstructured variation under the convolution prior requires estimates of marginal rather than conditional variances. One may obtain an empirical estimate of the marginal variances (Mollié, 2001) as

$$V_\epsilon = \sum_i (\epsilon_i - \bar{\epsilon})^2/(n-1).$$

This may be compared with the estimated marginal variance for the $u_i$,

$$V_u = \sum_i (u_i - \bar{u})^2/(n-1),$$

or with the total random variation

$$V_\epsilon/(V_\epsilon + V_u).$$

Each of these quantities may be calculated and monitored during MCMC sampling. There may be sensitivity to priors adopted on $\sigma_\epsilon^2$ and $\sigma_u^2$ in the convolution model (e.g. Bernardinelli et al., 1995a; Yan, 2007), and Lee (2011) shows limitations of the convolution prior in both weak and strong spatial correlation applications. Some studies report over-smoothing of relative risks under the BYM model, limiting the ability to detect localised increases in risk (Richardson et al., 2004), and leading to a high proportion of false negatives (Goovaerts, 2009).

An alternative to the convolution prior proposed by Leyland et al. (2000) retains two error terms, but assumes that underlying the spatial errors $\epsilon_i$ are unstructured errors $v_i$. Under this approach joint densities may be specified for $u_i$ and $v_i$, such as a bivariate normal. The averaged spatial effect is

$$\epsilon_i = \sum_j w_{ij} v_j,$$

where the $w_{ij}$ are row standardised spatial interactions. If the $w_{ij}$ are based on contiguity, then $w_{ij} = 1/d_i$ if areas $i$ and $j$ are adjacent, and

$$\epsilon_i = \frac{1}{d_i} \sum_{j \in \partial_i} v_j,$$

with $\partial_i$ denoting the neighbourhood of areas adjacent to $i$. The corresponding relative risk model is

$$\log(\theta_i) = X_i \beta + u_i + \sum_j w_{ij} v_j.$$

This form of prior extends multivariate error forms for $j = 1, \ldots, J$ different outcomes. For instance, a multivariate normal prior of dimension $2J$ allows correlation between outcome specific errors $u_{ij}$ and $v_{ij}$ (Congdon, 2002).

Spatial patterning may not be smooth, and commonly adopted spatial priors may not adequately model discontinuities in disease maps; for example, a low mortality area surrounded by high mortality areas may have a distorted relative risk estimate if the area random effect consists solely of a pure spatial prior such as the ICAR(1). In the convolution model an alternative to normal errors $u_i$ is a heavier tailed Student-$t$ distribution obtained through scale mixing, namely

$$u_i \sim N(0, \sigma_u^2/\kappa_i),$$

$$\kappa_i \sim G\left(\frac{\nu}{2}, \frac{\nu}{2}\right),$$

with the impact of spatial outliers apparent in low $\kappa_i$ (Parent and LeSage, 2008).

Forms of discrete mixture have also been proposed to represent discontinuities in high disease risk (Militino et al., 2001). Besag et al. (1991, p. 8) mention a double exponential (Laplace) prior

$$p(\epsilon_i|\epsilon_{[i]}) \propto \frac{1}{\kappa} \exp\left[\frac{1}{\kappa} \sum_{j \in \partial_i} |\epsilon_i - \epsilon_j|\right]$$

as robust to outliers or discontinuities in the risk surface. $\kappa$ is a scaling parameter, with smaller values implying smaller spatial variability. Lawson and Clark (2002) propose a mixture of the ICAR(1) and Laplace priors, with the mixture defined by a continuous (beta) density rather than a discrete mixture, leading to

$$\log(\theta_i) = \gamma + u_i + \eta_i \epsilon_i + (1 - \eta_i) f_i,$$

where $f_i$ follows the conditional Laplace form, and the beta prior on $\eta_i$ has fixed parameters, e.g. $\eta_i \sim \text{Beta}(1, 1)$, or an unknown hyperparameter $\eta_i \sim \text{Beta}(\varphi, \varphi)$. Another option (Congdon, 2007) puts emphasis on the unstructured component in 'discontinuous' areas, under the scheme

$$\log(\theta_i) = \gamma + \eta_i u_i + (1 - \eta_i)\epsilon_i,$$

$$\eta_i \sim \text{Beta}(\varphi, \varphi).$$

Congdon (2007) mentions a spatial switching prior, involving two regimes, a pure spatial scheme in $\epsilon$ (e.g. an ICAR(1)), and a convolution prior combining a pure spatial effect and an iid error $u$,

$$\log(\theta_i | R_i = 0) = X_i \beta + \epsilon_i,$$

$$\log(\theta_i | R_i = 1) = X_i \beta + \epsilon_i + u_i,$$

$$R_i \sim Ber(\pi_R),$$

where $\pi_R$ may be preset (e.g. to 0.5) or taken as unknown.

A particular focus is small area disease models is the probability of elevated risk in a particular area. For example, consider the BYM model with an intercept $\beta_0$ only, and let $\theta_i^{(t)} = \exp(\beta_0^{(t)} + u_i^{(t)} + \epsilon_i^{(t)})$ denote the relative risk of area $i$ at iteration $t \in (1, \ldots, T)$. Then one may define binary indicators $\Delta_i^{(t)} = I(\theta_i^{(t)} > t_\theta)$, where $t_\theta$ is some high risk threshold (e.g. $t_\theta = 1$, or 1.5), with posterior probabilities of elevated risk then estimated as $\pi_i = \frac{1}{T} \sum_{t=1}^{T} \Delta_i^{(t)}$. Areas with probabilities $\pi_i$ exceeding some threshold probability $t_\pi$ (e.g. $t_\pi = 0.8$ or 0.9) can be classified as high risk. Decision rules for suitable $(t_\theta, t_\pi)$ providing optimal trade off between false positive and false negative have been proposed; for example, Richardson et al. (2004) suggest $(t_\theta = 1, 0.7 < t_\pi < 0.8)$ based on a simulation study.

Assessment of areas as high risk is subject to type 1 errors, when areas are falsely rated as high risk, or generically are 'false discoveries'. Thus one may seek a decision rule for classifying areas as high risk subject to a relatively low false discovery rate (FDR), defined as the proportion of areas assessed as high risk which are actually not high risk. To this end, the above modelling approaches may be adapted to include an area-specific null hypothesis indicator $H_{0i}$ in the model for $\theta_i$, with the null hypothesis being that $\theta = 1$ (when $H_0 = 1$), with a one sided alternative $\theta > 1$ (when $H_0 = 0$). In terms of the area risk model, $H_{0i}$ is binary with $\Pr(H_{0i} = 1) = \phi_i$, with $\phi_i$ in turn taken to be uniform $U(0, 1)$. When $H_{0i} = 1$ is selected at a particular MCMC iteration, the area relative risk $\theta_i$ is set to 1, whereas when $H_{0i} = 0$, one has the usual model for $\theta_i$.

For example, the BYM prior becomes

$$\theta_i = 1^{H_{0i}} \exp(\beta_0 + u_i + s_i)^{1 - H_{0i}},$$

$$H_{0i} \sim \text{Bern}(\phi_i),$$

$$\phi_i \sim U(0, 1),$$

with priors on $u_i$ and $s_i$ as discussed above. Let $\pi_{0i} = \frac{1}{T} \sum_{t=1}^{T} H_{0i}^{(t)}$ denote the MCMC posterior mean of $H_{0i}$, namely the probability that area $i$ has a relative risk of 1. The alternative to the hypothesis is one sided, so the FDR is calculated only among the subset of observations with $y_i \geq E_i$. Thus define a high risk classification rule $d_i = 1$ if $\pi_{0i} < t_\pi$, and $d_i = 0$ otherwise. Then the FDR is estimated as $\sum_{i:y_i \geq E_i} \pi_{0i} d_i / \sum_{i:y_i \geq E_i} d_i$. Ventrucci et al. (2011, p. 61) propose a general recommendation for scenarios where true high-risk areas have a moderately high value (e.g. $\theta = 1.5$), namely an FDR of 20% yielding a moderate sensitivity without major loss of specificity. An alternative scheme discussed by Catelan *et al.* (2010) proposes

$$\theta_i = 1^{(G_i=1)} \exp(\beta_0 + u_i + s_i)^{I(G_i=2)},$$

$$G_i \sim \text{Categoric}(\pi_{1:2}),$$

$$\pi_{1:2} \sim D(w_{1:2})$$

where $w_j$ are prior weights.

## Example 8.3  Crime rates in English local authorities

Spatial models for count outcomes are illustrated by counts of notifiable offenses $y_i$ (over the financial year 2009-10) in 324 English local authorities (excluding the City of London and Scilly Islands). These are considered in relation to an offset consisting of total populations $P_i$ in 2009, namely $y_i \sim \text{Po}(P_i \theta_i)$. A log link is used so that the modelling is of the log relative crime risks in each area (local rates relative to the national rate), which have an empirical variance of 0.135.

Three models are considered, with spatial interaction $C$ defined by adjacency. The first (model 1) contains no predictors but uses random area effects according to the convolution prior; the second model uses an alternative conditional prior but still without predictors; while the third adds predictors to assess how far the random variance in crime rates is explained. The ICAR prior in model 1 uses the car.normal function in BUGS, which includes the useful identifying device of centring the $\epsilon$ effects. An intercept $\alpha$ is included though the actual mean may differ from $\alpha$, as the empirical mean of the iid effects $u_i$ is not necessarily the nominal mean of zero. The crime counts are relatively large, and the $u$ effects will model overdispersion and irregular area effects that may distort the smooth spatial signal. So model 1 is

$$y_i \sim \text{Po}(P_i \theta_i),$$

$$\log(\theta_i) = \alpha + u_i + \epsilon_i,$$

$$\epsilon_i \sim \text{ICAR}(W, \sigma_\epsilon^2), \qquad \sigma_\epsilon^2 \sim IG(0.5, 0.0005), \qquad \sigma_u^2 \sim IG(0.5, 0.0005),$$

where the IG(0.5, 0.0005) prior for variance terms follows Kelsall and Wakefield (1999). The average smoothed crime rate per 1000 population can be obtained as $1000\exp(\alpha + \bar{u} + \bar{\epsilon})$. The second half of a two chain run of 1 00 000 iterations in BUGS is used for inferences. One chain starts with $\{\sigma_\epsilon^{-2} = 2, \sigma_u^{-2} = 0.5\}$, while the other starts with $\{\sigma_\epsilon^{-2} = 0.5, \sigma_u^{-2} = 2\}$. Convergence is obtained early for the total variance $V_u + V_\epsilon$, but the ratio $V_\epsilon/(V_u + V_\epsilon)$ takes the first half of the run to stabilise. Based on the second half of the run around 49% of the variance is due to the spatial signal. The variance of $t_i = u_i + \epsilon_i$ is estimated as 0.126, slightly less than the empirical variance of 0.135, and implying a shrinkage (smoothing) in the modelled crime rates compared to crude observed rates.

The Leroux *et al.* (1999) scheme[1] is also applied (model 2), with

$$\log(\theta_i) = \alpha + \epsilon_i,$$

$$\epsilon_i|\epsilon_{[i]} \sim N\left(\frac{\rho}{1 - \rho + \rho d_i} \sum_{j \in \partial_i} \epsilon_j, \frac{\sigma_\epsilon^2}{1 - \rho + \rho d_i}\right).$$

$$\sigma_\epsilon^2 \sim IG(0.5, 0.0005),$$

$$\rho \sim U(0, 1).$$

One chain starts with $\{\sigma_\epsilon^{-2} = 0.5, \rho = 0.75\}$, while the other starts with $\{\sigma_\epsilon^{-2} = 2, \rho = 0.5\}$. Convergence in $V_\epsilon$ and $\rho$ is straightforward, and the second half of a two chain run of 1 00 000 iterations in BUGS is used for inferences. The LPML improves slightly under this model to $-2194$, as compared to $-2206$ under the convolution model. The $\rho$ parameter is estimated at 0.36 (with 95% interval from 0.22 to 0.51), with $V_\epsilon$ estimated at 0.126.

The preceding model is then extended to include four predictors, obtained by principal component analysis from a broader set of indicators. These predictors are measures of aspects of area social structure, namely $x_1$ = social capital, $x_2$ = social fragmentation, $x_3$ = urbanicity, and $x_4$ = socioeconomic deprivation. Scores on these constructs are standardised, so that $\beta$ coefficients are a measure of explanatory importance. So

$$\log(\theta_i) = \alpha + X_i\beta + \epsilon_i,$$

with diffuse priors on the $\beta$ coefficients, and a Leroux *et al.* (1999) prior on $\epsilon_i$. The second half of a 100 000 iteration run of two chains in BUGS gives posterior means (sd) for $\beta_1$ of $-0.15$ (0.019), and for $\beta_2$ of 0.11 (0.015). The other two predictors have less impact on crime, but are still significant with $\beta_3$ having 95% interval (0.045,0.106) and $\beta_4$ having 95% interval (0.063,0.109). Residual spatial variation is reduced to var$(\epsilon_i) = 0.035$, with $\rho$ having mean 0.80. This implies that although residual variation is considerably reduced, the remaining unobserved influences on crime risk are spatially clustered.

**Example 8.4    INLA spatial estimation: Self harm in English small areas**

To illustrate the application of R-INLA to spatial estimation, consider self harm hospitalisation counts $y_i$ over the five year period (financial years 2006/07 to 2010/11) in 6781 small areas (Middle Level Super Output Areas, MSOAs) across England. Self harm stays are those with external cause ICD10 codes X60 to X84. The offset $E_i$ denotes expected events, with $y_i \sim Po(E_i\theta_i)$, and three predictors are a socioeconomic deprivation score (dep), a social fragmentation score (frag), and a rurality score (rur). Spatial data is contained in a text file ENG_MSOA_SH.txt includes columns headed y,E,id,id.sp, dep, frag and rur. The identifiers id and id.sp both range from 1 to 6781. An adjacency file ENG_MSOA_ADJ.txt has lines

6781

1 13 192 193 371 575 576 618 808 809 878 884 889 890 977

2 6 3 468 474 752 763 769

...

6781    1    3950

with lines subsequent to the first containing the area identifier, its number of neighbours, and a list of the neighbour codes. The Scilly Isles is assumed adjacent to the closest mainland MSOA.

The model adopted also allows for smoothly varying predictor effects $g_{1t}$, $g_{2t}$, and $g_{3t}$ (following a first order random walk prior on differences in knot values in the predictors). Then a convolution model specifies

$$\log(\theta_i) = \beta_0 + g_1(dep) + g_2(frag) + g_3(rur) + u_i + \epsilon_i.$$

Default gamma parameters 1 (shape) and 0.001 (scale) are adopted for precisions of the predictor smooths, the CAR spatial errors $\epsilon_i$, and the white noise errors $u_i$.

The sequence of commands to read in data and specify and run the model (including the option to report the DIC) is

```
library(INLA)
D <- read.table("ENG_MSOA_SH.txt",header=T)
G="C:/ENG_MSOA_ADJ.txt"
formula = y~f(inla.group(dep), model="rw1")+ f(inla.group(frag),
            model="rw1")+ f(inla.group(rur), model="rw1")+
f(id.sp,model="besag", graph =G)+f(id)
m1 = inla(formula,family="poisson",E=E,data=D,
        control.compute=list(graph=T,dic=T)).
```

Summaries of overall fit, posterior density plots, the predictions (predicted relative risks $\theta_i$), and the two sets of area random effects are obtained by the commands

```
summary(m1); plot(m1)
F <- m1$summary.fitted.values
R1 <- m1$summary.random$id.sp
R2 <- m1$summary.random$id
```

The first summary reports a DIC of 51 975, and effective parameter total of 5166, and a marginal likelihood of -34 336. Plots of the smooth predictor effects can be obtained using the second and third columns of summaries obtained via the sequence

```
R3 <- m1$summary.random$'inla.group(dep)'
R4 <- m1$summary.random$'inla.group(frag)'
R5 <- m1$summary.random$'inla.group(rur)'.
```

In fact, non-linearity is clearly apparent in the effects of fragmentation and rurality. If interest is in related parameters, such as the variance of the spatial effects, this may be obtained via

```
prec.sp = m1$marginals.hyperpar$"Precision for id.sp"
var.sp <- inla.expectation(function(x) 1/x, prec.sp).
```

Marginal likelihood estimates may be affected by the level of prior information, and to guage this, more informative priors for the area effects were adopted, based on the procedure of Mollié (1996). Thus the variance of the observed log relative risks is $s^2 = 0.073$, and so prior means for the precisions of the white noise variance and spatial variance are $2/0.073 = 27.4$ and $\frac{2}{0.073 \times 5.8} = 4.7$, with 5.8 being the average number of neighbours. With the scale parameter set at 0.1, the corresponding priors on the precision of $u$ and the conditional precision of $\epsilon$ are taken as Ga(2.5,0.1) and Ga(0.5,0.1) respectively. In fact under the change in prior assumptions, the marginal likelihood now estimated as $-34\,312$, with the DIC also reduced to 51 955.

# 8.4    Spatial covariation and interpolation in continuous space

The preceding two sections have considered continuous and discrete outcomes for zones (also called 'lattice' data). For point based data, a fixed interaction matrix is generally inappropriate, and the influence of proximity needs to be estimated. Consider observations $y_i$ at points $s_i$ in two dimensional space, $s_i = (s_{1i}, s_{2i})$, where $s_{1i}$ is the easting (longitude) and $s_{2i}$ the northing (latitude). A starting point for estimating the effect of proximity is provided by the matrix of Euclidean interpoint distances, $d_{ij} = |s_i - s_j|$, though other distance metrics can be used.

The widely applied isotropic stationary Gaussian model (Cressie and Wikle, 2011) for such data may be stated as

$$y(s) = \mu(s) + \epsilon(s),$$

where the joint prior for errors $\epsilon(s)$ is multivariate normal,

$$\epsilon(s) \sim N_n(0, \Sigma(d)),$$

with spatial covariance in errors $\Sigma(d) = \sigma^2 R(d)$, depending on distance but independent of location. These assumptions are also known as the intrinsic hypothesis (Oliver et al., 1998). The mean $\mu(s) = E[y(s)] = X(s)\beta$ models any trend over space, or the impact of other covariates. Trend, sometimes known as large scale variation, is exemplified by the North West to South East gradient in cardiovascular disease in England (Richardson, 1992), or the linear trend identified by Diggle and Ribeiro (2007, chapter 2) in topological elevation data. Small scale (second order) variation represents features such as spatial correlation in neighbouring locations.

The correlation function $R(d)$ is defined to ensure that $r_{ii}(d_{ii}) = r_{ii}(0) = 1$ and that $R$ is positive definite (Fotheringham et al., 2000; Diggle and Ribeiro, 2007). Commonly used functions meeting these requirements include the exponential correlation function

$$r_{ij} = \exp(-d_{ij}/\phi),$$

where $\phi$ is the range, or inter-point distance at which spatial correlation ceases to be important. The Gaussian correlation function has

$$r_{ij} = \exp(-d_{ij}^2/\phi^2),$$

and the spherical (Mardia and Marshall, 1984) has $r_{ij} = 0$ for $d_{ij} > \phi$, and

$$r_{ij} = (1 - 1.5d_{ij}/\phi + 0.5d_{ij}^3/\phi^3),$$

for $d_{ij} < \phi$. The Matérn correlation function is

$$r_{ij} = \frac{1}{\Gamma(\kappa)2^{\kappa-1}} \left( 2\sqrt{\kappa}\frac{d_{ij}}{\phi} \right)^\kappa K_\kappa(2\sqrt{\kappa}\frac{d_{ij}}{\phi}),$$

where $\kappa$ is a smoothness or shape parameter, and $K_\kappa$ is the modified Bessel function of the second kind. Mixtures of covariance functions may be used (e.g. Pilz and Spöck, 2008).

Often there will be further iid variability (e.g. due to measurement error or microscale variation), leading to the linear Gaussian model with two error terms

$$y(s_i) = \mu(s_i) + \epsilon(s_i) + u_i,$$

where $u_i \sim N(0, \tau^2)$ are independent of location or distance, with covariance

$$\Sigma(d) = \sigma^2 R(d) + \tau^2 I.$$

The parameter $\tau^2$ is often denoted the nugget variance, with the limiting variance as $d_{ij}$ tends to zero being $\tau^2 + \sigma^2$ instead of $\sigma^2$. Diggle et al. (1998) and Diggle and Ribeiro (2007, Ch. 4) consider adaptations of the Gaussian linear model for Poisson or binomial data. For example, for count data $y_i$ assumed Poisson, the means $v(s_i)$ may include simply a spatial error

$$\log(v(s_i)) = X(s_i)\beta + \epsilon(s_i),$$

with $\epsilon(s) \sim N_n(0, \sigma^2 R(d))$. However, for overdispersed data, both spatial and iid errors may be relevant, namely

$$\log(v(s_i)) = X(s_i)\beta + \epsilon(s_i) + u_i,$$

where the errors $u_i$ are independently $N(0, \tau^2)$.

In geostatistical applications, the emphasis is often on prediction at locations $s.new_i = (s_{new1i}, s_{new2i})$ on the basis of observations made at points $s_i = (s_{1i}, s_{2i})$. An example of spatial interpolation or 'kriging' from a Bayesian perspective is provided by Handcock and Stein (1993), who consider prediction of topological elevations $y_{new}$ at unobserved locations on a hillside, given an observed sample of 52 elevations. Prediction of $y_{new}$ at a new point $s_{new}$ under a linear model involves the $n \times 1$ vector of covariances $g = \text{Cov}(s_{new}, s)$ between the new point and the sampled sites $s = (s_1, s_2 \dots s_n)$. For instance if $\Sigma(d) = \sigma^2 e^{-d/\phi}$, the covariance vector is obtained by plugging in to this parametric form the distances $d_{1new} = |s_{new} - s_1|$, $d_{2new} = |s_{new} - s_2|$, etc. The prediction $y_{new}$ is a weighted combination of values at observed locations with weights $\lambda_i$ determined by

$$\lambda = g\Sigma^{-1}.$$

Just as for time series models, prediction to new locations may be facilitated by data transformation to remove trend and induce stationarity, meaning the mean and variance are independent of location, and providing a basis for the covariance assumption $\text{cov}(\epsilon) = \sigma^2 R(d)$. An example of such data transformation known as row and median polish is illustrated by Cressie (1993) on data for percents coal ash. One may also use trend surface regression (Ecker and Gelfand, 1997; e.g. Gaudard $et\ al.$, 1999; Militino and Ugarte, 2001), where in the regression $y(s) = X(s)\beta + \epsilon(s)$, the $X(s)$ are functions of the spatial coordinates $(s_{1i}, s_{2i})$. For example, one might set $x_{1i} = s_{1i}$, $x_{2i} = s_{2i}$, take $x_{3i}$ and $x_{4i}$ as squared terms in $s_1$ and $s_2$, and $x_{5i}$ as an interaction between $s_{1i}$ and $s_{2i}$, with Bayesian predictor selection methods to obtain an optimal trend model. Simultaneous modelling of spatial effects in the mean and variance structure may lead to weakened identifiability, especially for small datasets (Ecker and Gelfand, 1997).

A technique often used to display the covariance structure focuses on distance based functions of dissimilarity between observations or regression residuals. Assume a stationary process such that any large scale trend has been modelled (e.g. Diggle and Ribeiro, 2007, p. 34). Whereas the covariance terms in

$$\Sigma(d) = \sigma^2 R(d),$$

diminish to zero for widely separated points and attain their highest levels as $d_{ij}$ tends to zero, the variogram function

$$\gamma(d) = \sigma^2 - \Sigma(d) = \sigma^2(I - R(d))$$

has value zero when $d_{ij} = 0$, and reaches its maximum at $\sigma^2$ as spatial covariation in $\Sigma(d)$ disappears. Hence $\sigma^2$ is known as the sill in geostatistics. For instance, the variogram for the exponential model is

$$\gamma(d) = \sigma^2(1 - e^{-d/\phi}).$$

In a model including measurement error at $d = 0$, one has

$$\gamma(d) = \tau^2 + \sigma^2[I - R(d)],$$

with sill $\tau^2 + \sigma^2$. For a stationary model, the expected value of the difference between $y(s_i)$ and $y(s_j)$ at separation $d_{ij}$ is $2\gamma(d_{ij})$ (Ecker and Gelfand, 1997). Estimation of the variogram using interpoint differences may be carried out within distance bands, such that for the $n(d)$ points within bands $\{d - 0.5L, d + 0.5L\}$ of length $L$, a point estimate of the variogram is (Reilly and Gelman, 2007)

$$\hat{\gamma}(d) = 0.5 \frac{1}{n(d)} \left[ \sum_{d_{ij} \, \epsilon \, (d-0.5L, d+0.5L)} (y_i - y_j)^2 \right].$$

The relation between $\gamma(d)$ and $d$ may be modelled over distance bands (e.g. via non-linear regression) involving parametric forms such as above.

However, there are drawbacks to such procedures (Diggle and Ribeiro, 2002), and Bayesian approaches to spatial covariance generally focus on likelihood based estimation and interpolation (Ecker and Gelfand, 1997). For the linear model

$$y(s_i) = \mu(s_i) + \epsilon(s_i) + u_i,$$

with covariance

$$V(d) = \sigma^2 R(d) + \tau^2 I,$$

the log-likelihood kernel is

$$-0.5 \log |V(d)| - 0.5(y(s) - \mu(s))' V(d)^{-1}(y(s) - \mu(s)).$$

Identifiability of parameters may be improved (and sampling reduced) by discrete priors for the range parameter $\phi$, or on the nugget to sill ratio $v^2 = \tau^2/\sigma^2$ in the reparameterized covariance matrix $V(d) = \sigma^2(R(d) + v^2 I)$ (Diggle and Ribeiro, 2007, chapter 7; Diggle et al., 2003, p. 65). Modifications of this likelihood to allow Box–Cox transformations of $y$ are considered by Pilz and Spöck (2008), and a Box–Cox transformation option (the trans-Gaussian model) is included in the geoR package for the linear Gaussian model.

## 8.4.1    Discrete convolution processes

High dimensional Gaussian process spatial covariance models may be subject to slow estimation, and underlying stationarity and isotropy assumptions violated. An alternative lower-dimensional representation, that can be adapted to spatial non-stationarity and anisotropy, is the discrete process convolution (DPC) approach (Higdon et al., 1999; Higdon, 2007). Although different convolution schemes are possible, the most basic convolves a continuous white noise process $x(u)$, with an isotropic smoothing kernel $K(s - u)$, with the spatial trend obtained as

$$z(s) = \int_R K(u - s)x(u)du,$$

where $R$ is the region of interest. If the $x(u)$ are white noise with variance one, then (Lee et al., 2005)

$$\text{cov}(z(s), z(s')) = \int_R K(u - s)K(u - s')du.$$

The underlying process $x(u)$ is in practice approximated by a discretized process defined over a regular grid containing the region, and the parameter requirements reflect the dimension of the underlying grid, not the number of observations (Calder, 2003, 2007).

So with $i = 1, \ldots, n$ observations at points $(s_1, \ldots, s_n)$ where $s_i = (s_{1i}, s_{2i})$, and with $x_j$ sampled at grid locations (or knots) $\{w_j, j = 1, \ldots, m\}$ with $w_j = (w_{1j}, w_{2j})$, one may define the discretized kernel smoother as

$$z(s_i) = \sum_{j=1}^{m} K(s_i - w_j | \phi) x_j,$$

where $K(d|\phi)$ is a kernel function with parameters $\phi$ (Lee *et al.*, 2005; Higdon, 2007). If both the $K$ function and $w$ series have unknown variances then there are potential identifiability issues. The $x_j$ may be taken as iid effects, such as normal or Student-$t$, or constrained to be positive, such as gamma or log-normal, with $z(s_i)$ then also positive. Semi-parametric approaches can also be used (Reich and Fuentes, 2007), for example kernel functions for each of $m$ potential clusters, with unknown centres $w_j = (w_{1j}, w_{2j})$, and cluster allocation probabilities for sites or areas $i$ at location $s_i = (s_{1i}, s_{2i})$ incorporating spatial information.

Among possible generalizations are multivariate DPCs (where for $R$ outcomes, there are $R$ underlying discretised processes), multivariate factor DPC models (with $Q<R$ underlying discretised processes), and space-time DPCs. For example, a space-time DPC might take a knot-time specific random effect $x_{jt}$ to be a low order random walk. Thus a first order random walk in $x_{jt}$ would lead to

$$z(s_i, t) = \sum_{j=1}^{m} K(s_i - w_j | \phi_t) x_{jt},$$

$$x_{jt} \sim N(x_{j,t-1}, \sigma_x^2), \qquad t = 2, \ldots, T,$$

which incorporates evolution also in the kernel parameters, $\phi_t$ (Sanso *et al.*, 2008).

**Example 8.5   Ozone readings**

The principles involved in deriving a variogram, correlation function estimation, and kriging are illustrated using a set of ozone readings from $n = 32$ sites in the Los Angeles area aggregated over one month (UCLA Academic Technology Services, 2011). In R the data are input in the order {ozone, longitude, latitude}, and the distances obtained in R using the dist function. Distance bands for the variogram are based on the profile of the distances, with 10 intervals of width 0.15 adopted. Thus

```
ozone <- read.table("c://ozone.csv", sep=",", header=T)
dists <- dist(ozone[,2:3])
quantile(dists, c(0.01,0.05,0.95,0.99))
```

where the percentile summary of interpoint distances is useful in deciding distance bands for variogram plotting. To use routines from geoR (Diggle and Ribeiro, 2007), the data may be recast using the command

```
ozone.geo <- as.geodata(ozone,coords.col=2:3, data.col=1),
```

and the empirical variogram estimated using

```
vgm <- variog(ozone.geo, breaks= seq(0,1.5,l=11)),
```

with 10 intervals up to a maximum interpoint distance of 1.5. One may also use different options in geoR for parametric variogram smoothing, illustrated by an exponential model

```
thvgm <- variofit(vgm,ini=c(5,1),cov.model="exp"),
```

which estimates the spatial variance $\sigma^2$ as 7.02, the nugget variance $\tau^2$ as 0, and the range $\phi$ as 0.47. The empirical and smoothed variograms may be compared in the same plot using

```
plot(vgm)
lines.variomodel(cov.model="exp",cov.pars = c(7.02,0.47),nug = 0,
max.dist = 1.5, lwd = 1).
```

Separate plots of the ozone readings against longitude and latitude (e.g. Diggle and Ribeiro, 2007, p. 31) suggest trend effects, stronger for longitude. However, the first model fitted in BUGS omits trend modelling (model 1). A hierarchically centred structure is used, coded as

```
model { for (i in 1:n) {y[i] ~dnorm(mu[i],inv.tau2)
mu[i] <- beta+eps[i]
mu.eps[i] <- 0}
eps[1:n] ~spatial.exp(mu.eps[], s1[], s2[], inv.sig2, delta, 1)
```

where the parameterisation of the exponential correlation function (spatial.exp) involves the inverse range $\delta = 1/\phi$. The observed inter-site distances range from 0.07 to 2.4, and a $U(0.001,2.5)$ prior on $\delta$ is adopted, corresponding to a minimum correlation of 0.002 {at distance $d = 2.4$, and with $\delta = 2.5$} and a maximum correlation of 0.9999 {for $d = 0.07$, and $\delta = 0.001$}. The estimates for the hyperparameters $\{\sigma^2, \tau^2, \phi\}$ are sensitive to alternative priors on the variances, as may be verified by running under three options contained in the code for this example: separate Ga(1,0.001) priors on the inverse variances, $1/\sigma^2$ and $1/\tau^2$; separate uniform U(0,10) priors on $\sigma$ and $\tau$; and a uniform shrinkage prior on $T = \frac{\tau^2}{\tau^2+\sigma^2}$ combined with a Ga(1,0.001) prior on $1/\sigma^2$. Under the first option, a two chain run of 10 000 iterations in OpenBUGS 3.1.2 (with inferences from the last 5000) provides posterior means for $\sigma^2, \tau^2$, and $\phi$ of 8.96, 0.07 and 0.57.

Interpolation at 5 new locations can be made using the spatial.pred or spatial.unipred options, and these are less sensitive to priors. The interpolations under BUGS may be compared with kriging predictions from geoR obtained using

```
locs <- c(-117.50,-117.43,-117.31,-117.29,-117.23,
                 34.00,33.93,33.94,34.05,33.90)
dim(locs) <- c(5,2)
krg <- krige.conv(ozone.geo, coords=ozone.geo$coords,
                      data=ozone.geo$data,locs,
krige=krige.control(cov.pars=c(8.96,0.57),
nug=0.07))
```

The geoR predictions are (8.56, 8.67, 9.61, 9.99, 9.55). Using the spatial.pred option in Open-BUGS and the code

```
for (j in 1:n.new) { y.new[j] ~dnorm(mu.new[j],inv.tau2)
mu.new[j] <- beta+eps.new[j]
mu.eps.new[j] <- 0}
```

```
# Predictions at new (unobserved) locations
eps.new[1:n.new] ~spatial.pred(mu.eps.new[], s1.new[], s2.new[], eps[])
```

leads to the predictions (8.54,8.66, 9.56, 9.97, 9.51).

A second model includes a linear trend but allows for predictor selection. Specifically

$$y_i \sim N(\beta_1 + \gamma_1\beta_2 s_{1i} + \gamma_2\beta_3 s_{2i} + \epsilon_i, \tau^2)$$

where $\gamma_j \sim \text{Bern}(0.5)$. Because the results obtained from predictor selection may depend on priors adopted, the $\beta_2$ and $\beta_3$ coefficients are assigned priors based on separate univariate regressions of ozone readings on $s_1$ and $s_2$. Specifically the prior means are the same as from the univariate regressions, while the prior precisions for $\beta_2$ and $\beta_3$ are obtained by dividing the linear regression precisions by $n = 32$, analogous to the procedure proposed by Shively *et al.* (1999). Priors on other parameters are as in model 1. Posterior means for the $\gamma_j$ (namely 0.44 and 0.40) from the second half of a two chain run of 10 000 iterations suggest no gain from including a linear trend. However, the other parameters produced by the model averaging inherent under this selection approach show an attenuation of the spatial covariance effects: $\sigma^2$ is reduced to 6.1, and $\phi$ increased to 0.64.

Bayesian estimation and prediction for the linear Gaussian model in geoR uses the function krige.bayes. This function sets priors by discretizing the distribution of the range parameter $\phi$ as well as the 'relative nugget' parameter $v^2 = \tau^2/\sigma^2$. The default discrete prior on $\phi$ is a sequence of 51 values between 0 and 2 times the maximum intersite distance. Here the discrete prior is over 21 values between 0 and the maximum intersite distance of 2.4, while the discrete prior on relative nugget has maximum value 1. These settings are combined with a Matern covariance prior (with smoothness parameter preset at 0.5) and a linear trend, namely:

```
require(geoR)
MC <- model.control(trend.d = "1st", trend.l = "1st", cov.model = "matern",
                    kappa = 0.5, aniso.pars = NULL, lambda = 1)
PC <-  prior.control(phi.discrete=seq(0, 2.4, l=21),phi.prior="reciprocal",
tausq.rel.discrete=seq(0, 1, l=11),tausq.rel.prior="unif")
OC <- output.control(moments=T, n.posterior = 1000)
# define 10x10 grid for which predictions required
locs <- expand.grid(seq(-118.6,-117.6,l=10),seq(33.8,34.4,l=10))
# estimation
skb <- krige.bayes(ozone.geo,loc=locs,model=MC,prior=PC,output=OC)
# parameter samples
summary(skb$posterior$sample$phi)
summary(skb$posterior$sample$sigmasq)
summary(skb$posterior$sample$tausq.rel)
# mean predictions at each point in 10x10 grid
pred.m <- numeric(100)
for (k in 1:100) {pred.m[k] <- mean(skb$predictive$simulations[k,500:1000])}
# plot of mean predictions
image(skb)
contour(skb, add=TRUE, nlev=21)
```

The resulting plot shows increasingly higher predicted ozone values in the north east parts of the region. Posterior mean for $\phi$ and $\sigma^2$ are 0.79 and 3.0.

### Example 8.6   Interpolation between different spatial frameworks

Health data (e.g. on prevalence patterns) may be collected by practitioner agencies (e.g. general practitioner or GP practices responsible for primary care) for the populations they are responsible for. Unless catchment areas are defined, these populations may be spatially dispersed over several neighbourhoods. However, spatial contrasts between residential neighbourhoods in prevalence may be of interest in assessing health equity. In the UK, population prevalence totals for a range of chronic diseases are maintained by general practices for their populations under a scheme called the Quality and Outcomes Framework. Only total disease counts for GP practices are publicly released, with no additional disaggregation of prevalence by patient demographic category, and no totals provided for small areas. In England there are around 8200 general practices with populations averaging 6600, and around 32 000 neighbourhoods with populations typically of 1500-2000 (census units known as Lower level super output areas, or LSOAs).

The analysis here concerns prevalence registers for psychotic illness (schizophrenia, bipolar disorder) for 190 GP practice populations in a sector of London, known as Outer North East London (ONEL), consisting of four London boroughs (Barking and Dagenham, Havering, Redbridge, Waltham Forest). These practice populations are generally dispersed over several neighbourhoods, with population centroids $s_i = (s_{1i}, s_{2i})$ based on weighted averages over the subpopulations in each LSOA. The same sector has 562 LSOA neighbourhoods, with population centroids $s_{nei,k} = (s_{nei,1k}, s_{nei,2k})$. A discrete process convolution model is applied to counts $y_i$ of prevalent psychosis, assumed Poisson with

$$y_i \sim Po(P_i\theta_i),$$

for $i = 1, \dots , 190$ GP practice populations of size $P_i$. The aim is to interpolate prevalence counts $y_{nei,k}$ for the $k = 1, \dots , 562$ residential neighbourhoods with populations $P_{nei,k}$.

The DPC model uses a grid of $m = 90$ points encompassing the outer NE London sector. The discrete grid effects $\{x_j, j = 1, \dots , m\}$ are assigned an intrinsic CAR prior based on rook contiguity of the grid points. A baseline model for the mean prevalence in each GP population includes a spatial effect $z(s_i)$ generated by the DPC prior, and an additional iid effect to represent overdispersion, and measurement variation (e.g. variable effectiveness in disease case finding between GP practices). The model for GP practice prevalence rates then has the form

$$\log(\theta_i) = \beta_0 + z(s_i) + u_i,$$

$$z(s_i) = \sum_{j=1}^{m} K(d_{ij})x_j,$$

$$x_{1:n} \sim ICAR(1, 1/\tau_x),$$

$$u_i \sim N(0, 1/\tau_u).$$

The random effects have precisions $\tau_x$ and $\tau_u$ that are assigned gamma $Ga(1, 0.001)$ priors. A normal kernel of known form, namely

$$K(d) = \frac{1}{2\pi} \exp(-0.5d^2),$$

is assumed. An alternative would be to take the $x_j$ to have known variance and the kernel function to have an unknown scaling parameter. Interpolated prevalence totals for neighbourhoods are generated according to

$$y_{nei,k} \sim Po(P_{nei,k}\theta_{nei,k})$$

$$\log(\theta_{nei,k}) = \beta_0 + z(s_{nei,k}),$$

$$z(s_{nei,k}) = \sum_{j=1}^{m} K(d_{kj})x_j,$$

where $d_{kj}$ are distances between LSOA population centroids $s_{nei,k}$ and the grid point locations $w_j = (w_{1j}, w_{2j})$.

A two chain run of 5000 iterations shows a posterior mean $\sigma_x$ of 0.063, with modelled GP practice psychosis prevalences rates per 1000 (i.e. $1000 \times \theta_i$) having 2.5th and 97.5th percentiles of 4 and 13.6 per 1000. Some variation between GP practices may reflect effectiveness in case finding or clinical variation, and the variation between interpolated neighbourhood rates (based on the smooth spatial signal underlying practice variations) is smaller, with 95% of neighbourhoods between 5.1 and 8.9 per 1000. The average deviance and DIC are respectively 214 and 365, with the LPML at $-764$.

A second model uses a socioeconomic deprivation score available both for GP practice populations and for neighbourhoods. Let $D_i$ denote the GP practice score, and $D_{nei,k}$ the neighbourhood score. Then a second model has

$$\log(\theta_i) = \beta_0 + \beta_1 D_i + z(s_i) + u_i,$$

$$\log(\theta_{nei,k}) = \beta_0 + \beta_1 D_{nei,k} + z(s_{nei,k}).$$

This model shows a significant deprivation effect on prevalence with 95% interval for $\beta_1$ of $(0.07,0.19)$. Variation in the discrete grid effects is reduced, with mean for $\sigma_x$ now at 0.043, but with a wider 95% interval for neighbourhood rates, from 4.9 to 10.4 per 1000. The average deviance and DIC are now respectively 210 and 355, with the LPML at $-736$.

## 8.5 Spatial heterogeneity and spatially varying coefficient priors

With geographical data, the preceding sections has shown how regression errors may well be spatially correlated and have sought to model spatial dependencies in errors. However, an alternative or at least complementary perspective on spatial outcomes is in terms of spatial heterogeneity or non-stationarity, including heteroscedastic variances, spatially varying regression effects, and varying spatial correlations (Paez and Scott, 2004; Fortin and Dale, 2005; López-Hernandez et al., 2010). A similar theme is raised by the concept of 'spatial regimes', meaning different intercepts or slopes for subsets of spatial units. Spatial heteroscedasticity can be handled by allowing the iid error (e.g. in the convolution prior) to have a variance that differs between individual areas, or perhaps governed by a spatial regime, involving $k < n$ subsets of areas, within which the areas have the same variance (Anselin, 2001).

While spatial heteroscedasticity may well be relevant in accounting for spatially correlated residuals, an alternative perspective on spatial heterogeneity focuses on the regression parameters themselves, since it is likely in fact in many applications that regression effects are not constant over the region of application (Casetti, 1992). Spatial correlation in residuals under spatially homogeneous predictor effects may reflect spatially varying predictor effects (e.g. Breitenecker and Harms, 2010). Classical approaches to spatial heterogeneity in regression coefficients (Section 8.5.1) have focused on the normal linear model (with modifications for heavy tailed errors via Student $t$ extensions), while Bayesian spatially varying coefficient approaches apply straightforwardly to both normal and to general linear models.

## 8.5.1   Spatial expansion and geographically weighted regression

The spatial expansion model (Casetti 1992, 2010; LeSage 1997) assumes that the impacts of one or more of regressors on a continuous outcome $y_i$ vary according to the locations $(s_{1i}, s_{2i})$ of the areas or points. For convenience, locations may be taken with reference to a central point $(0, 0)$, so that one or both elements in the grid reference may be negative. One may allow for fixed (i.e. non-spatially varying) impacts $\gamma_1, \gamma_2, \ldots, \gamma_P$ of regressors as well as spatially varying ones $b_{ji}$ ($j = 1, \ldots, P$), though the constant $\gamma_0$ does not have a parallel spatially varying effect. The latter choice is made by LeSage (1998, p. 106) on identifiability grounds. For a linear regression this leads to

$$y_i \sim N(\mu_i, \phi),$$

$$\mu_i = \gamma_0 + \gamma_1 x_{1i} + \gamma_2 x_{2i} + \ldots + \gamma_P x_{Pi} + b_{1i} x_{1i} + b_{2i} x_{2i} + \ldots + b_{Pi} x_{Pi},$$

with a linear spatial trend in the regression parameters modelled as

$$b_{ji} = s_{1i}\delta_{1j} + s_{2i}\delta_{2j}, \qquad j = 1, \ldots, P.$$

There are $3P + 1$ parameters to estimate: the constant, the fixed regression effects, and the $2P$ linear trend parameters, $\delta_{1j}$ and $\delta_{2j}$. The combined non-spatial and spatial impact of $x_j$ then varies by area:

$$\beta_{ji} = \gamma_j + b_{ji},$$

with average over all areas

$$\beta_j = \gamma_j + B_j.$$

This model may therefore capture variations in the regression relationships, especially if clusters of adjacent observations have similar regression behaviours, or if economic relationships vary according to distance from city centres. Non-linear spatial trends can be modelled, and non-locational predictors of $b_{ji}$ can also be used. Heteroscedasticity can also involve functions of the locations, as in

$$\phi_i = \exp(\eta_0 + \eta_1 s_{1i} + \eta_2 s_{2i}).$$

The method of geographically weighted regression (GWR) also makes regressions specific to the location of point or area $i$ (Brunsdon et al., 1996, 1998). Somewhat like cross-validation with single case omission, GWR essentially consists in re-using the data $n$ times, such that the $i$th regression regards the $i$th point as the origin. The coefficients $\{b_{1i}, \ldots b_{Pi}\}$ for the $i$th regression entail distance based weightings $a_{ik}$ of $n - 1$ remaining areas $k \neq i$, but with area $i$ as the centre. These weightings modify the precision parameters $\tau_{ik} = \phi_i a_{ik}$ for each area in a normal or Student t likelihood for the $i$th regression.

The distance weightings might be exponential, Gaussian, or spherical with the Gaussian form being

$$a_{ik} = \exp(-d_{ik}^2 / 2\eta^2), \qquad k \neq i,$$

with $a_{ii} = 1$. At small distances $a_{ik}$ is close to 1, with the decay effect increasing as $\eta$ tends to zero. In all these functions, a prior may be set on $\eta$, taking account of the maximum observed inter-point or inter-area distance. Other options are profile type analyses to maximise a fit criterion. For the $i$th regression (centred at area $i$), one has

$$y_k \sim N(\mu_{ik}, \tau_{ik}^{-1}), \qquad k = 1, \ldots, n,$$

$$\tau_{ik} = \phi_i a_{ik},$$

$$\mu_{ik} = \alpha + \beta_{1i} x_{1k} + \beta_{2i} x_{2k} + \dots + \beta_{Pi} x,$$

where random effects priors for the $\beta_{pi}$ may be used as a way to borrow strength and improve identifiability. The precision parameters $\phi_i$ may be assigned independent (e.g. gamma) priors, or possibly taken to follow a hierarchical prior, such as a hierarchical gamma mixture $\phi_i \sim Ga(a_\phi, b_\phi)$ where $a_\phi, b_\phi$ are extra unknowns. This format implies higher weighting in the $i$th regression for points $k$ close to $i$, so these have more influence on the estimates of $(\alpha_i, \beta_{1i}, \dots, \beta_{Pi})$.

Robustification against outlying areas may be obtained by taking a scale mixture using scaling factors

$$\kappa_{ik} \sim Ga(0.5v, 0.5v),$$

so that

$$\tau_{ik} = \phi_i \kappa_{ik} a_{ik},$$

though separate identifiability of $\eta$ and $v$ may require informative priors. For example, taking $v \sim Ga(8, 2)$ would be an informative prior consistent with expected spatial heteroscedasticity, and allow one to discriminate between non-constant variances over space as against non-constant regression relationships.

### 8.5.2    Spatially varying coefficients via multivariate priors

Some evaluations of GWR point to problems such as multicollinearity (Wheeler and Tiefelsdorf, 2005; Paez *et al.*, 2011). A Bayesian random effects alternative is provided by spatially varying coefficient (SVC) models (Assuncao, 2002; Gamerman *et al.*, 2003). For univariate responses $y_i$ of dimension $n \times 1$, let $\beta$ be a $nP \times 1$ vector of spatially varying regression coefficients for $P$ predictors contained in the $nP \times n$ block diagonal matrix $X_i$. SVC models may be based on geostatistical or autoregressive priors.

The normal linear SVC model based on geostatistical principles may be stated (Gelfand *et al.*, 2003) as

$$y(s_i) \sim N(X(s_i)\beta_i, \sigma^2 I),$$

with prior for $\beta = (\beta_1, \dots, \beta_n)$,

$$\pi(\beta | \mu_\beta, V_\beta) = N(1_{n \times 1} \otimes \mu_\beta, V_\beta),$$

where $\mu_\beta = (\mu_{\beta_1}, \dots, \mu_{\beta_P})'$ are prior mean coefficients and $V_\beta$ is $nP \times nP$. $V_\beta$ has form

$$V_\beta = C \otimes \Lambda,$$

where $\Lambda$ is a $P \times P$ matrix of within area covariances between regression coefficients, and $C = [c(s_i - s_j; \eta)]$ is a $n \times n$ correlation matrix modelling spatial interaction between areas or sites with locations $s_i = (s_{1i}, s_{2i})$ and inter-area distances $d_{ij}$. For example, under exponential spatial interaction

$$c_{ij} = \exp(-d_{ij}/\eta), \qquad \eta > 0.$$

Setting $\beta = 1_{n \times 1} \otimes \mu_\beta + b$ where $b \sim N(0, C(\eta) \otimes \Lambda)$, the geostatistical SVC model becomes

$$y = X'\mu_\beta + X'b + u,$$

where $u \sim N(0, \sigma^2 I)$.

To illustrate a conditional autoregressive scheme for SVC, let $\beta_i = (\beta_{1i}, \ldots, \beta_{Pi})$ denote area specific regression effects in the linear predictor

$$\eta_i = \sum_{p=1}^{P} \beta_{pi} x_{pi},$$

of a general linear model with mean $\mu_i = E(y_i)$, and link $g(\mu_i) = \eta_i$. It is possible to use separate spatial priors for each of the $P$ sets of spatially varying coefficients (Assuncao *et al.*, 2002). However, effects of different predictors may be correlated within areas. With $\beta = (\beta_1, \ldots, \beta_n)$, a multivariate ICAR(1) prior for spatial effects involves a joint pairwise difference prior for $\beta$ (Assuncao, 2003)

$$p(\beta|\Phi) \propto |\Phi|^{n/2} \exp\{-0.5 \sum_i \sum_j c_{ij}(\beta_i - \beta_j)'\Phi(\beta_i - \beta_j)\},$$

with $P \times P$ within area precision matrix $\Phi$. As noted above this prior is technically improper though propriety may be obtained empirically under a constraint $\sum_{i=1}^{n} \beta_{pi} = A_p$ where $A_p$ is a known $P$-vector (Assuncao, 2003). A practical strategy takes

$$\beta_i = \mu_\beta + b_i,$$

where the $\mu_\beta$ are fixed effects, and the $b_i$ follow the multivariate pairwise difference prior but are zero centred at each MCMC iteration.

Assuming separate regression effects for each area may lead to relatively imprecise coefficient estimates (especially for boundary areas), and possible alternative spatial regimes assume a discrete set of coefficient values based on an underlying continuous spatial prior. For example, a binary scheme for a particular coefficient $\beta_{pi}$ (e.g. a choice between relatively high regression effect in some areas and a weak effect in others) can be obtained by taking

$$(z_1, \ldots, z_n) \sim CAR(\rho),$$
$$\beta_{pi} = \beta_{p1}^* \qquad z_i > \eta,$$
$$\beta_{pi} = \beta_{p2}^* \qquad z_i \leq \eta.$$

A multinomial scheme with $K$ possible coefficient values $\beta_{pk}^*$ involves taking $K - 1$ cut-points on $z$.

It may be questioned whether exclusively spatial dependence in varying regression effects is always plausible. A multivariate convolution prior such as

$$\beta_i = \mu_\beta + b_i + u_i,$$

where $u_i$ is iid of dimension $P$ may be adopted instead. Alternatively schemes such as the proper CAR model or Leroux *et al.* (1999) model may be applied to SVC situations. SVC models allowing for a non-spatial component are also obtained under the prior of Leyland *et al.* (2000) and Langford *et al.* (1999). Consider a model with one independent variable and linear predictor

$$\eta_i = \alpha + u_i + \epsilon_i + \beta_i x_i,$$

where underlying the spatially structured effects $\epsilon_i$, $\beta_i$, are unstructured errors $v_{1i}$ and $v_{2i}$. With row standardised spatial interactions $w_{ij}$ taken to be the same across the two sets of effects, one has

$$\epsilon_i = \sum_{j \neq i} w_{ij} v_{j1},$$

$$\beta_i = \sum_{j \neq i} w_{ij} v_{j2},$$

$$u_i = v_{3i},$$

where one may set a multivariate prior of dimension 3 (e.g. MVN or MVt) on $(v_1, v_2, v_3)$, with mean zero for $v_1$ and $v_3$, but non-zero mean $\mu_\beta$ (with a separate prior) for $v_2$.

**Example 8.7   TB incidence: A comparison of GWR and SVC**

Here GWR methods and SVC models are considered for the TB incidence data in $n = 31$ London districts. Population centroids for each district are averages over neighbourhood (Lower Super Output Area, LSOA) centroids. For the GWR analysis, the regression with district $i$ as the centre has the form

$$y_k \sim N(\alpha + \beta_{1i}x_{1k} + \beta_{2i}x_{2k} + \ldots + \beta_{Pi}x_{Pk}, \tau_{ik}^{-1}), \qquad k = 1, \ldots, n,$$

$$\tau_{ik} = \phi_i a_{ik},$$

where $P = 2$, with $x_1 = \%$ non-white, $x_2 =$ area deprivation. An exponential distance decay function is assumed

$$a_{ik} = \exp(-\zeta d_{ik}/100)$$

where distances $d_{ik}$ are in kilometres, and a common unknown decay parameter $\zeta$ is assumed across the $n$ regressions. As in Example 8.1, the $y$ variables are incidence rates per 100 000 and normal linear regression is the underlying technique (though an alternative Poisson sampling scheme is considered in the SVC model for these data). The $y$ and $x$ variables are scaled to have a standard deviation of unity, so that the $\beta$ parameters are standardized regression effects. The priors for the regression coefficients $\beta_{pi}$ have variance 1, in line with a prior belief that standardized regression effects have 95% probability of being in an interval $\mu_{\beta p} \pm 2$. So $\pi(\beta_{pi}) = N(\mu_{\beta p}, 1)$. $N(0, 1)$ priors are assumed on the $\mu_{\beta p}$ themselves. The Bayesian sampling approach to estimation enables one to monitor the minimum and maximum $\beta_{pi}$ for predictor $p$ over all $i = 1, \ldots, n$ regressions, and also the probability that the $i$th regression provides the maximum or minimum coefficient. A grid prior on $\zeta$ is adopted with 100 equally spaced values between 0.1 and 10.

A two chain run of 5000 iterations with inferences from the last 4000 shows an insignificant deprivation effect, replicating Example 8.1, but a significant effect of percent non-white; the respective means and 95% intervals for $\mu_{\beta p}$ are 0.82 (0.47,1.16) and 0.09 (−0.25, 0.44). The $\zeta$ parameter has a well defined posterior density, with mean (95% interval) of 1.55 (0.40,2.80), though the resulting distance decay is relatively shallow within each regression. For example, district 16 (Hillingdon) is on the western edge of London and distances to the other 30 districts range from 7.7 to 44.5 km, with posterior means for $a_{ik}$ ranging from 0.89 to 0.53. The coefficients $\beta_{1i}$ on percent non-white range from 0.81 (with 95% interval 0.62 to 1.00) to 0.87 (0.69, 1.07). Figure 8.1 shows the geographic patterning of the non-white coefficients.

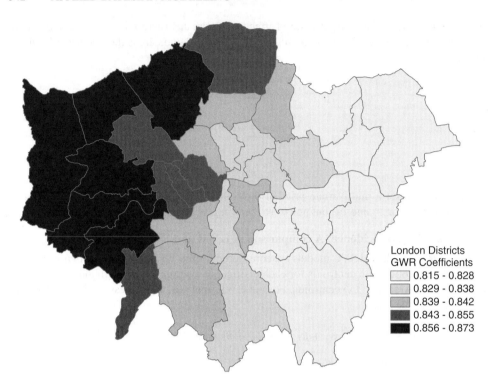

**Figure 8.1** Posterior mean GWR coefficients, TB rates and % non-white, linear regression.

For the SVC model, the incidence counts are assumed to be Poisson with means $\mu_i = P_i\theta_i$, where $P_i$ are populations (in units of 100 000), so that the $\theta_i$ are rates per 100 000. Before fitting the SVC model, a model with spatially homogeneous regression coefficients and Leroux-CAR errors is applied (a 'spatial errors' model). This will enable a goodness of fit comparison with the SVC model. So $y_i \sim Po(P_i\theta_i)$,

$$\log(\theta_i) = \eta_i = \alpha + \beta_1 x_{1i} + \beta_2 x_{2i} + \epsilon_i,$$

$$\epsilon_i|\epsilon_{[i]} \sim N(\frac{\rho}{1 - \rho + \rho d_i}\sum_{j \in \partial_i}\epsilon_j, \frac{\sigma_\epsilon^2}{1 - \rho + \rho d_i}),$$

where $d_i$ is the number of neighbours of borough $i$. A two chain run of 10 000 iterations is run, with inferences from the last 9000. It is interesting to note that this model produces significant impacts for both predictors, namely 0.033 (0.026,0.041) for percent non-white, and 0.022 (0.010,0.034) for deprivation, illustrating possible drawbacks in applying normal regression to a variable based on counts. A comparison with the GWR results on regression coefficients can be obtained by monitoring standardised coefficients (see Exercises). The LPML for this model is $-146.4$.

For the spatial heterogeneity model, a bivariate Leroux-CAR scheme is adopted for the area coefficients $b_i = (b_{1i}, b_{2i})$ in the model

$$\log(\theta_i) = \eta_i = \alpha + (\beta_1 + b_{1i})x_{1i} + (\beta_2 + b_{2i})x_{2i},$$

$$E(b_{pi}|b_{[i]}) = M_{pi} = \frac{\rho \Sigma_{j \in \partial_i} b_{pj}}{[1 - \rho + \rho d_i]},$$

$$\text{Prec}(b_i|b_{[i]}) = [1 - \rho + \rho d_i]\Phi,$$

$$\Phi \sim W(R, 10)$$

where $\Phi$ is $2 \times 2$, with $\text{diag}(R) = (0.002, 0.001)$. The total regression coefficients are $\beta_{pi}^* = \beta_p + b_{pi}$ with means $\mu_{\beta_p} = E(\beta_{pi}^*)$. Taking the estimates for $\beta_p$ from the spatial errors model as a guide, prior standard deviations of 0.014 =sqrt(0.002/10), and 0.01 for $b_{1i}$ and $b_{2i}$ are assumed. This corresponds to a prior belief that 95% of $\beta_{1i}^*$ will be within the interval (0.005,0.061), and 95% of the $\beta_{2i}^*$ will be in the interval (0.002,0.042). Negative effects of these predictors are implausible on substantive grounds. This model produces a higher LPML ($-137.6$) than the spatial errors model and the mean predictor effects $\mu_{\beta_p}$ are also slightly elevated, namely 0.036 (0.024,0.049) for percent non-white, and 0.036 (0.016,0.054) for deprivation. The map of mean $\beta_{1i}^*$ coefficients is shown in Figure 8.2, and has some similarity with that obtained by using GWR.

A varying coefficients model can also be estimated in R-INLA. This option assumes independent CAR priors on the zero mean variations $(b_{1i}, b_{2i})$ around the central fixed effects $(\beta_1, \beta_2)$. Identifiers for each random effect (id.sp.x1 and id.sp.x2) are needed in the input. The default Ga(1,0.001) priors on the precisions of these effects may be assumed. A code comparing a simple fixed effects model with an SVC model (and specifying relevant outputs) is then

```
D <- read.table("LONDONDIST_TB.txt",header=T);
G="LONDONDIST_ADJ.txt"
formula1 = y~x1+x2+f(id.sp.x1,x1,model="besag",graph =G)+
           f(id.sp.x2,x2,model="besag",graph =G)
m1 = inla(formula1,family="poisson",E=P,data=D, con-
trol.compute=list(graph=T, dic=T))
summary(m1)
FV1 <- m1$summary.fitted.values
R1 <- m1$summary.random$id.sp.x1
R2 <- m1$summary.random$id.sp.x2
formula2 = y~x1+x2
m2 = inla(formula2,family="poisson",E=P,data=D,
control.compute=list(graph=T, dic=T))
summary(m2) .
```

## 8.6   Spatio-temporal models

Economic or health data are often available as averages or counts $y_{it}$ for areas $i = 1, \dots , n$ and times $t = 1, \dots T$. For metric data $y_{it}$, one may generalise simultaneous spatial lag or spatial error schemes to include temporal effects (Anselin, 2010). For example, the space-time extension of the spatial errors model (Elhorst, 2010; Barufi et al., 2011) can be represented as

$$y_{it} = X_{it}\beta + \alpha_i + \epsilon_{it},$$

$$\epsilon_{it} = \rho \sum_j w_{ij}\epsilon_{jt} + u_{it},$$

$$E(u_{it}) = 0, \qquad E(u_{it}, u_{ir}) = \sigma^2,$$

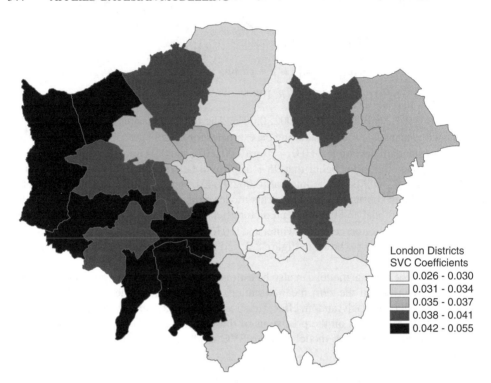

**Figure 8.2**    Posterior mean SVC coefficients, TB rates and % non-white, Poisson regression.

where the long term means $\alpha_i$ in area $i$ are either fixed or random effects (Kakamu and Wago, 2008). In matrix terms

$$\epsilon = \rho\tilde{W}\epsilon + u \Rightarrow \epsilon = (I - \rho\tilde{W})^{-1}u$$

where $u$ is a $(Tn \times 1)$ vector of $N(0, \sigma^2)$ effects, and $\tilde{W}$ is $Tn \times Tn$ block diagonal containing $T$ copies of a $n \times n$ interaction matrix $W = [w_{ij}]$. Among possible options, Anselin (1988, Section 4.3) mentions spatial and temporal heteroscedasticity in the $u_{it}$, and space-time varying regression coefficients $\beta_{it}$ as in

$$y_{it} = X_{it}\beta_{it} + \alpha_i + \epsilon_{it},$$

with special cases being coefficients correlated over areas, $\beta_i$, or over time, $\beta_t$. Variations on the spatio-temporal error lag model have been proposed to include both spatial and temporal lags in the errors (Baltagi *et al.*, 2007; Elhorst, 2008), as in

$$y_{it} = X_{it}\beta + \alpha_i + \epsilon_{it},$$

$$\epsilon_{it} = \rho\sum_j w_{ij}\epsilon_{jt} + v_{it},$$

$$v_{it} = \gamma v_{i,t-1} + u_{it}.$$

Among possible spatio-temporal versions of the spatial lag model is the specification

$$y_{it} = \lambda\sum_j w_{ij}y_{jt} + X_{it}\beta + \alpha_i + u_{it},$$

where the $u_{it}$ are iid, $u_{it} \sim N(0, \sigma^2)$. Assuming the $\alpha_i$ are random, $\alpha_i \sim N(0, \sigma_\alpha^2)$, the likelihood for period $t$ is

$$p(y_t | \lambda, \beta, \alpha, \sigma^2) = (2\pi\sigma^2)^{-n/2} |I - \lambda W| \exp(-0.5 e_t' e_t / \sigma^2)$$

where $e_{it} = y_{it} - \lambda \sum_j w_{ij} y_{jt} - X_{it}\beta - \alpha_i$. Debarsy *et al.* (2012) incorporate time-lags in the response for both own area and neighbouring areas, as well as contemporary spatial lags in both outcomes and errors, as in

$$y_{it} = \phi y_{i,t-1} + \lambda \sum_j w_{ij} y_{jt} + \theta \sum_j w_{ij} y_{j,t-1} + X_{it}\beta + \alpha_i + \epsilon_{it},$$

$$\epsilon_{it} = \rho \sum_j w_{ij} \epsilon_{jt} + u_{it},$$

while Elhorst (2012) additionally includes spatial and time lags in predictors.

Spatio-temporal extension of spatial probit or logit models (binary, ordinal or multinomial) using the augmented data representation have been proposed. For example, Chakir and Parent (2009) propose a spatial errors multinomial probit model for land use choices. A spatial lag model could include lags in both observed and latent data, for example with underlying utilities $z_{it}$ and binary responses $y_{it} = 1$,

$$z_{it} = X_{it}\beta + \lambda_1 \sum_j w_{ij} y_{jt} + \lambda_2 \sum_j w_{ij} z_{jt} + u_{it},$$

where the $u_{it}$ are iid with known variance. Dubin (1995) considers a dynamic logit model for innovations with $y_{it} = 1$ if the innovation is adopted at time $t$, and $y_{it} = 0$ otherwise, and with interaction matrix depending on distances $d_{ij}$ from earlier adopters according to $w_{ij} = \alpha_1 \exp(-\alpha_2 d_{ij})$.

## 8.6.1  Conditional prior representations

In area-time models involving binomial or Poisson count data, conditional space-time priors are more convenient, though these may also be specified for metric data (e.g. Wikle *et al.*, 1998). One may assume spatially structured random variation for both the level and the growth effect (analogous to a random intercept-random slope model for panel data), so that neighbouring areas have similar linear trends in relative risk (Bernardinelli *et al.*, 1995b). More complex functions of time may be relevant, e.g. to represent seasonal effects in climate models (Wikle *et al.*, 1998). Denote level effects $\omega_{1i}$ describing the initial relative risk pattern, and trend parameters $\omega_{2i}$ describing incremental changes in relative risk. An aggregate trend can additionally be represented by parameters $\delta_t$, which for $T$ small are probably better modelled as fixed effects with a corner constraint ($\delta_1 = 0$).

Then for Poisson data with equally spaced time points and offsets $E_{it}$ (expected events or populations), a spatial linear trend model has $y_{it} \sim Po(E_{it}\theta_{it})$, with

$$\log(\theta_{it}) = \alpha + \delta_t + \omega_{1i} + \omega_{2i}t,$$

where the two sets of effects $\omega_{1i}$ and $\omega_{2i}$ can be assigned separate conditional spatial priors, or a bivariate prior used with level and trend correlated. The priors on $(\omega_{1i}, \omega_{2i})$ can encompass extremes of complete spatial and iid dependence as under the Leroux *et al.* (1999) scheme, or a convolution form used both for level and trend (cf. Knorr-Held, 2000), namely

$$\log(\theta_{it}) = \alpha + \delta_t + \omega_{1i} + u_{1i} + (\omega_{2i} + u_{2i})t,$$

where $u_{1i}$ and $u_{2i}$ are iid effects, and the $\omega_{ji}$ are pure spatial, such as ICAR(1), effects. Setting $c_{ji} = \omega_{ji} + u_{ji}$ one has

$$\log(\theta_{it}) = \alpha + \delta_t + c_{1i} + c_{2i}t.$$

Even when there is some shuffling of spatial relativities in the outcome over time, one may nevertheless ask how persistent differentials across areas are, for example, via autoregressive dependence in $c_{it} = \omega_{it} + u_{it}$ where $\omega_{it}$ are ICAR(1) and $u_{it}$ are iid (Martinez-Beneito et al., 2008). Thus an AR1 scheme has for $t > 2$,

$$\log(\theta_{it}) = \alpha + \delta_t + c_{it} + \gamma(\log(\theta_{i,t-1}) - \alpha - \delta_{t-1}),$$

$$\gamma \in (-1, 1),$$

while for $t = 1$,

$$\log(\theta_{i1}) = \alpha + \delta_1 + \frac{c_{i1}}{(1 - \gamma^2)^{0.5}}.$$

Variations in persistence of spatial clustering can be taken as area specific, as in Congdon (2001), namely

$$\log(\theta_{it}) = \alpha + \delta_t + \omega_{it} + u_{it}, \qquad \omega_{it} = \rho_i \omega_{it-1} + v_{it}, \qquad t > 1,$$

$$\log(\theta_{i1}) = \alpha + \delta_1 + \omega_{i1} + u_{i1},$$

where the $\omega_{i1}$ follow a distinct spatial prior with precision $\tau_1$. The $\rho_i$ are taken to be spatially (i.e. ICAR) distributed with non-zero mean $\mu_\rho$, namely a form of spatially varying regression effect.

Area-time schemes for health outcomes may be extended to include ages $x = 1, \dots, X$, with observations $y_{ixt} \sim Po(E_{ixt}\theta_{ixt})$. The area-time model above may be extended to include separate trends for each age and area, namely

$$\log(\theta_{ixt}) = \alpha + \delta_t + \omega_{1i} + \gamma_{1x} + (\omega_{2i} + \gamma_{2x})t,$$

where $\{\gamma_{1x}, \gamma_{2x}\}$ can be assigned autoregressive or random walk priors. There may also be a case for age-area interaction effects (Congdon, 2006), while in the event of overdispersion one may additionally include unstructured age-area-time effects, $u_{ixt}$, so that

$$\log(\theta_{ixt}) = \alpha + \delta_t + \omega_{1i} + \gamma_{1x} + (\omega_{2i} + \gamma_{2x})t + u_{ixt}.$$

Rather than assuming linear trends in time or other age scales, more generality may be gained by assuming structured age-time interactions $\psi_{xt}$ or area-time interactions $\psi_{it}$, maybe combined with unstructured area-time or age-time effects. For example, it may be that the relative balance of unstructured random variation as against spatial clustering is changing over time. Prior precisions for structured interactions may be obtained as a Kronecker product of the structure matrices (which define the inverse covariance matrices) for the relevant dimensions (Clayton, 1996; Knorr-Held, 2000; Lagazio et al., 2001). Thus the ICAR(1) joint prior for spatial errors $s = (s_1, \dots, s_n)$, with interaction matrix $C$ based on contiguity of areas, is multivariate normal with precision matrix $\tau_s K_s$, where $\tau_s$ is a precision parameter, and off-diagonal terms $K_{s[ij]} = -1$ if areas $i$ and $j$ are adjacent, and $K_{s[ij]} = 0$ for non-adjacent areas. The diagonal terms in $K_s$ are $d_i$, the number of neighbours of area $i$. With $K_t$ as the structure matrices for times, an $RW1$ prior in time has a structure matrix with off-diagonal elements $K_{t[ab]} = -1$ if times $a$ and $b$ are adjacent, and $K_{t[ab]} = 0$ otherwise. Diagonal elements are 1 if $a = b = 1$ or $a = b = T$, and equal 2 otherwise.

Then an area-time interaction effect $\psi_{it}$ formed by crossing an $RW1$ time prior with a ICAR(1) spatial effect has a joint precision based on the Kronecker product

$$\tau_\psi K_s \otimes K_t,$$

and prior conditionals with precisions $\tau_\psi d_i$ when $t = 1$ or $t = T$, and $2\tau_\psi d_i$ otherwise. With $\partial_i$ denoting the neighbourhood of area $i$, the prior conditional means $M_{it}$ for $\psi_{it}$ are

$$M_{i1} = \psi_{i2} + \sum_{j\in\partial_i} \psi_{j1}/d_i - \sum_{j\in\partial_i} \psi_{j2}/d_i,$$

$$M_{it} = 0.5(\psi_{i,t-1} + \psi_{i,t+1}) + \sum_{j\in\partial_i} \psi_{jt}/d_i - \sum_{j\in\partial_i} (\psi_{j,t+1} + \psi_{j,t-1})/(2d_i), \qquad t = 2, \dots, T-1,$$

$$M_{iT} = \psi_{i,T-1} + \sum_{j\in\partial_i} \psi_{jT}/d_i - \sum_{j\in\partial_i} \psi_{j,T-1}/d_i.$$

For identification, the $\psi_{it}$ should be doubly centred at each iteration (over areas for a given $t$, and over times for a given area $i$).

### Example 8.8  Trends in TB incidence

The data here are as in Examples 8.1 and 8.7, but include two earlier periods, namely 2001-03 and 2004-06. One analysis treats the incidence rates $y_{it}$ for areas $i$ and $T = 3$ periods as continuous, and uses a dynamic spatial lag approach, as considered in Section 8.2. Two further models adopt Poisson sampling with offsets provided by area populations, and compare a Kronecker product approach (see above) with a discrete latent mixture model (Lawson et al., 2010; Kirby et al., 2011). The spatial lag model specifies a constant lag parameter combined with time varying regression effects, namely

$$y_{it} = \lambda \sum_j w_{ij} y_{jt} + X_{it}\beta_t + \alpha_i + u_{it},$$

$$u_{it} \sim N(0, \sigma^2), \alpha_i \sim N(0, \sigma_\alpha^2),$$

with log-likelihood for period $t$

$$\log[p(y_t | \lambda, \beta, \alpha, \sigma^2)] = -n/2\log(\sigma^2) + \log|I - \lambda W| - 0.5 e_t' e_t / \sigma^2,$$

where $e_{it} = y_{it} - \lambda \sum_j w_{ij} y_{jt} - X_{it}\beta_t - \alpha_i$. A grid prior of $R = 100$ values $\lambda_r$ from 0 to 0.99 is used, combined with precalculated determinant values $|I - \lambda_r W|$ in a non-standard likelihood. The last 4000 iterations of a two chain run of 5000 show that effects $\beta_{2t}$ of non-white ethnicity $(x_2)$ on incidence are increasing, while effects $\beta_{3t}$ of deprivation are declining. The spatial lag coefficient has posterior mean (95% CRI) of 0.27 (0.07, 0.43). Significant $\alpha_i$ occur for 8 of the 31 areas, with the largest in absolute terms being for area 22 (Lewisham) which has relatively low incidence, given both relatively high incidence in neighbouring areas and its own ethnic mix and deprivation level.

An alternative perspective is provided by a Poisson model (without covariates) for the period-specific counts, with $y_{it}$ now denoting counts of new TB cases[2]. Thus

$$y_{it} \sim Po(P_{it}\theta_{it}),$$

$$\log(\theta_{it}) = \beta_0 + \omega_i + \psi_{it}^*,$$

with $\psi_{it}^* = \psi_{it} - \psi_{i1}$, and area-time interactions $\psi_{it}$ have joint prior based on the Kronecker product of an ICAR(1) spatial effect and an RW1 time effect, as described above, to provide a precision matrix $\tau_\psi K_s \otimes K_t$. A main time effect is not included, but the trend can be inferred by averaging over the $\psi_{it}^*$ in each period. Following Lagazio et al. (2001) a corner constraint (namely $\psi_{i1}^* = 0$) is applied to the area-time interactions actually included in the model for $\log(\theta_{it})$. This constraint combined with centering of the $\omega_i$ at each iteration ensures that the intercept $\beta_0$ can be separately identified. From the last 4000 iterations of a two chain run of 5000, it appears that risk differences between areas are relatively stable in that $\tau_\psi$ is considerably larger than $\tau_\omega$. This model has a LPML of $-360$.

A latent mixture approach to spatial clustering is also considered, with $y_{it}$ again counts of new cases. This model is based on an evolving relative risk parameter set $\theta_{kt}$ for periods $t$ and $k = 1, \ldots, K$ groups. For example, an RW1 evolution, $\theta_{kt} \sim N(\theta_{k,t-1}, \sigma_k^2)$ is one option. The appropriate mix of $\theta_{kt}$ for area $i$ at time $t$ is based on a Dirichlet mixture of dimension $K$, with proportions $w_{itk} = g_{itk} / \sum_{k=1}^{K} g_{itk}$, in turn based on $K$ sets of positive variables $g_{itk} = \exp(\zeta_{itk})$, where the $\zeta_{itk}$ are spatially correlated over areas $i$. While correlation of the $\zeta$ over times or groups could be modelled, here the $\zeta_{ikt}$ are assumed to follow $TK$ separate ICAR(1) priors[3]. Then

$$\log(\theta_{it}) = \beta_0 + w_{itk}\theta_{kt}.$$

Although models with unknown $K$ could be considered, a common scenario in spatial health applications involves $K = 3$ area types defined by low risk, average risk, and high risk. To ensure consistent labelling of risk categories and assist convergence, the $\theta$ parameters are constrained within each period, namely $\theta_{3t} > \theta_{2t} > \theta_{1t}$. This is implemented by taking

$$\delta_{kt} \sim N(\delta_{k,t-1}, \sigma_k^2), \qquad k = 2, \ldots, K, \qquad t > 1,$$

$$\theta_{1t} = \delta_{1t},$$

$$\theta_{kt} = \theta_{k-1,t} + \exp(\delta_{kt}) \qquad k > 1.$$

A two chain run of 100 000 iterations shows convergence in the $\theta_{kt}$ parameters after 50 000 iterations. The LPML is lower than the above model at $-397$, but adding long term means $\omega_i$, as in

$$\log(\theta_{it}) = \beta_0 + \omega_i + w_{itk}\theta_{kt}.$$

could be considered as a way to improve fit.

## 8.7   Clustering in relation to known centres

The above modelling strategies have considered whether outcomes for aggregate areas, or individual events, exhibit a spatial structure without focussing on known locational influences on risk (e.g. pollution sources), or seeking to identify unknown cluster centres. In environmental epidemiology it is frequently the goal to assess the degree of focussed clustering in relation to one or more hazard sites (Biggeri et al., 1999; Ismaila et al., 2007; Lawson et al.,

2007; Maule *et al.*, 2007; Dreassi *et al.*, 2008; Lawson, 2009). The analysis may focus on the spatial locations of individual disease events in relation to one or more hazard sites, or on aggregate disease counts for small areas, again in relation to point sources. In the absence of individual or aggregate measures of exposure, the distance from the source is often used as a proxy for exposure.

For testing clustering of events around a specified point source (i.e. a putative hazard site) a number of testing procedures have been suggested (e.g. Tango, 2000; Rogerson, 2005), but not necessarily including features such as the impact of covariates, the presence of overdispersion, or generalised spatial dependence in the outcome beyond that associated with exposure to the hazard source. The basis of a more general parametric modelling strategy in relation to environmental hazard sites builds on the approaches set out by authors such as Diggle *et al.* (2003), Lawson (2001), Morris and Wakefield (2000), and Wakefield and Morris (2001).

Diggle (1990) suggested that the relative risk for disease events at location $s$ in relation to a point source at $s_0$ could be represented by a Poisson point process

$$\lambda(s) = R_0 g(s) a(d)$$

where $d = |s - s_0|$ is distance between the location and the point source, $R_0$ is the overall region-wide rate, $g(s)$ is the population at risk (or background risk) at location $s$. The main focus for risk assessment is a distance decay function $a(d)$, expressing the postulated decline in exposure at greater distances from the source, and also the relative risk of disease at the source as compared to the 'background' level.

For instance, possible distance decay functions, expressing monotonic lessening in risk as distance from the point source increases, involve the additive form

$$a(d) = 1 + \eta f(d),$$

where $\eta > 0$, and $f()$ might be a simple exponential function

$$f(d) = \exp(-\alpha d),$$

or a squared exponential

$$f(d) = \exp(-\alpha d^2).$$

Taking $\alpha > 0$, these functions have the property that $a()$ tends to 1 (the background risk) as $d$ tends to infinity. Also $1 + \eta$ can be interpreted as the relative risk at or close to the source itself (where $d$ is near 0). If in fact $\eta = 0$ or effectively so, there is no association between risk and distance from the source. Another type of model is 'hot spot' clustering when there is uniformly elevated risk in a neighbourhood around the focus, but background risk elsewhere. Thus for tracts at distances $d < \delta$ from the focus, the risk is $1 + \eta$, but for $d > \delta$, the risk is either just set at 1, or maybe follows a distance decay from the threshold distance:

$$a(d) = 1 + \eta \exp(-\alpha[d - \delta]^2).$$

Models where direction $\theta$, not just distance, from the point source to the case event or tract is relevant (e.g. when prevailing winds influence the potential spread of pollution) might include terms in $\sin(\theta)$ and $\cos(\theta)$. If there is a peak risk away from the source or other non-linearities (e.g. due to topographical effects), then a term in $\log(d)$ as well as $d$ can be included. Lawson and Williams (1994) suggest the additive form

$$a(d, \theta) = 1 + \eta f(d, \theta)/d,$$

so that risk still tends to 1 as $d \to \infty$, but $f()$ may include aspects of direction.

### 8.7.1  Areas or cases as data

In area studies with observations consisting of disease counts $y_i$, the population risk in area $i$ might be approximated by the expected disease total $E_i$, based on the population age structure in that area and the regional disease rate, while $d$ might be the distance between the area centroid and the point source. Additionally to account for the more usual sources of overdispersion or spatial correlation, the model for the Poisson mean $\mu_i$ might include unstructured effects $u_i$, and spatial effects $\epsilon_i$ (e.g. CAR distributed), as in preceding sections. If area characteristics $X_i$ also influence the disease risk one has a model such as

$$\mu_i = R_0 E_i \exp(X_i \beta + u_i + \epsilon_i)[1 + \eta f(d_i)],$$

or

$$\log(\mu_i) = \log(R_0) + \log(E_i) + X_i \beta + u_i + \epsilon_i + \log[1 + \eta f(d_i)].$$

Often observations consist of a set of cases and their locations. On the assumption that individuals are independent (in spatial terms) with regard to risk of disease, possibly after allowing for relevant risk factors or confounders, one may, however, model the likelihood of case events $i = 1, \ldots, n$ at locations $s_i$ in relation to a point source at $s_0$. The population density at $s_i$ may be modelled via kernel methods, typically using population data for areas within the region. An alternative is to proxy the population distribution using a control disease unrelated to the risk from the point source (e.g. cardiovascular conditions may be assumed independent of residence near high voltage electricity lines, but certain cancers may not be).

Let the $n_1$ disease cases be Bernoulli events $y_i = 1$, with $y_i \sim \mathrm{Bern}(\pi_i)$ and probability of caseness $\pi_i$. So if the $n_2$ control cases (with $y_i = 0$) have locations $s_i$, $i = n_1 + 1, \ldots, n_1 + n_2$, caseness can be modelled in terms of the decay function, and possibly individual risk factors $X_i$ (confounders in assessing the relation between risk and the pollution sources). Then the odds of caseness may be represented as

$$\pi_i/(1 - \pi_i) = R_0^*[1 + \eta f(d)] \exp(X_i \beta)$$

This conditional approach (Diggle and Rowlingson, 1994; Rogerson, 2006, 2010, Lawson, 2009) has the advantage of not requiring an estimate of $g(s)$. Note, however, the interpretation of $R_0^*$ depends on the selection of cases and controls; specifically, $R_0^* = (a/b)R_0$ where $a$ is the sampling proportion of cases (often 100%), $b$ the sampling proportion of controls, and $R_0$ the population odds of disease (Diggle *et al.*, 2003).

### 8.7.2  Multiple sources

Often there are multiple sources $k = 1, \ldots, K$ of environmental pollution, and assuming each site contributes to the risk independently of others, one may consider (Biggeri and Lagazio, 1999) odds of caseness (for individual event data)

$$\frac{\pi(s_i)}{1 - \pi(s_i)} = \exp(X_i \beta)[1 + \sum_{k=1}^{K} \eta_k f(s_i - s_0, \alpha_k)].$$

For instance, one option is

$$\frac{\pi(s_i)}{1 - \pi(s_i)} = \exp(X_i \beta)[1 + \sum_{k=1}^{K} \eta_k \exp(-\alpha_k |s_i - s_0|)],$$

though either or both of $\eta_k$ and $\alpha_k$ may be taken equal. Morris and Wakefield (2000, p. 171) propose a slightly different formulation with

$$\mu(s) = R_0 g(s) a(d, \varphi) = R_0 g(s) \prod_{k=1}^{K} [1 + \eta_k f(d, \alpha_k)].$$

Lawson (2009) and Lawson and Kulldorf (2000) discuss adaptation of the multiple source model to the case where both the geographic location of the cluster centres and their number are unknown. Other issues are relevant in multi-site analysis, for example, proximities between sites themselves: risk from two sources relatively close to one another may overlap.

## Example 8.9   Larynx cancer in Lancashire

As an example of an individual level outcome, consider event case data consisting of 58 larynx cancer cases that occurred in a part of Lancashire over 1974–83. The controls are provided by 978 lung cancer cases in the same study region. The point source implicated in the disease is a waste incinerator at location (355000, 414000). This data set has a relatively small case total and firm identification of exposure effects in terms of (say) distance decay may be difficult to obtain.

Allowing for effects of distance $d$ from the incinerator, and direction effects in angles $r$ (in radians) between source and case/control event, the intensity model is

$$\lambda(s, d) = R_0^* g(s)[1 + \eta f(d, r)],$$

or with individual confounders $X_i$

$$\lambda(s, d) = R_0^* \exp(X_i \beta) g(s)[1 + \eta f(d, r)]$$

but in the conditional logistic model (Diggle and Rowlinson, 1994) the population intensity $g(s)$ is 'conditioned out'. In the present example there are no confounders. So if $y_i = 1$ for larynx cancers and $y_i = 0$ for lung cancers, one has

$$y_i \sim \text{Bern}(\pi_i),$$
$$\pi_i/(1 - \pi_i) = R_0^*[1 + \eta f(d, r)/d],$$

where the term $1/d$ allows $f()$ to be density function (Lawson and Williams, 1994). Terms in $d$ and $\log(d)$ in $f()$ allow for possibly peaked exposure and incidence away from the source. Additionally, directional measures $\cos(r)$ and $\sin(r)$ are introduced. For instance, an event due North of the source would have angle $90°$ and radian value $r = 1.57$, while an event due South of the source would have angle and radian of $270°$ and 4.71 respectively.

An initial estimation stage specifies the distance-direction function

$$f(d, r) = \exp[-\alpha_1 d + \alpha_2 \log(d) + \alpha_3 \cos(r) + \alpha_4 \sin(r)]$$

Estimates of $\alpha$ and $\eta$ are sensitive to priors, and diffuse priors produce implausible estimates. Here $N(0,10)$ priors are adopted on the $\alpha_j$, and on $\delta = \log(\eta)$, with $\alpha_1$ constrained to be positive. To estimate the probability of intensity exceedance for individual cases one may monitor the indicators

$$I(\lambda_i^{(t)} > \lambda_H),$$

where $\lambda_i = 1 + \eta f(d,r)/d$, and $\lambda_H$ is some threshold (e.g. $\lambda_H = 1.2$ or $\lambda_H = 1.3$) judged as indicating high risk. Setting $\lambda_H = 1.2$, and using 0.9 as the significance threshold on the posterior probability estimated over all $T$ iterations, namely

$$\sum_{t=1}^{T} I(\lambda_i^{(t)} > \lambda_H)/T > 0.9,$$

one obtains an estimated 58 subjects with elevated risk, of whom 52 are non-cases.

Posterior means and precisions on $\alpha_j$ and $\delta$ (from a two chain run of 10 000 iterations in OpenBUGS) from the initial estimation stage are used to provide weakly data based priors for a second variable selection stage, using the approach of Shively et al. (1999) which involves downweighting the first stage precisions by the sample size $n = 1036$. The subsequent variable selection stage involves binary selection indices $\{G_j \sim \text{Bern}(0.5), j = 1, 5\}$ on $\{\eta, \alpha_1, \ldots, \alpha_4\}$, with retention of the $\alpha$ parameters conditional on $\eta$ being retained (i.e. on $G_1 = 1$). Thus retention of the $\alpha_j$ is based on binary indicators $K_j = G_1 G_{j+1} (j = 1, 4)$ in the scheme

$$\eta = G_1 \exp(\delta)$$

$$a(d,r) = [1 + \eta \exp\{-K_1\alpha_1 d + K_2\alpha_2 \log(d) + K_3\alpha_3 \cos(r) + K_4\alpha_4 \sin(r)\}/d].$$

The probability that $G_1 = 0$ amounts to a test that $a(d,r) = (1 + \eta f(d,r)/d) = 1$ (cf. Diggle, 1990, p. 355). The last 9000 iterations from a two chain run of 10 000 iterations show $\Pr(G_1 = 0|y) = 0.675$, compared to a prior probability of 0.5. This provides inconclusive evidence for or against a cancer risk from the source, though inferences from any such analysis are conditional on the form of $f(d,r)$ and the priors for the parameters.

# Exercises

**8.1.** Estimate a spatial errors model (without predictors) for the London TB data (Example 8.1) using a grid prior for the autocorrelation parameter, and compare its LPML with that under the pure spatial lag model.

**8.2.** A weighted grid prior for the spatial correlation coefficient in spatial lag and error models can be specified (Lesage, 1997) as

$$\pi(\lambda) \propto (s+1)\lambda^s I_{(0,1)}(\lambda),$$

where the positive coefficient $s$ is preset, and $I_{(0,1)}(\lambda)$ indicates that $\lambda$ is constrained between 0 and 1. The value $s = 0$ leads to a uniform (equal probability) prior over different $\lambda_g$ values in the grid $g = 1, \ldots, G$, while higher integer values for $s$ (e.g. $s = 5$) give higher weight to high correlations. Apply a prior with $s = 2$ to the pure spatial autoregression model (without predictors) for the London TB data (Example 8.1), and compare the estimated $\lambda$ and LPML with those obtained under a uniform grid prior.

**8.3.** Estimate the spatial error model for the Columbus crime data (Example 8.2) retaining the grid prior and special likelihood approach, but with Student $t$ distributed disturbances $u_i$. The precision for the $i$th neighbourhood is

$$\kappa_i/\sigma_u^2$$

where $\kappa_i \sim \text{Ga}(v/2, v/2)$ and $v$ is an unknown. Compare the LPML to that obtained under the homoscedastic model.

**8.4.**   Estimate the convolution model, together with predictors $X_1 - X_4$ for the English crime rate data (Example 8.3), namely

$$\log(\theta_i) = \alpha + X_i\beta + \epsilon_i + u_i,$$

where the $\epsilon_i$ are ICAR(1) and the $u_i$ are iid. Assess convergence in $V_\epsilon/(V_u + V_\epsilon)$, and the impact that including predictors has on the relative importance of spatial to iid random effects. Compare the LPML with that obtained for the predictor model combined with the Leroux *et al.* (1999) spatial prior.

**8.5.**   In Example 8.6, assess the impact on the profile of neighbourhood psychosis prevalence rates per 1000, $R_k = 1000\theta_{k,nei}$ (e.g. levels of skewness in such prevalence rates) and exceedance probabilities $\Pr(R_k > 7.75)$ (where 7.75 per 1000 is the regional prevalence rate), of using a standard Student $t$ kernel with 5 degrees of freedom, namely

$$K(d) = \frac{\Gamma(3)}{\Gamma(2.5)(5\pi)^{0.5}}[1 + 0.2d^2]^{-3}.$$

**8.6.**   For the GWR model of TB incidence in Example 8.7, find the posterior probability that district 9 (Ealing) has the highest coefficient for the effect of percent non-white.

**8.7.**   For the Leroux *et al.* (1999) CAR spatial errors model for TB incidence in Example 8.7 (Poisson sampling and homogenous regression effects) monitor the standardised regression coefficients $\beta_p sd(x_p)/sd(\eta)$ and compare them with the $\mu_{\beta_p}$ obtained under the GWR model.

**8.8.**   Compare the LPML of the spatial lag model fitted in Example 8.8 with a model assuming a time varying lag coefficient

$$y_{it} = \lambda_t \sum_j w_{ij} y_{jt} + X_{it}\beta_t + \alpha_i + u_{it},$$

and with a model with time-specific lag coefficient but excluding the long term area means $\alpha_i$, namely

$$y_{it} = \lambda_t \sum_j w_{ij} y_{jt} + X_{it}\beta_t + u_{it}.$$

**8.9.**   In the second model considered in Example 8.8, namely

$$y_{it} \sim \text{Po}(P_{it} v_{it}),$$

$$\log(v_{it}) = \beta_0 + \omega_i + \psi_{it}^*,$$

$$\psi_{it}^* = \psi_{it} - \psi_{i1},$$

monitor the marginal variances $\{V_\omega, V_\psi\}$ of the $\omega$ and $\psi$ parameters, together with the 'intra-cluster correlation' $V_\omega/(V_\omega + V_\psi)$. Also test whether any areas have significant monotonic trends, defined by $\Pr(\psi_{i2}^* > \psi_{i1}^*|y)$ and $\Pr(\psi_{i3}^* > \psi_{i2}^*|y)$ both exceeding 0.9. Finally assess the gain in fit through adding spatially varying regression effects (for borough ethnicity and deprivation), as in

$$\log(v_{it}) = \beta_0 + \omega_{1i} + \omega_{2i} X_{it} + \psi_{it}^*$$

where $(\omega_{1i}, \omega_{2i})$ follow a bivariate ICAR prior (implemented via the mv.car prior in BUGS).

**8.10.**  In Example 8.9, estimate a three parameter distance decay model for the cancer case-control data (without direction effects), namely

$$y_i \sim \text{Bern}(\pi_i),$$

$$\pi_i/(1 - \pi_i) = R_0^*[1 + \eta f(d)],$$

$$f(d) = \exp[-\alpha d^2],$$

and assess the probability that $\eta = 0$ using a variable selection method.

# Notes

1. The code for this model (model 2, Example 8.3) is

```
model { for (i in 1 : N) { y[i] ~dpois(mu[i]);
mu[i] <- nu[i]*P[i]; rate[i] <- 1000*nu[i]
LL[i] <- -mu[i]+y[i]*log(mu[i])-logfact(y[i]); G[i] <- 1/exp(LL[i])
# model for crime rate
log(nu[i]) <- alph+eps[i]}
# select area codes of areas adjacent to area i
for (i in 1:NN) {eps.neigh[i] <- eps[map[i]] }
# spatial effects
for (i in 1:N) {eps[i] ~dnorm(M[i],tau[i])
tau[i] <- tau.eps * (1-rho+rho*d[i])
M[i] <- rho*sum(eps.neigh[cum[i]+1:cum[i+1] ])/
(1-rho+rho*d[i])}
rho ~dunif(0,1); tau.eps ~dgamma(0.5,0.0005); alph ~dflat()
V.eps <- sd(eps[])*sd(eps[])
av.rate <- 1000*exp(alph+mean(eps[]))}
```

2. The code for this analysis is

```
model {for (i in 1:n) {for (t in 1:T) {
y[i,t] ~dpois(mu[i,t]); mu[i,t] <- P[i,t]*nu[i,t]
LL[i,t] <- y[i,t]*log(mu[i,t])-mu[i,t]-logfact(y[i,t]);
G[i,t] <- 1/exp(LL[i,t])
log(nu[i,t]) <- alph+omega[i]+psi.star[i,t]}}
# varying spatial effect
for (i in 1:nn) {wt[i] <- 1;
for (t in 1:T) { psi.adj[i,t] <- psi[adj[i],t]}}
for (i in 1:n) {for (t in 1:T){
psi.adj.M[i,t] <- mean(psi.adj[cum[i]+1:cum[i+1],t])
psi.prec[i,t] <- d.psi[i,t]*tau.psi
psi[i,t] ~dnorm(psi.M[i,t],psi.prec[i,t]);
psi.star[i,t] <- psi[i,t]-psi[i,1]}}
# persistent spatial effect
omega[1:n] ~car.normal(adj[],wt[],d[],tau.omega)
alph ~dnorm(0,0.001);
tau.psi ~dexp(1); tau.omega ~dexp(1);
# inferred time trend
delta[1] <- alph; for (t in 2:T) {delta[t] <- alph+mean(psi.star[,t])}
# conditional mean and precision weight by periods
for (i in 1:n) {psi.M[i,1] <- psi[i,2]+psi.adj.M[i,1]-psi.adj.M[i,2]
                d.psi[i,1] <- d[i];
for (t in 2:T-1){psi.M[i,t] <- (psi[i,t-1]+psi[i,t+1])/2+
```

```
psi.adj.M[i,t]  -  (psi.adj.M[i,t-1]+psi.adj.M[i,t+1])/2
                d.psi[i,t]  <-  2*d[i]}
psi.M[i,T]  <-  psi[i,T-1]+psi.adj.M[i,T]-psi.adj.M[i,T-1]
                d.psi[i,T]  <-  d[i]}}
```

3. The code for this model is

```
model {for (i in 1 : n) { for (t in 1:T) {
# model
nu[i,t]  <-  alph+sum(w.s[i,t,1:K])
for (k in 1:K) {w.s[i,t,k]  <-  w[i,t,k]*theta[k,t]
w[i,t,k]  <-  g[i,t,k]/sum(g[i,t,1:K])
g[i,t,k]  <-  exp(psi[t,k,i])}
# likelihood
y[i,t]  ~dpois(mu[i,t]);  mu[i,t]  <-  P[i,t]*exp(nu[i,t])
LL[i,t]  <-  y[i,t]*log(mu[i,t])-mu[i,t]-logfact(y[i,t]);
G[i,t]  <-  1/exp(LL[i,t])}}
alph  ~dnorm(0,0.001)
for (i in 1:nn) {wt[i]  <-  1}
for (k in 1:K) { for (t in 1:T) { tau.psi[t,k]  ~dexp(1)
theta.star[k,t]  <-  theta[k,t]-theta[k,1]
psi[t,k,1:n]  ~car.normal(adj[], wt[], d[], tau.psi[t,k])}}
for (k in 1:K) { tau.del[k]  ~dexp(1)
del[k,1]  ~dnorm(0,1)
for (t in 2:T) { del[k,t]  ~dnorm(del[k,t-1],tau.del[k])}}
for (t in 1:T) {theta[1,t]  <-  del[1,t]
for (k in 2:K) {theta[k,t]  <-  theta[k-1,t]+exp(del[k,t])}}}
```

# References

Abreu, M., de Groot, H. and Florax, R. (2005) Space and growth: a survey of empirical evidence and methods. *Région et Développement*, **21**, 13–44.

Anselin, L. (1988) *Spatial Econometrics: Methods and Models*. Kluwer Academic, Dordrecht.

Anselin, L. (2001) Spatial econometrics. In B. Baltagi (ed.), *A Companion to Theoretical Econometrics*, pp. 310–330. Blackwell, Oxford, UK.

Anselin, L. (2006) Spatial econometrics. In T.C. Mills and K. Patterson (eds), *Palgrave Handbook of Econometrics: Volume 1, Econometric Theory*, pp. 901–969. Palgrave Macmillan, Basingstoke, UK.

Anselin, L. (2010) Thirty years of spatial econometrics. *Papers in Regional Science*, **89**, 3–25.

Anselin, L. and Bera, A. (1998) Spatial dependence in linear regression models with an introduction to spatial econometrics. In A. Ullah and D. Giles (eds), *Handbook of Applied Economic Statistics*, pp. 237–289. Marcel Dekker, New York, NY.

Anselin, L. and Rey, S. (2010) *Perspectives on Spatial Data Analysis*. Springer, New York, NY.

Arbia, G. and Baltagi, B. (2009) *Spatial Econometrics: Methods and Applications*. Springer, New York, NY.

Assunção, R. and Krainski, E. (2009) Neighborhood dependence in Bayesian spatial models. *Biometrical Journal*, **51**(5), 851–869.

Assunção, R., Potter, J. and Cavenaghi, S. (2002) A Bayesian space varying parameter model applied to estimating fertility schedules. *Statistics in Medicine*, **14**, 2057–2076.

Auchincloss, A., Gebreab, S., Mair, C. and Diez Roux, A. (2012) A review of spatial methods in epidemiology, 2000–2010. *Annual Review of Public Health*, **33**, 107–22.

Bailey, T. and Gatrell, A. (1995) *Interactive Spatial Data Analysis*. Addison Wesley.

Bakar, K. and Sahu, S. (2013). spTimer: Spatio-temporal Bayesian modelling using R. *Technical Report*, University of Southampton. http://www.southampton.ac.uk/~sks/research/papers/spTimeRpaper.pd.

Baltagi, B. and Dong, L. (2002) Prediction in the panel data model with spatial correlation. In L. Anselin and R. Florax (eds), *Advances in Spatial Econometrics*. Springer-Verlag, Berlin.

Baltagi, B., Song, S., Jung, B. and Koh, W. (2007) Testing for serial correlation, spatial autocorrelation and random effects using panel data. *Journal of Econometrics*, **140**, 5–51.

Banerjee, S., Carlin, B. and Gelfand, A. (2004) *Hierarchical Modeling and Analysis for Spatial Data*. CRC, Boca Raton, FL.

Barufi, A., Haddad, E. and Paez, A. (2011) Infant mortality in Brazil, 1980–2000: a spatial panel data analysis. *BMC Public Health*, **12**, 181–, doi:10.1186/1471-2458-12-181.

Beale, L., Abellan, J., Hodgson, S. and Jarup, L. (2008) Methodologic issues and approaches to spatial epidemiology. *Environmental Health Perspectives*, **116**, 1105–1110.

Beckhurst, C., Evans, S., MacFarlane, A. and Packe, G. (2000) Factors influencing the distribution of tuberculosis cases in an inner London borough. *Communicable Disease and Public Health*, **3**, 28–31.

Bell, B. and Broemeling, L. (2000) A Bayesian analysis for spatial processes with application to disease mapping. *Statistics in Medicine*, **19**, 957–974.

Bernardinelli, L. and Montomoli, C. (1992) Empirical Bayes versus fully Bayesian analysis of geographical variation in disease risk. *Statistics in Medicine*, **11**, 983–1007.

Bernardinelli, L., Clayton, D. and Montomoli, C. (1995a) Bayesian estimates of disease maps: how important are priors? *Statistics in Medicine*, **14**, 2411–2432.

Bernardinelli, L., Clayton, D., Pascutto, C., Montomoli, C. and Ghislandi, M. (1995b) Bayesian analysis of space-time variations in disease risk. *Statistics in Medicine*, **11**, 983–1007.

Besag, J. and Kooperberg, C. (1995) On conditional and intrinsic autoregressions. *Biometrika*, **82**(4), 733–746.

Besag, J., York, J. and Mollie, A. (1991) Bayesian image restoration, with two applications in spatial statistics. *Annals of the Institute of Statistical Mathematics*, **43**, 1–20.

Biggeri, A. and Lagazio, C. (1999) Case-control analysis of risk around putative sources, In A. Lawson *et al.* (eds), *Disease Mapping and Risk Assessment for Public Health*, pp. 271–286. Wiley, Chichester, UK.

Breitenecker, R. and Harms, R. (2010) Dealing with spatial heterogeneity in entrepreneurship research. *Organizational Research Methods*, **13**, 176.

Brunsdon, C., Fotheringham, A. and Charlton, M. (1996) Geographically weighted regression: a method for exploring spatial nonstationarity. *Geographical Analysis*, **28**(4), 281–298.

Brunsdon, C., Fotheringham, A. and Charlton, M. (1998) Geographically weighted regression – modelling spatial non- stationarity. *Journal of the Royal Statistical Society D*, **47**(3), 431–443.

Calder, C. (2003) Exploring latent structure in spatial temporal processes using process convolutions. Ph.D. thesis, Duke University, Durham, NC 27708.

Calder, C. (2007) Dynamic factor process convolution models for multivariate space–time data with application to air quality assessment. *Environmental and Ecological Statistics*, **14**, 229–247.

Casetti, E. (1972) Generating models by the expansion method: applications to geographic research. *Geographical Analysis*, **4**, 81–91.

Casetti, E. (1992) Bayesian regression and the expansion method, *Geographical Analysis*, **24**, 58–74.

Casetti, E. (2010) Expansion method, dependency, and multimodeling. In M. Fischer and A. Getis (eds), *Handbook of Applied Spatial Analysis*, pp. 487–505. Springer, New York, NY.

Catelan, D. and Biggeri, A. (2008) A statistical approach to rank multiple priorities in environmental epidemiology: an example from high-risk areas in Sardinia, Italy. *Geospatial Health*, **3**, 81–9.

Catelan, D. and Biggeri, A. (2011) Selective inference in disease mapping. In B. Cafarelli (ed.), *Spatial2: Spatial Data Methods for Environmental and Ecological Processes*. Conference Proceedings, Spatial Data Methods for Environmental and Ecological Processes. Foggia, September 2011. http://hdl.handle.net/10446/25261

Catelan, D., Lagazio, C. and Biggeri, A. (2010) A hierarchical Bayesian approach to multiple testing in disease mapping. *Biometrical Journal*, **52**(6), 784–797.

Chakir, R. and Parent, O. (2009) Determinants of land use changes: a spatial multinomial probit approach. *Papers in Regional Science*, **88**, 327–344.

Clayton, D. (1996) Generalized linear mixed models. In W. Gilks, S. Richardson and D. Spiegelhalter (eds), *Markov Chain Monte Carlo in Practice*, pp. 275–301. Chapman and Hall, London, UK.

Congdon, P. (2001) Bayesian models for suicide monitoring. *European Journal of Population*, **15**, 1–34.

Congdon, P. (2002) A model for mental health needs and resourcing in small geographic areas: a multivariate spatial perspective. *Geographical Analysis*, **34**, 168–186.

Congdon, P. (2006) A model framework for mortality and health data classified by age, area and time. *Biometrics*, **61**, 269–278.

Congdon, P. (2007) Mixtures of spatial and unstructured effects for spatially discontinuous health outcomes. *Computational Statistics and Data Analysis*, **51**, 3197–3212.

Congdon, P. (2009) Modelling the impact of socioeconomic structure on spatial health outcomes. *Computational Statistics and Data Analysis*, **53**, 3047–3056.

Cressie, N. (1993) *Statistics for Spatial Data*. Wiley, Chichester, UK.

Cressie, N. and Wikle, C. (2011) *Statistics for Spatio-Temporal Data*. Wiley, New York, NY.

Debarsy, N., Ertur, C. and LeSage, J. (2012) Interpreting dynamic space–time panel data models. *Statistical Methodology*, **9**, 158–171.

De Graaff, T., Florax, R., Nijkamp, P. and Reggiani, A. (2001) A general misspecification test for spatial regression models: dependence, heterogeneity, and nonlinearity. *Journal of Regional Science*, **41**, 255–276.

De Oliveira, V. (2011) Bayesian analysis of conditional autoregressive models. *Annals of the Institute of Statistical Mathematics*, **64**(1), 107–133.

De Oliveira, V. and Song, J. (2008) Bayesian analysis of simultaneous autoregressive models. *Sankhya*, **70B**, 323–350.

Diggle, P.J. (1990) A point process modelling approach to raised incidence of a rare phenomenon in the vicinity of a prespecified point. *Journal of the Royal Statistical Society*, **153**, 349–362.

Diggle, P. and Ribeiro, P. (2002) Bayesian inference in Gaussian model based geostatistics. *Geographical and Environmental Modelling*, **6**, 129–146.

Diggle, P. and Ribeiro, P. (2007) *Model-based Geostatistics*. Springer, New York.

Diggle, P. and Rowlingson, B. (1994) A conditional approach to point process modelling of elevated risk. *Journal of the Royal Statistical Society A*, **157**, 433–440.

Diggle, P., Elliott, P., Morris, S. and Shaddick, G. (1997) Regression modelling of disease risk in relation to point sources. *Journal of the Royal Statistical Society A*, **160**, 491–505.

Diggle, P., Tawn, J. and Moyeed, R. (1998) Model-based geostatistics. *Journal of the Royal Statistical Society C*, **47**(3), 299–350.

Diggle, P., Morris, S. and Wakefield, J. (2000) Point-source modelling using matched case-control data. *Biostatistics*, **1**, 89–115.

Diggle, P., Ribeiro, P. and Christensen, O. (2003) An introduction to model-based geostatistics. In J. Møller (ed.), *Spatial Statistics and Computational Methods*, pp. 43–86. Springer-Verlag, Berlin.

Dreassi, E., Lagazio, C., Maule, M., Magnani, C. and Biggeri, A. (2008) Sensitivity analysis of the relationship between disease occurrence and distance from a putative source of pollution. *Geospatial Health*, **2**, 263–271.

Dubin, R. (1995) Estimating logit models with spatial dependence. In L. Anselin and R. Florax (eds), *New Directions in Spatial Econometrics*. Springer-Verlag, Berlin.

Ecker, M. and Gelfand, A. (1997) Bayesian variogram modeling for an isotropic spatial process. *Journal of Agricultural, Biological, and Environmental Statistics*, **2**, 347–369.

Elhorst, J. (2008) Serial and spatial error correlation. *Economics Letters*, **100**, 422–424.

Elhorst, J. (2010) Spatial panel data models. In M. Fischer and A. Getis (eds), *Handbook of Applied Spatial Analysis*, pp. 377–407. Springer, Berlin.

Elhorst, J. (2012) Dynamic spatial panels: models, methods, and inferences. *Journal of Geographical Systems*, **14**, 5–28.

Fortin, M.-J. and Dale, M. (2005) *Spatial Analysis: a Guide for Ecologists*. Cambridge University Press, Cambridge, UK.

Fotheringham, A., Brunsdon, C. and Charlton, M. (2000) *Quantitative Geography*. Sage, London, UK.

Franzese, R., Hays, J. and Schaeffer, L. (2010) Spatial, temporal, and spatiotemporal autoregressive probit models of binary outcomes: estimation, interpretation, and presentation. APSA 2010 Annual Meeting Paper. http://ssrn.com/abstract=1643867

Gaetan, C. and Guyon, X. (2009) *Spatial Statistics and Modeling*. Springer, Berlin.

Gamerman, D., Moreira, A. and Rue, H. (2003) Space-varying regression models: specifications and simulation. *Computational Statistics and Data Analysis*, **42**, 513 – 533.

Gaudard, M., Karson, M., Linder, E. and Sinha, D. (1999) Bayesian spatial prediction. *Environmental and Ecological Statistics*, **6**, 147–171.

Gelfand, A. (1996) Model determination using sampling-based methods. In W. Gilks, S. Richardson and D. Spiegelhalter (eds), *Markov Chain Monte Carlo in Practice*, pp. 145–161. Chapman Hall, London, UK.

Gelfand, A., Ghosh, S., Knight, J. and Sirmans, C. (1998) Spatio-temporal modeling of residential sales markets. *Journal of Business and Economic Statistics*, **16**, 312–321.

Gelfand, A., Kim, H.-J., Sirmans, C. and Banerjee, S. (2003) Spatial modeling with spatially varying coefficient processes. *Journal of the American Statistical Association*, **98**, 387–396.

Gelfand, A., Diggle, P., Guttorp, P. and Fuentes, M. (2010) *Handbook of Spatial Statistics*. CRC Press, Boca Raton, FL.

Getis, A. (1989) A spatial association model approach to the identification of spatial dependence. *Geographical Analysis*, **21**, 251–259.

Getis, A. (1995) Spatial filtering in a regression framework. In L. Anselin and R. Florax (eds), *New Directions in Spatial Econometrics*, pp. 172–185. Springer, Berlin.

Getis, A. (2010) Spatial filtering in a regression framework; examples using data on urban crime, regional inequality, and government expenditures. In M. Fischer, G. Hewings, P. Nijkamp and F. Snickars (eds), *Advances in Spatial Science*, pp. 191–202. Springer, Berlin.

Getis, A. and Aldstadt, J. (2004) Constructing the spatial weights matrix using a local statistic. *Geographical Analysis*, **36**, 90–104.

Getis, A. and Griffith, D. (2002) Comparative spatial filtering in regression analysis. *Geographical Analysis*, **34**, 130–140.

Getis, A., Mur, J. and Zoller, H. (2004) *Spatial Econometrics and Spatial Statistics*. Palgrave Macmillan, Basingstoke, UK.

Goovaerts, P. (2009) Medical geography: a promising field of application for geostatistics. *Mathematical Geosciences*, **41**, 243–264.

Goovaerts, P. and Gebreab, S. (2008) How does Poisson kriging compare to the popular BYM model for mapping disease risks? *International Journal of Health Geographics*, **7**, 6.

Graham, A., Atkinson, P. and Danson, F. (2004) Spatial analysis for epidemiology. *Acta Tropica*, **91**, 219–225.

Haining, R. (2003) *Spatial Data Analysis: Theory and Practice*. Cambridge University Press, Cambridge, UK.

Handcock, M. and Stein, M. (1993) A Bayesian analysis of Kriging. *Technometrics*, **35**, 403–410.

Higdon, D. (2007) A primer on space-time modelling from a Bayesian perspective. In Finkelstadt, Held and Isham (eds), *Statistical Methods for Spatio-Temporal Systems*. CRC, Boca Raton, FL.

Higdon, D., Swall, J. and Kern, J. (1999) Non-stationary spatial modeling. In *Bayesian Statistics 6*, pp. 761–768. Oxford University Press, Oxford, UK.

Hill, E., Ding, L. and Waller, L. (2000) A comparison of three tests to detect general clustering of a rare disease in Santa Clara County, California. *Statistics in Medicine*, **19**, 1363–1378.

Holloway, G., Shankar, B. and Rahman, S. (2002) Bayesian spatial probit estimation: a primer and an application to HYV rice adoption. *Agricultural Economics*, **27**, 383–402.

Hossain, M. and Lawson, A. (2006) Cluster detection diagnostics for small area health data. *Statistics in Medicine*, **25**, 771–786.

Ismaila, A., Canty, A. and Thabane, L. (2007) Comparison of Bayesian and frequentist approaches in modelling risk of preterm birth near the Sydney Tar Ponds, Nova Scotia, Canada. *BMC Medical Research Methodology*, **7**, 39.

Jaimovich, D. (2012) A Bayesian spatial probit estimation of free trade agreement contagion. *Applied Economics Letters*, **19**, 579–583.

Jerrett, M., Gale, S. and Kontgis, C. (2010) Spatial modeling in environmental and public health research. *International Journal of Environmental Research and Public Health*, **7**, 1302–1329.

Kakamu, K. and Wago, H. (2007) Bayesian spatial panel probit model with an application to business cycle in Japan. Working paper. http://www.mssanz.org.au/modsim05/proceedings/papers/kakamu_2.pdf

Kakamu, K. and Wago, H. (2008) Small-sample properties of panel spatial autoregressive models: comparison of the Bayesian and maximum likelihood methods. *Spatial Economic Analysis*, **3**, 305–319.

Kelsall, J. and Wakefield, J. (1999) Discussion of Best *et al.* 1999. In J. Bernardo, J. Berger, A. Dawid and A. Smith (eds), *Bayesian Statistics 6*, p. 151. Oxford University Press, Oxford, UK.

Kelsall, J. and Wakefield, J. (2002) Modeling spatial variation in disease risk: a geostatistical approach. *Journal of the American Statistical Association*, **97**(459), 692–701.

Kirby, R., Liu, J., Lawson, A., Choi, J., Cai, B. and Hossain, M. (2011) Spatio-temporal patterning of small area low birth weight incidence and its correlates: a latent spatial structure approach. *Spatial and Spatiotemporal Epidemiology*, **2**, 265–271.

Kissling, W. and Carl, G. (2008) Spatial autocorrelation and the selection of simultaneous autoregressive models. *Global Ecology and Biogeography*, **17**, 59–71.

Knorr-Held, L. (2000) Bayesian modelling of inseparable space-time variation in disease risk. *Statistics in Medicine*, **19**, 2555–2567.

Knorr-Held, L. and Rasser, G. (2000) Bayesian detection of clusters and discontinuities in disease maps. *Biometrics*, **56**, 13–21.

Kyriakidis, P. and Nagle, N. (2010). Geostatistical regression for areal data. In N. Tate and P. Fisher (eds), *9th International Symposium on Spatial Accuracy Assessment in Natural Resources and Environmental Sciences*. International Spatial Accuracy Research Association.

Lagazio, C., Dreassi, E. and Bernardinelli, A. (2001) A hierarchical Bayesian model for space-time variation of disease risk. *Statistical Modelling*, **1**(17), 29.

Langford, I., Leyland, A., Rasbash, J. and Goldstein, H. (1999) Multilevel modelling of the geographical distributions of rare diseases. *Journal of the Royal Statistical Society C*, **48**, 253–268.

Lawson, A. (2001) *Statistical Methods in Spatial Epidemiology*. Wiley, Chichester, UK.

Lawson, A. (2009) *Bayesian Disease Mapping: Hierarchical Modeling in Spatial Epidemiology*, 1st edn. CRC Press, New York, NY.

Lawson, A. and Clark, A. (2002) Spatial mixture relative risk models applied to disease mapping. *Statistics in Medicine*, **21**, 359–370.

Lawson, A. and Kulldorff, M. (2000) A review of cluster detection methods. In A. Lawson, A. Biggeri, D. Bohning, E. Lesaffre, J. Viel and R. Bertollini (eds), *Disease Mapping and Risk Assessment for Public Health*, chapter 7. Wiley, New York, NY.

Lawson, A. and Williams, F. (1994) Armadale: a case study in environmental epidemiology. *Journal of the Royal Statistical Society*, **157A**, 285–298.

Lawson, A., Biggeri, A., Lesaffre, E., Viel, J. and Bertollini, B. (1999) *Disease Mapping and Risk Assessment for Public Health*, Wiley, Chichester, UK.

Lawson, A., Biggeri, A. and Williams, F. (2000) A review of modelling approaches in health risk assessment around putative sources. In A. Lawson, A. Biggeri, D. Bohning, E. Lesaffre *et al.* (eds) *Disease Mapping and Risk Assessment for Public Health*. Wiley, Chichester, UK.

Lawson, A., Simeon, S., Kulldorff, M., Biggeri, A. and Magnani, C. (2007) Line and point cluster models for spatial health data. *Computational Statistics and Data Analysis*, **51**, 6027–6043.

Lawson, A., Song, H., Cai, B., Hossain, M. and Huang, K. (2011) Space-time latent component modeling of geo-referenced health data. *Statistics in Medicine*, **29**(19), 2012–27.

Lee, D. (2011) A comparison of conditional autoregressive models used in Bayesian disease mapping. *Spatial and Spatio-temporal Epidemiology*, **2**, 79–89.

Lee, H., Higdon, D., Calder, C. and Holloman, C. (2005) Efficient models for correlated data via convolutions of intrinsic processes. *Statistical Modelling*, **5**, 53–74.

Leroux, B., Lei, X. and Breslow, N. (1999) Estimation of disease rates in small areas: a new mixed model for spatial dependence. In M. Halloran and D. Berry (eds), *Statistical Models in Epidemiology, the Environment and Clinical Trials*, pp. 135–78. Springer-Verlag, New York, NY.

Lesage, J. (1997) Bayesian estimation of spatial autoregressive models. *International Regional Science Review*, **20**, 113–130.

Lesage, J. (1998) Spatial econometrics. http://www.spatial-econometrics.com/html/wbook.pdf

Lesage, J. (2000) Bayesian estimation of limited dependent variable spatial autoregressive models. *Geographical Analysis*, **32**(1), 19–35.

Lesage, J. (2001) A family of geographically weighted regression models. Manuscript, Department of Economics, University of Toledo.

LeSage, J. (2008) An introduction to spatial econometrics. *Revue d'Economie Industrielle*, **123**, 19–44.

LeSage, J. and Pace, R. (2009) *Introduction to Spatial Econometrics*. CRC, Boca Raton, FL.

Leyland, A., Langford, I., Rasbash, J. and Goldstein, H. (2000) Multivariate spatial models for event data. *Statistics in Medicine*, **19**(17–18), 2469–2478.

Liao, Y., Wang, J., Wu, J., Wang, J. and Zheng, X. (2010) A comparison of methods for spatial relative risk mapping of human neural tube defects. *Stochastic Environmental Research and Risk Assessment*, **25**, 99–106.

López-Hernandez, F., Mur, J. and Angulo, A. (2010) Local estimation of spatial autocorrelation processes. In A. Páez, J. Le Gallo and R. Buliung (eds), *Progress in Spatial Analysis: Advances in Spatial Science*, Part 1, pp. 93–116. Springer, New York, NY.

Mardia, K. (1988) Multi-dimensional multivariate Gaussian Markov random fields with application to image processing. *Journal of Multivariate Analysis*, **24**, 265–284.

Mardia, K. and Marshall, R. (1984) Maximum likelihood estimation of models for residual covariance in spatial regression. *Biometrika*, **71**, 135–146.

Marshall, E. and Spiegelhalter, D. (2007) Identifying outliers in Bayesian hierarchical models: a simulation-based approach. *Bayesian Analysis*, **2**(2), 409–444.

Martinez-Beneito, M., Lopez-Quilez, A. and Botella-Rocamora, P. (2008) An autoregressive approach to spatio-temporal disease mapping. *Statistics in Medicine*, **27**, 2874–2889.

Maule, M., Magnani, C., Dalmasso, P., Mirabelli, D., Merletti, F. and Biggeri, A. (2007) Modeling mesothelioma risk associated with environmental asbestos exposure. *Environmental Health Perspectives*, **115**, 1066–1071.

Messner, S. and Anselin, L. (2004) Spatial analyses of homicide with areal data. In M. Goodchild and D. Janelle (eds), *Spatially Integrated Social Science*, pp. 127–144. Oxford University Press, Oxford, UK.

Militino, A. and Ugarte, M. (2001) Assessing the covariance function in geostatistics. *Statistics and Probability Letters*, **52**, 199–206.

Militino, A., Ugarte, M. and Dean, C. (2001) The use of mixture models for identifying high rates in disease mapping. *Statistics in Medicine*, **20**, 2035–2049.

Mollié, A. (1996) Bayesian mapping of disease. In W. Gilks, S. Richardson and D.J. Spiegelhalter (eds), *Markov Chain Monte Carlo in Practice*, pp. 359–379. Chapman & Hall, London, UK.

Mollié, A. (2001) Bayesian mapping of Hodgkin's disease in France. In M. Elliott, J. Wakefield, N. Best and D. Briggs (eds), *Spatial Epidemiology: Methods and Applications*, pp. 267–286. Oxford University Press, Oxford, UK.

Morris, S.E. and Wakefield, J.C. (2000). Assessment of disease risk in relation to a pre-specified source. In M. Elliott, J. Wakefield, N. Best and D. Briggs (eds), *Spatial Epidemiology: Methods and Applications*, pp. 153–184. Oxford University Press, Oxford, UK.

Neumayer, E. and Pluemper, T. (2009) Spatial effects in dyadic data. *International Organization*, **64**, 145–166.

Oliver, M., Webster, R., Lajaunie, C., Muir, K., Parkes, S., Cameron, A., Stevens, M. and Mann, J. (1998) Binomial cokriging for estimating and mapping the risk of childhood cancer. *IMA Journal of Mathematics Applied in Medicine and Biology*, **15**, 279–297.

Ord, J. and Getis, A. (1995) Local spatial autocorrelation statistics – distributional issues and an application. *Geographical Analysis*, **27**(4), 286–230.

Pace, R. and LeSage, J. (2010) Spatial Econometrics. In A. Gelfand, P. Diggle, M. Fuentes and P. Guttorp (eds), *Handbook of Spatial Statistics*. CRC, Boca Raton, FL.

Páez, A. and Scott, D. (2004) Spatial statistics for urban analysis: a review of techniques with examples. *GeoJournal*, **61**, 53–67.

Páez, A., Farber, S. and Wheeler, D. (2011) A simulation-based study of geographically weighted regression as a method for investigating spatially varying relationships. *Environment and Planning A*, **43**, 2992–3010.

Parent, O. and LeSage, J. (2008) Using the variance structure of the conditional autoregressive spatial specification to model knowledge spillovers. *Journal of Applied Econometrics*, **23**, 235–256.

Pfeiffer, D., Robinson, T., Stevenson, M., Stevens, K., Rogers, D. and Clements, A. (2008) *Spatial Analysis in Epidemiology*. Oxford University Press, Oxford, UK.

Pilz, J. and Spöck, G. (2008) Why do we need and how should we implement Bayesian kriging methods. *Stochastic Environmental Research and Risk Assessment*, **22**, 621–632.

Plummer, M. (2008) Penalized loss functions for Bayesian model comparison. *Biostatistics*, **9**(3), 523–539.

Reich, B. and Fuentes, M. (2007) A multivariate semiparametric Bayesian spatial modeling framework for hurricane surface wind fields. *Annals of Applied Statistics*, **1**, 249–264.

Reilly, C. and Gelman, A. (2007) Weighted variogram estimation for data with clustering. *Technometrics*, **49**, 184–195.

Richardson, S. (1992) Statistical methods for geographical correlation studies. In P. Elliott, J. Cuzick, D. English and R. Stern (eds), *Geographical and Environmental Epidemiology: Methods for Small Area Studies*. Oxford University Press, Oxford, UK.

Richardson, S. and Monfort, C. (2000) Ecological correlation studies. In P. Elliott, J. Wakefield, N. Best and D. Briggs (eds), *Spatial Epidemiology: Methods and Applications*, pp. 205–220. Oxford University Press, Oxford, UK.

Richardson, S., Guihenneuc, C. and Lasserre, V. (1992) Spatial linear models with autocorrelated error structure. *The Statistician*, **41**, 539–557.

Ripley, B. (2004) *Spatial Statistics*, 2nd edn. Wiley, Chichester, UK.

Rodrigues, A. and Assuncao, R. (2008) Propriety of posterior in Bayesian space varying parameter models with normal data. *Statistics and Probability Letters*, **78**, 2408–2411.

Rogerson, P. (2005) A set of associated statistical tests for spatial clustering. *Environmental and Ecological Statistics*, **12**, 275–288.

Rogerson, P. (2006) Statistical methods for the detection of spatial clustering in case–control data. *Statistics in Medicine*, **25**, 811–823.

Rogerson, P. (2010) Health surveillance around prespecified locations using case-control data. In M. Fischer, G. Hewings, P. Nijkamp and F. Snickars (eds), *Advances in Spatial Science*, pp. 181–188. Springer, Berlin.

Rue, H. and Held, L. (2005). *Gaussian Markov Random Fields: Theory and Applications*. CRC, Boca Raton, FL.

Sanso, B., Schmidt, A. and Nobre, A. (2008) Spatio-temporal models based on discrete convolutions. *Canadian Journal of Statistics*, **36**, 239–258.

Sauleau, E. A., Hennerfeind, A., Buemi, A. and Held, L. (2007) Age, period and cohort effects in Bayesian smoothing of spatial cancer survival with geoadditive models. *Statistics in Medicine*, **26**, 212–229.

Schrödle, B. and Held, L. (2011) A primer on disease mapping and ecological regression using INLA. *Computational Statistics*, **26**, 241–258.

Seya, H., Tsutsumi, M. and Yamagata, Y. (2012) Income convergence in Japan: A Bayesian spatial Durbin model approach. *Economic Modelling*, **29**, 60–71.

Shi, W. (2009) *Principles of Modeling Uncertainties in Spatial Data and Spatial Analyses*. CRC Press/Taylor & Francis, Boca Raton, FL.

Shively, T., Kohn, R. and Wood, S. (1999) Variable selection and function estimation in additive nonparametric regression using a data-based prior. *Journal of the American Statistical Association*, **94**, 777–794.

Smith, T. and LeSage, J. (2004) A Bayesian probit model with spatial dependencies. In J. Lesage and R. Kelley Pace (eds), *Spatial and Spatiotemporal Econometrics (Advances in Econometrics, Volume 18)*, pp 127–160. Emerald Group Publishing, Bradford, UK.

Song, J.J., Ghosh, M., Miaou, S. and Mallick, B. (2006) Bayesian multivariate spatial models for roadway traffic crash mapping. *Journal of Multivariate Analysis*, **97**(1), 246–273.

Stern, H. and Cressie, N. (2000) Posterior predictive model checks for disease mapping models. *Statistics in Medicine*, **19**, 2377–2397.

Sun, D., Tsutakawa, R. and Speckman, P. (1999) Posterior distribution of hierarchical models using CAR(1) distributions. *Biometrika*, **86**, 341–350.

Tango, T. (2000) Comparison of general tests for spatial clustering. In A. Lawson *et al.*, *Disease Mapping and Risk Assessment for Public Health*, pp. 111–118. Wiley, Chichester, UK.

UCLA Statistical Consulting Group (2013) http://www.ats.ucla.edu/stat/r/faq/variogram.htm

Ventrucci, M., Scott, E. and Cocchi, D. (2011) Multiple testing on standardized mortality ratios: a Bayesian hierarchical model for FDR estimation. *Biostatistics*, **12**(1), 51–67.

Voss, P., Long, D., Hammer, R. and Friedman, S. (2006) County child poverty rates in the U.S.: a spatial regression approach. *Population Research and Policy Review*, **25**, 369–391.

Wakefield, J. and Morris, S. (2001) The Bayesian modeling of disease risk in relation to a point source. *Journal of the American Statistical Association*, **96**, 77–91.

Wakefield, J., Best, N. and Waller, L. (2000) Bayesian approaches to disease mapping. In P. Elliott, J. Wakefield, N. Best and D. Briggs (eds), *Spatial Epidemiology: Methods and Applications*, pp. 104–127. Oxford University Press, Oxford, UK.

Wall, M. (2004) A close look at the spatial structure implied by the CAR and SAR models. *Journal of Statistical Planning and Inference*, **121**, 311–324.

Waller, L. and Gotway, C. (2004) *Applied Spatial Statistics for Public Health Data*. Wiley, Chichester, UK.

Ward, M. and Gleditsch, K. (2008) *Spatial Regression Models*. Sage, London, UK.

Wheeler, D. and Tiefelsdorf, M. (2005) Multicollinearity and correlation among local regression coefficients in geographically weighted regression. *Journal of Geographical Systems*, **7**, 161–187.

Wikle, C., Berliner, L. and Cressie, N. (1998) Hierarchical Bayesian space-time models. *Environmental and Ecological Statistics*, **5**, 117–154.

Wilhelm, S. and Matos, M. (2013) Estimating spatial probit models in R. *The R Journal*, **5**, 130–143.

Yan, J. (2007) Spatial stochastic volatility for lattice data. *Journal of Agricultural, Biological, and Environmental Statistics*, **12**, 25–40.

# 9

# Latent variable and structural equation models

## 9.1 Introduction

Structural equation models (or SEM methods) describes multiple equation representations that include latent or unmeasured variables ('factors', 'constructs', or 'domains') for which multiple observed indicators (or 'items') are available (e.g. Kelloway, 1995; Bollen, 1998; Tu, 2009; Kline, 2011; Ullman and Bentler, 2012). The intention is often the representation of causal relationships (Song and Lee, 2012, Section 2.2.2). Particular types of SEM include confirmatory and explanatory factor analysis (Hurley *et al.*, 1997), latent class analysis and item response models (Fox, 2010).

Such methods have found a major application in areas such as psychology, education, marketing and sociology where underlying constructs (depression, product appeal, teacher style, anomie, authoritarianism, etc.) are not possible to measure directly. Among issues that often occur are the treatment of non-normality or discrete manifest indicators (Muthen, 1984; Quinn, 2004), missing data (Muthen *et al.*, 1987; Kamakura and Wedel, 2000; Allison, 2003), and invariance of measurement structures (Meredith and Teresi, 2006).

Bayesian approaches to factor analysis and structural equation models are discussed by Song and Lee (2012), Lee (2007), Muthen and Asparouhov, (2012), MacCallum *et al.* (2012), Palomo *et al.* (2007), and Ghosh and Dunson (2009), with applications exemplified by Arhonditsis *et al.* (2006), Wu *et al.* (2010) and Ho and Quinn (2008). The applications may be to cross-sectional data, multilevel data (Goldstein and Browne, 2002), to time series and longitudinal data models, and spatially configured data. Structural equation and factor model implementations in R include the packages sem, lavaan, ltm and tsfa (Fox, 2002; Rizopoulos, 2006; Rosseel, 2012), while specifically Bayesian approaches are adopted in the BFRM package (http://www.stat.duke.edu/research/software/west/bfrm/), the BFA package (http://cran.r-project.org/web/packages/bfa/bfa.pdf), the MCMCpack package (http://cran.r-project.org/web/packages/MCMCpack/MCMCpack.pdf), and MLIRT (Fox, 2010).

Generally the observed data on a large number of indicators are used to define (i.e. serve as proxies for) the underlying constructs, which are usually stipulated as fewer in number

*Applied Bayesian Modelling*, Second Edition. Peter Congdon.
© 2014 John Wiley & Sons, Ltd. Published 2014 by John Wiley & Sons, Ltd.

than the observed variables to ensure identifiability. One might assume that conditional on the constructs, the observed indicators are independent (e.g. Bartholomew *et al.*, 2008, Section 7.3), so that the constructs account for the observed correlations between the indicators (the conditional independence or local independence property). However, as Bollen (2002) points out, this is not an intrinsic feature of structural equation models, and a common alternative is to allow conditional residual correlations within subsets of the observed indicators (Stern and Jeon, 2004).

## 9.2   Normal linear structural equation models

### 9.2.1   Cross-sectional normal SEMs

Structural equation models include both measurement models, typically confirmatory factor analysis models, confined to representing the constructs as functions of the indicators, and models including both measurement and structural components, with the latter expressing interdependence between the constructs. The canonical structural equation model takes the LISREL form (Joreskog, 1973; Zhu and Lee, 2001; Ghosh and Dunson, 2009) with a structural model relating endogenous constructs $\psi$ to each other, and to exogenous constructs $\xi$, and a measurement model linking observed indicators $Y$ and $X$ to the latent variables.

Thus for subjects $i = 1, \ldots, n$, the structural model is

$$\psi_i = A + B\psi_i + \Gamma\xi_i + w_i$$

where $\psi_i$ is a $p \times 1$ vector of endogenous constructs, $\xi_i$ is a $q \times 1$ vector of exogenous constructs, $w_i$ is a $p \times 1$ vector of normally distributed errors on the endogenous constructs, $B$ is a $p \times p$ parameter matrix describing interrelations between endogenous constructs, $\Gamma$ is a $p \times q$ parameter matrix describing the effect of exogenous on endogenous constructs, and $A$ is a $p \times 1$ intercept. A structural model may also include measured explanatory indicators $Z_i$ of dimension $r$, sometimes known as formative indicators (Bollen and Bauldry, 2011), as in

$$\psi_i = A + B\psi_i + \Gamma\xi_i + HZ_i + w_i.$$

The links between observed indicators as responses and the constructs as latent predictors are defined by the measurement model:

$$Y_i = \kappa_Y + \Lambda_Y\psi_i + \epsilon_i,$$
$$X_i = \kappa_X + \Lambda_X\xi_i + \delta_i,$$

where both sets of errors are taken to be normal, $Y_i$ is an $M \times 1$ vector of indicators describing the endogenous construct vector $\psi$, and $X_i$ is an $L \times 1$ vector of indicators that proxy the exogenous construct vector $\xi$. The vectors $\kappa_Y$ and $\kappa_X$ are $M \times 1$ and $L \times 1$ intercepts, and $\Lambda_Y$ and $\Lambda_X$ are $M \times p$ and $L \times q$ matrices of loading coefficients on the scores $(\psi_i, \xi_i)$ of an individual on the latent constructs.

Many applications involve simply a measurement model, without any distinction between endogenous and exogenous factors, so that

$$Y_i = \kappa + \Lambda\psi_i + \epsilon_i, \tag{9.1}$$

where $\kappa$ is $M \times 1$, $\Lambda$ is $M \times p$, and the $\epsilon_i$ are normal. Constraints are applied on loadings and factor variances or both to ensure identifiability (see Section 9.2.2 for an example).

Since the scale of the latent variables is unknown, a common constraint involves assuming standardised factors with scale unity (Lopes and West, 2004), though an alternative is a unit loading constraint for one among the set of loadings relating a group of indicators to a particular construct (Bartholomew *et al.*, 2008, p. 296; Song and Lee, 2012, p. 22). For $p > 1$, identifiability also requires that $\Lambda$ under (9.1) be of full-rank lower triangular form (Ghosh and Dunson, 2009), implying $p(p-1)/2$ additional constraints, usually zero constraints, for example, setting the loading of $y_1$ on $\psi_2$ at 0, the loadings of $y_1$ and $y_2$ on $\psi_3$ at 0, etc.

Under a standardised factors assumption, there are then up to $K = Mp - p(p-1)/2$ unknown parameters in $\Lambda$ and $\text{cov}(\psi)$, collectively encompassing the latent model parameters. The latent model must be specified such that $K \leq M(M+1)/2 - M$, namely the total available parameters under a conditional independence assumption whereby

$$\text{cov}(\epsilon) = \Sigma = \text{diag}(\sigma_1^2, \sigma_2^2, \ldots, \sigma_M^2).$$

If additionally the constructs are assumed to be normally distributed, and uncorrelated such that $\text{cov}(\psi) = I_p$, the $Y_i$ have a marginal normal form

$$Y_i \sim N(\kappa, \Lambda\Lambda' + \Sigma).$$

However, if, as under unit-loading constraints, factor variances or covariances are unknown with $\text{cov}(\psi) = \Phi$, one has

$$Y_i \sim N(\kappa, \Lambda\Phi\Lambda' + \Sigma).$$

This framework may be extended to data $y_{img}$ stratified by group $g$, or to multiple items $y_{imt}$ observed over subjects, variables and times $t = 1, \ldots, T$ (Section 9.3).

As for all hierarchical models involving random effects, diffuse priors (e.g. on unknown factor variances or loadings) may impede identifiability and MCMC convergence. If postulated linkages from latent variables to indicators are defined parsimoniously on subject matter grounds, as in a confirmatory factor analysis where $\Lambda$ has a non-overlapping or 'simple structure', this may assist identification (Song and Lee, 2012, Section 2.2.4). Often a scaling of the observed indicators (e.g. centred or standardized $X$ and $Y$) is useful also in identification and convergence.

Priors on loadings may be defined to ensure consistent labelling (see Section 9.2.2). Assuming normal priors

$$\lambda_{mk} \sim N(\lambda_{0mk}, 1/\Delta_{mk}),$$

then relatively high precisions, e.g. $\Delta_{mk} = 1$ or $\Delta_{mk} = 2$, may be adopted if there is confidence that prior means $\lambda_{0mk}$ are not too far away from the true values (Song and Lee, 2012, Section 3.2.2). Informative priors may also be based on expert elicitation (Palomo *et al.*, 2007). Directional constraints may be imposed, as in item response modelling where a log-normal prior is adopted for item loadings (i.e. discrimination parameters) on latent ability (Sinharay, 2004).

Other possibilities oriented to parsimony are the automatic relevance determination (ARD) prior (Neal, 1996), whereby for the loading linking indicator $m$ to factor $k$

$$\lambda_{mk} \sim N(0, 1/\Delta_{mk}),$$

$$\Delta_{mk} \sim \text{Ga}(a_0, b_0)$$

with possibly diffuse settings for $a_0$ and $b_0$. Loading selection may be implemented using spike-slab priors (Knowles and Ghahramani, 2011), whereby

$$\lambda_{mk} \sim D_{mk}N(0, 1/\Delta_{mk}) + (1 - D_{mk})\delta_0(\lambda_{mk}),$$

$$D_{mk} \sim \text{Bern}(\pi_k),$$

$$\pi_k \sim \text{Be}(c_0, d_0),$$

$$\Delta_{mk} \sim \text{Ga}(a_0, b_0),$$

and $\delta_0$ is a delta function (point-mass) at 0.

## 9.2.2    Identifiability constraints

As an example of identifiability constraints in practice, suppose $M = 5$ continuous indicators $y_{i1}, \ldots, y_{i5}$ are available, under a normal linear model as in 9.1, to be measures of $p = 2$ constructs $\psi_1$ and $\psi_2$. There are a maximum of $M(M + 1)/2 = 15$ parameters available to represent the residual covariance matrix in the five indicator regressions (for $y$ variables), the indicator-construct loadings, and the construct covariances (Bartholomew *et al.*, 2008, p. 296). The indicator-construct loadings and the construct covariances may be denoted as the latent model parameters.

The indicator regressions specify the linear relations

$$y_{im} = \kappa_m + \lambda_{m1}\psi_{i1} + \lambda_{m2}\psi_{i2} + \epsilon_{im},$$

with residuals assumed normal with iid errors $\epsilon_{im} \sim N(0, \sigma_m^2)$, under a conditional independence assumption.

There are then a maximum of $M(M + 1)/2 - M = 10$ remaining parameters for defining indicator-construct loadings and construct covariances.

Suppose a simple structure confirmatory model is adopted with indicators $y_1, y_2, y_3$ (measures of spatial ability) having non-zero loadings $\lambda_1, \lambda_2$, and $\lambda_3$ on construct $\psi_1$, while indicators $y_4$ and $y_5$ (measures of linguistic skills) have non-zero loadings $\lambda_4, \lambda_5$ on $\psi_2$. It follows from simple structure that there are no loadings of $y_1, y_2$ and $y_3$ on $\psi_2$, or of $y_4$ and $y_5$ on $\psi_1$.

Since $\psi_1$ and $\psi_2$ have arbitrary location and scale, one constraint scheme to gain identifiability is the unit factor variance constraint (Ghosh and Dunson, 2009). So if correlation between the two constructs is allowed, $\psi_i = (\psi_{i1}, \psi_{i2})$ might be taken to be bivariate normal, with variances 1 in the diagonal of the covariance matrix, but with an unknown off-diagonal correlation parameter.

Under this predefined scale option, the loadings $\lambda_1, \lambda_2$ and $\lambda_3$ relating $y_1, y_2$ and $y_3$ to $\psi_1$, and $\lambda_4$ and $\lambda_5$ relating the verbal test scores to $\psi_2$, are unknowns. So

$$\psi_i \sim N_2, \left( \begin{bmatrix} 0 \\ 0 \end{bmatrix}, \begin{bmatrix} 1 & \rho \\ \rho & 1 \end{bmatrix} \right),$$

$$y_{i1} = \kappa_1 + \lambda_1\psi_{i1} + \epsilon_{i1},$$

$$y_{i2} = \kappa_2 + \lambda_2\psi_{i1} + \epsilon_{i2},$$

$$y_{i3} = \kappa_3 + \lambda_3\psi_{i1} + \epsilon_{i3},$$

$$y_{i4} = \kappa_4 + \lambda_4\psi_{i2} + \epsilon_{i4},$$

$$y_{i5} = \kappa_5 + \lambda_5\psi_{i2} + \epsilon_{i5}.$$

The inter-construct correlation $\rho$ is identified because each indicator loads on only one construct under a simple structure. In total, there are six unknowns in the latent model against a maximum of 10 available.

Since the constructs are intended to summarise information in the indicators relatively informative priors, e.g. $N(1, 1)$ or $N(0, 1)$, may be used for the loadings (e.g. Johnson and Albert, 1999; Song and Lee, 2012, p. 39). This form of assumption, however, does not preclude shifts in labels. Additionally constraining one or more loadings on each of $\psi_1$ and $\psi_2$ to be positive, as under exponential priors such as

$$\lambda_1 \sim E(1); \lambda_4 \sim E(1),$$

assists in ensuring consistent labelling of the factors. The labelling issue arises because a construct and associated loadings may simply switch sign, with the product $\lambda_{mk}\psi_{ik}$ indistinguishable from the product $(-\lambda_{mk})(-\psi_{ik})$ (Geweke and Zhou, 1996).

An alternative constraint scheme fixes the scale of the constructs by selecting one loading corresponding to each factor – here one among the loadings $\{\lambda_1, \lambda_2, \lambda_3\}$ and one among $\{\lambda_4, \lambda_5\}$ – and setting them to a predetermined non-zero value, usually 1 (Song and Lee, 2012, pp. 21–22). The variances of $\psi_1$ and $\psi_2$ are then free parameters and related to variances of observed indicators. For example, suppose $\lambda_1 = \lambda_4 = 1$, so that

$$y_{i1} = \kappa_1 + \psi_{i1} + \epsilon_{i1},$$
$$y_{i2} = \kappa_2 + \lambda_2\psi_{i1} + \epsilon_{i2},$$
$$y_{i3} = \kappa_3 + \lambda_3\psi_{i1} + \epsilon_{i3},$$
$$y_{i4} = \kappa_4 + \psi_{i2} + \epsilon_{i4},$$
$$y_{i5} = \kappa_5 + \lambda_5\psi_{i2} + \epsilon_{i5},$$

with the $\psi$ again being bivariate normal with zero means, but now with all three parameters in cov($\psi$) being unknown. There are again six unknowns.

The unit loading constraint has utility in assisting against labelling shifts of the construct scores $\psi_1$ during MCMC sampling. Since $y_1 - y_3$ in this example are positive measures of spatial ability, setting $\lambda_1 = 1$ means the construct $\psi_1$ will generally be estimated as a positive measure of this ability. However, when a large number of indicators load on a particular factor, additional constraints may be needed for consistent labelling.

### Example 9.1    Structural equation model for democracy and development

A dataset considered by Bollen (1989) and Palomo *et al.* (2007) is the basis of a normal linear structural model relating democracy (endogenous) to industrial development (exogenous). There are 75 observations on 11 indicators, with eight being indicators of democracy in 1960 and of democracy in 1965 (the endogenous constructs, denoted $\psi_1$ and $\psi_2$ respectively), and three measuring industrial stage in 1960 (the exogenous construct, denoted $\xi$).

Indicators for democracy in 1960 are: $y_1$, freedom of the press, 1960; $y_2$, freedom of political opposition, 1960; $y_3$, fairness of elections, 1960; and $y_4$, effectiveness of elected legislature, 1960. Indicators for democracy in 1965 are: $y_5$, freedom of the press, 1965; $y_6$, freedom of political opposition, 1965; $y_7$, fairness of elections, 1965; and $y_8$, effectiveness of elected legislature, 1965. Indicators for industrial development are: $x_1$, GNP per capita, 1960; $x_2$, energy consumption per capita, 1960; and $x_3$, percentage of labour force in industry, 1960.

The structural model specifies that $\xi$ influences $\psi_1$, and that both $\xi$ and $\psi_1$ influence $\psi_2$. So one has

$$x_{ik} = \kappa_{xk} + \lambda_{xk}\xi_i + \delta_{ik}, \qquad\qquad k = 1, \dots, 3,$$
$$y_{im} = \kappa_{ym} + \lambda_{ym}\psi_{i1} + \epsilon_{im}, \qquad\qquad m = 1, \dots, 4,$$

$$y_{im} = \kappa_{ym} + \lambda_{ym}\psi_{i2} + \epsilon_{im}, \qquad m = 5, \ldots, 8,$$

$$\psi_{i1} = \alpha_1 + \gamma_1 \xi_i + w_{i1},$$

$$\psi_{i2} = \alpha_2 + \gamma_2 \xi_i + \beta \psi_{i1} + w_{i2},$$

where $\xi_i \sim N(0, 1/\tau_\xi)$, $w_{i1} \sim N(0, 1/\tau_{w1})$, $w_{i2} \sim N(0, 1/\tau_{w2})$, $\epsilon_{im} \sim N(0, \sigma_{ym}^2)$, and $\delta_{ik} \sim N(0, \sigma_{xk}^2)$.

In the actual estimation, indicator values are standardised. Three unit loading constraints are adopted, whereby

$$\lambda_{x1} = \lambda_{y1} = \lambda_{y5} = 1,$$

and the three sets of construct scores have unknown variances. Remaining loadings are assigned $N(0, 1)$ priors, since there is no particular reason to expect widely differing loadings between indicators (as implied by flat priors), or to expect some of the remaining indicators to contribute disproportionately to defining the latent constructs, as this goes against the defining principles of latent variable analysis whereby indicators are chosen to have broadly comparable relevance in defining constructs. Normal priors with variance 1000 are adopted for the intercepts $\kappa_{xk}$ and $\kappa_{ym}$. Gamma priors with shape 1 and scale 0.001 are assigned to all precision parameters.

The second half of a two chain run of 50 000 iterations provides posterior summaries for important parameters as in Table 9.1. The strong autocorrelation in democracy scores between 1960 and 1965 is apparent in the estimate for $\beta$, whereas the 2.5% point of the 95% credible interval for $\gamma_2$ (the lagged impact on democracy of industrialisation five years earlier) is close to zero. The indicators are of similar relevance to defining the underlying constructs. This interpretation follows from standardisation of the reflexive indicators, so that $\lambda$ coefficients measure the relevance of each indicator to the underlying construct.

Posterior mean communalities for the $x$ variables are obtained for $x_k$ as

$$[\lambda_{xk}^2/\tau_\xi]/[\lambda_{xk}^2/\tau_\xi + \sigma_{xk}^2].$$

For the $y$ indicators the empirical variances $V_{\psi_1}$ and $V_{\psi_2}$ of the construct scores $\psi_{i1}$ and $\psi_{i2}$ are obtained at each MCMC iteration, with communalities for $y_1 - y_4$ then given by

$$[\lambda_{ym}^2 V_{\psi_1}]/[\lambda_{ym}^2 V_{\psi_1} + \sigma_{ym}^2].$$

**Table 9.1** Posterior summaries.

| Parameter | Mean | St devn | 2.5% | 97.5% |
|---|---|---|---|---|
| $\beta$ | 0.90 | 0.11 | 0.70 | 1.14 |
| $\gamma_1$ | 0.41 | 0.11 | 0.19 | 0.64 |
| $\gamma_2$ | 0.13 | 0.06 | 0.01 | 0.26 |
| $\lambda_{x2}$ | 1.13 | 0.07 | 0.99 | 1.28 |
| $\lambda_{x3}$ | 0.96 | 0.09 | 0.80 | 1.15 |
| $\lambda_{y2}$ | 0.91 | 0.12 | 0.68 | 1.15 |
| $\lambda_{y3}$ | 0.84 | 0.12 | 0.61 | 1.08 |
| $\lambda_{y4}$ | 1.03 | 0.11 | 0.82 | 1.26 |
| $\lambda_{y6}$ | 0.97 | 0.13 | 0.73 | 1.25 |
| $\lambda_{y7}$ | 1.02 | 0.13 | 0.79 | 1.30 |
| $\lambda_{y8}$ | 1.05 | 0.13 | 0.82 | 1.33 |

Communalities for the $x$ indicators range from 0.70 to 0.99, while those for the $y$ variables range from 0.57 (opposition freedom, 1960) to 0.73 (legislature effectiveness, 1960).

### Example 9.2   Confirmatory factor analysis and invariance, psychometric tests

Consider a confirmatory measurement model, as in equation 9.1, for psychometric item scores for $n = 301$ children classified by gender (1 = male, 2 = female), and school (1 = Pasteur, 2 = Grant–White). There are $M = 9$ items (Kolenikov, 2009), namely

1. VISUAL PERCEPTION

2. CUBES

3. LOZENGES

4. PARAGRAPH COMPREHENSION

5. SENTENCE COMPREHENSION

6. WORD MEANINGS

7. ADDITION

8. COUNTING DOTS

9. STRAIGHT CURVED CAPITALS

Three factors are postulated: spatial ability $\psi_1$, taken to explain $Y_1 - Y_3$; verbal ability $\psi_2$, explaining $Y_4 - Y_6$; and maths ability $\psi_3$, to explain $Y_7 - Y_9$.

The measurement model equations have the form

$$y_{im} \sim N(\mu_{im}, 1/\tau_{im}),$$

where, under conditional independence given constructs, the precision parameters for subject $i$ and indicator $m$ are defined as

$$\tau_{im} = T_m w_{im},$$

with $T_m = 1/\sigma_m^2$, and weights $w_{im}$ following $Ga(v_m/2, v_m/2)$ densities. Robustness issues regarding normality in the residuals for these data have been considered (Yuan and Bentler, 1998), and so a scale mixture form of the Student $t$ is adopted for the indicator likelihoods. The degrees of freedom parameters $v_m$ are drawn from an exponential prior with parameters $\eta_m$, sampled from a uniform (0.01, 1) prior, so that prior extreme $\eta_m$ correspond approximately to means 1 and 100 for $v_m$.

The means $\mu_{im}$ are given by

$$\mu_{im} = \kappa_m + \lambda_{m,C[m]}\psi_{i,C[m]}$$

where $C = \{1, 1, 1, 2, 2, 2, 3, 3, 3\}$ is the construct index for variable $m$. The scale of the constructs is defined by unit loading constraints

$$\lambda_{11} = \lambda_{42} = \lambda_{73} = 1,$$

and sufficient constraints on $cov(\epsilon)$ under a conditional independence assumption permit covariances between the constructs to be estimated. The six unknown loadings are assigned $N(0, 1)$ priors.

The model is estimated using standardised indicator scores, and this assists in achieving convergence. A posterior predictive check, based on comparing the observed square root of the mean square error to that obtained with replicated item scores, shows acceptable fit, averaging around 0.69 from the second half of a run of 20 000 iterations. The largest factor correlation, namely 0.46, is between factors 1 and 2. Posterior means (sd) for the unknown loadings are $\lambda_{21} = 0.52(0.12)$, $\lambda_{31} = 0.73(0.14)$, $\lambda_{52} = 1.01(0.06)$, $\lambda_{53} = 0.96(0.06)$, $\lambda_{83} = 1.25(0.18)$, and $\lambda_{93} = 1.10(0.18)$ (cf. Kolenikov, 2009). Estimates for $v_m$ show the WORD MEANINGS and COUNTING DOTS items as the most dubious in terms of normality, with posterior means 6.0 and 7.4. Of the 16 pupils with mean $w_{im}$ values under 0.5, eight are on WORD MEANINGS.

A second analysis extends Student $t$ robustification to the factor model[1], so that construct scores are multivariate $t$, using the scale mixture method (Nadarajah and Kotz, 2008) with unknown degrees of freedom $v_F$ and scaling factors $w_{Fi}$. The posterior median and mean $v_F$ (from second half of 10 000 iterations over two chains) are respectively 22 and 34, so some departure from multivariate normality is apparent. The degrees of freedom $v_m$ in the indicator Student $t$ likelihoods show the WORD MEANINGS and COUNTING DOTS items now have posterior means of 6.9 and 9.2. The $L$ criterion (Laud and Ibrahim, 1995), obtained using item score replicates, is reduced under this model, from 52.224 to 52.155.

### Example 9.3    Non-linear and interactive latent variable effects

The question of non-linear effects of latent variables, or of interactions between them, has been raised in applications of structural equation models (e.g. Zhu and Lee, 1999; Lee *et al.*, 2007; Wall and Amemiya, 2007; Mooijaart and Bentler, 2010). Defining exogenous constructs $\xi_{i1}, \ldots, \xi_{iq}$ by indicators $x_{i1}, \ldots, x_{iL}$, and endogenous constructs $\psi_{i1}, \ldots, \psi_{ip}$ by indicators $y_{i1}, \ldots, y_{iM}$, an example of a non-linear structural model would be if endogenous constructs are predicted by powers of, or interactions between, the $\xi_{ik}$. Non-linear construct effects may also occur in the measurement equations. While $q < L$ and $p < M$, the actual number of polynomial terms in a measurement equation or structural model may equal or exceed the number of indicators (Song and Lee, 2002). For instance, if M = 2, p = 1, one might have as measurement equations

$$y_{i1} = \kappa_1 + \lambda_{11}\psi_i + \lambda_{12}\psi_i^2 + \epsilon_{i1},$$

$$y_{i2} = \kappa_2 + \lambda_{21}\psi_i + \lambda_{22}\psi_i^2 + \epsilon_{i2},$$

namely two polynomial terms in each indicator model.

Consider simulated data for $n = 100$ subjects based on a measurement model relating $L = 5$ exogenous indicators to $q = 2$ constructs, $\xi_1$ and $\xi_2$, and a structural model for a single observed and centred endogenous indicator $y_i = \psi_i$ (Arminger and Muthen, 1998). Residuals $w_i$ for the $y$ regression are assumed normal with variance 0.6, with the regression including main terms in $\xi_1$ and $\xi_2$ and an interaction between them:

$$y_i = \gamma_1\xi_{i1} + \gamma_2\xi_{i2} + \gamma_3\xi_{i1}\xi_{i2} + w_i.$$

This type of model has applications in performance testing (Earley *et al.*, 1990), where $y$ represents performance, $\xi_1$ task complexity, and $\xi_2$ goal specificity, and where $\gamma_1$ and $\gamma_2$ are expected to be positive, but $\gamma_3$ negative on theoretical grounds.

The measurement model for the simulation involves loadings $\lambda_{mk}$ of items $x_{im}(m = 1, \ldots, 5)$ on factors $\xi_{ik}(k = 1, 2)$ as follows: $\lambda_{11} = 1$ (preset), $\lambda_{21} = 0.7$, $\lambda_{31} = -0.5$, $\lambda_{42} = 1$ (preset), and $\lambda_{52} = 1.6$. The coefficients in the structural model are

$$\{\gamma_1, \gamma_2, \gamma_3\} = (0.8, 1.7, -0.5).$$

The remaining parameters used to simulate the data are contained in the first code[2] for Example 9.3, with the dispersion matrix $\Phi_\xi$ including a positive correlation between the two constructs $\xi_{ik}$. The simulated data are generated in BUGS by checking syntax, then compiling, then generating inits randomly, and then using the steps info/node info for $y$ and $x$.

In the first analysis of these simulated data, unit loading constraints on $\lambda_{11}$ and $\lambda_{42}$ are combined with $N(0, 1)$ priors on unknown loadings $\lambda$, and on the structural coefficients $\gamma$. Inferences are based on the second half of a two chain run of 100 000 iterations. Posterior means (sd) on the structural coefficients $\{\gamma_1, \gamma_2, \gamma_3\}$ are 0.79 (0.18), 1.83 (0.30) and −0.58 (0.17), while those on loadings $\{\lambda_{21}, \lambda_{31}, \lambda_{52}\}$ are 0.78 (0.10), −0.48 (0.08) and 1.81 (0.24).

As an illustration of loading parameter selection the assumed structural model is now taken to include quadratic terms, which were not included in the data generation process. A spike-slab prior[3] on the structural model is adopted, so that

$$y_i = \gamma_{1\delta_1}\xi_{i1} + \gamma_{2\delta_2}\xi_{i2} + \gamma_{3\delta_3}\xi_{i1}\xi_{i2} + \gamma_{4\delta_4}\xi_{i1}^2 + \gamma_{5\delta_5}\xi_{i2}^2,$$

with

$$\gamma_{j\delta_j}|\delta_j = 1 \sim N(0, 0.01), \qquad j = 1, \ldots, 5$$

$$\gamma_{j\delta_j}|\delta_j = 2 \sim N(0, 1),$$

$$\delta_j \sim \text{Categoric}(\pi_\delta),$$

$$\pi_\delta = (0.5, 0.5).$$

Hence, conditional on $\delta_j = 1$, the loadings have a low variance around zero.

Estimates for this model (from the second half of two chains of 100 000 iterations) show posterior probabilities of rejection, $\Pr(\delta_j = 1|y, x)$, of 0.88 and 0.83 for $\gamma_4$ and $\gamma_5$, as against 0.005, 0 and 0.11 for $\{\gamma_1, \gamma_2, \gamma_3\}$. Posterior means (sd) on the structural coefficients $\{\gamma_1, \gamma_2, \gamma_3\}$ are now 0.71 (0.17), 1.82 (0.17) and −0.46 (0.18), while the remaining $\gamma$ coefficients have insignificant effects.

# 9.3   Dynamic factor models, panel data factor models and spatial factor models

Presentations of factor and structural equation models for individual subjects (e.g. in psychometrics) often assume factor scores are independent over cases, but for spatial or time series data it may be more realistic to allow for correlation between units.

## 9.3.1   Dynamic factor models

Thus dynamic factor models may be applied to multivariate financial data, such as exchange rates or stock returns, offering dimension reduction in representing high inter-series correlations and identifying common trends (Harvey and Shephard, 1993; Gilbert and Meijer, 2005). Bayesian dynamic factor models (e.g. Aguilar and West, 2000; Koop and Korobilis, 2010) have been mainly applied in econometrics, for example, to portfolio allocation. A more specific focus involves factor models to represent multivariate stochastic volatility (Asai et al., 2006).

The representation of latent constructs $\psi_t = (\psi_{t1}, \ldots, \psi_{tp})$ and $\xi_t = (\xi_{t1}, \ldots, \xi_{tq})$ in measurement or SEM models for time series data draws on methods discussed in Chapter 6.

Dynamic dependence may feature in either the measurement or structural equations, or both. For example, assuming $p = 1$ endogenous construct, let $B^d(\psi_t) = \psi_{t-d}$ and

$$\Lambda_m(B) = (\lambda_{m0} + \lambda_{m1}B + \lambda_{m2}B^2 + \ldots + \lambda_{mr}B^r).$$

Then autoregressive dependence in the measurement equation can be represented for continuous indicators $m = 1, \ldots, M$ as

$$y_{mt} = \Lambda_m(B)\psi_t + \epsilon_{mt},$$

where $\epsilon_{mt}$ and $\psi_t$ are normally distributed. Similarly, letting

$$\Gamma(B) = (1 - \gamma_1 B - \gamma_2 B^2 - \ldots - \gamma_r B^r),$$

dependence in the structural equation could be represented as

$$y_{mt} = \lambda_m \psi_t + \epsilon_{mt},$$
$$\Gamma(B)\psi_t = w_t,$$

where $w_t$ is normally distributed. Structural model time dependence might in practice involve simple forms such as first or second order random walks in $\psi_t$, with $\gamma$ coefficients preset. If $p > 1$, the structural equation could include lags across constructs.

As in other forms of latent variable model, issues may include assessing departures from measurement invariance (e.g. between sub-periods), or violation of the conditional independence assumption. Departures from conditional independence might involve introducing lagged dependencies in $y_{mt}$ or $\epsilon_{mt}$ into the measurement equations.

### Example 9.4 Canadian money measurement

This example considers $M = 6$ indicators of latent variables representing shifts in the financial assets of the population, available in the R package tsfa. Specifically the indicators are differences

$$y_{mt} = z_{mt} - z_{m,t-1}$$

over the period of $T = 214$ months for from February 1986 to November 2003, with $z_{m1}$ being indicators for January 1986 (cf. Gilbert and Meijer, 2005). The undifferenced indicators are currency ($z_1$), personal chequing deposits ($z_2$), non-bank chequing deposits ($z_3$), non-personal demand and notice deposits ($z_4$), non-personal term deposits ($z_5$), and investment ($z_6$). The indicators are per capita and in real terms.

Eigenvalues of the correlation matrix for the difference data show two eigenvalues over 1, with a third at 0.91. For $M = 6$ indicators, there are a maximum of $p = 3$ factors according to the Ledermann criterion (Ten Berge and Socan, 2007) namely that $(M - p)^2 \geq (M + p)$. An exploratory factor approach with $p > 1$ constructs is adopted.

Consider first the case $p = 2$, and consider a bivariate normal construct prior without autoregressive time dependence in the constructs. Thus

$$y_{mt} = \alpha_m + \lambda_{m1}\psi_{t1} + \lambda_{m2}\psi_{t2} + \epsilon_{mt},$$

with $\epsilon_{mt} \sim N(0, 1/\tau_m)$, and $\psi_t = (\psi_{t1}, \psi_{t2})$ following a standardised bivariate normal form with unknown correlation (Damien and Walker, 2001). Hence all loadings $\{\lambda_{m1}, \lambda_{m2}\}$ are unknowns, except for a single rotational invariance (RI) constraint.

Arbitrary selection of $y_1$ (differences in currency) for the RI constraint $\lambda_{12} = 0$ may not be the most suitable for representing substantive patterns. Instead, one might apply an initial model with $p = 1$, and ascertain the indicator with the highest posterior communality, and then under a $p = 2$ model assume a zero loading of that maximum communality indicator on the second construct. However, a variable selection procedure is considered here, without a zero RI constraint initially applied, but with selection of loadings on the second construct:

$$\lambda_{m2}|\delta_m = 0 \sim N(0, 0.01),$$

$$\lambda_{m2}|\delta_m = 1 \sim N(0, 1),$$

$$\delta_m \sim \text{Bern}(0.5).$$

Loadings $\lambda_{m1}$ on the leading factor are taken to be $N(0, 1)$, except for $\lambda_{11}$ which is assumed to be exponential, $\lambda_{11} \sim E(1)$, to ensure consistent factor labelling. However, the main goal of this preliminary analysis is the selection of a suitable indicator to which the RI constraint is applied, and labelling issues are not paramount. The second half of a two chain analysis with 10 000 iterations shows the highest rejection probability for the loading $\lambda_{m2}$ on $y_6$ (investment differences).

Accordingly, a time series EFA with $p = 2$ is applied, with the RI constraint $\lambda_{62} = 0$, and with $N(0,1)$ priors on unknown loadings except for the labelling constraints

$$\lambda_{11} \sim E(1); \quad \lambda_{12} \sim E(1).$$

Inferences from the second half of a two chain run of 50 000 iterations show $y_2$ having a particularly high loading of 0.88 on the first construct, and the second construct having high loadings on $y_4$ and $y_1$ (of 1.12 and 0.80 respectively).

Communalities are low for $y_5$ and $y_6$, with the posterior mean communalities for the six indicators being (0.57, 0.70, 0.28, 0.91, 0.12, 0.12). Possible extensions are to $p = 3$ constructs, or to extend the measurement models, e.g. to include autoregressive lags in the $y$ variables.

An alternative is to use the undifferenced data, as these show much higher correlations between indicators, indicating scope for dimension reduction. As an illustration of a time dependent factor structure, the undifferenced data for $T = 215$ months are accordingly considered with $p = 1$. It is assumed that successive construct scores[4] follow an RW1 scheme (implemented using the car.normal prior option in BUGS)

$$z_{mt} = \alpha_m + \lambda_m \psi_t + \epsilon_{mt},$$

$$\psi_t = \psi_{t-1} + w_t,$$

with $w_t$ normal. A unit loading constraint is applied, with $\lambda_1 = 1$, so that the construct variance is now unknown. With inferences from iterations 50 000–100 000 of a two chain run, the lowest communality is now 0.65 for $y_2$, with all other communalities exceeding 0.9.

## 9.3.2   Linear SEMs for panel data

The cross-sectional linear SEM of Section 9.2 may be extended to panel designs with observations consisting of time varying endogenous indicators $Y_{it} = \{y_{imt}\}$, and time varying exogenous indicators $X_{it} = \{x_{iht}\}$, which are respectively indicators for time varying endogenous constructs $\psi_{it}$ and time varying exogenous constructs $\xi_{it}$. The structural model may involve impacts of both contemporary or lagged constructs. Replication of manifest indicators over subjects means that certain parameters in the measurement model (e.g. residual variances, loadings) may be assumed time-varying.

For example, suppose $p = q = 1$, and that item-construct loadings are constant, such that for $t = 1, \ldots, T$,

$$y_{imt} = \kappa_{mt}^y + \lambda_{1m}\psi_{it} + \epsilon_{imt}^y, \qquad m = 1, \ldots, M$$

$$x_{iht} = \kappa_{ht}^x + \lambda_{2h}\xi_{it} + \epsilon_{iht}^x, \qquad h = 1, \ldots, L.$$

The associated structural model could involve both contemporaneous impacts of the exogenous construct, and lagged impacts of the endogenous construct, as in

$$\psi_{it} = \gamma_0 + \gamma_1\xi_{it} + \gamma_2\psi_{i,t-1} + w_{it}.$$

In a measurement model with indicators $\{y_{imt}, m = 1, \ldots, M\}$ representing only endogenous constructs, a normal linear state-space panel model with multivariate construct $\psi_{it} = (\psi_{i1t}, \psi_{i2t}, \ldots, \psi_{ipt})$ of dimension $p < M$, may be described by the model

$$y_{it} = \Lambda_t\psi_{it} + \epsilon_{it},$$

$$\Gamma(B)\psi_{it} = w_{it},$$

where $\Lambda_t$ is $M \times p$, $w_{it}$ is $p$-variate normal, and cross-construct lags may be included in the structural equation.

By analogy to the panel models considered in Chapter 7, replication over subjects implies that latent constructs may be designed to represent permanent subject effects. For example, a measurement model with constant latent effects $\psi_{i1}$, and time varying latent effects $\psi_{i2t}$, for multiple continuous indicators $y_{imt}$, would take the form

$$y_{imt} = \kappa_m + \lambda_{m1}\psi_{i1} + \lambda_{m2}\psi_{i2t} + \epsilon_{imt},$$

with $\epsilon_{imt}$ normal, subject to identifiability constraints (Longford and Muthen, 1992). For example, if no constraints are placed on the $M$ pairs of loadings $\{\lambda_{m1}, \lambda_{m2}\}$ then both construct variances must be fixed. If the measurement model included only stable traits $\psi_{i1}$, then one might consider changing loadings $\lambda_{m1t}$.

### Example 9.5    Two wave data on verbal and quantitative achievement

Consider data originally from the US Education Testing Service. There are $M = 5$ observed indicators over $T = 2$ waves, namely a cohort of $n = 383$ children observed at seventh and ninth grades. The data considered here are mean-centred and derived by simulation from the between the observed indicator covariance matrix (Bartholomew *et al.*, 2008, chapter 11). Any results reported may differ to some extent from those based on a direct analysis of the covariance matrix, without considering the data for individuals (an approach often adopted in frequentist analysis). The simulated data are obtained using the code

```
model { for (i in 1 :M2 ) {gamma[i] <- 0;
for (j in i + 1 : M2) {V[i,j] <- V[j,i]}}
Inv.Cov[1:M2, 1:M2] <- inverse(V[ , ])
for (i in 1:383) { X[i,1:M2] ~ dmnorm(gamma[], Inv.Cov[,]) }}
```

where M2 = 10, and with the covariance matrix V[,] as data input. The BUGS steps are check model, load data (i.e. the covariance matrix), compile and then 'gen inits' to obtain the simulated data.

Now the simulated data are regarded as observations with unknown DGP. There are $p = 2$ constructs, quantitative and verbal ability, postulated at each grade, with indicators from the earlier grade measuring the exogenous constructs $\xi = (\xi_1, \xi_2)$, with $\xi_1$ for quantitative ability in seventh grade, and $\xi_2$ for verbal ability in seventh grade. Indicators from the later wave provide the endogenous constructs: $\psi_1$ (quantitative ability), and $\psi_2$ (verbal ability). There is a departure from simple structure in that science achievement test scores (abbreviated as SCI7 and SCI9 for seventh and ninth grade respectively) are taken to load on both quantitative and verbal ability. Two repeated indicators (Scholastic Aptitude Test Quantitative, SCATQ7 and SCATQ9; and maths achievement tests, MATH7 and MATH9) load only on $\xi_1$ and $\psi_1$. Another two repeated indicators (Scholastic Aptitude Test Verbal, SCATV7 and SCATV9; and reading achievement tests, READ7 and READ9) load only on the verbal ability constructs, $\xi_2$ and $\psi_2$.

Assume the measured indicators (obtained by simulation as above) are standardised (i.e. $x_1, x_2, x_3, x_4, x_5$ being standardised versions of grade 7 scores SCATQ7, MATH7, SCI7, READ7 and SCAT7V, and similarly for $y$ indicators), as this is important for ensuring convergence. Assuming conditional independence, one then has for subjects $i$

$$x_{1i} \sim N(\lambda_{11}\xi_{1i}, \sigma_{11}^2),$$

$$x_{2i} \sim N(\lambda_{21}\xi_{1i}, \sigma_{12}^2),$$

$$x_{3i} \sim N(\lambda_{31}\xi_{1i} + \lambda_{32}\xi_{2i}, \sigma_{13}^2),$$

$$x_{4i} \sim N(\lambda_{42}\xi_{2i}, \sigma_{14}^2),$$

$$x_{5i} \sim N(\lambda_{52}\xi_{2i}, \sigma_{15}^2),$$

$$y_{1i} \sim N(\kappa_{11}\psi_{1i}, \sigma_{21}^2),$$

$$y_{2i} \sim N(\kappa_{21}\psi_{1i}, \sigma_{22}^2),$$

$$y_{3i} \sim N(\kappa_{31}\psi_{1i} + \kappa_{32}\psi_{2i}, \sigma_{23}^2),$$

$$y_{4i} \sim N(\kappa_{42}\psi_{2i}, \sigma_{24}^2),$$

$$y_{5i} \sim N(\kappa_{52}\psi_{2i}, \sigma_{25}^2).$$

Unit loading constraints $\lambda_{11} = \lambda_{42} = \kappa_{11} = \kappa_{42} = 1$ are used for identifiability. Cumulating the $M = 5$ indicators over both grades, there are $M(M + 1) - 2M - p(p - 1) = 18$ available parameters for the latent model (indicator-construct loadings, construct variances/covariances, and structural equation parameters).

In the measurement equations set out above, and with unit loading constraints, there are eight unknown loadings. Suppose one assumes uncorrelated constructs, so that cov($\xi$) and cov($\psi$) together contain four unknown construct variances. This leaves a maximum of six coefficients for the structural model linking $\psi$ to $\xi$. So with $\xi_1$ and $\xi_2$ representing quantitative and verbal ability at grade 7, and $\psi_1$ and $\psi_2$ representing these abilities at grade 9, a baseline structural model postulates dependence of each construct on its earlier values, namely

$$\xi_{1i} \sim N(0, \phi_1^2), \qquad \xi_{2i} \sim N(0, \phi_2^2),$$

$$\psi_{1i} \sim N(\beta_1\xi_{1i}, \phi_3^2), \qquad \psi_{2i} \sim N(\beta_2\xi_{2i}, \phi_4^2),$$

with two additional structural regression coefficients, and thus a total of 14 parameters in the latent model.

Adaptations of this model may be envisaged such as contemporaneous dependence of quantitative on verbal ability and vice versa, namely

$$\xi_{1i} \sim N(\alpha_1 \xi_{2i}, \phi_1^2), \qquad\qquad \xi_{2i} \sim N(\alpha_2 \xi_{1i}, \phi_2^2),$$
$$\psi_{1i} \sim N(\beta_{11}\xi_{1i} + \beta_{12}\psi_{2i}, \phi_3^2), \quad \psi_{2i} \sim N(\beta_{21}\xi_{2i} + \beta_{22}\psi_{1i}, \phi_4^2).$$

However, the appropriate extension reflects the form of any lack of fit. Frequentist estimation and evaluation typically focus on reproducing the covariance/correlation structure, and this can be one facet of a Bayesian evaluation. Lack of fit may require that one modify the measurement model instead of, or as well as, the structural model.

The above simple structural model (dependence of each grade 9 construct only on its earlier values) is estimated using $N(0,1)$ priors on unknown loadings and structural coefficients. A posterior predictive loss criterion or PPLC measure (Laud and Ibrahim, 1995) assesses fit at both subject and global level. A posterior predictive check also compares actual correlations and modelled lag correlations on the $M$ indicators: the actual correlations $c_m, m = 1, \dots, M$ of scores between grades 7 and 9 for SCATQ, MATHS, SCI, READ and SCATV are 0.67, 0.75, 0.74, 0.79 and 0.90. Note these are correlations for the simulated dataset.

The simple structural model combined with a conditional independence assumption (model 1 in the code for this example) indicates lack of fit in terms of reproducing these grade 7–grade 9 correlations (with inferences from the second half of a two chain sequence of 5000 iterations). This is especially the case for the correlations between SCI7 and SCI9, and between READ7 and READ9. A posterior predictive check produces essentially zero probabilities that $c_{m.\text{rep}}$ exceeds $c_m$, where $c_{m.\text{rep}}$ is the correlation between grades obtained using replicate indicator values sampled from the model.

A revised model (model 2) drops the conditional independence assumption, and allows for bivariate normal residual covariance in the measurement equations for SCATQ7 and SCATQ9, for MATHS7 and MATHS9, etc, through to the equations for SCATV7 and SCATV9. This results in five more parameters in the measurement model so that there are now 13 available parameters for the latent model (indicator-construct loadings, construct variances/covariances, and structural equation parameters).

There are eight unknown loadings in the indicator regressions. The structural model in model 2 remains as above, involving only a grade lag effect in the constructs,

$$\xi_{1i} \sim N(0, \phi_1^2), \qquad\qquad \xi_{2i} \sim N(0, \phi_2^2),$$
$$\psi_{1i} \sim N(\beta_1 \xi_{1i}, \phi_3^2), \qquad \psi_{2i} \sim N(\beta_2 \xi_{2i}, \phi_4^2),$$

and so involves six parameters in all.

To ensure the constraint on available (fixed effect) latent model parameters is ensured, one could reduce the number of loadings in the measurement or structural equations, e.g. by setting $\kappa_{32} = 0$, or $\beta_1 = \beta_2$. Instead of this, a random effects model is adopted for the eight unknown loadings in the measurement equation. A gamma prior on the precision of the loadings has shape 1 and scale 0.001.

Posterior predictive probabilities that $c_{m.\text{rep}}$ exceeds $c_m$ are all now satisfactory, with values between 0.4 and 0.7. However, overall fit as measured by the PPLC worsens slightly from 1955.6 to 1961.1: the predictive variance component increases from 1160.9 to 1226.6, while the fit component improves, falling from 794.7 to 734.5. Posterior means (sd) for the loading parameters are $\lambda_{21} = 1.11$ (0.08), $\lambda_{31} = 0.356$ (0.07), $\lambda_{32} = 0.62$ (0.06), $\lambda_{52} = 0.97$ (0.04), $\kappa_{21} = 1.13$ (0.07), $\kappa_{31} = 0.378$ (0.07), $\kappa_{32} = 0.59$ (0.06), and $\kappa_{52} = 0.92$ (0.04), while for the structural parameters, they are $\beta_1 = 0.94$ (0.06), and $\beta_1 = 0.96$ (0.03).

### 9.3.3   Spatial factor models

Consider multivariate spatial indicators $(y_{i1}, \dots, y_{iM})$ from an exponential family density observed over areas $i = 1, \dots, n$. Defining $M$-dimensional spatially structured random effects $s_i = (s_{i1}, \dots, s_{iM})$ measuring unmeasured spatially configured predictors, and measured predictors $X_i$ one has for the $m$th indicator

$$p(y_{im} | s_{im}, X_i) \propto \exp \left\{ \frac{y_{im}\,\theta_{im} - b\left(\theta_{im}\right)}{\phi_{im}} + c(y_{im}, \phi_{im}) \right\}$$

where $\theta_{im}$ is the canonical parameter, and $\phi_{im}$ is a known scale. Denoting regression terms as $\eta_{im} = g(\theta_{im})$ with link $g$, a widely applied regression scheme (Chapter 8) is

$$\eta_{im} = \alpha_m + \beta_m X_i + s_{im},$$

where the spatial effects $s_i$ follow a multivariate spatial prior. For example, the joint density for a normal $M$-variate intrinsic conditional autoregressive (ICAR) spatial prior with $nM \times nM$ precision matrix $Q$ may be expressed

$$p(s|Q) = \left( \frac{1}{2\pi} \right)^{nP/2} |Q|^{0.5} \sum_{ij} \exp[s_i Q_{ij} s_j],$$

where $Q$ is block diagonal with $M \times M$ sub-matrices $Q_{ij}$ that are non-zero (zero) if area $j$ is (is not) a neighbour of area $i$. The corresponding full conditional densities are

$$s_i | s_{[i]} \sim N\left( -Q_{ii}^{-1} \sum_{j \neq i} Q_{ij} s_j, \frac{1}{Q_{ii}} \right),$$

with conditional precision matrices $\mathrm{Prec}(s_i | s_{[i]}) = Q_{ii} = \Delta_i$. Equivalently define $M \times M$ matrices $B_{ij} = -Q_{ij}/Q_{ii}$, with $B_{ii} = 0$, and $\Delta_i = Q_{ii}$. Then

$$E(s_i | s_{[i]}) = \sum_{j \neq i} B_{ij} s_j; \qquad \mathrm{Prec}(s_i | s_i) = \Delta_i.$$

A valid joint density exists (Banerjee et al., 2004; Rue and Held, 2005) when $\Delta_i B_{ij} = \Delta_j B_{ji}$. For example, setting $B_{ij} = [w_{ij}/d_i] I_{P \times P}$, where $d_i$ is the number of areas adjacent to $i$, and

$$\Delta_i = d_i \Omega$$

where $\Omega$ is a $M \times M$ within area precision matrix, will ensure a valid joint density.

When high correlations are evident in $\Omega^{-1}$, common spatial factor models may be more parsimonious leading to regressions of the form

$$\eta_i = \alpha + \beta X_i + \Lambda F_i,$$

where $\eta_i = (\eta_{i1}, \ldots, \eta_{iM})'$, $\Lambda$ is of dimension $M \times p$, and the factor scores

$$F_i = (F_{il}, \ldots, F_{ip})'$$

are both spatially dependent over areas $i$, and mutually intercorrelated. If the spatial prior is specified, as under the ICAR prior, in terms of differences $F_i - F_j$ (and an overall location is not explicit), then the location may be fixed by centring each of the $p$ sets of spatial factor scores at each MCMC iteration. The spatial scheme may also use a proper prior, such as the multivariate version of the Leroux *et al.* (1999) prior. The scale of the spatial constructs may be determined as usual either by standardising the factor scores or by unit loading constraints. If there is more than one spatial construct, additional loadings would need to be fixed to avoid rotational indeterminacy, such as $\lambda_{mk} = 0$ for $k > m$.

### Example 9.6   Air quality and asthma admissions

This example considers the spatial relationship between hospitalisations for asthma $y_i$ and a latent index of air quality, based on $M = 5$ air pollutant indicators $x_{im}$. The air quality index $\xi_i$ is regarded as an exogenous construct in a structural model with a single endogenous response, namely hospitalisation risks for asthma. The spatial framework consists of $n = 562$ neighbourhoods (lower super output areas) in four boroughs in outer NE London. The hospitalisation counts are for 2011–2012, all patient ages, and include an asthma diagnosis among the precipitating conditions (counts under 8 are treated as missing), while the pollution indicators (for 2011) are NO2 annual mean, NOx annual mean, PM10 annual mean, PM10 exceedances, and PM2.5 exceedances. The original readings for annual means are in micrograms (one-millionth of a gram) per cubic meter (or $\mu g/m^3$), and all pollution indicators are first log transformed and then standardised. Additionally, Student $t$ likelihoods are adopted in the measurement model linking the pollution indicators to the latent index $\xi_i$.

The latent air quality index $\xi$ is assumed to be spatially correlated according to the Leroux *et al.* (1999) scheme, so avoiding an assumption that the index is necessarily spatially structured: the Leroux *et al.* scheme encompasses fully spatial structured and unstructured random effects as extremes. A measure of area deprivation, denoted $D_i$, and an additional spatial error denoted $s_i$, are included along with the exogenous factor to explain variations in the logarithms of relative hospitalisation risk for asthma, $\rho_i$. To allow for the explanatory role of deprivation in area pollution differences (e.g. Crouse *et al.*, 2009), a standardised version $D_{zi}$ of the deprivation measure is used as a 'formative' indicator in defining $\xi_i$.

With $E_i$ denoting expected hospitalisations (namely region-wide hospitalisation rates multiplied by populations), and $w_{ih}$ being binary spatial adjacency indicators, a full statement of the assumed model[5] is

$$y_i \sim \text{Po}(E_i \rho_i), \qquad i = l, \ldots, n$$

$$\log(\rho_i) = \beta_1 + \beta_2 D_i + \beta_3 \xi_i + s_i$$

$$x_{im} \sim t(\mu_{im}, \sigma_m^2, v_m), \qquad m = 1, \ldots, M$$

$$\mu_{im} = \alpha_m + \lambda_m F_i,$$

$$F_i | F_{[i]} \sim N\left( M_i, \frac{\omega_1}{1 - \kappa_1 + \kappa_1 \sum_{h \neq i} w_{ih}} \right),$$

$$e_h = F_h - \gamma D_{zh},$$

$$M_i = \gamma D_{zi} + \kappa_1 \sum_{h \neq i} w_{ih} e_h / \left[1 - \kappa_1 + \kappa_1 \sum_{h \neq i} w_{ih}\right],$$

$$s_i | s_{[i]} \sim N\left(S_i, \frac{\omega_2}{1 - \kappa_2 + \kappa_2 \sum_{h \neq i} w_{ih}}\right),$$

$$S_i = \kappa_2 \sum_{h \neq i} w_{ih} s_h / \left[1 - \kappa_2 + \kappa_2 \sum_{h \neq i} w_{ih}\right].$$

The spatial mix parameters $\kappa_j$ are taken as $U(0, 1)$, and the $\xi$ scores are assumed to be standardised, with $\omega_1 = 1$, and the unknown loadings $\lambda_m$ constrained to be positive,

$$\lambda_m \sim N(1, 1) \, I(0, ).$$

Iterations 10 000–25 000 of a two chain run provide a posterior summary of parameters as in Table 9.2. The $\beta$ coefficients are standardised, namely $\beta_2^* = \beta_2 s_D / s_\psi$, and $\beta_3^* = \beta_3 s_\xi / s_\psi$, where $\psi_i = \log(\rho_i)$, and where standard deviations $s_\psi$ and $s_\xi$ are obtained at each iteration. From the estimated loadings it can be seen that the indicators have similar relevance to defining the overall index, except that PM10 annual means play a greater role than PM10 exceedances. Furthermore, the effect $\gamma$ of deprivation on the pollution index is significant, so that both the reflexive and formative indicators are relevant to obtaining the latent index. However, while the direct effect of deprivation on asthma hospitalisations is highly significant, the impact of air pollution itself on asthma hospital admissions is not significant. The assumption of spatial dependence in the latent pollution index is supported, with the 95% interval for $\kappa_1$ essentially including the value of unity, under which the Leroux *et al.* prior reduces to the intrinsic CAR prior (Lee, 2011).

**Table 9.2**    Asthma and pollution, summary parameter estimates.

| Parameter | Mean | St devn | 2.5% | 97.5% |
|-----------|------|---------|------|-------|
| $\beta_2^*$ | 0.48 | 0.03 | 0.41 | 0.53 |
| $\beta_3^*$ | −0.01 | 0.04 | −0.12 | 0.07 |
| $\gamma$ | 0.10 | 0.03 | 0.04 | 0.16 |
| $\kappa_1$ | 0.96 | 0.04 | 0.87 | 1.00 |
| $\kappa_2$ | 0.07 | 0.04 | 0.01 | 0.16 |
| $\lambda_1$ | 1.42 | 0.09 | 1.29 | 1.64 |
| $\lambda_2$ | 1.46 | 0.09 | 1.34 | 1.68 |
| $\lambda_3$ | 1.43 | 0.09 | 1.32 | 1.65 |
| $\lambda_4$ | 1.28 | 0.08 | 1.17 | 1.47 |
| $\lambda_5$ | 1.43 | 0.09 | 1.32 | 1.65 |
| $v_1$ | 6.0 | 1.7 | 3.7 | 10.3 |
| $v_2$ | 8.5 | 3.5 | 4.5 | 17.9 |
| $v_3$ | 3.1 | 0.6 | 2.2 | 4.5 |
| $v_4$ | 2.1 | 0.1 | 2.0 | 2.4 |
| $v_5$ | 6.1 | 1.2 | 4.2 | 8.9 |

## 9.4   Latent trait and latent class analysis for discrete outcomes

### 9.4.1   Latent trait models

A similar framework to the normal linear measurement and SEM models may be postulated for observations on $M$ discrete items (e.g. binary or ordinal data), which are to be explained by $p$ metric factors. For example, suppose there are $M$ binary exam items and $p = 1$ (a single latent trait). Then a typical item response theory (IRT) model (e.g. Fox, 2010) has the form

$$y_{im} \sim \text{Bern}(\pi_{im}), \qquad i = 1, \dots, n;\ m = 1, \dots, M$$

$$\pi_{im} = \Phi(\omega_m + \lambda_m \psi_i),$$

for examinees $i$, with $\psi_i$ representing standard normal factor scores. If instead the assumption $\psi_i \sim \text{Logist}(0, 1)$ is made, the loadings are $\kappa_m \approx (\sqrt{3/\pi})\lambda_m$. A logit link may also be used, and the scheme can be extended to examinees clustered into groups (Azevedo $et\ al.$, 2012). Under conditional independence the joint success probability given $\psi_i$ is a product of Bernoulli likelihoods,

$$\Pr(y_{i1} = 1, y_{i2} = 1, \dots, y_{iM} = 1 | \psi_i) = \Pr(y_{i1} = 1 | \psi_i)\Pr(y_{i2} = 1 | \psi_i) \dots \Pr(y_{iM} = 1 | \psi_i).$$

The IRT model may also be estimated by invoking latent continuous variables underlying the observations (Albert and Chib, 1993). Thus continuous $z_{i1}, z_{i2}, \dots, z_{iM}$ generate responses to $M$ observed binary items $y_{i1}, y_{i2}, \dots, y_{iM}$, while traits $\psi_{i1}, \dots, \psi_{ip}$ explain variation in the $z$ variables. Correlations between the $z$ variables may be introduced in situations where conditional independence is not supported (Asparouhov and Muthen, 2011).

If the items are positive measures of ability, then the $\psi$ scores will measure overall ability, provided the $\lambda_m$ are constrained to prevent label switching (i.e. ensure a consistent direction for the $\psi$ scores), for example, by using a positive prior on the loadings (Albert and Ghosh, 2000; Sinharay, 2004). If $p > 1$, only lower triangular loadings $\lambda_{mk}$ may be non-zero, since otherwise orthogonal transformation of the $\lambda_{mk}$ leaves the likelihood unchanged. For example, for $p = 2$, one may set $\lambda_{12} = 0$.

Analysis with a single trait ($p = 1$) typically seeks both to rank the abilities of examinees and also establish the effectiveness of different items in measuring the ability trait. The item response probability may be written

$$\pi_{im} = \Phi(\lambda_m \psi_i - \delta_m),$$

so that $\delta_m$ measures the difficulty of item $m$, while $\lambda_m$ measures an item's power to detect ability differences between examinees. For a given difference in $\psi$ scores, the larger the absolute size of $\lambda_m$ the greater the difference in the probability of a positive response, $y_{im} = 1$, as abilities vary. The item response probability may also include a guessing parameter $0 \le \gamma_m \le 1$ (Sinharay, 2005), such that

$$\pi_{im} = \gamma_m + (1 - \gamma_m)\Phi(\lambda_m \psi_i - \delta_m)$$

under a probit link or

$$\pi_{im} = \gamma_m + (1 - \gamma_m) \exp(\lambda_m \psi_i - \delta_m)/[1 + \exp(\lambda_m \psi_i - \delta_m)]$$

under a logit link.

A potential complication in such models is caused by differential item functioning (DIF) when, for given ability, the probability of a positive response may depend on cultural factors. Thus let $x_i = 0$ for a reference group and $x_i = 1$ for a minority group, then DIF is indicated if the group-differentiated model

$$\pi_{im} = \Phi(\lambda_m \psi_i - \delta_m + x_i(\kappa_m \psi_i - \eta_m))$$

has better support than the standard model.

Latent trait models and item response model formulations apply for other forms of discrete observation such as ordinal responses (e.g. Shi and Lee, 2000; Cagnone et al., 2009; Huber et al., 2009). For example, suppose the $M$ indicators contain M-H continuous variables $(y_{H+1}, \ldots, y_M)$, and $H$ ordinal outcomes with $R_1, R_2, \ldots, R_H$ categories respectively. To model correlation among these variables or introduce regression effects, one may define latent continuous variables $z_{im} (m = 1, \ldots, H)$, and $R_m - 1$ cut-points $\delta_{mk}$ on their range, such that

$$y_{im} = k \quad \text{if} \quad \delta_{m,k-1} \le z_{im} < \delta_{mk}$$

with cut-points constrained according to

$$-\infty \le \delta_{m1} \le \ldots \le \delta_{m,R_{m-1}} \le \infty.$$

The $z_{im}$ might be taken to be multivariate normal or Student $t$ of dimension $H$. With a large number of ordinal or binary outcomes, observed together with metric outcomes, one may consider representing the data in terms of a smaller number of constructs. Then the combined set of variables $(z_1, \ldots, z_H, y_{H+1}, \ldots, y_M)$ is expressed in terms of the normal linear measurement model (Section 9.2), but with variance structures for the $z$ variables defined by identifiability.

## 9.4.2    Latent class models

In some circumstances it may be more plausible to treat latent variables themselves as categoric rather than metric. Thus subjects are classified into one of $K$ classes for a single latent categorical variable $\psi$, or cross-classified into one of $K_1 \times K_2$ classes if there are $p = 2$ categoric constructs, rather than being located on a continuous scale or scales. For example, Langeheine (1994) considers a longitudinal setting where observed items relate to children at different ages, with items taken to represent stage theories of developmental psychology which postulate distinct stages (i.e. discrete categories) of intellectual development. In other circumstances there may be no substantive rationale for preferring a latent trait or latent class model, but both provide adequate fit to the observed data – so leading to model indeterminacy (Bartholomew and Knott, 1999, chapter 6).

Let the prior probabilities on the $K$ classes of a single latent category $\psi$ be denoted $\eta_k$, with $\sum_k \eta_k = 1$. Viewed as an 'independent variable' predicting observed indicators, $\psi$ is comparable to a categorical factor in regression models, with the first category $\psi_i = 1$ providing a reference category under a 'corner constraint'. So for $M$ binary items the impact of a single latent class $\psi$ may be expressed

$$y_{im} \sim \text{Bern}(\pi_{im}),$$

$$g(\pi_{im}) = \omega_m + \lambda_{m2} I[\psi_i = 2] + \ldots + \lambda_{mK} I[\psi_i = K],$$

where $g$ is a probit or logit link, and $I[\psi_i = k] = 1$ if the $i$th subject is allocated to the $k$th category of $\psi$. Depending on whether $K = 2$ or $K > 2$, subjects are allocated to classes according to

$$\psi_i \sim \text{Categoric}(1 - \eta, \eta),$$

with $\eta$ assigned a beta prior, or according to

$$\psi_i \sim \text{Categorical}(\eta_1, \ldots, \eta_K),$$

with $\eta = (\eta_1, \ldots, \eta_K)$ assigned a Dirichlet prior. If there are formative or group indicators $G_i$ (e.g. gender) that explain the latent category then subject level $\eta_{ik}$ may be predicted by logit or probit regression.

An equivalent framework takes

$$y_{im} \sim \text{Bern}(\rho_{\psi_i, m}),$$

where $\psi_i$ has $K$ categories of 'caseness', and $\rho_{km}$ are the probabilities $\text{Pr}(y_{im} = 1)$ for items m according to the caseness of subject $i$. With priors on $\psi_i$ as above, beta priors may then be assigned to the $\rho_{km}$, or a link function used such as

$$\rho_{km} = \exp(\theta_{km})/[1 + \exp(\theta_{km})],$$

where normal priors on $\theta_{km}$, perhaps constrained to produce consistent labels during MCMC sampling.

Latent categories may be useful in medical diagnosis where the observed indicators are various diagnostic tests or criteria of illness (or different clinicians), with none being certain or 'gold standard' indicators of the presence of a disease. One may assess whether a single latent categorisation (e.g. if $K = 2$, the latent categories might be ill vs. not ill) underlies several observed binary diagnostic items, none of which provide a 'gold standard' test (Rindskopf, 2002).

**Example 9.7    Psychiatric caseness**

As an example of latent class analysis applied to clinical diagnosis, consider responses to three binary diagnostic items for $n = 103$ patients (Dunn, 1999). The binary responses are taken by dichotomising more extensive scales, and are denoted CIS (Clinical Interview Schedule), GHQ (General Health Questionnaire) and HADS (Hospital Anxiety and Depression Scale). Categories 1 of CIS, GHQ and HADS define less ill patients and categories 2 of CIS, GHQ and HADS define more ill patients. Let $G_{hjk}$ denote the observed counts of patients in category $h$ of CIS, $j$ of GHQ, and $k$ of HADS.

A latent class model is initially considered, with $m = 1$ for CIS, $m = 2$ for GHQ and $m = 3$ for HADS. Thus for binary versions of the items ($y_{im} = 0$ for category 1, $y_{im} = 1$ for category 2)

$$y_{im} \sim \text{Bern}(\pi_{im}), \qquad m = 1, \ldots, 3,$$
$$\text{logit}(\pi_{im}) = \omega_m + \lambda_m I[\psi_i = 2],$$

where $\psi_i = 2$ for 'caseness', and

$$\psi_i \sim \text{Categoric}(1 - \eta, \eta),$$

**Table 9.3**   Latent class analysis of diagnosis.

|  |  | Mean | St devn | 2.5% | Median | 97.5% | Actual |
|---|---|---|---|---|---|---|---|
| Predicted tabulation $G_{hjk}$ over items | | | | | | | |
|  | Cell (1,1,1) | 28.2 | 4.7 | 19 | 28 | 37 | 35 |
|  | Cell (1,1,2) | 7.7 | 3.4 | 2 | 7 | 15 | 5 |
|  | Cell (1,2,1) | 5.9 | 3.0 | 1 | 6 | 12 | 3 |
|  | Cell (1,2,2) | 7.1 | 3.2 | 2 | 7 | 14 | 6 |
|  | Cell (2,1,1) | 8.4 | 3.5 | 2 | 8 | 16 | 6 |
|  | Cell (2,1,2) | 9.5 | 3.7 | 3 | 9 | 18 | 8 |
|  | Cell (2,2,1) | 7.4 | 3.3 | 2 | 7 | 15 | 6 |
|  | Cell (2,2,2) | 28.8 | 4.9 | 19 | 29 | 38 | 34 |
| Marginal tables (items by LC) | | | | | | | |
| LC by CIS | Both no | 41.4 | 2.4 | 36 | 42 | 46 | |
|  | LC no, CIS yes | 7.2 | 2.8 | 2 | 7 | 13 | |
|  | LC yes, CIS no | 7.6 | 2.4 | 3 | 7 | 13 | |
|  | LC yes, CIS yes | 46.8 | 2.8 | 41 | 47 | 52 | |
| LC by GHQ | Both no | 44.2 | 2.9 | 38 | 44 | 50 | |
|  | LC no, GHQ yes | 4.4 | 2.3 | 1 | 4 | 10 | |
|  | LC yes, GHQ no | 9.8 | 2.9 | 4 | 10 | 16 | |
|  | LC yes, GHQ yes | 44.6 | 2.3 | 39 | 45 | 48 | |
| LC by HAD | Both no | 42.3 | 2.5 | 37 | 42 | 47 | |
|  | LC no, HAD yes | 6.3 | 2.7 | 2 | 6 | 12 | |
|  | LC yes, HAD no | 7.7 | 2.5 | 3 | 8 | 13 | |
|  | LC yes, HAD yes | 46.7 | 2.7 | 41 | 47 | 51 | |
| Prob($\psi_i = 2$) | $\eta$ | 0.53 | 0.06 | 0.40 | 0.53 | 0.65 | |
| Loadings | $\lambda_1$ | 3.1 | 0.6 | 2.1 | 3.1 | 4.3 | |
|  | $\lambda_2$ | 3.2 | 0.6 | 2.1 | 3.2 | 4.4 | |
|  | $\lambda_3$ | 3.2 | 0.6 | 2.1 | 3.2 | 4.4 | |

with a beta prior assumed on $\eta$. To ensure consistent labelling, the loadings $\lambda_m$ are assigned $N(1, 1)$ priors constrained to positive values.

If the latent categorisation (LC) by $\psi_i$ is regarded as the 'true' diagnosis, then the sensitivity of each observed item can be obtained with regard to the true diagnosis. For instance, in the LC by GHQ table (Table 9.3, based on the second half of a two chain run of 20 000 iterations) the LC variable classifies 55 patients as cases (using posterior medians), 45 of whom are also classified as cases by the GHQ, with the sensitivity of the GHQ obtained as 45/55 = 82%.

The LCA model underpredicts subtotals in $G_{hjk}$ where all items agree, suggesting possible violation of the conditional independence assumption. The observed frequencies with all items having value 1 and all items having value 2 are respectively 35 and 34.

Alternatively, a latent trait analysis with a single continuous factor is applied (in Open-BUGS), namely

$$y_{im} \sim \text{Bern}(\pi_{im}), \qquad i = 1, \ldots, n; \ m = 1, \ldots, M$$

$$\text{logit}(\pi_{im}) = \kappa_m + \lambda_m \psi_i,$$

with priors on $\lambda_m$ constrained to positive values, and $\psi_i \sim N(0, 1)$. Sampling new data $y_{new,im} \sim \text{Bern}(\pi_{im})$, one may obtain predicted frequencies in the 8 cells, and compare them to actual frequencies $G_{hjk}$ by a predictive fit criterion (specifically with $k = 1$ in equation 6 of Gelfand and Ghosh, 1998). The latent trait model allows choice of $K \geq 2$ cut points on the latent trait $\psi_i$; for instance taking $K = 3$ might correspond to the divisions: well, some symptoms, and definitely ill. Here $K = 2$ is chosen with the cut point at zero, and $H[1 : 103]$ is the resulting classifier in the code for this example.

This model has a clear advantage over the latent class model, with a predictive fit criterion of 87 compared to 164 under the LCA model, based on more accurate and more precise predictions. The advantage of the latent trait model (as compared to latent class analysis) lies in providing distinct scores for each of the eight multinomial categories formed by the three binary items, with the highest $\psi_i$ for the 34 subjects with category 2 on all items.

From the second half of a 20 000 iteration two chain sample, the posterior means (sd) for $(\lambda_1, \lambda_2, \lambda_3)$ are 3.39 (1.22), 3.86 (1.42), 3.69 (1.32). Using a zero cut point on the continuous latent trait (note that such binarisation implies information loss) leads to similar marginal tables as those in Table 9.3 based on the categorical factor, though slightly fewer patients are classed as cases (namely those with positive $\psi_i$).

### Example 9.8    Government services and interventions

To illustrate the application of a latent trait model to ordinal responses, consider data on $M = 5$ questions asked in the 1996 British Attitudes Survey regarding government activity, and involving $n = 786$ respondents. The questions are:

1. Should the government provide a job for everyone;

2. Should the government keep prices under control;

3. Should the government provide a decent standard of living for the unemployed;

4. Should the government reduce income differences between the rich and poor;

5. Should the government provide decent housing for those who cannot afford it.

Each question has $R = 4$ ranked response categories: $1 =$ definitely should; $2 =$ probably should; $3 =$ probably should not; $4 =$ definitely should not.

Initially a single latent variable $\psi_i$ is used to explain covariation between responses to the observed questions. Observed values on the ordinal indicators $y_{im}$ are linked to underlying continuous scales with $R - 1$ cut-points $\delta_{mr}$. Thus

$$y_{im} \sim \text{Categorical}(\pi_{i,m,1:R}),$$

where

$$\pi_{im1} = \gamma_{im1},$$

$$\pi_{imr} = \gamma_{imr} - \gamma_{im,r-1}, \qquad r = 2, \dots , R - 1$$

$$\pi_{imR} = 1 - \gamma_{im,R-1},$$

and

$$\text{logit}[\gamma_{imr}] = \delta_{mr} - \lambda_m \psi_i.$$

The first loading is assumed positive with mean 1, to ensure consistent labelling, with other loadings assigned a $N(1, 1)$ prior.

An important aspect of model success is in terms of predictions effectively reproducing the $M(M - 1)/2$ two-way frequency tabulations between the items. Let $\{G_{mh}, m \neq h\}$ be tabulations (with cells $g_{mhrs}$) containing $R \times R$ frequency totals between indicators $m$ and $h$. One may then obtain chi-square statistics $\chi^2_{mh}$ comparing $g_{mhrs}$ to expected values $e_{mhrs}$. Let $G_{rep,mh}$ similarly be the frequency cross-tabulation based on replicate indicator samples $y_{rep,im}$ and $y_{rep,ih}$ from the model. Then one may compare $\chi^2_{rep,mh}$ with $\chi^2_{mh}$ using a posterior predictive p-value. Another assessment tool is the proportion of observations on each item that are correctly classified, namely the concordancy rate (over all subjects) between predicted and actual ordinal categories.

The second half of a two chain run of 5000 iterations produces estimated loadings and predictive checks as in Table 9.4. It can be seen that the single factor model has particular deficits in reproducing $G_{12}, G_{15}, G_{23}$ and $G_{35}$ (cf. Bartholomew et al., 2008, p. 264).

A two factor model is then estimated with loading constraint $\lambda_{12} = 0$, where $\lambda_{12}$ is the loading of item 1 (government provide job for everyone) on factor 2. Additionally $\lambda_{22}$ is

**Table 9.4**    Factor models for ordinal items.

| Statistic | One Construct | | Two Constructs | |
|---|---|---|---|---|
| | Mean | St devn | Mean | St devn |
| Predictive check for $\chi^2, G_{12}$ | 0.00 | 0.04 | 0.26 | 0.44 |
| Predictive check for $\chi^2, G_{13}$ | 0.48 | 0.50 | 0.11 | 0.31 |
| Predictive check for $\chi^2, G_{14}$ | 0.31 | 0.46 | 0.62 | 0.49 |
| Predictive check for $\chi^2, G_{15}$ | 0.98 | 0.14 | 0.58 | 0.49 |
| Predictive check for $\chi^2, G_{23}$ | 0.95 | 0.22 | 0.86 | 0.35 |
| Predictive check for $\chi^2, G_{24}$ | 0.11 | 0.31 | 0.24 | 0.43 |
| Predictive check for $\chi^2, G_{25}$ | 0.92 | 0.28 | 0.65 | 0.48 |
| Predictive check for $\chi^2, G_{34}$ | 0.76 | 0.43 | 0.62 | 0.48 |
| Predictive check for $\chi^2, G_{35}$ | 0.02 | 0.15 | 0.32 | 0.47 |
| Predictive check for $\chi^2, G_{45}$ | 0.45 | 0.50 | 0.27 | 0.44 |
| Concordancy rate, indicator 1 | 0.45 | 0.02 | 0.53 | 0.03 |
| Concordancy rate, indicator 2 | 0.48 | 0.02 | 0.51 | 0.02 |
| Concordancy rate, indicator 3 | 0.58 | 0.02 | 0.61 | 0.03 |
| Concordancy rate, indicator 4 | 0.49 | 0.02 | 0.48 | 0.02 |
| Concordancy rate, indicator 5 | 0.62 | 0.02 | 0.69 | 0.03 |
| $\lambda_1$ | 1.93 | 0.14 | | |
| $\lambda_2$ | 1.32 | 0.12 | | |
| $\lambda_3$ | 2.53 | 0.21 | | |
| $\lambda_4$ | 2.30 | 0.18 | | |
| $\lambda_5$ | 2.44 | 0.21 | | |
| $\lambda_{11}$ | | | 2.74 | 0.31 |
| $\lambda_{21}$ | | | 1.59 | 0.15 |
| $\lambda_{31}$ | | | 1.48 | 0.47 |
| $\lambda_{41}$ | | | 1.79 | 0.26 |
| $\lambda_{51}$ | | | 1.31 | 0.56 |
| $\lambda_{22}$ | | | 0.06 | 0.06 |
| $\lambda_{32}$ | | | 2.10 | 0.35 |
| $\lambda_{42}$ | | | 0.82 | 0.17 |
| $\lambda_{52}$ | | | 2.61 | 0.41 |
| $\rho$ | | | 0.42 | 0.19 |

constrained to be positive. The correlation $\rho$ between the standardised factors is assigned a $U(-1, 1)$ prior.

It can be seen from Table 9.4 that a two factor model no longer has predictive deficits in reproducing item two-way cross-tabulations. Concordance rates also generally improve. The second factor loads particularly highly on items 3 and 5, and may be interpreted as support for social security spending. There may be some sensitivity to which item is chosen for the rotation invariance constraint, for example, if the items were sorted so that item for government controlling prices is set as item 1 (see Exercise 9.9).

## 9.5    Latent trait models for multilevel data

The principle of multivariate data reduction extends to hierarchically structured data (Rabe-Hesketh *et al.*, 2007; Bartholomew *et al.*, 2008). Thus consider cross-sectional indicators $y_{ijm}$ with subjects $i$ at level 1 (e.g. students) nested within clusters $j$ at level 2 (e.g. schools), and with latent variables potentially operating at each level. Correlations between multiple indicators at level 1 may reflect both latent student ability (i.e. a level 1 factor), and abilities of other students in the same school (i.e. a level 2 or cluster factor), with cluster effects reflecting school attributes and selection effects (Longford and Muthen, 1992).

Assume a two level model with a $p_1$ dimensional factor vector $\psi_{ij}$ at level 1, and a $p_2$ dimensional factor vector $\varphi_j$ at level 2 (clusters), and continuous indicators $y_{ij} = (y_{ij_1}, \ldots, y_{ijM})$ for clusters $j = 1, \ldots, J$, individuals $i = 1, \ldots, n_j$ within clusters, and variables $m = 1, \ldots, M$. Then a normal linear two level factor model (i.e. a measurement model) can be written as

$$y_{ij} = \alpha_j + \Gamma\psi_{ij} + \epsilon_{1ij},$$

$$\alpha_j = \kappa + \Lambda\varphi_j + \epsilon_{2j},$$

where $\Gamma$ is of dimension $M \times p_1$ and $\Lambda$ is $M \times p_2$. For indicator $m$ one has

$$y_{ijm} = \alpha_{jm} + \Gamma_m\psi_{ij} + \epsilon_{1ijm},$$

$$\alpha_{jm} = \kappa_m + \Lambda_m\varphi_j + \epsilon_{2jm}.$$

Communalities are defined for both levels in terms of the proportions of variation in $\epsilon_{1ijm}$ and $\epsilon_{2jm}$ accounted for.

Assume for illustration that $p_1 = p_2 = 2$. Then one has

$$y_{ijm} = \kappa_m + \lambda_{m1}\varphi_{1j} + \lambda_{m2}\varphi_{2j} + \gamma_{m1}\psi_{1ij} + \gamma_{m2}\psi_{2ij} + \epsilon_{2jm} + \epsilon_{1ijm},$$

where subject errors $\epsilon_{1ijm}$ are normal with variances $\sigma_{1m}^2$, cluster errors $\epsilon_{2jm}$ are normal with variances $\sigma_{2m}^2$, and constructs $\psi = (\psi_{1ij}, \psi_{2ij})$ and $\varphi = (\varphi_{1j}, \varphi_{2j})$ are normal with respective covariance matrices $\Phi_1$ and $\Phi_2$. The model in this case can equivalently be written

$$y_{ijm} = \alpha_{jm} + \gamma_{m1}\,\psi_{1ij} + \gamma_{m2}\,\psi_{2ij} + \epsilon_{1ijm},$$

$$\alpha_{jm} = \kappa_m + \lambda_{m1}\,\varphi_{1j} + \lambda_{m2}\,\varphi_{2j} + \epsilon_{2jm}.$$

Priors adopted for $\Phi_1$ and $\Phi_2$ depend partly on assumed relationships between the loadings at different levels. Thus if factors are standardised and uncorrelated at each level, and level 2 loadings $\Lambda$ are estimated independently of level 1 loadings $\Gamma$, $\Phi_1$ and $\Phi_2$ reduce to identity matrices in an exploratory factor analysis (Longford and Muthen, 1992). With $\Lambda$ estimated independently of $\Gamma$, there are $p_1(p_1 - 1)/2$ rotation invariance constraints needed on $\Lambda$, and $p_2(p_2 - 1)/2$ on $\Gamma$ (e.g. for $p_1 = p_2 = 2$ standardised factors rotation invariance constraints might be $\gamma_{12} = 0$ and $\lambda_{12} = 0$). Setting structural relationships between the loadings at different levels (or setting extra loadings to fixed values) makes certain dispersion parameters estimable. For example, one might take $\Lambda = \Gamma$, or centre priors for $\lambda_{mk}$ around $\gamma_{mk}$.

## Example 9.9   Nested data on science tests

Consider data on $M = 4$ tests of science knowledge over $N = 2439$ pupils in $J = 99$ schools (Bartholomew *et al.*, 2008). The tests relate to earth science, biology, physics and an additional biology test for a subsample of 1222 pupils. Missing results on the latter are assumed to be missing at random. A single student level latent ability $\psi_{ij}$, and a single school effect $\varphi_j$, are assumed. Then

$$y_{ijm} = \kappa_m + \lambda_m \varphi_j + \gamma_m \psi_{ij} + \epsilon_{2jm} + \epsilon_{1ijm},$$

where $\epsilon_{1ijm} \sim N(0, \sigma_{1m}^2)$, $\epsilon_{2jm} \sim N(0, \sigma_{2m}^2)$. The constructs $\psi_{ij}$ and $\varphi_j$ are taken as standard normal, initial loadings $\{\lambda_1, \gamma_1\}$ are constrained to be positive, and remaining loadings assigned $N(0, 1)$ priors. To assess possible gender impacts, average $\psi$ scores by gender[6] are calculated at each MCMC iteration. Since the factor scores are standardised, communalities at levels 1 and 2 are routinely obtained as $\gamma_m^2/(\gamma_m^2 + \sigma_{1m}^2)$ and $\lambda_m^2/(\lambda_m^2 + \sigma_{2m}^2)$.

The second half of a two chain run of 20 000 iterations provides parameter estimates as in Table 9.5, and a DIC of 21 051. A gender difference is apparent in the level 1 scores. Loadings at both levels show a higher impact of physics and biology on the overall construct than earth sciences. The model explains cluster variation better than student variation.

**Table 9.5**   Multilevel factor model, posterior summary.

|  |  | Mean | St devn | 2.5% | 97.5% |
|---|---|---|---|---|---|
| Average $\psi$ | Boys | −0.089 | 0.024 | −0.137 | −0.041 |
|  | Girls | 0.092 | 0.024 | 0.045 | 0.139 |
|  | ES | 0.170 | 0.020 | 0.132 | 0.211 |
| Level 1 communalities | BIOL | 0.457 | 0.038 | 0.384 | 0.534 |
|  | PHYS | 0.381 | 0.033 | 0.318 | 0.448 |
|  | BIOL(S) | 0.041 | 0.014 | 0.017 | 0.072 |
|  | ES | 0.460 | 0.095 | 0.267 | 0.635 |
| Level 2 communalities | BIOL | 0.959 | 0.057 | 0.799 | 0.999 |
|  | PHYS | 0.848 | 0.074 | 0.710 | 0.997 |
|  | BIOL(S) | 0.654 | 0.198 | 0.281 | 0.979 |
| $\gamma_1$ | ES | 0.378 | 0.024 | 0.330 | 0.426 |
| $\gamma_2$ | BIOL | 0.598 | 0.028 | 0.543 | 0.653 |
| $\gamma_3$ | PHYS | 0.541 | 0.026 | 0.489 | 0.593 |
| $\gamma_4$ | BIOL(S) | 0.193 | 0.035 | 0.127 | 0.262 |
| $\lambda_1$ | ES | 0.280 | 0.044 | 0.195 | 0.368 |
| $\lambda_2$ | BIOL | 0.477 | 0.045 | 0.393 | 0.570 |
| $\lambda_3$ | PHYS | 0.458 | 0.047 | 0.371 | 0.554 |
| $\lambda_4$ | BIOL(S) | 0.199 | 0.039 | 0.124 | 0.278 |

A second model introduces gender as an explanatory variable (formative indicator) for level 1 scores. Thus the assumption $\psi_{ij} \sim N(0, 1)$ is replaced by

$$\psi_{ij} \sim N(\alpha_1 + \alpha_2 x_{ij}, 1)$$

where $x_{ij}$ represents gender (= 1 for females). The parameter $\alpha_2$ has 95% credible interval $(-0.47, -0.25)$ and the DIC is reduced to 21 018. Level 2 communalities now stand at 0.47, 0.97, 0.84 and 0.70, so contextual effects are mostly better explained.

## 9.6    Structural equation models for missing data

In structural equation models, including confirmatory factor models, it may be that latent variables, rather than (or as well as) observed indicators, contribute to predicting or understanding missingness mechanisms, and shared random effect approaches to missingness have become more widely applied (e.g. Yang and Shoptaw, 2005; Albert and Follmann, 2007). Sample selection or selective attrition that lead to missing data may be more clearly related to constructs than to any combination of the multiple indicators for such constructs. Denote the full set of observed and missing indicator data as $Y = \{Y_{obs}, Y_{mis}\}$, with the observed substantive indicators amplified by an $n \times M$ matrix of binary indicators $R_{im}$ corresponding to whether $y_{im}$ is missing ($R_{im} = 1$) or observed ($R_{im} = 0$).

As noted by Arbuckle (1996), missing data methods for structural equation applications are often based on the missing at random (MAR) assumption, or assume special patterns of missingess, such as monotone patterns. Under the MAR assumption the distribution of $R$ depends only on the observed data, so

$$P(R|Y, \omega) = P(R|Y_{obs}, \omega).$$

In many situations non-response may be non-monotone and non-random (informative). A potential benefit of factor and SEM models is in accounting for missingness via latent constructs $\psi$ based on the expanded observations $(Y_{obs}, R)$. Thus under a selection approach to missing data, informative missingness would imply the sequence

$$P(R, Y|\theta, \omega, \psi) = P(R|Y, \omega, \psi)P(Y|\theta, \psi),$$

with parameters $(\omega, \theta)$, and constructs $\psi$ relevant to explaining missingness as well as correlations between the substantive indicators. Conditional independence of substantive indicators $Y$ and missingness indicators $R$ applies if $P(R|Y, \omega, \psi) = P(R|\omega, \psi)$ (e.g. Yang and Shoptaw, 2005).

**Example 9.10    Alienation over time**

This example considers adaptations of the data used in a structural equation model of alienation over time (Wheaton *et al.*, 1977), originally with $n = 932$ subjects. In a reworked analysis of the simulations of Muthen *et al.* (1987), there are six indicators of two constructs (social status and alienation) at time 1, and three indicators of alienation at time 2. A slightly smaller number of subjects ($n = 600$) is assumed.

Denote social status and alienation at time 1 by $\xi_1$ and $\psi_1$, and alienation at time 2 by $\psi_2$. The indicators for $i = 1, \ldots, 600$ subjects are standardised, with three positive status indicators $X_1 - X_3$ at time 1 related to the social status (exogenous) construct:

$$x_{i1} = \lambda_{11}\xi_i + \delta_{i1},$$

$$x_{i2} = \lambda_{21}\xi_i + \delta_{i2},$$

$$x_{i3} = \lambda_{31}\xi_i + \delta_{i3},$$

where $\delta_1, \delta_2$, and $\delta_3$ are independent univariate normal. The three indicators of alienation are denoted $Y_{11}, Y_{21}$ and $Y_{31}$ at time 1 and $Y_{12}, Y_{22}$ and $Y_{32}$ at time 2. They are related to the alienation construct at times 1 and 2 as follows:

$$y_{11i} = \lambda_{12}\psi_{1i} + \epsilon_{i1},$$

$$y_{21i} = \lambda_{22}\psi_{1i} + \epsilon_{i2},$$

$$y_{31i} = \lambda_{32}\psi_{1i} + \epsilon_{i3},$$

$$y_{12i} = \lambda_{13}\psi_{2i} + \epsilon_{i4},$$

$$y_{22i} = \lambda_{23}\psi_{2i+}\epsilon_{i5},$$

$$y_{32i} = \lambda_{33}\psi_{2i} + \epsilon_{i6}.$$

The constructs themselves are related in a structural model with cross-sectional dependence at time 1, namely

$$\psi_{1i} = \beta_{11}\xi_i + w_{1i},$$

and with longitudinal dependence relating alienation at time 2 to earlier alienation and social status, namely

$$\psi_{2i} = \beta_{21}\xi_i + \beta_{22}\psi_{1i} + w_{2i}.$$

Missingness is assumed confined to second wave indicators of alienation, and to then apply to all three $y$-items together. Thus let $R_i$ be a binary indicator of whether a subject is missing at wave 2 (i.e. unit rather than item non-response), with $R_i = 1$ for response missing, and $R_i = 0$ for response present. Underlying the binary missingness indicator is a latent continuous variable $R_i^*$, so that lower $R_i^*$ lead to raised chances of being classed as missing, namely $R_i = 1$ if $R_i^* < \tau$.

Different missingness schemes may be devised. A MAR missingness model assumes $R_i^*$ related only to fully observed (i.e. first wave) data $x_{ki}$ and $y_{k1i}$, as in the illustrative scheme (Muthen et al., 1987)

$$R_i^* = 0.667\omega^*(x_{i1} + x_{i2} + x_{i3}) - 0.333\omega^*(y_{11i} + y_{21i} + y_{31i}) + v_i, \tag{9.2}$$

where $v_i \sim N(0, 1)$, $\omega^* = 0.329$, and $\tau = -0.675$. This leads to a missingness rate of around 25%. Another choice is to make missingness depend on all wave 1 and 2 outcomes, whether observed or not, so that missingness is non-random. Thus, again with $\tau = -0.675$,

$$R_i^* = 0.667\omega^*(x_{i1} + x_{i2} + x_{i3}) - 0.333\omega^*(y_{11i} + y_{21i} + y_{31i} + y_{12i} + y_{22i} + y_{32i}) + v_i. \tag{9.3}$$

A further option, also non-random, makes missingness depend on latent constructs so that

$$R_i^* = 0.667\omega^*\xi_i - 0.333\omega^*(\psi_{1i} + \psi_{2i}) + v_i, \tag{9.4}$$

with $\omega^* = 0.619$. This is non-random because $\psi_2$ is defined both by observed and missing data at phase 2.

Accordingly a complete dataset $\{X_j, j = 1, 3; Y_{j_1}, j = 1, 3; Y_{j_2}, j = 1, 3\}$ is generated (from the observed covariance or correlation matrix), but then a subset of the wave 2 data is removed according to one of the above missingness schemes. One may subsequently compare (a) parameter estimates using the complete dataset; (b) parameters obtained under MAR or MCAR missingness; and (c) parameters obtained under a MNAR missingness model based on the constructs. Here a complete dataset is generated using the $9 \times 9$ correlation matrix for the full data, and missingness then applied according to scheme (9.3), with $\omega^* = 0.27$ (see the first program code for this example). Wave 2 data is removed (regarded as missing) when $R_i^*$ is under the threshold $-0.675$. There are 169 of 600 observations with wave 2 missingness, a rate of 28%. Under the selected response mechanism, attrition is greater for lower status persons and more alienated persons: so missingess might be expected to be greater for subjects with higher scores on $\psi_1$ and $\psi_2$.

Consider now the simulated data, regarded as observations with an unknown DGP. In model 1 in the code, the response model relates $\pi_i = \Pr(R_i = 1)$ to the factor scores, namely,

$$\text{logit}(\pi_i) = \omega_0 + \omega_1 \xi_i + \omega_2 \psi_{1i} + \omega_3 \psi_{2i}. \tag{9.5}$$

Since $\psi_2$ is defined both by observed and missing data at phase 2, this corresponds to assuming missingness not at random.

From the second half of a two chain sample of 5000 iterations (in OpenBUGS), the expected positive impacts on missingness of alienation ($\psi_1$ and $\psi_2$), and a negative impact of status $\xi$, are obtained (see Table 9.6). The impact of $\psi_2$ is, as might be expected, less precisely estimated than that of $\psi_1$. Other models relating $\pi_i$ to (say) just $\xi$ and $\psi_2$ might be tried (see Exercise 9.10). The coefficients of the structural model ($\beta$ coefficients in Table 9.6) are close to parameters obtained from the complete dataset, though the negative impact $\beta_{21}$ of social status $\xi$ on time 2 alienation ($\psi_2$) is enhanced.

**Table 9.6**  Alienation study, shared factor missingness, parameter summary.

|  | Mean | St devn | 2.50% | Median | 97.50% |
|---|---|---|---|---|---|
| Missingness coefficients |  |  |  |  |  |
| $\omega_0$ | −1.12 | 0.12 | −1.35 | −1.11 | −0.90 |
| $\omega_1$ | −0.72 | 0.25 | −1.23 | −0.72 | −0.23 |
| $\omega_2$ | 0.64 | 0.32 | 0.03 | 0.64 | 1.28 |
| $\omega_3$ | 0.43 | 0.46 | −0.44 | 0.42 | 1.37 |
| Structural model coefficients |  |  |  |  |  |
| $\beta_{11}$ | −0.56 | 0.06 | −0.68 | −0.56 | −0.46 |
| $\beta_{21}$ | −0.31 | 0.07 | −0.46 | −0.31 | −0.16 |
| $\beta_{22}$ | 0.54 | 0.07 | 0.40 | 0.54 | 0.69 |
| Measurement model coefficients |  |  |  |  |  |
| $\lambda_{11}$ | 1.00 |  |  |  |  |
| $\lambda_{21}$ | 1.05 | 0.07 | 0.92 | 1.05 | 1.19 |
| $\lambda_{31}$ | 0.70 | 0.06 | 0.60 | 0.70 | 0.82 |
| $\lambda_{12}$ | 1.00 |  |  |  |  |
| $\lambda_{22}$ | 1.00 | 0.06 | 0.89 | 1.00 | 1.13 |
| $\lambda_{32}$ | 0.73 | 0.05 | 0.62 | 0.72 | 0.83 |
| $\lambda_{13}$ | 1.00 |  |  |  |  |
| $\lambda_{23}$ | 0.90 | 0.09 | 0.73 | 0.89 | 1.08 |
| $\lambda_{33}$ | 0.56 | 0.08 | 0.41 | 0.56 | 0.73 |

Instead of MNAR missingness, one might assume a model (see Allison, 2003) with no information to predict missingess (i.e. MCAR), with

$$\text{logit}(\pi_i) = \omega_0.$$

This is specified as pi.1[] in the model 1 code. In the present case, and with the particular sample of data from the correlation matrix, this option produces very similar estimates of structural and measurement coefficients to the non-random missingness model.

Both models in turn provide similar parameter estimates to those based on the fully observed data set of $600 \times 9$ variables (model 2 in the code). For example, the structural coefficients based on the complete data have posterior means (sd) of $\beta_{11} = -0.57(0.06)$, $\beta_{21} = -0.28(0.07)$ and $\beta_{22} = 0.54(0.06)$.

The model (9.5) allowing for non-random missingness provides an estimate for $\lambda_{23}$ slightly closer to the complete data estimate, but $\beta_{21}$ is better estimated under MCAR missingness. So for this particular sampled data set there is no benefit in using a missingness model linked to values on the latent constructs. However, to draw firm conclusions about the benefits of random vs. non-random missingess it would be necessary to repeat this analysis with a large number of replicate data sets.

## Exercises

**9.1.** Re-estimate the democracy-industrialisation analysis, Example 9.1, using a standardised factor scores constraint (i.e. a unit variance rather than unit loading constraint for identifiability). It is suggested to constrain the loadings $\lambda_{x1}$, $\lambda_{y_1}$ and $\lambda_{y5}$ to take only positive values.

**9.2.** In Example 9.2, repeat the posterior predictive checks applied in model 1, but with the added step

$$w_{im,\text{new}} \sim \text{Ga}(v_m/2, v_m/2),$$

so that $\tau_{im,\text{new}} = T_m w_{im,\text{new}}$ and $y_{im,\text{new}} \sim N(\mu_{im}, 1/\tau_{im,\text{new}})$.

**9.3.** An important question in confirmatory factor analysis is the invariance of structural relationships across groups: for example, should the effect of each indicator on the underlying constructs be allowed to vary by group. In Example 9.2, first assess whether significant differences in factor score averages exist between four pupil groups defined by gender and school (see the vector mnF.G in the code). Find which group has the highest median rank on each of the three factors (this can be assessed using the BUGS rank monitor). Then assess invariance (assuming a multivariate normal factor model) using group specific intercepts, loadings and factor covariance matrices.

**9.4.** In Example 9.3, assess how posterior rejection/retention rates for $\gamma$ coefficients under the extended model with slab/spike prior are affected by more diffuse priors under the retention option ($\delta_j = 2$). The extended model is

$$y_i = \gamma_{1\delta_1}\xi_{i1} + \gamma_{2\delta_2}\xi_{i2} + \gamma_{3\delta_3}\xi_{i1}\xi_{i2} + \gamma_{4\delta_4}\xi_{i1}^2 + \gamma_{5\delta_5}\xi_{i2}^2$$

with

$$\gamma_{j\delta_j}|\delta_j = 1 \sim N(0, 0.01), \qquad j = 1, \dots, 5$$

$$\gamma_{j\delta_j}|\delta_j = 2 \sim N(0, V_2),$$

Thus consider impacts on retention of the settings $V_2 = 10$ and $V_2 = 100$.

**9.5.** Examine whether a second order random walk in the latent factor improves fit for the un-differenced monetary data (Example 9.4).

**9.6.** Re-estimate model 2 in Example 9.5 (verbal and quantitative ability), namely the specification with bivariate normal residual covariance for the indicator pairs $(x_1, y_1)$ through to $(x_5, y_5)$, and also with fixed effects priors for the eight unknown indicator-construct loadings. However, to ensure identifiability adopt a reduced structural model without autoregressive parameters,

$$\xi_{1i} \sim N(0, \phi_1^2), \qquad \xi_{2i} \sim N(0, \phi_2^2),$$
$$\psi_{1i} \sim N(\xi_{1i}, \phi_3^2), \qquad \psi_{2i} \sim N(\xi_{2i}, \phi_4^2).$$

How does this option affect the posterior predictive loss criterion and the predictive checks comparing the $M = 5$ observed and predicted indicator correlations ($c_m$ and $c_{m,\text{rep}}$) between grades 7 and 9.

**9.7.** Assess fit of models 1 and 2 in Example 9.7 (LCA and latent trait models for diagnosis) using a posterior predictive check (Sinharay, 2005, p. 379) of the observed score distribution, which in this application is the total number of patients with category 2 scores on the three manifest items (unwell diagnosis). For example, some patients have zero items with category 2, while some patients have all items registering category 2. For values $i = 0$ through to $i = 3$ of category 2 scores, let $p_i$ be the observed proportions of $n = 103$ subjects with $i = 0, \ldots, 3$, and $q_i$ be the proportions based on replicate data. Then one may compare multinomial diversity indices for observed and replicate data, namely $G_p = 1 - \sum_i p_i$ and $G_q = 1 - \sum_i q_i$ (Gini indices), or $E_p = -\sum_i p_i \log(p_i)$ and $E_q = -\sum_i q_i \log(q_i)$ (entropy indices). Diversity can also be measured by $\exp(E_p)$ and $\exp(E_q)$ (Jost, 2006).

**9.8.** Apply a 3PL model (including guessing parameters) to the Law School Admission Test (LSAT) data in the R program ltm. The 2PL model may be fitted with the code

```
model { for (i in 1:1000) { F[i] ~ dnorm(0,1)
 for (m in 1:5) {y[i,m] ~ dbern(pi[i,m])
logit(pi[i,m]) <- lam[m]*F[i] - delta[m]}}
for (m in 1:5) {lam[m] ~ dexp(1); delta[m] ~ dnorm(0,0.001)}}
```

Suitability of the two models in reproducing the data may be assessed by comparing the 10 actual and predicted $2 \times 2$ frequency tables between the five items, with predicted tables based on aggregating subject level replicates

```
yrep[i,m] ~ dbern(pi[i,m]).
```

A posterior predictive check may be based on comparing chi-square statistics using replicate and actual frequency tables.

**9.9.** In Example 9.8 (involving ordinal indicators of the perceived desirability of various government actions) consider how the two factor solution is changed if the item 'Should the government keep prices under control' becomes item 1, and the loading constraint $\lambda_{12} = 0$ is retained.

**9.10.** In Example 9.10, modify model 1 to assess sensitivity of structural coefficients to a MNAR missingness model (for wave 2 unit missingness), so that $\pi_i$ depends only on $\xi$ (social status) and $\psi_2$ (wave 2 alienation).

**9.11.** In Example 9.10, generate data with wave 2 missingness using the scheme 9.3 but with threshold $\tau = -0.1$ (leading to higher missingness). Then compare inferences (e.g. for structural model coefficients) under MCAR and MNAR missingness models, where the MNAR model for $\pi_i = \Pr(R_i = 1)$ is

$$\text{logit}(\pi_i) = \omega_0 + \omega_1 \xi_i + \omega_2 \psi_{1i} + \omega_3 \psi_{2i}.$$

# Notes

1. The code for the multivariate t factor model in Example 9.2 is

```
model {# prior on d.f. for Student t (by indicator m=1 to M)
for (m in 1:M) { nu[m] ~ dexp(eta[m]) I(1,);     eta[m] ~ dunif(0.01,1)
                 nu2[m] <- nu[m]/2
                 T[m] ~ dgamma(1,0.001)
for   (i in 1:n) {#  Student t likelihoods for indicators
                 z[i,m] <- (y[i,m]-mean(y[,m]))/sd(y[,m])
                 z[i,m] ~ dnorm(mu[i,m],tau[i,m])
# predictions
                 znew[i,m] ~ dnorm(mu[i,m],tau[i,m])
# regrn errors
                 e2[i,m] <- pow(z[i,m]-mu[i,m],2)
                 e2.new[i,m] <- pow(znew[i,m] - mu[i,m],2)
                 tau[i,m] <- T[m]*w[i,m];
                 w[i,m] ~ dgamma(nu2[m],nu2[m])}}
for (i in 1:n) {for (m in 1:3) {mu[i,m] <- kappa[m]+lambda[m]*F[i,1]}
for (m in 4:6) {mu[i,m] <- kappa[m]+lambda[m]*F[i,2]}
for (m in 7:9) {mu[i,m] <- kappa[m]+lambda[m]*F[i,3]}}
# Posterior Predictive check
MSE <- sqrt(sum(e2[1:n,1:M])/(n-M))
MSE.new <- sqrt(sum(e2.new[1:n,1:M])/(n-M))
PPC <- step(MSE.new-MSE)
# Fixed effect priors
for (m in 1:9) {kappa[m] ~ dflat()}
# Priors for Latent variable model
# Correlated Factors
nu.F ~ dexp(eta.F) I(1,);     eta.F ~ dunif(0.01,1);     nu2.F <- nu.F/2
for   (i in 1:n) { w.F[i] ~ dgamma(nu2.F,nu2.F)
                 F[i,1:p] ~ dmnorm(nought[],P.F[i,1:p,1:p])
for (j in 1:p) {  for (k in 1:p) { P.F[i,j,k] <- w.F[i]*P[j,k]}}}
# pupil group indicators
 for   (i in 1:n) { G[i] <- (sch[i]-1)*2+gend[i]
 for (g in 1:4) {IndG[i,g] <- equals(G[i],g)
 for (k in 1:p) { F.G[i,g,k] <- F[i,k]*equals(G[i],g)}}}
# Factor score averages by group
 for (k in 1:p) {for (g in 1:4) {mnF.G[g,k] <- sum(F.G[1:n,g,k])
                                              /sum(IndG[1:n,g])}}
# Precision matrix of factor scores
P[1:p,1:p]  ~ dwish(R[,],3); Phi[1:p,1:p] <- inverse(P[,]);
for (j in 1:p) {  for (k in 1:p) {
            CorrF[j,k] <- Phi[j,k]/sqrt(Phi[ j,j]*Phi[k,k])}}
# unit loading constraint
lambda[1] <- 1; lambda[4] <- 1; lambda[7] <- 1
for (j in 2:3) {lambda[j] ~ dnorm(0,1)}
for (j in 5:6) {lambda[j] ~ dnorm(0,1)}
for (j in 8:9) {lambda[j] ~ dnorm(0,1)}}
```

2. The code to generate the data in Example 9.3 is

```
model {for (i in 1:100) {y[i] ~ dnorm(mu.y[i],1.667)
mu.y[i] <- gamma[1]*xi[i,1]+gamma[2]*xi[i,2]+gamma[3]*xi[i,1]*xi[i,2]
xi[i,1:2] ~ dmnorm(nought[],P[,])
x[1,i] ~ dnorm(mu.x[i,1],5)
x[2,i] ~ dnorm(mu.x[i,2],3.33)
x[3,i] ~ dnorm(mu.x[i,3],2.5)
x[4,i] ~ dnorm(mu.x[i,4],2)
x[5,i] ~ dnorm(mu.x[i,5],1.67)
mu.x[i,1] <- lam[1,1]*xi[i,1]
mu.x[i,2] <- lam[1,2]*xi[i,1]
mu.x[i,3] <- lam[1,3]*xi[i,1]
mu.x[i,4] <- lam[2,1]*xi[i,2]
mu.x[i,5] <- lam[2,2]*xi[i,2]}
# Factor Means and Precision Matrix
nought[1] <- 0; nought[2] <- 0; P[1:2,1:2] <- inverse(Phi.xi[,])
Phi.xi[1,1] <- 1.2; Phi.xi[2,2] <- 0.7; Phi.xi[1,2] <- 0.1;
                                        Phi.xi[2,1] <- 0.1
gamma[1] <- 0.8; gamma[2] <- 1.7; gamma[3] <- -0.5
# Indicator to factor loadings
lam[1,1] <- 1; lam[1,2] <- 0.7; lam[1,3] <- -0.5; lam[2,1]
                                        <- 1; lam[2,2] <- 1.6}
```

3. The code for the loading selection model (Example 9.3) is

```
model {  pi.del[1] <- 0.5; pi.del[2] <- 0.5;
for (j in 1:5) {gamma[j] ~ dnorm(0,inv.tau2.gam[j,del[j]])
# loading retention probability
pr.rej[j] <- equals(del[j],1)
del[j] ~dcat(pi.del[1:2])
inv.tau2.gam[j,1] <- 100
inv.tau2.gam[j,2] <- 1}
 # Structural Model
for (i in 1:100) { y[i] ~ dnorm(mu.y[i],tau.y)
                        ynew[i] ~ dnorm(mu.y[i],tau.y)
mu.y[i] <- gamma[1]*xi[i,1]+gamma[2]*xi[i,2]+gamma[3]
   *xi[i,1]*xi[i,2]+gamma[4]*xi[i,1]*xi[i,1]+gamma[5]*xi[i,2]*xi[i,2]}
# Measurement Model
for (i in 1:100) {for (j in 1:5) {x[i,j] ~ dnorm(mu.x[i,j],tau.x[j])
                        xnew[i,j] ~ dnorm(mu.x[i,j],tau.x[j])}
mu.x[i,1] <-             xi[i,1]
mu.x[i,2] <- lam[1,2]*xi[i,1]
mu.x[i,3] <- lam[1,3]*xi[i,1]
mu.x[i,4] <-             xi[i,2]
mu.x[i,5] <- lam[2,2]*xi[i,2]}
# priors
for (j in 1:5) {tau.x[j] ~ dgamma(0.5,0.01) }
                tau.y ~ dgamma(0.5,0.01)
# coefficients
lam[1,2] ~ dnorm(0,1);    lam[1,3] ~ dnorm(0,1);    lam[2,2] ~ dnorm(0,1)
# precision matrix for constructs
P[1 : 2, 1 : 2] ~ dwish(R[ , ], 2)
R[1,1] <- 0.01;   R[2,2] <- 0.01;  R[2,1] <- 0;       R[1,2] <- 0
nought[1] <- 0;     nought[2] <- 0
Phi.xi[1:2,1:2] <- inverse(P[,])
```

```
# Prior on factors
for (i in 1:100) {xi[i,1:2] ~ dmnorm(nought[],P[,])}}
```

4. The code for the RW1 construct applied to undifferenced financial indicators in Example
9.4 is

```
model {lam[1] <- 1; for (j in 2:6){lam[j] ~ dnorm(0,1)}
 # standardised indicators
for (t in 1:T) {for (j in 1:6){
z.st[t,j] <- (z[t,j] - mean(z[,j]))/sd(z[,j]);
z.st[t,j] ~ dnorm(mu[t,j], tau[j]);
mu[t,j] <- alph[j] + lam[j]*F[t]}}
# communalities
for (j in 1:6) {com-
mun.y[j] <- (pow(lam[j],2)*varF)/ (pow(lam[j],2)*varF+1/tau[j])}
# Priors
for (j in 1:6) {alph[j] ~ dnorm(0,0.001); tau[j] <- 1/(s[j]*s[j]);
                                         s[j]~dunif(0,1000)}
# RW1 prior adjacencies and weights
wt[1] <- 1;                  adj[1] <- 2;                      n[1] <- 1
wt[(T-2)*2+2] <- 1; adj[(T-2)*2+2] <- T-1; n[T] <- 1
for (t in 2:T-1) {wt[2+(t-2)*2] <- 1;    adj[2+(t-2)*2] <- t-1
                 wt[3+(t-2)*2] <- 1;     adj[3+(t-2)*2] <- t+1;
                                                      n[t] <- 2}
tauF ~ dgamma(1,0.001); varF <- sd(F[])*sd(F[])
F[1:T] ~ car.normal(adj[], wt[], n[], tauF)}
```

5. Code for the spatial factor model (Example 9.6) is

```
model { for (i in 1:n) { y[i]   ~  dpois(mu[i]);
mu[i]   <- E[i]*rh[i] ; logrh[i] <- log(rh[i])
log(rh[i]) <- b[1] +b[2]*dep[i] +b[3]*F[i] + s[i]
# pollution indicators
        z[1,i] <- (NO2a[i]-mean(NO2a[]))/sd(NO2a[])
        z[2,i] <- (NOXa[i]-mean(NOXa[]))/sd(NOXa[])
        z[3,i] <- (PM10a[i]-mean(PM10a[]))/sd(PM10a[])
        z[4,i] <- (PM10e[i]-mean(PM10e[]))/sd(PM10e[])
        z[5,i] <- (PM25e[i]-mean(PM25e[]))/sd(PM25e[])
for (j in 1:5) {z[j,i] ~ dt(mu.z[j,i],tauz[j],nu[j])
                mu.z[j,i] <- a[j] + lam[j]*F[i]}}
# Priors
for (j in 1:5) {lam[j] ~ dnorm(1,1) I(0,)}
                gam ~ dnorm(0,0.001)
for (j in 1:5) {a[j] ~ dnorm(0,0.001); tauz[j] ~ dgamma(1,0.001);
                inv.nu[j] ~ dunif(0.005,0.25); nu[j]
                                    <- 0.5/inv.nu[j]}
tauF <- 1; taus ~ dgamma(1,0.001);
for (j in 1:2) {kap[j] ~ dunif(0,1)}
# standardised beta's
sx[1] <- sd(dep[]); sx[2] <- sd(F[])
for (j in 1:3) {b[j] ~ dnorm(0,1)}
for (j in 1:2) {b.st[j] <- b[j+1]*sx[j]/sd(logrh[])}
# spatial prior on F
for (i in 1:NN) {e.adj[i] <- e[adj[i]]; s.adj[i] <- s[adj[i]] }
for (i in 1:n) {F[i]   ~ dnorm(muF[i],tau.F[i])
```

```
z.dep[i] <- (dep[i]-mean(dep[]))/sd(dep[])
eta[i] <- gam*z.dep[i]
e[i] <- F[i]-eta[i]
tau.F[i]    <-   tauF *  (1-kap[1]+kap[1]*num[i])
muF[i] <- eta[i]+kap[1]*sum(e.adj[cum[i]+1:cum[i+1] ])
                         /(1-kap[1]+kap[1]*num[i])}
# spatial prior on s
for (i in 1:n) {s[i]  ~ dnorm(mu.s[i],tau.s[i])
tau.s[i]    <-   taus *  (1-kap[2]+kap[2]*num[i])
mu.s[i] <- kap[2]*sum(s.adj[cum[i]+1:cum[i+1] ])
                       /(1-kap[2]+kap[2]*num[i])}}
```

6. The code for first model applied to the multilevel science test data is as follows:

```
model { for (i in 1:n) {for (m in 1:4) {y[i,m] ~ dnorm(mu[i,m],tau1[m])
mu[i,m] <- gam0[m]+gam1[m]*F1[i]+lam[m]*F2[sch[i]]+u2[sch[i],m]}
F1[i] ~ dnorm(0,1)
F1mn[i,1] <- equals(fem[i],1)*F1[i]
F1mn[i,2] <- equals(fem[i],0)*F1[i]}
# average construct scores by gender
for (g in 1:2) {F1mn.g[g] <- sum(F1mn[,g])/totg[g]}
totg[1] <- 1243; totg[2] <- 1196;
gam1[1] ~ dexp(1); for (j in 2:4) {gam1[j] ~ dnorm(0,1)}
lam[1] ~ dexp(1); for (j in 2:4) {lam[j] ~ dnorm(0,1)}
for (j in 1:4) {tau1[j] ~ dgamma(1,0.001); sig2.1[j] <- 1/tau1[j];
                tau2[j] ~ dgamma(1,0.001); sig2.2[j] <- 1/tau2[j];
     commun1[j] <- (gam1[j]*gam1[j])/(gam1[j]*gam1[j]+sig2.1[j])
     commun2[j] <- (lam[j]*lam[j])/(lam[j]*lam[j]+sig2.2[j])
                gam0[j] ~ dnorm(0,0.001)}
for (j in 1:J) {F2[j] ~ dnorm(0,1)
for (m in 1:4) {u2[j,m] ~ dnorm(0,tau2[m])}}}.
```

Average $\psi$ scores by gender are obtained by monitoring F1mn.g. Data are input in a student level dataset including school indicators sch[], with the first few records being

| sch[] | y[,1] | y[,2] | y[,3] | y[,4] | fem[] |
|-------|-------|-------|-------|-------|-------|
| 1 | 1.03 | −0.04 | −0.59 | 1.37 | 1 |
| 1 | −1.10 | −0.58 | −0.10 | NA | 0 |
| 1 | −0.04 | 0.51 | −1.07 | 1.37 | 1 |
| 1 | 1.03 | −0.58 | 0.38 | NA | 0 |

# References

Aguilar, O. and West, M. (2000) Bayesian dynamic factor models and portfolio allocation. *Journal of Business and Economic Statistics*, **18**(3), 338–357.

Albert, J. and Chib, S. (1993) Bayesian analysis of binary and polychotomous response data. *Journal of the American Statistical Association*, **88**(422), 669–679.

Albert, J. and Ghosh, M. (2000) Item response modeling. In D. Dey, S. Ghosh and B. Mallick (eds), *Generalized Linear Models: a Bayesian Perspective*, pp. 173–193. Marcel-Dekker, New York, NY.

Albert, P. and Follmann, D. (2007) Random effects and latent processes approaches for analyzing binary longitudinal data with missingness: a comparison of approaches using opiate clinical trial data. *Statistical Methods in Medical Research*, **16**(5), 417–439.

Allison, P. (2003) Missing data techniques for structural equation modeling. *Journal of Abnormal Psychology*, **112**(4), 545–557.

Arbuckle, J. (1996) Full information estimation in the presence of incomplete data. In G. Marocoulides and R. Schumacker (eds), *Advanced Structural Equation Modelling*. Lawrence Erlbaum, Mahwah, NJ.

Arhonditsis, G., Stow, C., Steinberg, L., Kenney, M., Lathrop, R., McBride, S. and Reckhow, K. (2006) Exploring ecological patterns with structural equation modeling and Bayesian analysis. *Ecological Modelling*, **192**(3), 385–409.

Arminger, G. and Muthen, B. (1998) A Bayesian approach to nonlinear latent variable models using the Gibbs sampler and the Metropolis-Hastings algorithm. *Psychometrika*, **63**, 271–300.

Asai, M., McAleer, M. and Yu, J. (2006) Multivariate stochastic volatility: a review. *Econometric Reviews*, **25**(2–3), 145–175.

Asparouhov, T. and Muthen, B. (2011) Using Bayesian priors for more flexible latent class analysis. In *Proceedings of the 2011 Joint Statistical Meeting, Section on Government Statistics*, pp. 4979–4993. American Statistical Association, Alexandria, VA.

Azevedo, C., Andrade, D. and Fox, J. (2012) A Bayesian generalized multiple group IRT model with model-fit assessment tools. *Computational Statistics and Data Analysis*, **56**, 4399–4412.

Banerjee, S., Carlin, B. and Gelfand, A. (2004) *Hierarchical Modeling and Analysis for Spatial Data*. CRC, Boca Raton, FL.

Bartholomew, D. and Knott, M. (1999) *Latent Variable Models and Factor Analysis*. Kendall's Library of Statistics, 7.

Bartholomew, D., Steele, F., Galbraith, J. and Moustaki, I. (2008) *Analysis of Multivariate Social Science Data*. CRC, Boca Raton, FL.

Bollen, K. (1989) *Structural Equations with Latent Variables*. Wiley, Chichester, UK.

Bollen, K. (1998) Structural equation models. In P. Armitage and T. Colton (eds), *Encyclopaedia of Biostatistics*, pp. 4363–4372. Wiley, Chichester, UK.

Bollen, K. (2002) Latent variables in psychology and the social sciences. *Annual Review of Psychology*, **53**, 605–634.

Bollen, K. and Bauldry, S. (2011) Three Cs in measurement models: causal indicators, composite indicators, and covariates. *Psychological Methods*, **16**, 265–284.

Cagnone, S., Mignani, S. and Moustaki, I. (2009) Latent variable models for ordinal data. In M. Bini (ed.), *Statistical Methods for the Evaluation of Educational Services and Quality of Products*. Springer-Verlag, Berlin.

Crouse, D., Ross, N. and Goldberg, M. (2009) Double burden of deprivation and high concentrations of ambient air pollution at the neighbourhood scale in Montreal, Canada. *Social Science and Medicine*, **69**(6), 971–981.

Damien, P. and Walker, S. (2001) Sampling truncated normal, beta, and gamma densities. *Journal of Computational and Graphical Statistics*, **10**(2), 206–215.

Dunn, G. (1999) *Statistics in Psychiatry*. Arnold, London.

Earley, P., Lee, C. and Hanson, L. (1990) Joint moderating effects of job experience and task component complexity: relations among goal setting, task strategies and performance. *Journal of Organizational Behaviour*, **11**, 3–15.

Fox, J. (2002) An R and S-PLUS Companion to Applied Regression. SAGE Publications, London, UK.

Fox, J.-P. (2010) *Bayesian Item Response Modeling: Theory and Applications*. Springer, Berlin.

Gelfand, A.E. and Ghosh, S. K. (1998) Model choice: A minimum posterior predictive loss approach. *Biometrika*, **85**(1), 1–11.

Geweke, J. and Zhou, G. (1996) Measuring the pricing error of the arbitrage pricing theory. *Review of Financial Studies*, **9**, 557–587.

Ghosh, J. and Dunson, D. (2009) Default prior distributions and efficient posterior computation in Bayesian factor analysis. *Journal of Computational and Graphical Statistics*, **18**(2), 306–320.

Gilbert, P. and Meijer, E. (2005) Time series factor analysis with an application to measuring money. Research Report 05F10 University of Groningen http://irs.ub.rug.nl/ppn/289322812.

Goldstein, H. and Browne, W. (2002) Multilevel factor analysis modelling using markov chain monte carlo estimation. In G. Marcoulides and I. Moustaki (eds), *Latent Variable and Latent Structure Models*, pp. 225–244. Lawrence Erlbaum, Mahwah, NJ.

Harvey, A. and Shephard, N. (1993) Structural time series models. In G. Maddala, C. Rao and H. Vinod (eds), *Handbook of Statistics*, Vol. **11**, pp 261–302. Elsevier, Oxford, UK.

Ho, D. and Quinn, K. (2008) Improving the presentation and interpretation of online ratings data with model-based figures. *The American Statistician*, **62**(4), 279–288.

Huber, P., Scaillet, O. and Victoria-Feser, M. (2009) Assessing multivariate predictors of financial market movements: A latent factor framework for ordinal data. *Annals of Applied Statistics*, **3**(1), 249–271.

Hurley, A., Scandura, T., Schriesheim, C., Brannick, M., Seers, A., Vandenberg, R. and Williams, L. (1997) Exploratory and confirmatory factor analysis: guidelines, issues, and alternatives. *Journal of Organizational Behavior*, **18**(6), 667–683.

Johnson, V. and Albert, J. (1999) *Ordinal Data Modeling*. Springer-Verlag, New York, NY.

Joreskog, K. (1973) A general method for estimating as linear structural equation system. In A.S. Goldberger and O.D. Duncan (eds), *Structural Equation Models in the Social Sciences*, pp. 85–112. Seminar Press, New York, NY.

Jost, L. (2006). Entropy and diversity. *Oikos*, **113**(2), 363–375.

Kamakura, W. and Wedel, M. (2000) Factor analysis and missing data. *Journal of Marketing Research*, **37**, 490–498.

Kelloway, E.K. (1995) Structural equation modelling in perspective. *Journal of Organizational Behavior*, **16**(3), 215–224.

Kline, R.B. (2010) *Principles and Practice of Structural Equation Modelling*, 3rd edn. The Guilford Press, New York, NY.

Knowles, D. and Ghahramani, Z. (2011) Nonparametric Bayesian sparse factor models with application to gene expression modeling. *Annals of Applied Statistics*, **5**(2B), 1534–1552.

Kolenikov, S. (2009) Confirmatory factor analysis using confa. *Stata Journal*, **9**(3), 329–373.

Koop, G. and Korobilis, D. (2010) Bayesian multivariate time series methods for empirical macroeconomics. *Foundations and Trends in Econometrics*, **3**(4), 267–358.

Langeheine, R. (1994) Latent variable Markov models. In A. Von Eye and C. Clogg (eds), *Latent Variables Analysis: Applications for Developmental Research*, pp. 373–395. Sage, London, UK.

Laud, P. and Ibrahim, J. (1995) Predictive model selection. *Journal of the Royal Statistical Society B*, **57**, 247–262.

Lee, D. (2011) A comparison of conditional autoregressive models used in Bayesian disease mapping. *Spatial and Spatio-temporal Epidemiology*, **2**, 79–89.

Lee, S.-Y. (2007) *Structural Equation Modelling: A Bayesian Approach*. Wiley, Chichester, UK.

Lee, S.-Y., Song, X.-Y. and Tang, N.-S. (2007) Bayesian methods for analyzing structural equation models with covariates, interaction, and quadratic latent variables. *Structural Equation Modeling*, **14**, 404–434.

Leroux, B., Lei, X. and Breslow, N. (1999) Estimation of disease rates in small areas: a new mixed model for spatial dependence. In M. Halloran and D. Berry (eds), *Statistical Models in Epidemiology, the Environment and Clinical Trials*, pp. 135–178. Springer-Verlag, New York, NY.

Levy, R. (2011) Bayesian data-model fit assessment for structural equation modeling. *Structural Equation Modeling*, **18**(4), 663–685.

Longford, N. and Muthen, B. (1992) Factor analysis for clustered observations. *Psychometrika*, **57**(4), 581–597.

Lopes, H. and West, M. (2004) Bayesian model assessment in factor analysis. *Statistica Sinica*, **14**, 41–67.

MacCallum, R., Edwards, M. and Cai, L. (2012) Hopes and cautions in implementing Bayesian structural equation modeling. *Psychological Methods*, **17**(3), 340–345.

Meredith, W. and Teresi, J. (2006) An essay on measurement and factorial invariance. *Medical Care*, **44**(11 Suppl 3), S69–77.

Mooijaart, A. and Bentler, P. (2010) An alternative approach for nonlinear latent variable models. *Structural Equation Modeling*, **17**(3), 357–373.

Muthen, B. (1984) A general structural equation model with dichotomous, ordered categorical, and continuous latent variable indicators. *Psychometrika*, **49**, 115–132.

Muthen, B. and Asparouhov, T. (2012) Bayesian structural equation modeling: a more flexible representation of substantive theory. *Psychological Methods*, **17**, 313–335.

Muthen, B., Kaplan, D. and Hollis, M. (1987) On structural equation modeling with data that are not missing completely at random. *Psychometrika*, **52**, 431–462.

Nadarajah, S. and Kotz, S. (2008) Estimation methods for the multivariate t distribution. *Acta Applicandae Mathematicae*, **102**(1), 99–118.

Neal, R. (1996) *Bayesian Learning for Neural Networks*. Springer-Verlag, Berlin.

Palomo, J., Dunson, D. and Bollen, K. (2007) Bayesian structural equation modeling. In S.-Y. Lee (ed.), *Handbook of Latent Variable and Related Models*, pp. 163–87. Elsevier, Boston, MA.

Quinn, K. (2004) Bayesian factor analysis for mixed ordinal and continuous responses. *Political Analysis*, **12**(4), 338–353.

Rabe-Hesketh, S., Skrondal, A. and Zheng, X. (2007) Multilevel structural equation modeling. In S.-Y. Lee (ed.), *Handbook of Latent Variable and Related Models*, pp. 209–228. Elsevier, Boston, MA.

Rindskopf, D. (2002) The use of latent class analysis in medical diagnosis. In *Joint Statistical Meetings – Social Statistics Section*, pp. 2912–2916. American Statistical Association, Alexandria, VA.

Rizopoulos, D. (2006) ltm: An R package for latent variable modeling and item response theory analyses. *Journal of Statistical Software*, **17**(5), 1–25.

Rosseel, Y. (2012) lavaan: An R package for structural equation modeling. *Journal of Statistical Software*, **48**(2), 1–36.

Rue, H. and Held, L. (2005) *Gaussian Markov Random Fields: Theory and Applications*. CRC, Boca Raton, FL.

Shi, J. and Lee, S.-Y. (2000) Latent variable models with mixed continuous and polytomous data. *Journal of the Royal Statistical Society B*, **62**, 77–87.

Sinharay, S. (2004) Experiences with Markov chain Monte Carlo convergence assessment in two psychometric examples. *Journal of Educational and Behavioral Statistics*, **29**(4), 461–488.

Sinharay, S. (2005) Assessing fit of unidimensional item response theory models using a Bayesian approach. *Journal of Educational Measurement*, **42**(4), 375–394.

Song, X.-Y. and Lee, S.-Y. (2002) A Bayesian approach for multigroup nonlinear factor analysis. *Structural Equation Modeling*, **9**(4), 523–553.

Song, X.-Y. and Lee, S.-Y. (2012) *Basic and Advanced Bayesian Structural Equation Modeling: with Applications in the Medical and Behavioral Sciences*. Wiley, Chichester, UK.

Stern, H. and Jeon, Y. (2004) Applying structural equation models with incomplete data. In A. Gelman and X. Meng (eds), *Applied Bayesian Modeling and Causal Inference from Incomplete-Data Perspectives*, pp. 331–342. Wiley, Chichester, UK.

Ten Berge, J. and Socan, G. (2007) The set of feasible solutions for reliability and factor analysis. In S.-Y. Lee (ed.), *Handbook of Latent Variable and Related Models*, pp. 303–320. North Holland/Elsevier, Amsterdam.

Tu, Y.-K. (2009) Commentary: Is structural equation modelling a step forward for epidemiologists? *International Journal of Epidemiology*, **38**, 549–551.

Ullman, J. and Bentler, P. (2012) *Structural Equation Modeling. Handbook of Psychology*, 2nd edn. Wiley, Chichester, UK.

Wall, M. and Amemiya, Y. (2007) Nonlinear structural equation modeling as a statistical method. In S.-Y. Lee (ed.), *Handbook of Latent Variable and Related Models*, pp. 321–344. Elsevier, The Netherlands.

Wheaton, B., Muthen, B., Alwin, D. and Summers, G. (1977) Assessing reliability and stability in panel models. In D. R. Heise (ed.), *Sociological Methodology 1977*, pp. 84–136. Jossey-Bass, San Francisco, CA.

Wu, X.-L., Heringstad, B. and Gianola, D. (2010) Bayesian structural equation models for inferring relationships between phenotypes: a review of methodology, identifiability, and applications. *Journal of Animal Breeding and Genetics*, **127**(1), 3–15.

Yang, X. and Shoptaw, S. (2005) Assessing missing data assumptions in longitudinal studies: an example using a smoking cessation trial. *Drug and Alcohol Dependence*, **77**(3), 213–225.

Yuan, K.-H. and Bentler, P. (1998) Structural equation modeling with robust covariances. *Sociological Methodology*, **28**, 363–96.

Zhu, H. and Lee, S.-Y. (1999) Statistical analysis of nonlinear factor analysis models. *British Journal of Mathematical and Statistical Psychology*, **52**, 225–242.

Zhu, H. and Lee, S.-Y. (2001) A Bayesian analysis of finite mixtures in the LISREL model. *Psychometrika*, **66**(1), 133–152.

# 10

# Survival and event history models

## 10.1 Introduction

Processes (economic, demographic) in the life cycle of individuals may be represented as event histories. These record the timing of changes of state, and associated durations of stay, in series of events such as marriage and divorce, job quits and promotions. Event histories may also describe mechanical operating times (Hamada *et al.*, 2008), and changes in political regimes (Box-Steffensmeier and Jones, 1997). Many applications of event history models are to non-renewable events such as mortality, and this type of application is often called survival analysis, with the stochastic variable being the time from entry into observation until the event in question. For renewable events the dependent variable is the duration between the previous event and the following event. Interest may focus on the instantaneous rate at which the event occurs (the hazard rate), or on average inter-event times. Heterogeneity in event rates or inter-event durations may be between population sub-groups, or between individuals defined by combinations of risk factors or therapies. A Bayesian inferential overview of survival models is provided by Ibrahim *et al.* (2001a), with computing options including BUGS (Mostafa, 2012), R-INLA (Martino *et al.*, 2011), R2BayesX, BMA (http://cran .r-project.org/web/packages/BMA/index.html), and ddpsurvival (De Iorio *et al.*, 2009).

Parametric representations of duration of stay and survival times are important in many applications (e.g. Ananda *et al.*, 1996). However, flexibility to accommodate a variety of differently shaped failure time distributions may be greater with semiparametric models, where the shape of the hazard function is essentially left unspecified (e.g. Pennell and Dunson, 2006; Wang *et al.*, 2012). These include the Cox proportional hazards model (Cox, 1972), and Bayesian implementations of the Cox model, including gamma process priors on the integrated hazard or hazard itself (Kalbflesich, 1978; Clayton, 1991; Chen *et al.*, 2006).

Among the major problems that occur in survival and inter-event time modelling is a form of data missingness known as censoring. A duration is right censored if a respondent withdraws from a study, or ceases being observed for reasons other than the terminating event, or if a subject does not undergo the event before the end of the observation period. The observed incomplete duration is necessarily less than the unknown full duration till the event. Other types of censoring are left censoring and interval censoring. In the first, subjects are known to have undergone the event but the time at which it occurred is unknown, while in the second it is known only that an event occurred within an interval, not the exact time within the

*Applied Bayesian Modelling*, Second Edition. Peter Congdon.
© 2014 John Wiley & Sons, Ltd. Published 2014 by John Wiley & Sons, Ltd.

interval. A widely adopted simplifying assumption (Leung *et al.*, 1997; Siannis *et al.*, 2005) is that censoring is non-informative, i.e. that censoring of failure times is not associated with the time that would have been observed.

Another complication arises through unobserved variations in the propensity to experience the event. These variations are often known as 'frailty' in medical and mortality applications (Duchateau and Janssen, 2008; Wienke, 2010). If repeated durations are observed on an individual, such as durations of stay in a series of jobs, or multiple event times for patients (Sinha and Dey, 1997), then the cluster is the individual employee or patient. The unobserved heterogeneity is then analogous to the constant subject effect in a panel model. Given the nature of the dependent variable, namely the length of time till an event occurs, unmeasured differences lead to a selection effect. For non-renewable events such as human mortality, high risk individuals die early and the remainder will tend to have lower risk. This will mean the hazard rate will rise less rapidly than it would in the absence of frailty variation.

A third major complication occurs in the presence of time varying covariates or time varying predictor effects. However, piecewise exponential and counting process models (Fleming and Harrington, 1991; Andersen *et al.*, 1993; Lindsey, 2001) are relatively flexible in incorporating such effects.

Survival model assessment from a Bayesian perspective has been considered by Ibrahim *et al.* (2001a), Kuo and Peng (2000), Hanagal and Dabade (2013) and Hanson *et al.* (2011), who use the log pseudo marginal likelihood (based on estimated conditional predictive ordinates), whereas Gu *et al.* (2011), and Ibrahim *et al.* (2001b) consider predictive loss criteria based on sampling new data. Model averaging (over different predictor subsets) for proportional hazard regression is discussed by Volinsky *et al.* (1997) and included in the R package BMA, while Volinsky and Raftery (2000) consider a BIC model choice criterion with multiplier $\log(n_d)$ for the number of parameters, where $n_d$ denotes total uncensored cases. Predictive loss methods may be illustrated by an adaptation of the Gelfand and Ghosh (1998) approach; thus let $t_i$ be the observed times, whether uncensored or censored, with density $f(t|\theta)$, and $t_{new,i}$ be predicted failure times sampled from the posterior predictive density $f(t_{new}|\theta)$. Denoting $v_i$ and $\varsigma_i$ as the mean and variance of $t_{new,i}$, then one criterion for any $w > 0$ is (Sahu *et al.*, 1997)

$$C(w) = \sum_{i=1}^{n} \varsigma_i + \frac{w}{(w+1)} \sum_{i=1}^{n} (v_i - u_i)^2,$$

where $u_i = \max(v_i, s_i)$ if $s_i$ is a censored time, and $u_i = t_i$ if the time is uncensored.

## 10.2   Continuous time functions for survival

Suppose event or failure times $T$ are recorded in continuous time. Then the density $f(t)$ of these times defines the probability that an event occurs in the interval $(t, t + dt)$, namely

$$f(t) = \lim_{dt \to 0} \frac{\Pr(t \le T \le t + dt)}{dt},$$

with cumulative density

$$F(t) = \int_0^t f(u)\mathrm{d}u.$$

From this density, the information contained in the failure times can be represented in two different ways. The first involves the chance of surviving till at least time $t$ (or not undergoing the event before duration $t$), namely

$$S(t) = \Pr(T \geq t) = 1 - F(t) = \int_t^\infty f(u)du.$$

The other way of representing the information involves the hazard rate, measuring the intensity of the event as a function of time,

$$h(t) = f(t)/S(t),$$

and in probability terms, representing the chance of an event in the interval $(t, t + dt)$ given survival till $t$. From $h(t)$ is obtained the cumulative hazard

$$H(t) = \int_0^t h(u)du,$$

and one may also write the survivor function as

$$S(t) = \exp(-H(t)).$$

Let $\delta_i = 1$ if the failure time $t_i$ is observed, and $\delta_i = 0$ for a subject with a right censored failure time. Then the likelihood for that subject is

$$L_i = f(t_i)^{\delta_i} S(t_i)^{(1-\delta_i)} = h(t_i)^{\delta_i} S(t_i).$$

Let $h_0(t)$, $H_0(t)$ and $S_0(t)$ denote the hazard, cumulative hazard and survival functions when there are no covariates. With covariates $X_i = (x_{i1}, \ldots, x_{ip})$ excluding the constant term, a common assumption is that their effect on the hazard rate is independent of $t$, known as the proportional hazards assumption (e.g. Dellaportas and Smith, 1993). One then has conditional hazard and survival rates as follows:

$$h(t|X_i) = h_0(t)e^{X_i\beta},$$
$$S(t|X_i) = \exp(-e^{X_i\beta} H_0(t_i)).$$

In practice, the parameter in $h_0(t)$ defining the average level of failure may be incorporated in an expanded regression term. Frailty effects may also be included in the conditional hazard rate (see Section 10.5).

Additional counting process functions (Andersen *et al.*, 1993) may be defined. The random variable $N(t)$ over $(0, t]$ (with $N(0) = 0$) defines a counting process if $N(s) \leq N(t)$ for $s < t$ and increments $dN(t) = N(t) - N(t-)$ are either 1 or 0. For a non-repeatable events, the counting process $N(t) = I(T \leq t)$ reduces to an indicator of whether the event has occurred by $t$. Let $A(t-)$ denote the event sequence up to but not including $t$. Then conditional on $A(t-)$, the probability that $dN(t) = 1$ can be written in terms of an intensity process $\lambda(t)$, namely,

$$\Pr\{N(t + \delta) - N(t-) = 1|A(t-)\} \simeq \lambda(t)\delta.$$

Equivalently

$$\Pr\{dN(t) = 1|A(t-)\} \simeq d\Lambda(t),$$

where $\Lambda(t) = \int_0^t \lambda(u)du$ denotes the integrated intensity, with $\Lambda(t) = E(N(t))$.

The intensity function is equal to the hazard while the subject is still being observed, but becomes zero when a sequence of (repeatable) events has finished, or when an event has happened (for non-repeatable events). So with $Y(t) = I(T \geq t)$ denoting an indicator of observation status, one has

$$\lambda(t) = Y(t)h(t).$$

The intensity function generalizes to include predictors and random frailties. Under a proportional hazards assumption, predictor effects can be included via

$$\lambda(t_i|X_i) = Y(t_i)h_0(t_i)\exp(X_i\beta).$$

One may compare observed and predicted counting process increments via the Martingale residual at $t$, defined as

$$M_i(t) = N(t_i) - \Lambda_0(t_i|X_i) = N(t_i) - \int_0^{t_i} Y_i(u)\exp(X_i\beta)dH_0(u).$$

The total residual $M_i = M_i(\infty)$ for a subject with observation time $t_i$ is obtainable for a non-repeatable event as

$$M_i = \delta_i - \Lambda_0(t_i|X_i).$$

Deviance residuals $r_i$ are obtained as

$$r_i = sgn(M_i)\sqrt{2\left[M_i - N_i(\infty)\log\left(\frac{N_i(\infty) - M_i}{N_i(\infty)}\right)\right]}.$$

While counting process functions are expressed in continuous time, in empirical work the observation of event times is often effectively discrete, and made at specific intervals, with no indication how the intensity changes within intervals. So the likelihood is a step function at the observed event times. In applications the observed increments $dN(t)$ are typically modelled as Poisson outcomes, with mean given by the intensity function. The intensity may be expressed

$$\lambda(t_i|X_i) = Y(t_i)dH_0(t_i)\exp(X_i\beta),$$

with a prior set on jumps in the integrated hazard $dH_0$. The conjugate prior for the Poisson mean is gamma, so a natural prior for $dH_0$ has the form

$$dH_0 \sim Ga(bH^*(t), b)$$

where $H^*(t)$ contains prior information or knowledge regarding the hazard rate per unit time, and $b > 0$ is higher for stronger beliefs (Sinha and Dey, 1997). One may also model $h_0(t)$ in $\lambda(t_i|X_i) = Y(t_i)h_0(t_i)\exp(X_i\beta)$ parametrically (see Example 10.3).

Another possibility in counting process applications to event histories is the modelling of the impact of previous events or durations in a subject's history. Thus the intensity for the next event could be made dependent on the number of previous events, in what are termed birth models (Lindsey, 1995).

## 10.2.1    Parametric hazard models

Often accumulated evidence supports a parametric form for the distribution of survival times. A baseline survival model is provided by the exponential density $f(t|\lambda) = \lambda\exp(-\lambda t)$, under which the hazard rate is constant $h_0(t) = \lambda$, and the survival function is $S_0(t) = \exp(-\lambda t)$. If there are covariates $X_i$ excluding the intercept, one may specify

$$h(t|X_i) = \lambda e^{X_i \beta},$$

$$S(t|X_i) = \exp(-\lambda e^{X_i \beta}),$$

with $\lambda$ a positive parameter (e.g. assigned a gamma prior). Alternatively

$$h(t|X_i) = e^{\beta_0 + X_i \beta},$$

$$S(t|X_i) = \exp(-e^{\beta_0 + X_i \beta}),$$

with $\lambda = e^{\beta_0}$.

To code survival models with time dependence according to a parametric density $f$ in BUGS (assuming the density $f$ is available), one creates a set of minimum possible values for censored times, say t.min[i], which equal the observed (right censored) durations $t_i$ for subjects with $\delta_i = 0$, but equal 0 for individuals whose failure times are observed. Another vector, say t.obs[i], of actual survival times, are set equal to $t_i$ if $\delta_i = 1$, but set as unobserved (that is NA in terms of BUGS data input) for individuals whose times are censored. So censored survival times are regarded as missing data. To code survival models in INLA, the inputs are times $t_i$ and event indicators $\delta_i$.

Assume exponential survival, $f(t|\lambda) = \lambda \exp(-\lambda t)$, with unknown parameter $\lambda$ and no covariates. The BUGS coding for right censored data, with input data t.obs[] and t.min[], is

```
t.obs[i] ~ dexp(lambda) I(t.min[i],),
```

so that potential survival times for right censored cases have lower bound t.min[i]. For left censored data this would be replaced by

```
t.obs[i] ~ dexp(lambda) I(,t.min[i]).
```

If there are covariates, one could code the exponential model for right censored data as

```
t.obs[i] ~ dexp(mu[i]) I(t.min[i],);
mu[i] <- lambda*exp(inprod(beta[], x[i,1:p])),
```

with lambda a positive parameter, and the covariates excluding a constant. Alternatively one may expand the regression, such that $\lambda = e^{\beta_0}$, namely

```
mu[i] <- exp(beta0+inprod(beta[], x[i,1:p])).
```

In INLA the command sequence for exponential survival, with a single predictor x, data file containing columns headed x,t, and delta, would be

```
D <- read.table("data.txt",header=T)
formula=inla.surv(t,delta)~x
m1=inla(formula, family ="exponential", data=D,control.compute=list
                                         (cpo=T,dic=T)).
```

A commonly adopted parameterised model for time dependence uses a Weibull distribution for durations $t_i \sim Wei(\lambda, \gamma)$, where $\lambda$ and $\gamma$ are scale and shape parameters respectively (Abrams et al., 1996; Kim and Ibrahim, 2000). The Weibull hazard is defined by the following functions

$$h(t) = \lambda \gamma t^{\gamma-1},$$

$$S(t) = \exp(-\lambda t^{\gamma}),$$

$$f(t) = \lambda \gamma t^{\gamma-1} \exp(-\lambda t^{\gamma}),$$

with hazard that is monotonically increasing or decreasing according as $\gamma > 1$, or $\gamma < 1$. The value $\gamma = 1$ leads to exponentially distributed durations with parameter $\lambda$. If there are covariates $X_i$ excluding a constant term, the Weibull hazard function becomes

$$h(t|X_i) = \lambda e^{X_i \beta} \gamma t^{\gamma-1}.$$

Equivalently setting $\lambda_i = e^{\beta_0 + X_i \beta}$, one has

$$h(t|X_i) = \lambda_i \gamma t^{\gamma-1}.$$

Thus in BUGS the Weibull density for failure times and covariates is routinely implemented as

```
t[i]~dweib(gamma, lambda[i]) I(t.min[i],),
```

where the log of lambda[i] (or possibly some other link) is expressed as a function of an intercept $\beta_0$ and covariate effects, and gamma is the shape parameter. In INLA, and with a single predictor x, one would simply specify

```
formula=inla.surv(t,delta)~x
m1=inla(formula, family ="weibull", data=D,control.compute=list
                                          (cpo=T,dic=T))
```

While the Weibull hazard is monotonic with regard to duration $t$, a non-monotonic alternative such as the log-logistic may be advantageous, and this may be achieved in BUGS by taking a logistic model for $y = \log(t)$. Here $t$ are observed durations, censored or complete. Thus

$$y_i \sim \text{Logistic}(\eta_i, \kappa),$$

where $\kappa$ is a scale parameter, and $\eta_i$ is the location of the $i$th subject. The location may be parameterised in terms of covariate impacts $\eta_i = X_i \beta^*$ on the mean length of log survival (rather than the hazard rate),

$$y_i = \log(t_i) = \eta_i + u_i/\kappa,$$

with $p(u) = e^u/(1 + e^u)^2$, $F(u) = e^u/(1 + e^u)$, and $S(u) = 1 - F(u) = 1/(1 + e^u)$. The variance of $y$ is obtained as $\pi^2/(3\kappa^2)$. The survivor function in the $y$ scale is

$$S(y) = \left[1 + \exp\left(\frac{y - \eta}{\sigma}\right)\right]^{-1},$$

where $\sigma = 1/\kappa$. In the original time scale, the survivor and hazard functions are

$$S(t_i|X_i) = \frac{1}{1 + (t_i \theta_i)^{\kappa}},$$

$$h(t_i|X_i) = \frac{\theta_i \kappa (t_i \theta_i)^{\kappa-1}}{1 + (t_i \theta_i)^{\kappa}},$$

where $\theta_i = e^{-\eta_i}$.

## 10.2.2  Semi-parametric hazards

Choice of a suitable parametric form for the baseline hazard $h_0(t)$ may be difficult, and semi-parametric approaches adopted instead. Consider a partition of the response time scale into $J$ intervals (e.g. Sahu *et al.*, 1997), with $J + 1$ knot points $a_j$, namely $(a_0, a_1], \dots, (a_{J-1}, a_J]$, where $a_0 = 0$, and $a_J$ equals the largest observed time, censored or uncensored. Intermediate knot points may be based on distinct failure times (Breslow, 1974), or by siting knots $a_j$ at selected points in the range $(t_{\min}, t_{\max})$ (Kalbfleisch and Prentice, 1973).

Piecewise exponential priors (Friedman, 1982; Brezger *et al.*, 2008) assume the baseline hazard is constant within each interval with parameter $h_{0j}$, possibly combined with interval specific regression parameters $\beta_j$. The hazard within interval $j$ is then

$$h(t_i \in (a_{j-1}, a_j]|X_i, t_i > a_{j-1}) = h_{0j} \exp(X_i \beta_j),$$

where $X_i$ excludes a constant term. Friedman (1982) recommends that analysis starts with $J$ set moderate (e.g. $J = 5$), with the estimated $h_{0j}$ and their credible intervals used to identify sharp changes in the underlying hazard rate. For a subject exiting after interval $j$ (i.e. with $t_i > a_j$), the likelihood contribution during interval $j$ is

$$\exp(-h_{0j}(a_j - a_{j-1})e^{X_i \beta_j}).$$

For a subject with $a_{j-1} < t_i \leq a_j$, either failing in interval $j (\delta_i = 1)$, or censored but with time $t_i$ in the $j$th interval $(\delta_i = 0)$, the likelihood contribution is

$$[h_{0j}e^{X_i \beta_j}]^{\delta_i} \exp[-h_{0j}(t_i - a_{j-1})e^{X_i \beta_j}].$$

A Poisson likelihood over subjects and intervals may be applied (Holford, 1980; Li *et al.*, 2012), with responses $d_{ij}$ defined by the event status in each interval

$$d_{ij} = \delta_i I(t_i > a_j) \, I(a_{j+1} > t_i),$$

and with means $O_{ij} h_{0j} \exp(X_i \beta_j)$, where offsets $O_{ij} = [\min(t_i, a_{j+1}) - a_j] \, I(t_i > a_j)$ are defined by the length of overlap of $t_i$ with interval $j$.

The parameters $h_{0j}$ may be taken as unrelated fixed effects. However these parameters, or transforms such as $\alpha_j = \log(h_{0j})$, will generally be correlated with earlier values, and possible hierarchical schemes include random walks in the $\alpha_j$, for example

$$\alpha_j \sim N(\alpha_{j-1}, \sigma_\alpha^2 \Delta_j), \qquad j > 1,$$

with $\Delta_j = (a_j - a_{j-1})$, and $\alpha_1$ a separate fixed effect. The $\beta_j$ may also be assumed autocorrelated through intervals, or random around a central parameter, such as $\beta_j \sim N(\beta_\mu, \sigma_\beta^2)$.

Semi-parametric approaches may also be applied to the cumulative baseline hazard $H_0$ (Kalbfleisch, 1978). Consider a non-repeatable outcome with event indicators $\delta_i$. Then a counting process approach with observation status indicators $Y_{ij} = I(t_i \geq a_j)$, will have counting process increments $dN_{ij} = Y_{ij} \delta_i I(a_j \leq t_i < a_{j+1})$. One may apply a Poisson likelihood, allowing for interval specific regression effects $\beta_j$, to the binary responses $dN_{ij}$ with mean intensities

$$\lambda_{ij} = Y_{ij} \exp(X_i \beta_j) dH_{0j}.$$

A gamma prior for the increments $dH_{0j}$ may then be adopted, namely

$$dH_{0j} \sim \text{Ga}(cm_j, c),$$

where $m_j$ is a prior estimate of the hazard rate per unit time, and $c$ expresses confidence in the estimate (lower $c$ for lesser confidence). An alternative additive Gamma-polygonal prior for $dH_{0j}$ is presented by Beamonte and Bermúdez (2003), and Mostafa and Ghorbal (2011).

### Example 10.1    Gompertz parametric model for mice mortality

As an example of parametric survival, and use of a non-standard likelihood in OpenBUGS, consider times of death $x$ (in days) from irradiation of 39 mice, as in Ananda *et al.* (1996). These times are all uncensored and range from 40 to 763 days. Consider the two parameter Gompertz density

$$f(x) = bc^x \exp\left[\frac{b}{\log(c)}(1 - c^x)\right], \qquad x > 0,$$

with $b > 0$, $c > 1$, and hazard rate

$$h(x) = bc^x.$$

This likelihood in OpenBUGS is specified as

```
model { for (i in 1:39) {LL[i] <- log(b)+x[i]*log(c)+b*(1-pow(c,x[i]))/log(c)
z[i] <- 0; z[i] ~dloglik(LL[i])}}
```

with Ga(1,0.001) priors on $b$ and $c_0 = c - 1$. Of particular interest is comparison of observed as against expected deaths in intervals of 100 days, which involves evaluating the survivor function at $x = 100, x = 200, \ldots x = 800$. This involves the code

```
S[1] <- 39
for (t in 1:8) {x.t[t] <- t*100
d[t] <- S[t]-S[t+1]
S[t+1] <- 39*exp(b*(1-pow(c,x.t[t]))/log(c))}
```

A two chain run to 50 000 iterations (with 1000 burn in) leads to an estimated parameters (posterior means) $b = 0.00063$ and $c = 1.0044$. Table 10.1 compares actual and expected deaths in intervals of 100 days. A chi-square statistic of 3.353 (betwen actual deaths and posterior mean predicted deaths) compares to a similar $\chi^2$ under the best fitting procedure of Ananda *et al.* (1996).

**Table 10.1**    Observed vs predicted deaths by time since irradiation.

| Time interval | Observed deaths | Expected deaths | | | |
|---|---|---|---|---|---|
| | | Mean | 2.5% | Median | 97.5% |
| 0–100 | 4 | 2.90 | 1.43 | 2.78 | 5.00 |
| 100–200 | 2 | 3.95 | 2.41 | 3.91 | 5.71 |
| 200–300 | 6 | 5.19 | 3.87 | 5.19 | 6.42 |
| 300–400 | 5 | 6.37 | 5.30 | 6.37 | 7.40 |
| 400–500 | 7 | 7.00 | 5.32 | 7.00 | 8.68 |
| 500–600 | 6 | 6.41 | 4.45 | 6.40 | 8.45 |
| 600–700 | 7 | 4.41 | 2.80 | 4.40 | 6.22 |
| 700–800 | 2 | 2.03 | 0.81 | 2.01 | 3.53 |

### Example 10.2   Parametric and semi-parametric models for cancer survival

The E1684 melanoma clinical trial (e.g. Ibrahim *et al.*, 2001a; Herring and Ibrahim, 2002) was carried out to assess the impact of interferon alpha-2b in chemotherapeutic treatment of melanoma. Consider data on 255 subjects with times of death $t_i$ (in years) and right censoring indicators ($\delta_i = 0$ if subject survives). Predictors are a binary treatment indicator, the log of the Breslow measure of the vertical tumor depth, age and sex. This is a subset of the total data on 286 subjects, excluding subjects with missing covariate values.

To illustrate standard parametric options available in BUGS, exponential and Weibull survival models are compared. Two chain runs of 10 000 iterations (with inferences from the last 9000) show a 90% credible interval ($-0.54$, $-0.02$) for the treatment variable under an exponential model, and ($-0.54$, $0.00$) under a Weibull model. The 95% credible interval for this covariate straddles zero under both assumptions. The Weibull time parameter $\gamma$ has 95% interval (0.69, 0.90), so the constant hazard model is not appropriate. This is also demonstrated in the LPML statistics of $-705$ for the exponential model, as compared to $-681$ for the Weibull model.

To illustrate models which avoid parametric assumptions, a piecewise exponential model is also applied with $J = 5$. Interior knots $(a_1, a_2, a_3, a_4)$ are set at the quintiles of all observed times, whether uncensored or censored. A random walk prior $\alpha_j \sim N(\alpha_{j-1}, \sigma_\alpha^2)$ is adopted on $\alpha_j = \log(h_{0j})$, with $\alpha_1$ assigned a $N(0,1000)$ prior, and $1/\sigma_\alpha^2$ assigned a $Ga(1,0.001)$ prior. The intercept is omitted as it can be estimated as the average of the $\alpha_j$.

The second half of a two chain run of 100 thousand iterations shows an LPML raised to $-539$. This representation shows clear variation in the baseline hazard, apparent in the posterior means (sd) for $\{\alpha_1, \dots \alpha_5\}$ of $-1.60 \,(0.25)$, $-1.59\,(0.26)$, $-2.19\,(0.25)$, $-2.94\,(0.30)$, and $-3.04\,(0.34)$. The code for this model is

```
model { # Priors
alph[1] ~dnorm(0,0.001);
for (j in 2:J) {alph[j] ~dnorm(alph[j-1],invsig2.alpha)}
invsig2.alpha ~dgamma(1,0.001)
for(k in 1:4) { beta[k] ~dnorm(0, 0.001)}
# event-time indicators
for (i in 1:N) {for (j in 1:J){
d[i,j] <- del[i]*step(t[i]-a[j])*step(a[j+1] - t[i])
# offset term
O[i,j] <- (min(t[i],a[j+1])-a[j])*step(t[i]-a[j]);
log(theta[i,j]) <- alph[j]+beta[1]*trt[i]+beta[2]*log(BRTH[i])+beta[3]*age[i]
                                                           +beta[4]*sex[i]
# Poisson likelihood
d[i,j] ~dpois(mu[i,j]); mu[i,j] <- O[i,j]*theta[i,j]
LL[i,j] <- -mu[i,j]+d[i,j]*log(mu[i,j])-logfact(d[i,j]);
G[i,j] <- 1/exp(LL[i,j])}}}.
```

This model can also be fitted in R-INLA using the coxph likelihood. The code is

```
D = read.table("E1684.txt",header=T)
# cutpoints
a=c(0,0.88,1.99,4.66,6.85,9.65)
formula = inla.surv(t, delta) ~trt+ln.BRTH+age+sex
# proper prior on intercept (precision = 0.001)
m1 = inla(formula, family = "coxph", data = D,control.fixed =list(prec.
                                                 intercept = 0.001),
control.hazard=list(model="rw1",cutpoints=a),control.compute=list(dic=T))
summary(m1); plot(m1).
```

The preceding plot command shows the baseline hazard changing most at intermediate durations. The $\alpha$ random parameters are constrained to sum to zero and a separate intercept $\beta_0$ is estimated, with posterior mean (sd) of $-2.23$ (0.33). Hazard parameters comparable to those from the BUGS code are obtained as sums of $\beta_0$ and the respective random effects, which may be obtained as R1 <- m1$summary.random.

### Example 10.3    Bladder cancer

To illustrate counting process models applied to repeated event data, consider a bladder cancer study conducted by the US Veterans Administrative Cooperative Urological Group. This involved $n = 116$ patients randomly allocated to one of three groups: a placebo group, a group receiving vitamin B6, and a group undergoing installation of thiotepa into the bladder. On follow up visits during the trial, incipient tumours were removed, so that an event history (with repeated observations on some patients) is obtained, with $N = 292$ events (or censorings) accumulated over the 116 patients. Times between recurrences are recorded in months, with many patients not experiencing recurrences (i.e. being censored). The counting process increments are defined over $J - 1 = 30$ intervals, based on distinct observed survival times. A beneficial effect of thiotepa would be apparent in a more negative impact on the recurrence rate than the two other treatment options. We compare Weibull and piecewise hazards and also allow a history effect, namely the impact of a count of previous recurrences within subjects.

Summaries are based on the last 2000 iterations of two chain runs of 2500 iterations after early convergence. The first model shows the Weibull parameter to have 95% interval including the the exponential null value ($\gamma = 1$). However, there is a clear influence of previous events on the chance of a further one. There is no apparent treatment benefit. The LPML for this model is $-704$. For a non-parametric hazard analysis, one may either set one of the treatment effects to a null value, or one of the piecewise coefficients. Here the first treatment effect is set to zero. Results for the history and treatment effects are similar to the parametric model, but the LPML worsens to $-721$, showing no gain in fit from adopting this type of hazard estimation. The rates $d\Lambda_0(t)$ for the first 18 months are precisely estimated and suggest a decline, albeit irregular, in the exit rate over this period.

## 10.3    Accelerated hazards

The proportional hazards model with

$$h(t_i|X_i) = h_0(t_i)e^{X_i\beta},$$

can be regarded as a special case of a general model

$$h(t_i|X_i) = L(h_0(t_i), e^{X_i\beta}),$$

where $L$ is a known function. For example, in an additive hazard one has

$$h(t_i|X_i) = h_0(t_i) + e^{X_i\xi},$$

while in an accelerated failure time (AFT) model one has

$$h(t_i|X_i) = e^{X_i\alpha}h_0(t_ie^{X_i\alpha}).$$

Thus the explanatory variates act multiplicatively on time, and affect the 'rate of passage' to the event; so in a clinical example, they might influence the speed of progression of a disease. Define $X_i$ excluding an intercept. If there is a single binary covariate (e.g. $x_i = 1$ for treatment group, 0 otherwise) then $\eta_i = X_i\alpha = \alpha$ when $x_i = 1$. Setting $\phi = e^\alpha$, the hazard for a treated patient is

$$\phi h_0(\phi t),$$

and the survivor function is $S_0(\phi t)$. The multiplier $\phi$ is sometimes termed the acceleration factor.

For example, under a Weibull hazard with $h_0(t) = \lambda \gamma t^{\gamma - 1}$, the AFT representation is

$$h(t_i | X_i) = e^{\eta_i} \lambda \gamma (t_i e^{\eta_i})^{\gamma - 1} = (e^{\eta_i})^{\gamma} \lambda \gamma t_i^{\gamma - 1}.$$

The median survival time under a Weibull AFT model is

$$t.50 = [\log 2 / \{\lambda e^{\gamma \eta_i}\}] / \gamma,$$

and under a Bayesian approach, priors for the regression parameters may be indirectly expressed in terms of their impact on median survival times (Bedrick et al., 2000).

The AFT model can be equivalently represented as a linear regression of $y = \log(t)$ on $X_i$,

$$\log(t_i) = X_i \alpha^* + \sigma u_i,$$

where different assumptions on $u$ correspond to different parametric survival assumptions. Taking $\sigma = 1$, and $p(u) = \exp[u - \exp(u)]$ leads to exponential survival, while taking $\sigma = 1/\gamma$ and $p(u) = \gamma \exp[\gamma u - \exp(\gamma u)]$ leads to Weibull survival. Taking $\psi_i = \exp(-X_i \alpha^*)$, one obtains the accelerated conditional survival function as $S(t_i | X_i) = S_0(t_i \psi_i)$.

The likelihood involves a product over observed failures of the density function $f(t_i | X_i)$ of event times, and a product over censored failure times of $S(t_i | X_i)$. These can be represented (Zhang and Lawson, 2011) in terms of the density and survival functions of the errors $u$, $f_0(u)$ and $S_0(u)$. With $u_i = [\log(t_i) - X_i \alpha^*] / \sigma$ as the realized residual, one has

$$f(t_i | X_i) = \frac{1}{\sigma t_i} f_0[u_i],$$

$$S(t_i | X_i) = S_0[u_i],$$

with likelihood

$$L_i = f(t_i | X_i)^{\delta_i} S(t_i | X_i)^{1 - \delta_i}.$$

## Example 10.4   AFT survival regression after bone marrow transplant

Consider data on survival after bone marrow transplant (Avalos et al.1, 1993), with time to death or relapse in days, and predictors graft type (1=allogenic, 0=autologous), disease type (0=Non-Hodgkin lymphoma, 1=Hodgkins disease), Karnofsky score, and waiting time to transplant in months. Normal and Weibull assumptions for $t$ are compared. For the normal errors assumption, with $t$ lognormal, and with $u$ as the standard residual

$$f(t) = \frac{1}{\sigma t} \frac{1}{\sqrt{2\pi}} \exp[-u^2 / 2],$$

$$S(t) = 1 - \Phi(u).$$

For the Weibull error assumption, with Weibull time parameter $\gamma = 1/\sigma$,

$$f(t) = \frac{1}{\sigma t} \gamma \exp[\gamma u - \exp(\gamma u)],$$

$$S(t) = \exp[-\exp(\gamma u)].$$

Using the BUGS option for non-standard likelihoods, the coding for the Weibull option is

```
model {for (i in 1:43) {u[i] <- (log(t[i])-eta[i])/sigma
eta[i] <- beta0+beta[1]*x1[i]+beta[2]*x2[i]+beta[3]*(x3[i]-mean(x3[]))
                                           +beta[4]*(x4[i]-mean(x4[]))
f0[i] <- gam*exp(gam*u[i]-exp(gam*u[i]))
S0[i] <- exp(-exp(gam*u[i]))
# log-likelihood
LL[i] <- delta[i]*log(f0[i]/(sigma*t[i]))+(1-delta[i])*log(S0[i])
invL[i] <- 1/exp(LL[i]);
nLL[i] <- -LL[i];
z[i] <- 0; z[i] ~dpois(nLL[i])}
# priors
for (j in 1:4) {beta[j] ~dnorm(0,0.0001)}
beta0 ~dflat(); sigma ~dunif(0,100); gam <- 1/sigma}
```

Convergence is delayed, possibly due to the small sample and correlation between the two binary predictors, and inferences are based on the second halves of two chain runs of 200,000 iterations.

It is interesting to note that substantive inferences are affected by the assumed form for $t$, with the effect of disease type being significant for the Weibull analysis but not the normal one: the Weibull analysis provides a negative impact of Hodgkins disease status on survival time with 95% interval $(-2.9, -0.1)$. The Weibull time parameter $\gamma$ is estimated as 0.81 (0.67, 0.95). However, the LPML is slightly higher for the normal errors option, namely $-172.5$ as against $-174$.

## 10.4  Discrete time approximations

Although events may actually occur in continuous time, event histories may only record time in discrete units, generally called periods or intervals, during which an event may only occur once. The discrete time framework includes population life tables, clinical life table methods such as the Kaplan–Meier method, and discrete time survival regressions. Applications of the latter include times to degree attainment (Singer and Willett, 1993), and the chance of exit from unemployment (Fahrmeir and Knorr-Held, 1997). Consider a discrete partition using $L$ knots of the positive real line, namely $0 < a_1 < a_2 \ldots < a_L < \infty$, and let $A_j$ denote the interval $(a_{j-1}, a_j]$, with the first interval being $(0, a_1]$. The discrete distributions analogous to those above are

$$f_j = \Pr(t \in A_j) = S(a_{j-1}) - S(a_j) = S_j - S_{j+1},$$

where (e.g. Aitkin et al., 2009),

$$S_j = \Pr(t > a_{j-1}) = f_j + f_{j+1} + \cdots + f_L.$$

The survivor function at $a_{j-1}$ is $S_j$ and at $a_j$ is $S_{j+1}$, with the first survivor rate at time 0 being $S_1 = 1$. The $j$th interval hazard rate is then

$$h_j = \Pr(t \in A_j | t > a_{j-1}) = f_j/S_j.$$

It follows that $h_j = (S_j - S_{j+1})/S_j$, and also that

$$S_{j+1}/S_j = 1 - h_j.$$

So the chance of surviving through $r$ successive intervals, which is algebraically

$$S_{r+1} = \prod_{j=1}^{r} S_{j+1}/S_j,$$

can be obtained as the product

$$S_{r+1} = \prod_{j=1}^{r} (1 - h_j).$$

In practice, the likelihood is defined over subjects $i$ and periods $j$, and an event variable $w_{ij}$ is coded for the end point $a_j$ of each interval $(a_{j-1}, a_j]$ and each subject. Conside a non-recurrent event. If a subject's observation sequence ends with an event within the interval $(a_{j-1}, a_j]$, the event variable would be coded 0 for preceding periods, while $w_{ij} = 1$. A subject still censored at the end of the study would have indicators $w_{ij} = 0$ throughout.

For subjects considered in aggregate, let the number at risk at the beginning of the $j$th interval be denoted $N_j$. This is composed of individuals still alive at $a_{j-1}$, and still under observation (i.e. neither failed nor lost to follow-up in previous intervals). For these $N_j$ individuals still at risk (for whom $R_{ij} = 1$), the total deaths in the $j$th interval are

$$\sum_{R_{ij}=1} w_{ij} = d_j.$$

The Kaplan–Meier estimate of the survival curve is applicable to studies where subjects may be censored through non-follow-up, and is based on comparing $N_j - d_j$ subjects surviving each interval from totals $N_j$. The cumulative probability of surviving to any point is estimated as a product of (moment estimated) probabilities of surviving each of the preceding time intervals.

### Example 10.5    Colorectal cancer: Kaplan–Meier method

To illustrate the Kaplan–Meier procedure but from a Bayesian perspectve, consider survival data among colon cancer patients, with times in weeks following an oral treatment (Ansfield *et al.*, 1977). For 45 of the 52 patients, times were uncensored (ranging from 6 to 142 weeks), while 7 patients had censored survival times. The data contain $L + 1 = 33$ distinct times of death. Totals at risk $N_j$ (for whom the at risk indicators $R_{ij} = 1$) are defined according to survival or withdrawal prior to the start of the $j$th interval. A technique useful in many survival applications involves reformulating the likelihood to reveal a Poisson kernel (Aitkin and Clayton, 1980; Lindsey, 1995; Fahrmeir and Tutz, 2001). Here a Poisson likelihood is adopted for the outcome indicators $w_{ij}$ with means

$$\theta_{ij} = R_{ij} h_j,$$

where $h_j$ is the hazard rate in the $j$th interval, and has a prior proportional to the width of that interval, namely $h_j \sim Ga(c\Delta_j, c)$, where $\Delta_j = (a_j - a_{j-1})$, and $c$ represents strength of prior belief. Taking $c = 0.001$, and two chains to 5000 iterations (with 1000 burn in), one may derive posterior mean survival rates $S_j$ for the distinct failure times, together with the hazard rates $h_j$.

To obtain an estimate of the underlying smooth survival curve, one may assume (Leonard *et al.*, 1994) an equally weighted mixture of survival functions with $K$ components

$$S(t, \xi, \eta) = K^{-1} \sum_{k=1}^{K} S(t, \xi_k, \eta),$$

where each $S(t, \xi_k, \eta)$ is a specific survivor function (e.g. exponential, Weibull), $\eta$ denotes parameters of that function not varying over the mixture, and $\xi_k$ are components that do vary. The equally weighted mixture is analogous to kernel estimation, and smooths the underlying density, without assuming that the density is from a parametric family. Here a Weibull mixture analysis with $K = 15$ components is applied, with the unknown shape parameter $\eta$ assigned an $E(1)$ prior. The location parameters $\xi_k$ are constrained to be increasing, namely $\xi_k = \xi_{k-1} + \Delta_k$, with $\Delta_k \sim Ga(0.01, 0.01)$. The smoothed survivor curve $S(u)$ is defined over a grid of values in weeks $u = 1, 2, \ldots, 250$. The last 4000 of a two chain run of 5000 iterations show $\eta$ estimated at 2.18, with 95% interval from 1.21 to 2.94. Figure 10.1 shows the Kaplan–Meier and smoothed survivor curve up to 250 weeks.

## 10.4.1 Discrete time hazards regression

Discrete time hazard regressions typically assume an underlying continuous time model but adapted to survival time observations grouped into intervals, such that durations or failure times between $a_{j-1}$ and $a_j$ are recorded as a single value. Assume that the underlying continuous time model is of proportional hazard form

$$h(t_i | X_i) = h_0(t_i) \exp(X_i \beta),$$

with integrated hazard $H_0(t_i) = \int_0^{t_i} h_0(u) du$ and survivor function

$$S(t_i | X_i) = \exp(-e^{X_i \beta} H_0(t_i)).$$

For discretized times, the conditional probability of surviving through $j - 1$ intervals is

$$S_{ij} = \exp[-e^{X_i \beta} H_0(a_{j-1})].$$

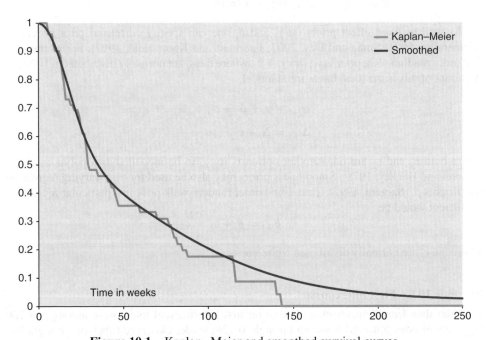

**Figure 10.1**   Kaplan–Meier and smoothed survival curves.

The corresponding hazard rate in the $j$th interval is

$$h_{ij} = 1 - S_{i,j+1}/S_{i,j} = 1 - \exp[-e^{X_i\beta}\{H_0(a_j) - H_0(a_{j-1})\}] = 1 - \exp(-e^{X_i\beta+\gamma_j}),$$

where $\gamma_j = \log[H_0(a_j) - H_0(a_{j-1})]$.

The likelihood of an event in the $j$th interval given survival till then, can be written (Kalbfleisch and Prentice, 1980; Fahrmeir and Tutz, 2001) as

$$h_{ij}S_{ij} = [1 - \exp(-e^{X_i\beta+\gamma_j})] \prod_{k=1}^{j-1} \exp(-e^{X_i\beta+\gamma_k}).$$

Let $w_{ij} = 1$ for an event in the $j$th interval for $j = 1, \ldots, J_i$ (where $J_i$ is the last interval subject $i$ is observed), and $w_{ij} = 0$ otherwise, and let $X_i[a_j]$ be potentially time varying predictors. Also let $\beta_j$ denote regression effects, fixed within intervals, but that may vary between intervals. If the predictors themselves are time specific, one may introduce lagged as well as contemporaneous effects (Fahrmeir and Tutz, 2001). The log-likelihood contribution for an individual surviving $j-1$ intervals till either an event or censoring is then

$$w_{ij}\log[1 - \exp(-\exp\{\gamma_j + X_i[a_j]\beta_j\})] - \sum_{k=1}^{j-1} \exp\{\gamma_k + X[a_k]\beta_k\}.$$

This likelihood reduces to Bernoulli sampling over individuals and intervals with probabilities $\pi_{ij}$ of the events $w_{ij}$ modeled via a complementary log-log link. Thus for a subject observed for $J_i$ intervals until either an event or censoring

$$w_{ij} \sim \text{Bernoulli}(\pi_{ij}), \qquad i = 1, \ldots, n; \qquad j = 1, \ldots, J_i,$$

$$\log\{-\log(1 - \pi_{ij})\} = \gamma_j + X_i[a_j]\beta_j.$$

As well as fixed effect priors on $\gamma_j$ and $\beta_j$ one can specify correlated prior processes (Gamerman, 1991; Sinha and Dey, 1997; Fahrmeir and Knorr-Held, 1997). For example, a first order random walk prior is $\gamma_{j+1} = \gamma_j + \epsilon_j$, where the $\epsilon_j$ are normally distributed iid effects. A variant of this is the local linear trend model

$$\gamma_{j+1} = \gamma_j + \Delta_j + \epsilon_{1j},$$

$$\Delta_{j+1} = \Delta_j + \epsilon_{2j},$$

where both $\epsilon_1$ and $\epsilon_2$ are iid. Another option is for $\gamma_j$ to be modelled as a polynomial in $j$ (Mantel and Hankey, 1978). Smoothness priors may also be used for time varying regression coefficients $\beta_j$ (Sargent, 1997). Thus a first order random walk prior in a particular regression coefficient would be

$$\beta_{j+1} = \beta_j + \epsilon_j,$$

where the $\epsilon_j$ are normally distributed white noise.

## Example 10.6    First intercourse

Consider data from Capaldi *et al.* (1996) on first heterosexual intercourse among $n = 180$ high-school boys followed from 7th through to 12th grade. Observed times (in year grades) between 7 and 12 are transfomed to times ranging from 1 to 6. Censored subjects account for around 30% of the sample. Predictors are a binary indicator (PT) of a parenting transition during the boy's early life, and a standardized measure (PAS) of parents' antisocial behavior during the child's formative years.

It is possible to expand the dataset to person-years form (Singer and Willett, 2003), but this is avoided by defining the likelihood in terms of risk and event status. A logit link and a fixed effects prior for the intercepts $\gamma_j$ is illustrated by the code

```
model { for (i in 1 : n) {for (j in 1 : J) {
# in risk set for jth interval if t[i] > a[j-1]
R[i,j] <- step(t[i] - a[j] - eps)
# event indicators
w[i, j] <- R[i,j]*step(a[j+1] - t[i]+eps)*delta[i]
w[i, j] ~dbern(p[i, j])
p[i, j] <- R[i,j]*exp(eta[i,j])/(1+exp(eta[i,j]))
eta[i,j] <- gam[j]+beta[1]*PT[i]+beta[2]*PAS[i]}}
            for (k in 1:2) {beta[k] ~dnorm(0,0.001)}
            for (j in 1: J) { gam[j] ~dnorm(0,0.001)}}}
```

A two chain run of 5000 iterations (and the last 4000 for inference) provides posterior means (sd) for $\{\gamma_1, \ldots \gamma_6\}$ of $-2.93$ (0.32), $-3.68$ (0.43), $-2.18$ (0.29), $-1.71$ (0.27), $-1.55$ (0.28), and $-1.02$ (0.28). Both predictors have significant positive effects, with posterior means (sd) for parenting transition of 0.68 (0.23), and 0.30 (0.13) for parental antisocial behavior. The DIC is 644.9.

## 10.5   Accounting for frailty in event history and survival models

Whether the event history is in discrete or continuous time, unobserved differences between subjects (e.g. due to unobserved covariates) may be confounded with the estimated survival curve and estimated impacts of observed covariates. Consider sub-populations defined by different levels of frailty or risk. Frail subjects will tend to undergo the event earlier, and with time the overall hazard rate will tend towards that of the lowest frailty sub-group. The observed hazard rate may thereby decline even though the hazard rates for the sub-groups are constant. Consider two sub-groups with hazard rates $h_1(t)$ and $h_2(t)$, survivorship rates $S_j(t) = \exp[-\int_0^t h_j(u)du]$, and initial sub-group fractions $p_1(0)$ and $p_2(0)$, with $p_1(0) + p_2(0) = 1$. Then the proportion of survivors at time t from the each sub-group is

$$p_j(t) = \frac{p_j(0)S_j(t)}{p_1(0)S_1(t) + p_2(0)S_2(t)}$$

and the observed population hazard rate, namely $h(t) = p_1(t)h_1(t) + p_2(t)h_2(t)$ will tend towards the hazard rate for the lower frailty sub-group. An example of a model with two sub-groups is the mover-stayer model, whereby a susceptible sub-population has high risk of the event (e.g. migration, job change), and another has zero or low risk; similar binary group models have been applied in modelling poverty persistence (Devicienti, 2011). Even if the hazard rate for the susceptible group is increasing, the observed hazard for the entire population may at first rise and then fall. One impact of neglected heterogeneity is that covariate effects may be estimated too precisely or subject to bias. For example, in mortality analysis if risks are not homogeneous over individuals, then the effect of age on individual mortality is greater than that on population mortality.

Unobserved differences may be summarised in a randomly distributed multiplicative frailty $\omega_i$, with the conditional hazard rate becoming

$$h(t_i|X_i, \omega_i) = \omega_i h_0(t_i) \exp(X_i \beta),$$

and with conditional survival function $S(t_i|X_i, \omega_i) = \exp(-\omega_i e^{X_i \beta} H_0(t_i))$. Inferences may be sensitive to the specification of unobserved heterogeneity, and choice of frailty distribution is important for estimating the dependence structure (Duchateau and Janssen, 2008). A common assumption is that the $\omega_i$ are gamma distributed, specifically with prior mean 1 to ensure the regression intercept is identified, namely $\omega_i \sim Ga(\varphi, \varphi)$ where $\varphi$ is unknown. Other parametric densities used for modelling frailty include the log-normal (Gustafson, 1997) and log skew-normal (Callegaro and Iacobelli, 2012), though non-parametric approaches to estimating the frailty density are also used (Heckman and Singer, 1984). These involve discrete mixture models with finite support at a small number $K$ of points, so that the conditional hazard can be represented as

$$h(t_i|X_i, G_i) = h_0(t_i) \exp(\alpha_{G_i} + X_i \beta),$$

where $G_i$ is a multinomial with $K$ categories. Non-parametric approaches to frailty can also include Dirichlet process and Polya tree priors (Zhao *et al*, 2009).

A frailty term can be used for univariate survival data without any clustering (Wienke, 2003), but also provides a way to account for within-cluster correlations in event history outcomes (Guo and Rodriguez, 1992), or for interdependence in multivariate survival times where a common underlying influence is present (Keiding *et al.*, 1997). Suppose subjects (patients, children) are arranged within aggregate units or clusters (hospitals, families) and event times are affected by cluster characteristics, known and unknown, as well as by the characteristics of individuals. Then frailty effects at cluster level represent unmeasured contextual influences on the subject level outcome.

**Example 10.7    Institutional frailty effects**

This analysis considers parametric, discrete mixture and mixed Dirichlet process options for institutional frailty in an analysis of mortality (survival times in days) among advanced lung cancer patients, conducted by the North Central Cancer Treatment Group (Loprinzi *et al.*, 1994). The $K = 18$ institutions enrolling subjects onto the trial are heterogeneus (including both community practices and a large tertiary care centre), and differences in the baseline risk may occur. There are 227 subjects with institution observed, and covariates gender (F = 1, M = 0), ph.karno (physician's estimate of Karnofsky score, an alternative to ECOG performance score), and pat.karno (patient's estimate of Karnofsky score). The lower the Karnofsky score, the worse are chances of survival. The latter two predictors are subject to slight missingness, which is accommodated by a regression scheme conditioning on ph.karno (missing for one subject). The data can be represented as multilevel form, with subjects $i = 1, \ldots, m_k$, nested in $k = 1, \ldots, K$ institutions. Then one might consider random fraility effects $\omega_k$ in the hazard

$$h(t_{ik}|X_{ik}, \omega_k) = h_0(t_{ik}) \exp(\beta_0 + X_{ik} \beta + \omega_k).$$

Under the parametic frailty option it is assumed that $\omega_k \sim N(0, 1/\tau_\omega)$, with Weibull survival and goodness of fit assessed using the LPML. The BUGS code is

```
model {for (i in 1 : 227) {t[i] ~dweib(gamma, mu[i]) I(t.min[i],)
L[i] <- pow(f[i],delta[i])*pow(S[i],1-delta[i]); LL[i] <- log(L[i]);
invL[i] <- 1/L[i]
```

```
S[i] <- exp(-mu[i]*pow(t[i],gamma));
f[i] <- mu[i]*gamma*pow(t[i],gamma-1)*S[i]
log(mu[i]) <- b0 + b[1]*female[i]+b[2]*ph.karno[i]+b[3]*pat.karno[i]
+omeg[inst[i]]
# conditional model for missingness
ph.karno[i] ~dnorm(mu.karno1,tau.karno[1])
pat.karno[i] ~dnorm(mu.karno2[i],tau.karno[2])
mu.karno2[i] <- c[1]+c[2]*ph.karno[i]}
# frailty model
for (j in 1:18) {omeg[j] ~dnorm(0,tau.omeg)}
# other priors
gamma ~dgamma(1,0.001);   tau.omeg ~dgamma(1,0.001)
for (j in 1:2) {tau.karno[j] ~dgamma(1,0.001)}
mu.karno1 ~dflat()
b0 ~dflat(); for (j in 1:3) {b[j] ~dnorm(0,0.001)}
for (j in 1:2) {c[j] ~dnorm(0,0.001)}}
```

The second half of a two chain run of 10 000 iterations is used for inference. Female sex and patient Karnofsky score have significant negative effects on mortality. However, the estimates of institutional frailty effects $\omega_k$ contain no significant outliers: institutions 15, 11, and 3 show the lowest residual effects on daily hazard rates (after controlling for gender and Karnofsky scores), while institutions 1 and 14 have the highest residual effects.

This model may be estimated in INLA after excluding subjects with missing predictor values and rescaling the time variable to avoid numeric errors. A mildly informative prior for the precision of the random effects is also assumed. The code including summary of the institutional effects is

```
D = read.table("Ex10_7_lung.txt",header=T); D$t <- D$time/100
f = inla.surv(t, delta) ~sex+ph.karno+pat.karno+f(inst, param=c(1,0.1))
m = inla(f, family = "weibull", data = D, control.fixed = list(prec.
          intercept = 0.001,prec=0.001), control.compute=list(dic=T))
summary(m); plot(m)
m$summary.random.
```

The discrete mixture model postulates two groups of institutions, with low and high frailty respectively, and represented by binary indicators $G_k$. Thus

$$h(t_{ik}|X_{ik}, \omega_{G_k}) = h_0(t_{ik}) \exp(X_{ik}\beta + \omega_{G_k}),$$

$$G_k \sim \text{Categoric}(\pi_{1:2})$$

$$\pi_{1:2} \sim D(\alpha_{1:2})$$

$$\alpha_1 = \alpha_2 = 0.5.$$

To achieve identifiability, a separate intercept is omitted, and the frailty effects are constrained so that $\omega_2 > \omega_1$. The code for this model is

```
model {for (i in 1 : 227) {t[i] ~dweib(gamma, mu[i])I(t.min[i],)
L[i] <- pow(f[i],delta[i])*pow(S[i],1-delta[i]); LL[i] <- log(L[i]);
invL[i] <- 1/L[i]
S[i] <- exp(-mu[i]*pow(t[i],gamma));
f[i] <- mu[i]*gamma*pow(t[i],gamma-1)*S[i]
# conditional model for missingness
ph.karno[i] ~dnorm(mu.karno1,tau.karno[1])
```

```
pat.karno[i] ~dnorm(mu.karno2[i],tau.karno[2])
mu.karno2[i] <- c[1]+c[2]*ph.karno[i]
# two group frailty model
log(mu[i]) <- b[1]*female[i]+b[2]*ph.karno[i]+b[3]*pat.karno[i]
+omeg[G[inst[i]]]}
# constrained frailty effects
omeg[1] ~dnorm(0,0.001); omeg[2] <- omeg[1]+del; del ~dexp(1)
# group allocation model for institutions
for (j in 1:18) {G[j] ~dcat(pi[1:2]); Gcat[j] <- equals(G[j],1)}
pi[1:2] ~ddirch(alph[1:2])
for (j in 1:2) {alph[j] <- 0.5}
# other priors
gamma ~dgamma(1,0.001);     mu.karno1 ~dflat()
for (j in 1:2) {tau.karno[j] ~dgamma(1,0.001)}
for (j in 1:3) {b[j] ~dnorm(0,0.001)}
for (j in 1:2) {c[j] ~dnorm(0,0.001)}}
```

As would be expected, posterior probabilities of belonging to the low frailty group 1 (obtained as the posterior means of Gcat[]) are lowest for institutions 1 and 14, namely 0.38 and 0.33. The LPML for this model is slightly higher at $-1300.1$ than that obtained (namely $-1302.3$) under the normal frailty model above.

Under the mixed DP prior, it is assumed that

$$h(t_{ik}|X_{ik}, \omega_k) = h_0(t_{ik}) \exp(X_{ik}\beta + \omega_k)$$

$$\omega_k \sim N(v_{G_k}, 1/\tau_{G_k})$$

$$G_k \sim \text{Categoric}(\pi_{1:M})$$

$$v_m \sim N(v_0, 1/\tau_0); \tau_m \sim \text{Ga}(1, 0.001); m = 1, \dots, M$$

$$v_0 \sim N(0, 1000), \tau_0 \sim \text{Ga}(1, 0.001).$$

The concentration parameter for the stick-breaking prior is set at $\alpha = 5$. Following Ohlssen et al. (2007), the number of potential mass points $M$ is taken large enough to ensure the final probability $p_M$ in the stick-breaking prior is approximately 0.01. The code is

```
model{ for (i in 1 : N) {t[i] ~dweib(gamma, mu[i]) I(t.min[i],)
L[i] <- pow(f[i],delta[i])*pow(S[i],1-delta[i]); LL[i] <- log(L[i]);
invL[i] <- 1/L[i]
S[i] <- exp(-mu[i]*pow(t[i],gamma));
f[i] <- mu[i]*gamma*pow(t[i],gamma-1)*S[i]
ph.karno[i] ~dnorm(mu.karno1,tau.karno[1])
pat.karno[i] ~dnorm(mu.karno2[i],tau.karno[2])
mu.karno2[i] <- c[1]+c[2]*ph.karno[i]
log(mu[i]) <- b[1]*female[i]+b[2]*ph.karno[i]+b[3]*pat.karno[i]+omeg[inst[i]]}
gamma ~dgamma(1,0.001);
for (j in 1:2) {tau.karno[j] ~dgamma(1,0.001)}
mu.karno1 ~dflat(); for (j in 1:2) {c[j] ~dnorm(0,0.001)}
for (j in 1:3) {b[j] ~dnorm(0,0.001)}
# Cluster Allocation under DPP
for (j in 1:K) {G[j] ~dcat(p[1:M])
# frailty in institution j
omeg[j] ~dnorm(nu[G[j]],tau.omeg[G[j]])
for (k in 1:M) {post[j,k] <- equals(G[j],k)}}
# baseline prior
nu0 ~dnorm(0,0.001); tau0 ~dgamma(1,0.001)
```

```
for (i in 1:M) {nu[i] ~dnorm(nu0,tau0)
                tau.omeg[i] ~dgamma(1,0.001); }
# truncated Dirichlet process
alpha <- 5; V[M] <- 1; p[1] <- V[1]
for (k in 1:M-1){ V[k] ~dbeta(1,alpha)}
for (j in 2:M) { p[j] <- V[j]*(1-V[j-1])*p[j-1]/V[j-1]}
# total clusters
 TClus <- sum(NonEmp[])
   for (j in 1:M) {NonEmp[j] <- step(sum(post[,j])-1)}}
```

Taking $M = 25$, a two chain run of $10\,000$ iterations (second half for inferences) provides posterior mean for $p_M$ of 0.013, with an average number of clusters (TClus in the code) of 8. A kernel plot of the estimated $\omega_k$ suggests some positive skew. The LPML deteriorates slightly to $-1306.5$.

## 10.6    Further applications of frailty models

Shared frailty models for survival data have a particular relevance to situations where longitudinal data on subjects is also observed. A typical scenario is in a follow-up study where repeated clinical biomarker observations are made, while the subject is followed to the event concerned (e.g. relapse, death) or to censoring. The question of interest is how the earlier longitudinal profile of a subject, or changes in that profile, relate to the risk of the subsequent event. As mentioned by Ibrahim *et al.* (2010) joint models for longitudinal-survival data provide more efficient estimates of treatment effects on both times to the event and on the longitudinal biomarkers, and reduce bias in the estimates of the overall treatment effect. Thus consider a 'shared trajectory model' for a continuous longitudinal marker with normal iid residuals

$$Y_{ij} \sim N(W_{ij}, \sigma^2),$$

for subjects $i$ measured at occasions $j$ and times $t_{ij}$. Then the trajectory function $W_{ij}$ may be taken to have a random intercepts and slopes form

$$W_{ij} = b_{1i} + b_{2i}t_{ij} + \beta X_{ij} + \delta Z_i,$$

where $Z_i$ is a treatment indicator, $X_{ij}$ are risk factors, and $b_i = (b_{1i}, b_{2i})$ are bivariate normal random effects (frailties). The survival component for event or censoring times $T_i$ consists of a parametric or semi-parametric model, such as a Weibull or piecewise exponential model. The longitudinal and survival components are linked through the trajectory function, so that the hazard is

$$h(T_i|W_{iT_i}, R_i, Z_i) = h_0(T_i)\exp(\theta R_i + \phi Z_i + \alpha W_{i,T_i}),$$

where $R_i$ are risk factors for survival, and $\alpha$ is denoted the association parameter. The overall treatment effect on survival is measured by $\alpha\delta + \phi$.

Henderson *et al.* (2000) propose a form of common factor strategy for linking longitudinal and event data. Continuing the above notation and assuming continuous longitudinal biomarker responses with normal iid residuals, this strategy can be represented as

$$Y_{ij} \sim N(\beta X_{ij} + \delta Z_i + U_{ij}, \sigma^2),$$

$$U_{ij} = b_{1i} + b_{2i}t_{ij},$$

$$h(T_i|R_i, Z_i, b_i) = h_0(T_i)\exp(\theta R_i + \phi Z_i + \lambda_1 b_{1i} + \lambda_2 b_{2i} + \lambda_3(b_{1i} + b_{2i}T_i) + b_{3i}),$$

where $b_{3i}$ represents unshared frailty relevant only to the survival process. The $\lambda$ parameters amount to loadings on the shared factor scores (random effects). Reduced versions of this model form can be considered such as

$$h(T_i|R_i, Z_i, b_i) = h_0(T_i)\exp(\theta R_i + \phi Z_i + \lambda_1 b_{1i} + \lambda_2 b_{2i} T_i),$$

and to ensure consistent labelling and identifiability, a positivity constraint on the $\lambda$ parameters can be imposed in line with the form of the biomarkers. In clinical applications, assuming the biomarkers are positive indicators of morbidity, and the event defining the hazard is expected to increase with morbidity, then the subject matter based constraints $\{\lambda_1 > 0, \lambda_2 > 0\}$ can be applied. If this constraint is inappropriate it will be apparent in posterior density profiles for the $\lambda$ parameters, such as a spike close to zero. Note that the $b_{2i}$ are higher for patients whose biomarker readings are rising more rapidly, with faster rising morbidity assuming positive morbidity measures, and so having raised chances of earlier mortality.

### Example 10.8    Primary biliary cirrhosis

Consider clinical trial data for treatment primary biliary cirrhosis (PBC) of the liver via the drug D-penicillamine (denoted $Z$). As in Crowther *et al.* (2013) the focus is on the additional information in longitudinal clinical readings that is potentially relevant to eventual death or survival. There are $n = 312$ subjects with varying longitudinal histories, with total readings for each subject $J_i$ ranging from 1 to 16. The total number of longitudinal data observations $N = \sum_i J_i$ is 1945, with BUGS input in stacked form. Times are considered in year units, as in Crowther *et al.* (2013), although some versions of the data (e.g. http://lib.stat.cmu.edu/datasets/pbcseq) contain times in days. Of particular interest are log transfomed readings $y_{ij}$ (for subjects $i$ and occasions $j$) of the biomarker serum bilirubin, of which higher levels indicate greater morbidity.

Under a shared trajectory approach, these readings are taken to have means $W_{ij}$ defined by a treatment effects, subject level random intercepts and trends in times $t_{ij}$,

$$W_{ij} = \beta_1 + \beta_2 t_{ij} + \delta Z_i + b_{1i} + b_{2i} t_{ij},$$
$$b_i \sim N(0, \Sigma_b),$$

with Wishart prior for the inverse covariance of the random effects, $\Sigma_b^{-1} \sim W(I, 2)$. From the covariance matrix, the parametric correlation $\rho_b$ between the two sets of random effects is obtained.

The survival component assumes a Weibull model with shape parameter $\gamma$, and the dloglik option in OpenBUGS is used with survival log-likelihood, which is defined by observed death times (observed or censored) denoted $T_i$, and event indicators $\delta_i$ (=1 for deaths). The log likelihood for the survival component is

$$\log(L_i) = \delta_i[\log\gamma + \log\lambda_i + (\gamma - 1)\log(T_i)] - \lambda_i T_i^\gamma,$$

with the regression term

$$\log(\lambda_i) = \theta_1 + \alpha W_{iT_i} + \phi Z_i,$$

with $\alpha$ denoting the association parameter, $\phi$ the direct treatment effect on mortality, and $W_{iT_i} = \beta_1 + \beta_2 T_i + \delta Z_i + b_{1i} + b_{2i} T_i$.

A two chain run of 35 000 iterations (with inferences from the last 10 000) provides summaries of major parameters in Table 10.2. There is a non-significant direct effect of the drug treatment on the biomarker, with $\delta$ having 95% interval $(-0.30, 0.0.08)$. Similarly there

**Table 10.2**  PBC joint longitudinal and survival model.

|            | Mean   | St devn | MC error | 2.5%   | 97.5%  |
|------------|--------|---------|----------|--------|--------|
| $\alpha$   | 0.62   | 0.07    | 0.004    | 0.47   | 0.76   |
| $\beta_1$  | 0.55   | 0.08    | 0.008    | 0.37   | 0.69   |
| $\beta_2$  | 0.18   | 0.01    | 0.001    | 0.15   | 0.20   |
| $\gamma$   | −0.13  | 0.10    | 0.011    | −0.30  | 0.08   |
| $\omega$   | 1.19   | 0.09    | 0.007    | 1.02   | 1.38   |
| $\theta_1$ | −3.97  | 0.27    | 0.021    | −4.51  | −3.48  |
| $\phi$     | −0.14  | 0.16    | 0.014    | −0.47  | 0.17   |
| $\rho_b$   | 0.37   | 0.07    | 0.002    | 0.22   | 0.50   |

is a non-significant direct treatment effect on survival, with $\phi$ having 95% interval $(−0.47, 0.17)$. However, the $\alpha$ parameter, with 95% interval $(0.47, 0.76)$ demonstrates the association of higher levels of the biomarker with increased eventual mortality. Despite this, the overall treatment effect defined as $\alpha\delta + \phi$ is non-significant with 95% interval $(−0.59, 0.15)$. The DIC is 490.6.

The reduced version of the common factor model is also applied, namely

$$Y_{ij} \sim N(\delta Z_i + U_{ij}, \sigma^2),$$

$$U_{ij} = b_{1i} + b_{2i}t_{ij},$$

$$h(T_i | R_i, Z_i, b_i) = h_0(T_i) \exp(\theta R_i + \phi Z_i + \lambda_1 b_{1i} + \lambda_2 b_{2i} T_i),$$

with positivity constraints on the $\lambda$ parameters, and a Weibull form for $h_0$. Inferences are based on the second half of a two chain run of 20 000 iterations. This model has a DIC of 495, and the treatment effect parameter on mortality $\phi$ with a 95% interval $(−0.53, 0.35)$. The $\lambda_1$ and $\lambda_2$ parameters have respective 95% intervals $(0.92, 1.33)$ and $(0.51, 0.90)$, confirming the morbidity effect on subsequent mortality.

## 10.7  Competing risks

Competing risks analysis is relevant when there are more than one type of failure, and one cause precludes the others (e.g. Pintilie, 2006; Klein, 2010). In event histories (e.g. job or marital histories) the types of failure refer to different possible destinations, while in clinical studies of treatment effects for particular diseases (e.g. effectiveness in preventing cancer relapse or death from cancer) it is also necessary to allow for death from other causes. Let $t_i$ be an event time and $C_i \in \{1, \ldots J\}$ be the observed failure type (with $C_i = 0$ for right censored subjects), where failure types are mutually exclusive and exhaustive. Then the survival process can be characterised by a cause specific hazard, also called a subhazard

$$h_j(t_i) = \lim_{\delta t \to 0} \frac{\Pr(t_i < T \leq t_i + \delta t, C_i = j | T > t_i)}{\delta t}.$$

The overall hazard, assuming causes are independent, is obtained by summing over failure types

$$h(t_i) = \sum_{j=1}^{J} h_j(t_i).$$

with the overall survivor function $\Pr(T > t)$ obtained as

$$S(t_i) = \exp\left[-\int_0^{t_i} h(u)\mathrm{d}u\right] = \exp\left[-\sum_{j=1}^{J}\int_0^{t_i} h_j(u)\mathrm{d}u\right].$$

The cause specific density or subdensity function

$$f_j(t_i) = \lim_{\delta t \to 0} \frac{\Pr(t_i < T \le t_i + \delta t, C_i = j)}{\delta t},$$

governing times till the $j$th failure type, is therefore equal to $h_j(t_i)S(t_i)$.

The likelihood encompasses all individuals and all possible causes, with censoring indicators $\delta_{ij} = 1$ if $C_i = j \in (1, \ldots , J)$, and $\delta_{ij} = 0$ otherwise. For an individual with $\delta_{ij} = 1$, survival times on causes $k \ne j$ are censored. For an individual whose exit is not observed ($C_i = 0$), survival times on all causes are censored.

Alternative approaches to competing risks (Pintilie, 2006) include the latent failure time interpretation (where observed times are taken as the minimum of $J$ possible latent failure times), and competing risks as a bivariate random variable with joint probability $F_j(t) = \Pr(T \le t, C = j)$. The latter function represents cumulative probability of failure from a specific cause over time, known as the cumulative incidence function. Regression analysis can be applied using either cause-specific hazard functions (Prentice *et al.*, 1978), or cumulative incidence functions (Fine and Gray, 1999; Jeong and Fine, 2006).

Impacts of covariates $X_{ij}$ are likely to be differentiated according to type of move or exit, with a proportional hazard assumption leading to

$$h_j(t_i|X_{ij}, \beta_j) = h_{0j}(t_i)\exp(\beta_j X_{ij}).$$

If the risks are conditionally independent, the likelihood for subject $i$ with exit type $C_i = j$ ($j > 0$) is

$$L_i = f_j(t_i|X_{ij}, \beta_j)\prod_{k \ne j} S_k(t_i|X_{ik}, \beta_k).$$

The independent risks assumption may be modified by introducing cause specific frailty effects to generate dependent competing risks; for example, positive dependence may occur if there is comorbidity between causes in clinical applications (e.g. Gordon, 2002). Thus using multivariate normal effects, one has

$$h_j(t_i|b_{ij}, X_{ij}, \beta_j) = h_{0j}(t_i)\exp(\beta_j X_{ij} + b_{ij}),$$

$$b_{i,1:J} \sim N_J(0, \Sigma).$$

### Example 10.9   Follicular cell lymphoma

Consider data on response to alternative therapies among patients with follicular cell lymphoma, with a median follow-up time of 5.5 years (Pintilie, 2006). There are 541 patients in all, treated either with radiation alone (chemo=0), or a combination of radiation and chemotherapy (chemo = 1). The competing risks are relapse or no response (cause 1, with 272 subjects), or death without relapse (cause 2, with 76 subjects), with 193 censored subjects. Predictors are treatment type (chemo), age, haemoglobin levels (hgb) and clinical stage (=1 for stage II, 0 for stage I).

A counting process formulation of the Cox model is adopted, with event times transformed to reduce the number of distinct exit times (over both causes of exit). Event times over one year are rounded to the nearest year, and apart from the smallest observed exit times, times under one year are expressed with one decimal place only. There are then $T_d=35$ distinct exit times (either relapse/no response or death without relapse).

A first model assumes independent competing risks. Mildly informative $N(0,1)$ priors are adopted are the age and haemoglobin effects for numeric stability. The BUGS code (for J=2) is

```
model {for (i in 1:N) { for (j in 1:J) {delta[i,j] <- equals(C[i], j)}
for (j in 1:Td){R[i,j] <- step(t[i] - a[j] + eps)
for (k in 1:J) {dN[i,j,k] <- R[i,j]*step(a[j+1] - t[i] - eps)*delta[i,k]
                dN[i,j,k] ~dpois(lambda[i, j, k])
lambda[i, j, k] <- R[i,j]*exp(beta0[k] + beta1[k]*stage[i] + beta2[k]*age[i]
+beta3[k]*chemo[i]+beta4[k]*hgb[i])*dL0[k,j]}}}
for (j in 1:Td) {for (k in 1:J) {
dL0[k,j] ~dgamma(mu[j,k], c)
mu[j,k] <- dL0.star[j,k]*c
dL0.star[j,k] <- max(0.1,r[k]*(a[j+1] - a[j]))}}
for (j in 1:J){ beta0[j] ~dnorm(0,0.001)
beta1[j] ~dnorm(0,0.001); beta2[j] ~dnorm(0,1)
beta3[j] ~dnorm(0,0.001); beta4[j] ~dnorm(0,1)}}
```

A two chain run of 5000 iterations (with inferences from the last 2500) shows that the stage and age effects are both positive risk factors for the first cause, with the stage coefficient having mean (95% CRI) of 0.53 (0.31, 0.78). The combination treatment effect on events due to the disease (cause 1) has a 95% credible interval including zero, but the 90% interval (−0.59, −0.04) is entirely negative. These effects are similar to those reported by Scheike and Zhang (2011) using cumulative incidence function regression. A second model includes bivariate normal cause-specific frailty, so inducing dependent risks. This model shows a marginally negative (non-significant) correlation between the frailties $b_{i1}$ and $b_{i2}$ for the two risks, and no change to inferences regarding the treatment effect.

# Exercises

**10.1.** In Example 10.1, apply a Weibull rather than Gompertz survival model to the mice death times. Compare the log pseudo marginal likelihoods of the two models and the chi-square statistic comparing actual deaths and posterior mean predicted deaths in the eight intervals of 100 days. The chi-square statistic under the Weibull should deteriorate, as early deaths are overpredicted and later deaths underpredicted.

**10.2.** In Example 10.2, adapt the BUGS code to include an intercept and using a car.normal prior for centred $\alpha_j$ effects (constrained to sum to zero).

**10.3.** In Example 10.2, modify the piecewise exponential model to allow an interval specific treatment effect, $\beta_{trt,j}$. A random walk prior, analogous to that used for $\alpha_j = \log(h_{0j})$, may also be used as the prior on this varying treatment effect. Report evidence on the profile of the estimated $\beta_{trt,j}$ and the changed LPML.

**10.4.** In Example 10.4 (marrow transplant survival), assess covariate effects and LPML resulting from a logistic assumption on the errors $u$ (log-logistic for $t$).

**10.5.** In Example 10.6 (first intercourse) compare the existing model with one allowing interval varying effects $\beta_{1j}$ for parental transition status. Assume both $\gamma_j$ and $\beta_{1j}$ to be normally distributed random effects following a first order random walk.

**10.6.** In Example 10.7 (institutional frailty) the Karnofsky scores might be regarded simply as morbidity proxies rather than risk factors per se. Therefore assess institutional frailty variations using the discrete mixture model with three rather than two groups (and with three constrained parameters) and with predictors age and sex. The data can be obtained using the commands library(survival) and data(lung) in R.

**10.7.** In Example 10.7 adapt the mixed DP frailty model to treat the concentration parameter $\alpha$ as an unknown. Also include commands to assess the probability that the frailty effect for each institution $\omega_k$ is higher than the average of the K=18 realised frailty values (effectively equal to an estimate of the intercept).

# References

Abrams, K., Ashby, D. and Errington, D. (1996) A Bayesian approach to Weibull survival models – application to a cancer clinical trial. *Lifetime Data Analysis*, **2**(2), 159–174.

Aitkin, M. and Clayton, D. (1980) The fitting of exponential, Weibull and extreme value distributions to complex censored survival data using GLIM. *Journal of the Royal Statistical Society C* **29**, 156–163.

Aitkin, M., Francis, B., Hinde, J. and Darnell, R. (2009) *Statistical Modelling in R*. Oxford University Press, Oxford, UK.

Ananda, M., Dalpatadu, R. and Singh, A. (1996) Adaptive Bayes estimators for parameters of the Gompertz survival model. *Applied Mathematics and Computation*, **75**(2–3), 167–177.

Andersen, P., Borgan, Ø., Gill, R. and Keiding, N. (1993) *Statistical Models Based on Counting Processes*. Springer-Verlag, Berlin.

Ansfield, F., Klotz, J., Nealon, T., Ramirez, G., Minton, J., Hill, G., Wilson, W., Davis, H. and Cornell, G. (1977) A phase III study comparing the clinical utility of four regimens of 5-fluorouracil: a preliminary report. *Cancer*, **39**(1), 34–40.

Avalos, B., Klein, J., Kapoor, N., Tutschka, P., Klein, J. and Copelan, E. (1993) Preparation for marrow transplantation in Hodgkin's and non-Hodgkin's lymphoma using Bu/CY. *Bone Marrow Transplant*, **12**(2), 133–138.

Beamonte, E. and Bermúdez, J. (2003) A Bayesian semiparmetric analysis for additive hazards models with censored observations. *Test*, **12**(2), 101–117.

Bedrick, E., Christensen, R. and Johnson, W. (2000) Bayesian accelerated failure time analysis with application to veterinary epidemiology. *Statistics in Medicine*, **19**(2), 221–237.

Box-Steffensmeier, J. and Jones, B. (1997) Time is of the essence: event history models in political science. *American Journal of Political Science*, **41**(4), 1414–1461.

Callegaro, A. and Iacobelli, S. (2012) The Cox shared frailty model with log-skew-normal frailties. *Statistical Modelling*, **12**(5), 399–418.

Capaldi, D., Crosby, L. and Stoolmiller, M. (1996) Predicting the timing of first sexual intercourse for adolescent males. *Child Development*, **67**, 344–359.

Chen, M., Ibrahim, J. and Shao, Q. (2006) Posterior propriety and computation for the Cox regression model with applications to missing covariates. *Biometrika*, **93**(4), 791–807.

Clayton, D. (1991) A Monte Carlo method for Bayesian inference in frailty models. *Biometrics*, **47**, 467–485.

Cox, D. (1972) Regression models and life-tables (with discussion). *Journal of the Royal Statistical Society B*, **34**, 187–220.

Crowther, M., Abrams, K. and Lambert, P. (2013) Joint modelling of longitudinal and survival data. *The Stata Journal*, **13**(1), 165–184.

De Iorio, M., Johnson, W., Müller, P. and Rosner, G. (2009) Bayesian nonparametric nonproportional hazards survival modeling. *Biometrics*, **65**, 762–771.

Dellaportas, P. and Smith, A. (1993) Bayesian inference for generalized linear and proportional hazards model via Gibbs sampling. *Applied Statistics*, **42**, 443–60.

Devicienti, F. (2011) Estimating poverty persistence in Britain. *Empirical Economics*, **40**(3), 657–686.

Duchateau, L. and Janssen, P. (2008) *The Frailty Model*. Springer, New York, NY.

Fahrmeir, L. and Knorr-Held, L. (1997) Dynamic discrete-time duration models: estimation via Markov Chain Monte Carlo. In A. Raftery (ed.), *A Sociological Methodology*. American Sociological Association.

Fahrmeir, L. and Tutz, G. (2001) *Multivariate Statistical Modelling based on Generalized Linear Models*, 2nd edn. Springer-Verlag, Berlin.

Fine, J. and Gray, R. (1999) A proportional hazards model for the subdistribution of a competing risk. *Journal of the American Statistical Association*, **94**, 496–509.

Fleming, T. and Harrington, D. (1991) *Counting Processes and Survival Analysis*. Wiley, Chichester, UK.

Gamerman, D. (1991) Dynamic Bayesian models for survival data. *Journal of the Royal Statistical Society C*, **40**(1), 63–79.

Gelfand, A. and Ghosh, S. (1998) Model choice: A minimum posterior predictive loss approach. *Biometrika*, **85**, 1–11.

Gordon, S. (2002) Stochastic dependence in competing risks. *American Journal of Political Science*, **46**, 200–217.

Gu, Y., Sinha, D. and Banerjee, S. (2011) Analysis of cure rate survival data under proportional odds model. *Lifetime Data Analysis*, **17**(1), 123–134.

Guo, G. and Rodriguez, G. (1992) Estimating a multivariate proportional hazards model for clustered data using the EM-algorithm, with an application to child survival in Guatemala. *Journal of the American Statistical Association*, **87**, 969–976.

Gustafson, P. (1997) Large hierarchical *Bayesian Analysis* of multivariate survival data. *Biometrics*, **53**, 230–242.

Hamada, M., Wilson, A., Reese, C. and Martz, H. (2008) *Bayesian Reliability*. Springer, New York, NY.

Hanagal, D. and Dabade, A. (2013) A comparative study of shared frailty models for kidney infection data with generalized exponential baseline distribution. *Journal of Data Science*, **11**, 109–142.

Hanson, T., Branscum, A. and Johnson, W. (2011) Predictive comparison of joint longitudinal-survival modeling: a case study illustrating competing approaches. *Lifetime Data Analysis*, **17**, 3–28.

Henderson, R., Diggle, P. and Dobson, A. (2000) Joint modelling of longitudinal measurements and event time data. *Biostatistics*, **1**(4), 465–480.

Herring, A. and Ibrahim, J. (2002) Maximum likelihood estimation in random effects cure rate models with nonignorable missing covariates. *Biostatistics*, **3**(3), 387–405.

Holford, T. (1980) The analysis of rates and survivorship using log-linear models. *Biometrics*, **36**, 299–305.

Ibrahim, J., Chen, M. and Sinha, D. (2001a) *Bayesian Survival Analysis*. Springer, New York, NY.

Ibrahim, J., Chen, M. and Sinha, D. (2001b) Criterion-based methods for Bayesian model assessment. *Statistica Sinica*, **11**, 419–443.

Ibrahim, J., Chu, H. and Chen, L. (2010) Basic concepts and methods for joint models of longitudinal and survival data. *Journal of Clinical Oncology*, **28**(16), 2796–2801.

Jeong, J. and Fine, J. (2006) Direct parametric inference for cumulative incidence function. *Journal of the Royal Statistical Society C*, **55**, 187–200.

Kalbfleisch, J. (1978) Non-parametric Bayesian analysis of survival time data. *Journal of the Royal Statistical Society B*, **40**, 214–221.

Kalbfleisch, J. and Prentice, R. (1973) Marginal likelihoods based on Cox's regression and life model. *Biometrika*, **60**(2), 267–278.

Kalbfleisch, J. and Prentice, R. (1980) *The Statistical Analysis of Failure Time Data*. Wiley, New York, NY.

Kay, R. (1977) Proportional hazard regression models and the analysis of censored survival data. *Applied Statistics*, **26**, 227–237.

Keiding, N., Andersen, P. and John, J. (1997) The role of frailty models an accelerated failure time models in describing heterogeneity due to omitted covariates. *Statistics in Medicine*, **16**, 215–225.

Kim, S. and Ibrahim, J. (2000) On Bayesian inference for proportional hazards models using noninformative priors. *Lifetime Data Analysis*, **6**, 331–341.

Klein, J. (2010) Competing risks. *Wiley Interdisciplinary Reviews: Computational Statistics*, **2**(3), 333–339.

Kuo, L. and Peng, F. (2000) A mixture model approach to the analysis of survival data. In D. Dey, S. Ghosh and B. Mallick (eds), *Generalized Linear Models; a Bayesian Perspective*. Marcel Dekker, New York, NY.

Laud, P., Damien, P. and Smith, A. (1998) Bayesian nonparametric and semiparametric analysis of failure time data. In D. Dey *et al.* (eds), *Practical Nonparametric and Semiparametric Bayesian Statistics. Lecture Notes in Statistics 133*. Springer, New York, NY.

Leonard, T., Hsu, J., Tsui, K. and Murray, J. (1994) Bayesian and likelihood inference from equally weighted mixtures. *Annals of the Institute of Statistical Mathematics*, **46**, 203–220.

Leung, K., Elashoff, R. and Afifi, A. (1997) Censoring issues in survival analysis. *Annual Review of Public Health*, **18**, 83–104.

Li, Y., Gail, M., Preston, D., Graubard, B. and Lubin, J. (2012) Piecewise exponential survival times and analysis of case-cohort data. *Statistics in Medicine*, **31**, 1361–1368.

Lindsey, J. (1995) Fitting parametric counting processes by using log-linear models. *Journal of the Royal Statistical Society C*, **44**(2), 201–221.

Lindsey, J. (2001) A general family of distributions for longitudinal dependence, with special reference to event histories. *Statistics in Medicine*, **20**, 1625–1638.

Loprinzi, C., Laurie, J., Wieand, H., Krook, J., Novotny, P., Kugler, J., Bartel, J., Law, Bateman, M., Klatt, N., *et al.* (1994) Prospective evaluation of prognostic variables from patient-completed questionnaires. North Central Cancer Treatment Group. *Journal of Clinical Oncology*, **12**(3), 601–607.

Mantel, N. and Hankey, B. (1978) A logistic regression analysis of response-time data where the hazard function is time dependent. *Communications in Statistics*, **7A**, 333–348.

Martino, S., Akerkar, R. and Rue, H. (2011) Approximate Bayesian inference for survival models. *Scandinavian Journal of Statistics*, **38**(3), 514–528.

Mostafa, A. (2012) *Bayesian Non-parametric Estimation in Proportional Hazard Models: Survival Data Analysis with WinBUGS*. Lambert Academic Publishing, Saarbrücken.

Mostafa, A. and Ghorbal, A. (2011) Using WinBUGS to Cox model with changing from the baseline hazard function. *Applied Mathematical Sciences*, **5**(45–48), 2217–2240.

Ohlssen, D., Sharples, L. and Spiegelhalter, D. (2007) Flexible random-effects models using Bayesian semi-parametric models: applications to institutional comparisons. *Statistics in Medicine*, **26**(9), 2088–2112.

Pennell, M. and Dunson, D. (2006) Bayesian semiparametric dynamic frailty models for multiple event time data. *Biometrics*, **62**, 1044–1052.

Pintilie, M. (2006) *Competing Risks: A Practical Perspective*. Wiley, Chichester, UK.

Prentice, R., Kalbfleisch, J., Peterson A., Flournoy N., Farewell, V. and Breslow, N. (1978) The analysis of failure times in the presence of competing risks. *Biometrics*, **34**, 541–54.

Rolin, J. (1998) Bayesian survival analysis. In *Encyclopedia of Biostatistics*, pp. 271–286, Wiley, Chichester, UK.

Sahu, S., Dey, D., Aslanidou, H. and Sinha, D. (1997) A Weibull regression model with gamma frailties for multivariate survival data. *Lifetime Data Analysis*, **3**(2), 123–137.

Sargent, D. (1997) A flexible approach to time-varying coefficients in the Cox regression setting, *Lifetime Data Analysis*, **3**, 13–25.

Sastry, N. (1997) A nested frailty model for survival data, with an application to the study of child survival in Northeast Brazil. *Journal of the American Statistical Association*, **92**, 426–435.

Scheike, T. and Zhang, M.-J. (2011) Analyzing competing risk data using the R timereg package. *Journal of Statistical Software*, **38**(2), 1–15.

Siannis, F., Copas, J. and Lu, G. (2005) Sensitivity analysis for informative censoring in parametric survival models. *Biostatistics*, **6**(1), 77–91.

Singer, J. and Willett, J. (1993) It's about time: using discrete-time survival analysis to study duration and timing of events. *Journal of Educational Statistics*, **18**, 155–195.

Singer, J. and Willett, J. (2003) *Applied Longitudinal Data Analysis: Modeling Change and Event Occurrence*. Oxford University Press, New York, NY.

Sinha, D. and Dey, D. (1997) Semiparametric Bayesian analysis of survival data. *Journal of the American Statistical Association*, **92**(439), 1195–1121.

Volinsky, C. and Raftery, A. (2000) Bayesian information criterion for censored survival models. *Biometrics*, **56**, 256–262.

Volinsky, C., Madigan, D., Raftery, A. and Kronmal, R. (1997) Bayesian model averaging in proportional hazard models: assessing the risk of a stroke. *Applied Statistics*, **46**, 433–448.

Wang, C., Fan, Z., Chang, H. and Douglas, J. (2012) A semiparametric model for jointly analyzing response times and accuracy in computerized testing. *Journal of Educational and Behavioral Statistics*, **38**, 381–417.

Wienke, A. (2011) *Frailty Models in Survival Analysis*. Chapman and Hall/CRC, Boca Raton, FL.

Zhang, J. and Lawson, A. (2011) Bayesian parametric accelerated failure time spatial model and its application to prostate cancer. *Journal of Applied Statistics*, **38**(3), 591–603.

Zhao, L., Hanson, T. and Carlin, B. (2009) Mixtures of Polya trees for flexible spatial frailty survival modelling. *Biometrika*, **96**(2), 263–276.

# Index

*Applied Bayesian Modelling*, Second Edition. Peter Congdon.
© 2014 John Wiley & Sons, Ltd. Published 2014 by John Wiley & Sons, Ltd.

# WILEY SERIES IN PROBABILITY AND STATISTICS
ESTABLISHED BY WALTER A. SHEWHART AND SAMUEL S. WILKS

Editors: *David J. Balding, Noel A. C. Cressie, Garrett M. Fitzmaurice, Geof H. Givens, Harvey Goldstein, Geert Molenberghs, David W. Scott, Adrian F. M. Smith, Ruey S. Tsay, Sanford Weisberg*
Editors Emeriti: *J. Stuart Hunter, Iain M. Johnstone, J. B. Kadane, Jozef L. Teugels*

The *Wiley Series in Probability and Statistics* is well established and authoritative. It covers many topics of current research interest in both pure and applied statistics and probability theory. Written by leading statisticians and institutions, the titles span both state-of-the-art developments in the field and classical methods.

Reflecting the wide range of current research in statistics, the series encompasses applied, methodological and theoretical statistics, ranging from applications and new techniques made possible by advances in computerized practice to rigorous treatment of theoretical approaches.

This series provides essential and invaluable reading for all statisticians, whether in academia, industry, government, or research.

[†] ABRAHAM and LEDOLTER · Statistical Methods for Forecasting

AGRESTI · Analysis of Ordinal Categorical Data, *Second Edition*

AGRESTI · An Introduction to Categorical Data Analysis, *Second Edition*

AGRESTI · Categorical Data Analysis, *Third Edition*

ALSTON, MENGERSEN and PETTITT (editors) · Case Studies in Bayesian Statistical Modelling and Analysis

ALTMAN, GILL, and McDONALD · Numerical Issues in Statistical Computing for the Social Scientist

AMARATUNGA and CABRERA · Exploration and Analysis of DNA Microarray and Protein Array Data

ANDĚL · Mathematics of Chance

ANDERSON · An Introduction to Multivariate Statistical Analysis, *Third Edition*

[*] ANDERSON · The Statistical Analysis of Time Series

ANDERSON, AUQUIER, HAUCK, OAKES, VANDAELE, and WEISBERG · Statistical Methods for Comparative Studies

ANDERSON and LOYNES · The Teaching of Practical Statistics

ARMITAGE and DAVID (editors) · Advances in Biometry

ARNOLD, BALAKRISHNAN, and NAGARAJA · Records

[*] ARTHANARI and DODGE · Mathematical Programming in Statistics

AUGUSTIN, COOLEN, DE COOMAN and TROFFAES (editors) · Introduction to Imprecise Probabilities

[*] BAILEY · The Elements of Stochastic Processes with Applications to the Natural Sciences

BAJORSKI · Statistics for Imaging, Optics, and Photonics

BALAKRISHNAN and KOUTRAS · Runs and Scans with Applications

BALAKRISHNAN and NG · Precedence-Type Tests and Applications

BARNETT · Comparative Statistical Inference, *Third Edition*

BARNETT · Environmental Statistics

BARNETT and LEWIS · Outliers in Statistical Data, *Third Edition*

BARTHOLOMEW, KNOTT, and MOUSTAKI · Latent Variable Models and Factor Analysis: A Unified Approach, *Third Edition*

BARTOSZYNSKI and NIEWIADOMSKA-BUGAJ · Probability and Statistical Inference, *Second Edition*

BASILEVSKY · Statistical Factor Analysis and Related Methods: Theory and Applications

BATES and WATTS · Nonlinear Regression Analysis and Its Applications

BECHHOFER, SANTNER, and GOLDSMAN · Design and Analysis of Experiments for Statistical Selection, Screening, and Multiple Comparisons

BEIRLANT, GOEGEBEUR, SEGERS, TEUGELS, and DE WAAL · Statistics of Extremes: Theory and Applications

BELSLEY · Conditioning Diagnostics: Collinearity and Weak Data in Regression

† BELSLEY, KUH, and WELSCH · Regression Diagnostics: Identifying Influential Data and Sources of Collinearity

BENDAT and PIERSOL · Random Data: Analysis and Measurement Procedures, *Fourth Edition*

BERNARDO and SMITH · Bayesian Theory

BHAT and MILLER · Elements of Applied Stochastic Processes, *Third Edition*

BHATTACHARYA and WAYMIRE · Stochastic Processes with Applications

BIEMER, GROVES, LYBERG, MATHIOWETZ, and SUDMAN · Measurement Errors in Surveys

BILLINGSLEY · Convergence of Probability Measures, *Second Edition*

BILLINGSLEY · Probability and Measure, *Anniversary Edition*

BIRKES and DODGE · Alternative Methods of Regression

BISGAARD and KULAHCI · Time Series Analysis and Forecasting by Example

BISWAS, DATTA, FINE, and SEGAL · Statistical Advances in the Biomedical Sciences: Clinical Trials, Epidemiology, Survival Analysis, and Bioinformatics

BLISCHKE and MURTHY (editors) · Case Studies in Reliability and Maintenance

BLISCHKE and MURTHY · Reliability: Modeling, Prediction, and Optimization

BLOOMFIELD · Fourier Analysis of Time Series: An Introduction, *Second Edition*

BOLLEN · Structural Equations with Latent Variables

BOLLEN and CURRAN · Latent Curve Models: A Structural Equation Perspective

BOROVKOV · Ergodicity and Stability of Stochastic Processes

BOSQ and BLANKE · Inference and Prediction in Large Dimensions

BOULEAU · Numerical Methods for Stochastic Processes

* BOX and TIAO · Bayesian Inference in Statistical Analysis

BOX · Improving Almost Anything, *Revised Edition*

*Now available in a lower priced paperback edition in the Wiley Classics Library.

†Now available in a lower priced paperback edition in the Wiley–Interscience Paperback Series.

CORNELL · Experiments with Mixtures, Designs, Models, and the Analysis of Mixture Data, *Third Edition*

COX · A Handbook of Introductory Statistical Methods

CRESSIE · Statistics for Spatial Data, *Revised Edition*

CRESSIE and WIKLE · Statistics for Spatio-Temporal Data

CSÖRGÖ and HORVÁTH · Limit Theorems in Change Point Analysis

DAGPUNAR · Simulation and Monte Carlo: With Applications in Finance and MCMC

DANIEL · Applications of Statistics to Industrial Experimentation

DANIEL · Biostatistics: A Foundation for Analysis in the Health Sciences, *Eighth Edition*

\* DANIEL · Fitting Equations to Data: Computer Analysis of Multifactor Data, *Second Edition*

DASU and JOHNSON · Exploratory Data Mining and Data Cleaning

DAVID and NAGARAJA · Order Statistics, *Third Edition*

DAVINO, FURNO and VISTOCCO · Quantile Regression: Theory and Applications

\* DEGROOT, FIENBERG, and KADANE · Statistics and the Law

DEL CASTILLO · Statistical Process Adjustment for Quality Control

DEMARIS · Regression with Social Data: Modeling Continuous and Limited Response Variables

DEMIDENKO · Mixed Models: Theory and Applications

DENISON, HOLMES, MALLICK and SMITH · Bayesian Methods for Nonlinear Classification and Regression

DETTE and STUDDEN · The Theory of Canonical Moments with Applications in Statistics, Probability, and Analysis

DEY and MUKERJEE · Fractional Factorial Plans

DE ROCQUIGNY · Modelling Under Risk and Uncertainty: An Introduction to Statistical, Phenomenological and Computational Models

DILLON and GOLDSTEIN · Multivariate Analysis: Methods and Applications

\* DODGE and ROMIG · Sampling Inspection Tables, *Second Edition*

\* DOOB · Stochastic Processes

DOWDY, WEARDEN, and CHILKO · Statistics for Research, *Third Edition*

DRAPER and SMITH · Applied Regression Analysis, *Third Edition*

DRYDEN and MARDIA · Statistical Shape Analysis

DUDEWICZ and MISHRA · Modern Mathematical Statistics

DUNN and CLARK · Basic Statistics: A Primer for the Biomedical Sciences, *Fourth Edition*

DUPUIS and ELLIS · A Weak Convergence Approach to the Theory of Large Deviations

EDLER and KITSOS · Recent Advances in Quantitative Methods in Cancer and Human Health Risk Assessment

\* ELANDT-JOHNSON and JOHNSON · Survival Models and Data Analysis

\*Now available in a lower priced paperback edition in the Wiley Classics Library.
†Now available in a lower priced paperback edition in the Wiley–Interscience Paperback Series.

---

\*Now available in a lower priced paperback edition in the Wiley Classics Library.
†Now available in a lower priced paperback edition in the Wiley–Interscience Paperback Series.

*Now available in a lower priced paperback edition in the Wiley Classics Library.
†Now available in a lower priced paperback edition in the Wiley–Interscience Paperback Series.

MARKOVICH · Nonparametric Analysis of Univariate Heavy-Tailed Data: Research and Practice

MARONNA, MARTIN and YOHAI · Robust Statistics: Theory and Methods

MASON, GUNST, and HESS · Statistical Design and Analysis of Experiments with Applications to Engineering and Science, *Second Edition*

McCULLOCH, SEARLE, and NEUHAUS · Generalized, Linear, and Mixed Models, *Second Edition*

McFADDEN · Management of Data in Clinical Trials, *Second Edition*

* McLACHLAN · Discriminant Analysis and Statistical Pattern Recognition

McLACHLAN, DO, and AMBROISE · Analyzing Microarray Gene Expression Data

McLACHLAN and KRISHNAN · The EM Algorithm and Extensions, *Second Edition*

McLACHLAN and PEEL · Finite Mixture Models

McNEIL · Epidemiological Research Methods

MEEKER and ESCOBAR · Statistical Methods for Reliability Data

MEERSCHAERT and SCHEFFLER · Limit Distributions for Sums of Independent Random Vectors: Heavy Tails in Theory and Practice

MENGERSEN, ROBERT, and TITTERINGTON · Mixtures: Estimation and Applications

MICKEY, DUNN, and CLARK · Applied Statistics: Analysis of Variance and Regression, *Third Edition*

* MILLER · Survival Analysis, *Second Edition*

MONTGOMERY, JENNINGS, and KULAHCI · Introduction to Time Series Analysis and Forecasting

MONTGOMERY, PECK, and VINING · Introduction to Linear Regression Analysis, *Fifth Edition*

MORGENTHALER and TUKEY · Configural Polysampling: A Route to Practical Robustness

MUIRHEAD · Aspects of Multivariate Statistical Theory

MULLER and STOYAN · Comparison Methods for Stochastic Models and Risks

MURTHY, XIE, and JIANG · Weibull Models

MYERS, MONTGOMERY, and ANDERSON-COOK · Response Surface Methodology: Process and Product Optimization Using Designed Experiments, *Third Edition*

MYERS, MONTGOMERY, VINING, and ROBINSON · Generalized Linear Models. With Applications in Engineering and the Sciences, *Second Edition*

NATVIG · Multistate Systems Reliability Theory With Applications

† NELSON · Accelerated Testing, Statistical Models, Test Plans, and Data Analyses

† NELSON · Applied Life Data Analysis

NEWMAN · Biostatistical Methods in Epidemiology

NG, TAIN, and TANG · Dirichlet Theory: Theory, Methods and Applications

OKABE, BOOTS, SUGIHARA, and CHIU · Spatial Tesselations: Concepts and Applications of Voronoi Diagrams, *Second Edition*

OLIVER and SMITH · Influence Diagrams, Belief Nets and Decision Analysis

---

*Now available in a lower priced paperback edition in the Wiley Classics Library.
†Now available in a lower priced paperback edition in the Wiley–Interscience Paperback Series.

*Now available in a lower priced paperback edition in the Wiley Classics Library.
†Now available in a lower priced paperback edition in the Wiley–Interscience Paperback Series.

SILVAPULLE and SEN · Constrained Statistical Inference: Inequality, Order, and Shape Restrictions

SINGPURWALLA · Reliability and Risk: A Bayesian Perspective

SMALL and McLEISH · Hilbert Space Methods in Probability and Statistical Inference

SRIVASTAVA · Methods of Multivariate Statistics

STAPLETON · Linear Statistical Models, *Second Edition*

STAPLETON · Models for Probability and Statistical Inference: Theory and Applications

STAUDTE and SHEATHER · Robust Estimation and Testing

STOYAN · Counterexamples in Probability, *Second Edition*

STOYAN and STOYAN · Fractals, Random Shapes and Point Fields: Methods of Geometrical Statistics

STREET and BURGESS · The Construction of Optimal Stated Choice Experiments: Theory and Methods

STYAN · The Collected Papers of T. W. Anderson: 1943–1985

SUTTON, ABRAMS, JONES, SHELDON, and SONG · Methods for Meta-Analysis in Medical Research

TAKEZAWA · Introduction to Nonparametric Regression

TAMHANE · Statistical Analysis of Designed Experiments: Theory and Applications

TANAKA · Time Series Analysis: Nonstationary and Noninvertible Distribution Theory

THOMPSON · Empirical Model Building: Data, Models, and Reality, *Second Edition*

THOMPSON · Sampling, *Third Edition*

THOMPSON · Simulation: A Modeler's Approach

THOMPSON and SEBER · Adaptive Sampling

THOMPSON, WILLIAMS, and FINDLAY · Models for Investors in Real World Markets

TIERNEY · LISP-STAT: An Object-Oriented Environment for Statistical Computing and Dynamic Graphics

TROFFAES and DE COOMAN · Lower Previsions

TSAY · Analysis of Financial Time Series, *Third Edition*

TSAY · An Introduction to Analysis of Financial Data with R

TSAY · Multivariate Time Series Analysis: With R and Financial Applications

UPTON and FINGLETON · Spatial Data Analysis by Example, Volume II: Categorical and Directional Data

† VAN BELLE · Statistical Rules of Thumb, *Second Edition*

VAN BELLE, FISHER, HEAGERTY, and LUMLEY · Biostatistics: A Methodology for the Health Sciences, *Second Edition*

VESTRUP · The Theory of Measures and Integration

VIDAKOVIC · Statistical Modeling by Wavelets

---

---

*Now available in a lower priced paperback edition in the Wiley Classics Library.
†Now available in a lower priced paperback edition in the Wiley–Interscience Paperback Series.